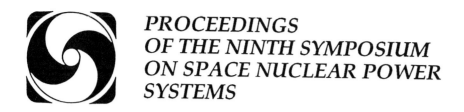

PROCEEDINGS OF THE NINTH SYMPOSIUM ON SPACE NUCLEAR POWER SYSTEMS

EDITORS

Mohamed S. El-Genk
University of New Mexico

Mark D. Hoover
Inhalation Toxicology
Research Institute

INSTITUTE FOR SPACE NUCLEAR POWER STUDIES
Chemical and Nuclear Engineering Department
The University of New Mexico
Albuquerque, NM 87131
(505) 277-2813, 277-2814

Co-sponsored by:

NATIONAL AERONAUTICS AND SPACE
ADMINISTRATION
    HEADQUARTERS
    LEWIS RESEARCH CENTER
STRATEGIC DEFENSE INITIATIVE ORGANIZATION
UNITED STATES DEPARTMENT OF ENERGY
    ARGONNE NATIONAL LABORATORY
    BROOKHAVEN NATIONAL LABORATORY
    IDAHO NATIONAL ENGINEERING LABORATORY
    LOS ALAMOS NATIONAL LABORATORY
    SANDIA NATIONAL LABORATORIES
UNITED STATES AIR FORCE
    PHILLIPS LABORATORY
    WRIGHT LABORATORY

In cooperation with:

AMERICAN NUCLEAR SOCIETY
    ANS TRINITY SECTION
    ANS ENVRONMENTAL SCIENCES DIVISION
    ANS NUCLEAR REACTOR SAFETY DIVISION
AMERICAN SOCIETY OF MECHANICAL ENGINEERS
    NUCLEAR ENGINEERING DIVISION
    HEAT TRANSFER DIVISION
ASTM, COMMITTEE-10 ON NUCLEAR TECHNOLOGY
AND APPLICATIONS
INTERNATIONAL ASTRONAUTICAL FEDERATION
NEW MEXICO ACADEMY OF SCIENCE

Industry Affiliates:

BABCOCK & WILCOX COMPANY
GRUMMAN CORPORATION
ROCKWELL INTERNATIONAL CORPORATION
    ROCKETDYNE DIVISION
WESTINGHOUSE ELECTRIC CORPORATION

Albuquerque Convention Center
Albuquerque, New Mexico
January 12-16, 1992

AMERICAN INSTITUTE OF PHYSICS

NEW YORK

INTRODUCTION

We are pleased to introduce the Proceedings of the Ninth Symposium on Space Nuclear Power Systems. These volumes contain the reviewed and edited papers that are being presented at the Ninth Symposium in Albuquerque, New Mexico, 12-16 January 1992. The objective of the symposium, and hence these volumes, is to summarize the state of knowledge in the area of space nuclear power and propulsion and to provide a forum at which the most recent findings and important new developments can be presented and discussed.

As in past meetings, the program is proceeded by two-day short courses. This year's topics are Heat Pipe Technology for Space Power Systems and Thermionic Power System Technology. They provide excellent overviews of these fields for symposium attendees, including graduate engineering students. Continuing education has always been one of the most valuable parts of the symposium.

The meeting program formally begins with a full day of plenary presentations which include congressional views and budget prospectives; an update on the Space Exploration Initiative; goals and issues for missions in space; and updates on plans for the SP-100 Program, the U.S. Thermionic Space Nuclear Program, Radioisotope Systems for Exploration of Space, and the use of Nuclear Propulsion for Future Space Missions. The plenary sessions are always a stimulating part of the program, and we are grateful to the congressional and agency leaders who are participating in those sessions. We are grateful to the members of Symposium Advisory Committee for helping to organize the plenary sessions.

At the same time that we and our professional colleagues are learning about the latest views and plans for space nuclear power and propulsion, 200 high school students and 20 teachers from around the state of New Mexico will learn about some of the exciting scientific and technical advances that are being made and about the career opportunities that await them. Irene El-Genk, chairwoman, and Rose Thome, co-chairwoman, are to be commended for their efforts and vision in organizing this important special session. We are grateful to NASA Lewis Research Center for providing the personalized space science folders for the students. We wish the attendees well, and encourage them to carry back their new found insights and enthusiasm to their schools.

This year's proceedings include nearly 200 papers prepared by over 400 authors from more than 50 organizations from many countries. We are grateful to the authors for their cooperation, enthusiasm, and sincere desire to contribute to a clear, useful, and technically sound publication. We gratefully acknowledge the contributions of the members of the Technical Program Committee in helping to organize the technical sessions and review the submissions. Reliable scientific literature cannot be produced without this kind of dedicated effort.

Many other individuals and organizations worked hard and contributed generously to sponsor, plan, organize, and manage this Symposium. They are also listed in the committee sections of these proceedings. They deserve credit for this ninth annual success. We gratefully acknowledge the competent and unflagging help of Mary Bragg and members of the staff of the Institute for Space Nuclear Power Studies in the preparation of these proceedings, and in the professional coordination of many functions and meetings for the preparation of the symposium. In addition, our families continue to provide us with their understanding, patience, and encouragement throughout the publication process. We deeply appreciate that.

In closing, we are sad to note the untimely death on 28 October 1991 at age 55 of Dr. Dudley G. McConnell of the National Aeronautics and Space Administration. He was a dynamic individual who had a distinguished professional career and who lived a caring and productive life as a human being. His leadership of the interagency efforts to launch the Galileo and Ulysses missions have made a lasting contribution to space exploration. Dr. McConnell's picture and obituary are included in the symposium program. He will be missed.

Mohamed S. El-Genk
Technical Chairman

Mark D. Hoover
Publications Chairman

SPACE NUCLEAR POWER PUBLICATIONS

Transactions of 1st Symposium ....out of print

Publications available from UNM's Institute for Space Nuclear Power Studies

Transactions of the 2nd Symposium $10.00
Transactions of the 3rd Symposium $10.00
Transactions of the 4th Symposium $10.00
Transactions of the 5th Symposium $10.00
Transactions of the 6th Symposium $20.00
Proceedings of the 7th Symposium $25.00

Publications available from the American Institute of Physics, c/o AIDC, 64 Depot Road, Colchester, VT 05446, 1-800-445-6638

Proceedings of the 8th Symposium (3-volume hardback set)
    ISBN # 0-88318-838-4 AIP Conference Proceedings #217...................................... $175.00
Proceedings of the 9th Symposium .... (to be determined)

Publications available from Orbit Book Company, P.O. Box 9542, Melbourne, FL 32902-9542, Phone: (407) 724-9542

Space Nuclear Power Systems 1984 .... $125.00
Space Nuclear Power Systems 1985 .... $125.00
Space Nuclear Power Systems 1986 .... $125.00
Space Nuclear Power Systems 1987 .... $125.00
Space Nuclear Power Systems 1988 .... $125.00
Space Nuclear Power Systems 1989 .... (in preparation)

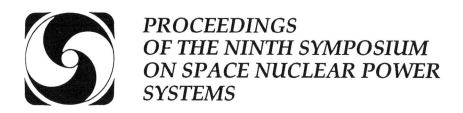

PROCEEDINGS OF THE NINTH SYMPOSIUM ON SPACE NUCLEAR POWER SYSTEMS

ORGANIZING COMMITTEE

HONORARY CHAIRMAN
Honorable Pete V. Domenici
United States Senate (R, NM)

GENERAL CHAIRMAN
Colonel Peter J. Marchiando
Commander, Phillips Laboratory

GENERAL CO-CHAIRMEN

Gary L. Bennett
NASA Headquarters

Wade Carroll
United States Department of Energy

TECHNICAL CHAIRMAN
Mohamed S. El-Genk
Institute for Space Nuclear Power Studies
University of New Mexico

TECHNICAL CO-CHAIRMEN

James H. Lee
Sandia National Laboratories

Ernest D. Herrera
Phillips Laboratory

Donald H. Roy
Babcock & Wilcox Company

PUBLICATIONS CHAIRMAN
Mark D. Hoover
Inhalation Toxicology Research Institute

ADMINISTRATIVE CHAIRWOMAN
Mary Bragg
Institute for Space Nuclear Power Studies
University of New Mexico

ADMINISTRATIVE CO-CHAIRWOMAN

Janice Holloman-Canales
Institute for Space Nuclear Power Studies
University of New Mexico

Maureen Alaburda
Institute for Space Nuclear Power Studies
Univeristy of new Mexico

SCHOLARSHIP AND HIGH SCHOOL PUBLIC RELATIONS COORDINATOR
Irene L. El-Genk
Education and Outreach Advisory Board
Institute for Space Nuclear Power Studies
University of New Mexico

SECONDARY SCHOOL SPACE SCIENCE COMPETITION COORDINATOR
David Kauffman
College of Engineering
University of New Mexico

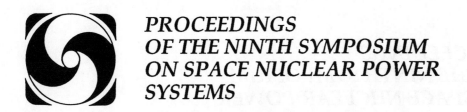

PROCEEDINGS OF THE NINTH SYMPOSIUM ON SPACE NUCLEAR POWER SYSTEMS

STEERING COMMITTEE

CHAIRMAN
Honorable Pete V. Domenici
United States Senate (R, NM)

MEMBERS

J. Sam Armijo
Program General Manager, Space Power
General Electric Company

Thomas J. Hirons
Division Leader, Nuclear Technology
and Engineering
Los Alamos National Laboratory

Richard A. Johnson
Director, Advanced Space Power
Rockwell International/Rocketdyne Division

Stephen J. Lanes
Deputy Assistant Secretary,
Space and Defense Power Programs
United States Department of Energy

Peter J. Marchiando
Commander, Phillips Laboratory

Paul North
Acting Associate Director
Idaho National Engineering Laboratory

Gregory M. Reck
Director, Space Technology
NASA Headquarters

Paul G. Risser
Provost and Vice President for
Academic Affairs
University of New Mexico

Honorable Steven H. Schiff
United States House of Representatives (R,NM)

ADVISORY COMMITTEE

CHAIRMAN
Mohamed S. El-Genk
University of New Mexico

MEMBERS

H. Sterling Bailey
General Electric Company

William J. Barattino
Phillips Laboratory

Gary Bennett
NASA Headquarters

Samit K. Bhattacharyya
Argonne National Laboratory

David L. Black
Westinghouse Electric Corporation

Richard J. Bohl
Los Alamos National Laboratory

David Buden
Idaho National Engineering Laboratory

Wade Carroll
United States Department of Energy

Edmund P. Coomes
Pacific Northwest Laboratory

Roy H. Cooper
Oak Ridge National Laboratory

Richard C. Dahlberg
General Atomics

Tony Gallegos
Senator Domenici's Office

Steve Harrison
National Space Council

Robert Haslett
Grumman Corporation

Ernest D. Herrera
Phillips Laboratory

Mark D. Hoover
Inhalation Toxicology Research Institute

Steven Howe
Los Alamos National Laboratory

Walter Kato
Brookhaven National Laboratory

Ehsan U. Khan
BDM International, Inc.

E. Tom Mahefkey
Wright Laboratory

Patrick J. McDaniel
Sandia National Laboratories

Thomas J. Miller
NASA Lewis Research Center

Joseph C. Mills
Rockwell International Corporation
        Rocketdyne Division

M. Frank Rose
Auburn University

Joseph A. Sholtis, Jr.
United States Air Force

R. Joseph Sovie
NASA Lewis Research Center

Earl J. Wahlquist
United States Department of Energy

Jack Walker
Sandia National Laboratories

Frank L. Williams
University of New Mexico

David M. Woodall
Idaho National Engineering Laboratory

Richard A. Zavadowski
Babcock & Wilcox Company

TECHNICAL PROGRAM COMMITTEE

TECHNICAL CHAIRMAN
Mohamed S. El-Genk
University of New Mexico

TECHNICAL CO-CHAIRMEN

Ernest Herrera
Phillips Laboratory

James H. Lee
Sandia National Laboratories

Donald H. Roy
Babcock & Wilcox

MEMBERS

Julio C. Acevedo
NASA Lewis Research Center

Douglas Allen
Strategic Defense Initiative Organization

Wayne R. Amos
EG&G Mound Applied Technologies, Inc.

Joseph Angelo
Science Applications International Corp.

H. Sterling Bailey
General Electric Company

Russell M. Ball
Babcock & Wilcox Company

C. Perry Bankston
Jet Propulsion Laboratory

Lester L. Begg
General Atomics

Kenneth J. Bell
Oklahoma State University

Gary L. Bennett
NASA Headquarters

Robert Bercaw
NASA Lewis Research Center

John A. Bernard
Massachusetts Institute of Technology

Frederick R. Best
Texas A&M University

Samit K. Bhattacharyya
Argonne National Laboratory

Walter Bienert
DTX Corporation

David L. Black
Westinghouse Electric Corporation

James B. Blackmon
McDonnell Douglas Astronautics
    Corporation

Richard J. Bohl
Los Alamos National Laboratory

Stanley K. Borowski
NASA Lewis Research Center

Edward J. Britt
Space Power, Inc.

Wayne M. Brittain
Teledyne Energy Systems

John Brophy
Jet Propulsion Laboratory

Neil W. Brown
General Electric Company

R. William Buckman
Refractory Metals Technology

David Buden
Idaho National Engineering Laboratory

David C. Byers
NASA Lewis Research Center

Colin Caldwell
Babcock & Wilcox Company

Wade Carroll
United States Department of Energy

A. Thomas Clark
United States Department of Energy

John S. Clark
NASA Lewis Research Center

TECHNICAL PROGRAM COMMITTEE *(CONTINUED)*

Edmund P. Coomes
Pacific Northwest Laboratory

Roy H. Cooper
Oak Ridge National Laboratory

Jeffery E. Dagle
Pacific Northwest Laboratory

Richard C. Dahlberg
General Atomics

Vijay K. Dhir
University of California-Los Angeles

Nils J. Diaz
University of Florida

James E. Dudenhoefer
NASA Lewis Research Center

Dale S. Dutt
Westinghouse Hanford Company

Amir Faghri
Wright State University

Gerald H. Farbman
Starpath, Inc.

Mario Fontana
Oak Ridge National Laboratory

Elzie E. Gerrels
General Electric Company

Toni L. Grobstein
NASA Lewis Research Center

Janelle W. Hales
Westinghouse Hanford Company

Bill Harper
Allied Signal Aerospace

Steve Harrison
National Space Council

Richard B. Harty
Rockwell International
    Corporation/Rocketdyne Division

Edwin Harvego
Idaho National Engineering Laboratory

Jack Heller
Sverdrup Technology, Inc.

Eugene Hoffman
United States Department of Energy

Robert S. Holcomb
Oak Ridge National Laboratory

Mark D. Hoover
Inhalation Toxicology Research Institute

Steven Howe
Los Alamos National Laboratory

Maribeth E. Hunt
Rockwell International
    Corporation/Rocketdyne Division

DeWayne L. Husser
Babcock & Wilcox Company

John R. Ireland
Los Alamos National Laboratory

Mohammad Jamshidi
University of New Mexico

Richard A. Johnson
Rockwell International Corporation
    Rocketdyne Division

Albert J. Juhasz
NASA Lewis Research Center

Michael Kania
Oak Ridge National Laboratory

Ehsan U. Khan
BDM International, Inc.

Donald B. King
Sandia National Laboratories

Andrew C. Klein
Oregon State University

Arvind Kumar
University of Missouri-Rolla

James H. Lee
Sandia National Laboratories

E. Tom Mahefkey
Wright Laboratory

Gregory Main
Office of Naval Research

John Mankins
NASA Headquarters

Albert C. Marshall
Sandia National Laboratories

Charles R. Martin
United States Air Force

John Martinell
EG&G Idaho, Inc.

L. David Massie
Wright Laboratory

Edward F. Mastal
United States Department of Energy

L. David Massie
Wright Laboratory

Edward F. Mastal
United States Department of Energy

Donald N. Matteo
General Electric Company

R. Bruce Matthews
Los Alamos National Laboratory

William H. McCulloch
Sandia National Laboratories

Patrick J. McDaniel
Sandia National Laboratories

Arthur Mehner
United States Department of Energy

Michael A. Merrigan
Los Alamos National Laboratory

Raymond A. Meyer
General Electric Company

Thomas J. Miller
NASA Lewis Research Center

Joseph C. Mills
Rockwell International Corporation
    Rocketdyne Division

Peyton Moore
Oak Ridge National Laboratory

Ira T. Myers
NASA Lewis Research Center

George F. Niederauer
Los Alamos National Laboratory

William E. Osmeyer
Teledyne Energy Systems

Lewis Peach
NASA Headquarters

Gary Polansky
Sandia National Laboratories

James Polk
Jet Propulsion Laboratory

James R. Powell
Brookhaven National Laboratory

William A. Ranken
Los Alamos National Laboratory

John W. Rice, Jr.
Idaho National Engineering Laboratory

Peter J. Ring
General Electric Company

M. Frank Rose
Auburn University

Richard Rovang
Rockwell International Corporation/
    Rocketdyne Division

Don Roy
Babcock & Wilcox Company

Paul H. Sager
General Dynamics Space Systems
    Division

Michael J. Schuller
Phillips Laboratory

Gene E. Schwarze
NASA Lewis Research Center

Stephen Seiffert
Benchmark Environmental Corp.

Joel C. Sercel
Jet Propulsion Laboratory

Harold A. Shelton
General Dynamics Space Systems
    Division

TECHNICAL PROGRAM COMMITTEE *(CONTINUED)*

Joseph A. Sholtis, Jr.
United States Air Force

James S. Sovey
NASA Lewis Research Center

R. Joseph Sovie
NASA Lewis Research Center

Marland L. Stanley
Idaho National Engineering Laboratory

Walter A. Stark
Los Alamos National Laboratory

Anthony Sutey
Boeing Aerospace Company

Melvin Swerdling
TRW, Inc.

Mark I. Temme
General Electric Aerospace Company

Charles W. Terrell
Phillips Laboratory

Frank V. Thome
Sandia National Laboratories

Robert H. Titran
NASA Lewis Research Center

James A. Turi
United States Department of Energy

Ronald F. Tuttle
United States Air Force Institute
of Technology

Jan W. Vandersande
Jet Propulsion Laboratory

Charles Vesely
Grumman Corporation

Carl E. Walter
Lawrence Livermore National
Laboratory

John W. Warren
United States Department of Energy

Joseph A. Weimer
Wright Laboratory

Joseph R. Wetch
Space Power, Inc.

Richard D. Widrig
Pacific Northwest Laboratory

Michael Wiemers
Westinghouse Hanford Company

Ken A. Williams
Science Applications International Corporation

David M. Woodall
Idaho National Engineering Laboratory

Thomas Wuchte
Phillips Laboratory

Francis J. Wyant
SandiaLaboratory Laboratories

SCHREIBER-SPENCE SPACE ACHIEVEMENT AWARD COMMITTEE

CHAIRMAN
Nils J. Diaz
University of Florida/INSPI

MEMBERS

Dale S. Dutt
Westinghouse Hanford Company

Thomas J. Hirons
Los Alamos National Laboratory

Richard A. Johnson
Rockwell International/Rocketdyne
    Division

Thomas J. Miller
NASA Lewis Research Center

Donald H. Roy
Babcock & Wilcox Company

Joseph A. Sholtis, Jr.
United States Air Force

MANUEL LUJAN, JR. STUDENT PAPER
AWARD COMMITTEE

CHAIRMAN
Frederick R. Best
Texas A&M University

MEMBERS

Hatice Cullingford
NASA Johnson Space Center

Philip Pluta
General Electric Company

Andrew Klein
Oregon Sate University

Michael Hall
Los Alamos National Laboratory

Ehsan U. Khan
General Electric Company

EDUCATION AND OUTREACH ADVISORY BOARD
(EOAB)

MEMBERS

HIGH SCHOOL SPECIAL SESSION
Education and Outreach Advisory Board (EOAB)

Co-sponsored by
INSTITUTE FOR SPACE NUCLEAR POWER STUDIES,
University of New Mexico
NASA LEWIS RESEARCH CENTER
NEW MEXICO SPACE GRANT CONSORTIUM
AMERICAN NUCLEAR SOCIETY - TRINITY SECTION

SESSION I

Irene L. El-Genk, Chairwoman
EOAB member
Albuquerque, NM

Rose Thome, Co-Chairwoman
EOAB member
Albuquerque, NM

Welcome
Irene L. El-Genk, Education Outreach Advisory Board, University of New Mexico, Albuquerque, NM

Introduction of Space Design Contest Winners
David Kauffman, Associate Dean, College of Engineering, University of New Mexico, Albuquerque, NM

Stellar Evolution
John A. Bernard, Ph.D., Massachusetts Institute of Technology, Cambridge, MA

Teamwork Makes It Happen
Sidney Gutierrez, NASA Astronaut, NASA Johnson Space Center, Houston, TX

SESSION II

Joanne Metzler, Chairwoman
EOAB member
Gallup/McKinley County Schools
Gallup, NM

Laura Reeves, Co-Chairwoman
EOAB member
Manzano High School
Albuquerque, NM

UNM College of Engineering Student Programs
Gail Ward, Ph. D., Director of Student Programs, College of Engineering, University of New Mexico, Albuquerque, NM

Medical and Biological Aspects of Space Travel
Barbara Lujan, NASA Headquarters, Washington, DC

Space Experiments
F. Andrew Gaffney, MD, Cardiologist and Payload Specialist on the Space Lab Mission, Southwest Medical School, Dallas, TX

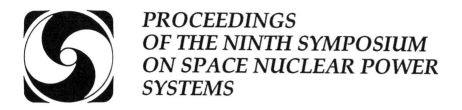

PROCEEDINGS OF THE NINTH SYMPOSIUM ON SPACE NUCLEAR POWER SYSTEMS

TABLE OF CONTENTS - PART ONE

TABLE OF CONTENTS - PART ONE

PLENARY SESSION IV
SPACE NUCLEAR PROGRAMS

Wade Carroll, Chairman
United States Department of Energy
Germantown, MD

Gary L. Bennett, Co-Chairman
NASA Headquarters
Washington, DC

POSTER SESSION

Gerald Farbman, Chairman
Starpath, Inc.
Pittsburgh, PA

Mark D. Hoover, Co-Chairman
Inhalation Toxicology Research Institute
Albuquerque, NM

TABLE OF CONTENTS - PART ONE

[1] SPACE EXPLORATION I

Gary L. Bennett, Chairman
NASA Headquarters
Washington, DC

Joe T. Howell, Co-Chairman
NASA Headquarters
Washington, DC

[2] REACTORS AND SHIELDING

Russell Ball, Chairman
Babcock & Wilcox Company
Lynchburg, VA

Carl Walter, Co-Chairman
Lawrence Livermore National
    Laboratory
Livermore, CA

TABLE OF CONTENTS - PART ONE

[3] MATERIALS I

J. P. Moore, Chairman
Oak Ridge National Laboratory
Oak Ridge, TN

R. William Buckman, Co-Chairman
Refractory Metal Alloys
Pittsburgh, PA

[4] RADIOISOTOPE POWER SYSTEMS

Arthur Mehner, Chairman
United States Department of Energy
Germantown, MD

Richard D. Rovang, Co-Chairman
Rockwell International Corporation/
   Rocketdyne Division
Canoga Park, CA

TABLE OF CONTENTS - PART ONE

[5] SPACE EXPLORATION II

Thomas J. Miller, Chairman
NASA Lewis Research Center
Cleveland, OH

Richard B. Harty, Co-Chairman
Rockwell International/
Rocketdyne Division
Canoga Park, CA

[6] RADIATION AND TEMPERATURE EFFECTS ON ELECTRONICS

Gene E. Schwarze, Chairman
NASA Lewis Research Center
Cleveland, OH

Donald B. King, Co-Chairman
Sandia National Laboratories
Albuquerque, NM

TABLE OF CONTENTS - PART ONE

[7] MATERIALS II

Steven L. Seiffert, Chairman
Benchmark Environmental Corp.
Albuquerque, NM

Robert H. Titran, Co-Chairman
NASA Lewis Research Center
Cleveland, OH

TABLE OF CONTENTS - PART ONE

[8] THERMOELECTRIC ENERGY CONVERSION

Jan W. Vandersande, Chairman
Jet Propulsion Laboratory
Pasadena, CA

Donald N. Matteo, Co-Chairman
General Electric Astro-Space
Valley Forge, PA

TABLE OF CONTENTS - PART ONE

[9] KEY NUCLEAR TECHNOLOGIES
FOR HUMAN EXPLORATION OF THE SOLAR SYSTEM

John Mankins, Chairman
NASA Headquarters
Washington, DC

Joseph Angelo, Co-Chairman
Science Applications International
Corporation
Melbourne, FL

[10] SPACE POWER ELECTRONICS I

Frank V. Thome, Chairman
Sandia National Laboratories
Albuquerque, NM

Joseph A. Weimer, Co-Chairman
Wright Laboratory
Wright Patterson AFB, OH

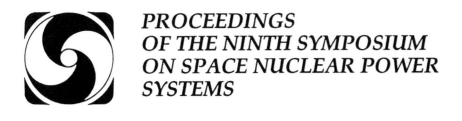

PROCEEDINGS OF THE NINTH SYMPOSIUM ON SPACE NUCLEAR POWER SYSTEMS

TABLE OF CONTENTS - PART TWO

[11] REACTOR AND POWER SYSTEMS CONTROL I

Samit K. Bhattacharyya, Chairman
Argonne National Laboratory
Argonne, IL

Francis J. Wyant, Co-Chairman
Sandia National Laboratories
Albuquerque, NM

[12] THERMIONIC ENERGY CONVERSION I

Richard A. Johnson, Chairman
Rockwell International Corporation
Canoga Park, CA

Richard J. Bohl, Co-Chairman
Los Alamos National Laboratory
Los Alamos, NM

TABLE OF CONTENTS - PART TWO

[13] SPACE MISSIONS AND POWER NEEDS

Steve Harrison, Chairman
National Space Council
Washington, DC

David Buden, Co-Chairman
Idaho National Engineering
Laboratory

[14] SPACE POWER ELECTRONICS II
(CANCELED)

[15] REACTOR AND POWER SYSTEM CONTROL II

Francis J. Wyant, Chairman
Sandia National Laboratory
Albuquerque, NM

Samit Bhattacharyya, Co-Chairman
Argonne National Laboratory
Argonne, IL

TABLE OF CONTENTS - PART TWO

[16] THERMIONIC ENERGY CONVERSION II

Richard Bohl, Co-Chairman
Los Alamos National Laboratory
Los Alamos, NM

Richard A. Johnson, Chairman
Rockwell International Corporation
Canoga Park, CA

[17] KEY ISSUES IN NUCLEAR POWER AND PROPULSION

Nils Diaz, Chairman
University of Florida
Gainesville, FL

Edmund P. Coomes, Co-Chairman
Pacific Northwest Laboratory
Richland, WA

TABLE OF CONTENTS - PART TWO

[18] NUCLEAR THERMAL PROPULSION I

John S. Clark, Chairman
NASA Lewis Research Center
Cleveland, OH

James R. Powell, Co-Chairman
Brookhaven National Laboratory
Upton, NY

[19] MANUFACTURING AND PROCESSING

Wayne R. Amos, Chairman
EG & G Mound Technologies, Inc.
Miamisburg, OH

Wayne M. Brittain, Co-Chairman
Teledyne Energy Systems
Timonium, MD

TABLE OF CONTENTS - PART TWO

[20] THERMAL MANAGEMENT I

Albert J. Juhasz, Chairman
NASA Lewis Research Center
Cleveland, OH

Maribeth E. Hunt, Co-Chairwoman
Rockwell International/
Rocketdyne Division
Canoga Park, CA

[21] SPACE NUCLEAR SAFETY I: POLICY AND REQUIREMENTS

George F. Niederauer, Chairman
Los Alamos National Laboratory
Los Alamos, NM

Ronald F. Tuttle, Co-Chairman
Air Force Institute of Technology
Wright-Patterson AFB, OH

TABLE OF CONTENTS - PART TWO

[22] NUCLEAR THERMAL PROPULSION II

Don Roy, Chairman   **Gary Polansky, Co-Chairman**
Babcock and Wilcox Company   Sandia National Laboratories
Lynchburg, VA   Albuquerque, NM

[23] NUCLEAR TESTING AND PRODUCTION FACILITIES

Michael Wiemers, Chairman   **Mario Fontana, Co-Chairman**
Westinghouse Hanford Company   Oak Ridge National Laboratory
Richland, WA   Oak Ridge, TN

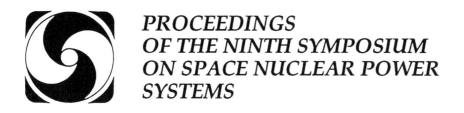

PROCEEDINGS OF THE NINTH SYMPOSIUM ON SPACE NUCLEAR POWER SYSTEMS

TABLE OF CONTENTS - PART THREE

TABLE OF CONTENTS - PART THREE

[26] NUCLEAR THERMAL PROPULSION III

Patrick J. McDaniel, Chairman
Sandia National Laboratories
Albuquerque, NM

David Woodall, Co-Chairman
Idaho National Engineering
Laboratory
Idaho Falls, ID

[27] SIMULATION AND MODELING I

Edwin A. Harvego, Chairman
Idaho National Engineering Laboratory
Idaho Falls, ID

Andrew C. Klein, Co-Chairman
Oregon State University
Corvallis, OR

TABLE OF CONTENTS - PART THREE

[28] HEAT PIPE TECHNOLOGY I

E. Tom Mahefkey, Chairman
Wright Laboratory
Wright-Patterson AFB, OH

Amir Faghri, Co-Chairman
Wright State University
Dayton, OH

[29] FLIGHT QUALIFICATION AND TESTING

Thomas J. Wuchte, Chairman
Phillips Laboratory
Kirtland AFB, NM

David Woodall, Co-Chairman
Idaho National Engineering
Laboratory
Idaho Falls, ID

TABLE OF CONTENTS - PART THREE

[30] NUCLEAR THERMAL PROPULSION IV

Steven Howe, Chairman
Los Alamos National Laboratory
Los Alamos, NM

John Martinell, Co-Chairman
EG & G Idaho, Inc.
Idaho Falls, ID

[31] SIMULATION AND MODELING II

Andrew C. Klein, Chairman
Oregon State University
Corvallis, OR

Edwin A. Harvego, Co-Chairman
Idaho National Engineering
Laboratory
Idaho Falls, ID

TABLE OF CONTENTS - PART THREE

[32] HEAT PIPES TECHNOLOGY II

Walter B. Bienert, Chairman
DTX Corporation
Cockeysville, MD

Michael A. Merrigan, Co-Chairman
Los Alamos National Laboratory
Los Alamos, NM

[33] APPLIED TECHNOLOGY I: CORE MATERIALS

Janelle Hales, Chairman
Westinghouse Hanford Company
Richland, WA

Walter Stark, Co-Chairman
Los Alamos National Laboratory
Los Alamos, NM

[34] NUCLEAR ELECTRIC PROPULSION I

James Sovey, Chairman
NASA Lewis Research Center
McLean, VA

James Polk, Co-Chairman
Jet Propulsion Laboratory
Pasadena, CA

TABLE OF CONTENTS - PART THREE

[35] MICROGRAVITY TWO PHASE FLOW

Zenen I. Antoniak, Chairman
Pacific Northwest Laboratory
Richland, WA

Frederick R. Best, Co-Chairman
Texas A & M University
College Station, TX

[36] SPACE POWER AND PROPULSION TECHNOLOGY I

Dan Mulder, Chairman
Phillips Laboratory
Kirtland AFB, NM

Yuri N. Niikolayev, Co-Chairman
Scientific and Industrial Association
(LUCH) Podlosk, USSR

TABLE OF CONTENTS - PART THREE

[37] APPLIED TECHNOLOGY II: FUEL MATERIALS

Dale S. Dutt, Chairman
Westinghouse Hanford Company
Richland, WA

Dewayne L. Husser, Co-Chairman
Babcock & Wilcox Company
Lynchburg, VA

[38] NUCLEAR ELECTRIC PROPULSION II

John Brophy, Chairman
Jet Propulsion Laboratory
Pasadena, CA

Jeffrey E. Dagle, Co-Chairman
Pacific Northwest Laboratory
Richland, WA

TABLE OF CONTENTS - PART THREE

SP-100 REACTOR WITH BRAYTON CONVERSION FOR LUNAR SURFACE APPLICATIONS

Lee S. Mason, Carlos D. Rodriguez, and Barbara I. McKissock
NASA Lewis Research Center
Cleveland, OH 44135
(216) 977-7106, 977-7116, 433-6102

James C. Hanlon
Sverdrup Technology, Inc.
Brookpark, OH 44142
(216) 977-7118

Brian C. Mansfield
University of Dayton
Dayton, OH 45409
(513) 223-6438

Abstract

This study examines the potential for integration of Brayton cycle power conversion with the SP-100 reactor for lunar surface power system applications. Two designs were characterized and modeled. The first design integrates a 100 kWe SP-100 Brayton power system with a lunar lander. This system is intended to meet early lunar mission power needs while minimizing on-site installation requirements. Man-rated radiation protection is provided by an integral multi-layer, cylindrical LiH/W shield encircling the reactor vessel. Design emphasis is on ease of deployment, safety and reliability while utilizing relatively near term technology. The second design combines Brayton conversion with the SP-100 reactor in an erectable 550 kWe power plant concept intended to satisfy later phase lunar base power requirements. This system capitalizes on experience gained from operation of the initial 100 kWe module while incorporating some technology improvements. For this system, the reactor is emplaced in a lunar regolith excavation to provide man-rated shielding and the Brayton engines and radiators are mounted on the lunar surface and extend radially from the central reactor. Design emphasis is on performance, safety, long life, and operational flexibility.

INTRODUCTION:

Recent studies examining approaches for lunar base missions have suggested a need for nuclear reactor power systems (*America at the Threshold*, Synthesis Group Report). Most mission development strategies suggest a phased approach for meeting mission objectives. Power requirements for permanent occupancy lunar surface missions range from 10 s of kilowatts for the early emplacement phases to 100 s of kilowatts for the latter operational phases. Nuclear reactor power systems provide a low mass, long life option for meeting these requirements.

One strategy for satisfying lunar base power requirements is through a centralized utility. Power could be generated by multiple systems and provided to a central user-common switching station. From the switching station, electric power would be distributed to the various users as shown in Figure 1. The stated power requirements are commensurate with results from NASA's 90 day study of the Moon and Mars. Within the power generation area, an initial nuclear reactor system could be emplaced which would have the capacity to meet near term power requirements associated with the emplacement phase. Principal power users for this phase of the mission might include the crew habitat, science platforms, rover recharging facilities, and in-situ resource utilization (ISRU) demonstrations. A photovoltaic and regenerative fuel cell (PV/RFC) system might also be utilized to provide redundant power to the habitat life support systems. If the lunar base grows and power requirements increase to accommodate laboratory modules, constructible habitats, liquid oxygen (LLOX) plants, launch and landing servicing facilities, and expanded science, a subsequent larger nuclear reactor system could be delivered to compliment and eventually replace the original system.

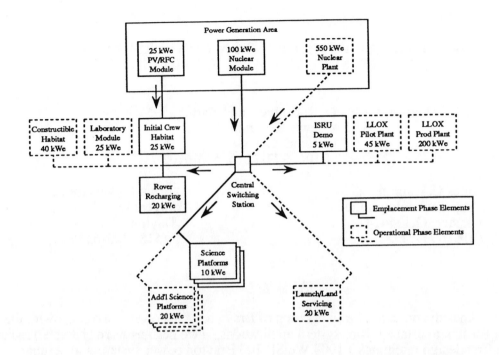

FIGURE 1.   Representative Base Layout and Power Distribution Network.

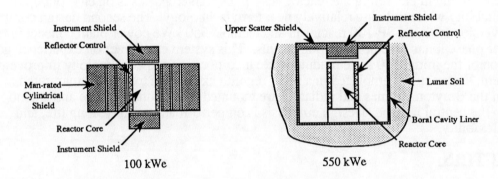

100 kWe   550 kWe

FIGURE 2.   Reactor Radiation Shielding Strategies to Achieve Man Rating.

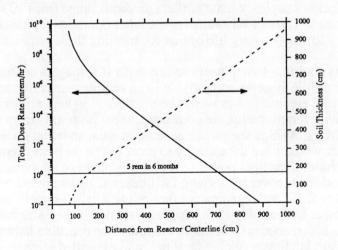

FIGURE 3.   Radiation Attenuation of a SP-100 Reactor Located in a Lunar Excavation.

GROUNDRULES AND DESIGN GUIDELINES:

This study characterizes two different SP-100 Brayton systems. The first system is sized for 100 kWe and is designed to be self-deployable. It was assumed that this system would be one of the first elements delivered to the lunar surface and therefore could rely on no equipment or manpower for its installation. It was concluded that a design in which the power system is integrated with a lunar lander would best satisfy this requirement. This concept was explored as an option for an easily deployable, self contained power system for a periodically tended lunar observatory mission (Hickman, Bloomfield 1989). An integrated LiH/W enclosure shield is included for crew radiation protection. While it was assumed that no infrastructure would be available for the deployment of the system, it was assumed that crew members would be available for the final electrical power transmission line connections and system start-up. A 10 year equivalent full power system lifetime was assumed. Operation of the reactor at less than full thermal power could potentially extend the system's service life.

The second system is sized for 550 kWe. This system is assumed to be delivered to the Moon when power requirements have increased to account for a full operational LLOX production facility. Because this system would be delivered once the base is established, it was assumed that a crew would be available for its installation. The basic concept consists of a single SP-100 reactor located in a cylindrical hole with surface mounted Brayton engines and radiators. Man-rated radiation protection is provided through the emplacement of the reactor in the excavation. Despite the crew availability, this power plant is designed for quick and easy assembly. It would be delivered to the site with all interface piping pre-connected and would only require the placement of the reactor in the excavation, installation of the radiator panels and final electrical transmission line connections. As in the 100 kWe case, a 10 year equivalent full power system life was assumed.

Reactor and Primary Heat Transport System

In both cases, SP-100 reactor technology is assumed. SP-100 is a joint DOE, NASA, DOD program to develop a space reactor power system. The reactor subsystem consists of UN fuel pins, reflector controls, safety rods, pressure vessel, auxiliary coolant loop, and instrumentation and control. The thermal power of the reactor was allowed to vary based on the electrical power output and the power conversion efficiency. In addition to the man-rated shielding, the 100 kWe and 550 kWe systems also include an instrument radiation shield directly above the reactor for the reflector control actuators, safety rod drives and reactor instrumentation. The 100 kWe system employs an additional instrument shield below the reactor to reduce ground scattering. Since SP-100 is a liquid lithium cooled reactor, a liquid metal to gas heat exchanger is required for the Brayton. Both systems utilize a segmented heat exchanger with redundant inlet and outlet manifolds for the reactor and one for each of the Brayton engines. Included in the primary heat transport system is the heat exchanger, inlet and outlet piping, and electro-magnetic pumps.

Man-rated Shielding

The radiation shielding strategies for both the 100 kWe and 550 kWe systems are shown in Figure 2. The 100 kWe system utilizes an integral man-rated shield for protection of the lunar base crew members. The shield circumferentially surrounds the reactor core and consists of alternating layers of tungsten and lithium hydride. The circumferential shield, while primarily intended to attenuate radiation directed towards the crew habitat, permits crew excursions and base expansion in any direction around the system. The dose constraints for sizing this shield were conservatively selected to be less than 5 rem in 6 months at a separation distance of 1000 m. This dose rate would vary with the inverse square of the separation distance. The 5 rem dose limit is consistent with the OSHA annual dose limit for terrestrial nuclear power plant workers and 6 months has been suggested as a typical tour of duty for a lunar base crew member. This limit is far below the 50 rem annual limit for astronauts proposed in NCRP Report Number 98. The conservative dose constraint was chosen to minimize the reactor's contribution to the cumulative radiation environment (including natural and other man-made sources) imposed on the crew.

The 550 kWe system relies on an excavated cylindrical hole to provide man-rated shielding. The surrounding lunar regolith acts to attenuate radiation in all directions around the reactor. Figure 3 shows that radiation from a 2.5 MWt SP-100 type reactor enclosed in a 3.5 m deep, 1.5 m diameter hole can be attenuated to less than 5 rem in 6 months at a distance of 7 m from the reactor centerline. This shielding approach eliminates the need for the massive circumferential shield used in the 100 kWe system designs. The excavation shield also permits relatively close, human proximity operations and the potential for repair of surface mounted power system equipment. For this design, the same 3.5 m deep and 1.5 m diameter excavation is assumed. In addition, the reactor assembly is completely enclosed in a multi-material bulkhead consisting of a 10 cm thick LiH upper scatter shield and a 2.5 cm thick Boral hole liner. The bulkhead is included to reduce the neutron scatter current out the top and through the sides of the excavation while enabling easier handling, delivery, and emplacement of the reactor assembly. The thermal conditions within the bulkhead result in the need for a dedicated cavity cooling system. This is accomplished through coolant tanks surrounding the bulkhead which feed surface mounted radiators.

Brayton Cycle Power Conversion

Since the 100 kWe system would likely be one of the earliest delivered lunar base elements, it was decided to utilize a non-refractory Brayton system and limit the turbine inlet temperature to 1144 K as a "baseline". Brayton rotating units operating at turbine inlet temperatures in this range underwent successful testing during the late 1960 s and early 1970 s. Since the mission start date is undefined, a high temperature system was also evaluated which fully utilized the 1350 K outlet temperature of the baseline SP-100 reactor. This "advanced" design point has a 1300 K turbine inlet temperature. The 550 kWe system would be delivered somewhat later in the development of the lunar base. For this reason, it was assumed that a refractory Brayton system could be available. Therefore, the turbine inlet temperature was based on the outlet temperature of the SP-100 reactor. The higher temperature can further be justified by considering the 100 kWe, 1144 K system as an evolutionary step toward the design and implementation of the 550 kWe, 1300 K system. The cycle state points have been chosen in both cases to represent a compromise between the minimum mass design point and the minimum radiator area design point. Table 1 provides a list of the parameters which were varied in determining the final design point.

For this study, the Brayton engines were assumed to operate in a parallel mode with at least one redundant engine. Parallel operation assumes that all engines operate initially at partial power. If an engine failure occurs, the power of the remaining engines can be increased to account for the loss. The alternative would be to operate the required engines at full power and maintain the redundant engines in cold standby. The advantage of the parallel mode is that there are no cold engine start-up problems and no complicated switching schemes in the primary and secondary heat transport loops. The disadvantage of parallel operation is that the engines must be able to operate over a wide range of operating characteristics (i.e. off-design). However, this disadvantage is not as extreme in the case of Brayton engines which are able to tolerate off-design conditions with minimal performance penalty.

Heat Rejection

The heat rejection system for the 100 kWe system is designed to fit within the lander system and be self-deployable. Previous designs for systems of this nature have shown a conical, deployable radiator which extends above the lander and is located within the instrument shield half angle to eliminate back scattering (Hickman, Bloomfield 1989). This is probably not the best approach because the radiator panels view downward toward the lunar surface rather than outward toward the colder sink of deep space. Analyses for the radiator design focused on configurations which minimize radiator area through favorable view factor geometries. Waste heat from all the Brayton engines is transferred to a single, gas to liquid NaK heat exchanger. This system uses a shared radiator system in which the NaK is pumped through an armored manifold to a number of heat pipe radiator panels connected in series. Since this manifold is vulnerable to single point failures, an independent stand-by loop was included.

For the 550 kWe system, each Brayton engine has its own dedicated radiator panel capable of rejecting the waste heat associated with peak engine power. During nominal operation, only a percentage of the radiator is used to reject the required heat load. This design approach allows failure of a radiator panel (or Brayton engine) without compromising full power output. Each Brayton engine has a waste heat, gas to liquid metal heat exchanger and a secondary pumped loop manifold with NaK as the working fluid. Heat pipe, heat rejection was selected for both the 100 kWe and 550 kWe systems because of the inherent reliability associated with the large number of heat pipes in any particular radiator segment. In both cases, mass estimates were determined from detailed analyses rather than an assumed mass per unit radiator area.

Brayton cycle conversion intrinsically rejects waste heat over a wide temperature range. For this reason, multiple heat pipe fluids may be required depending on cycle state point conditions. Another approach which may better accommodate the wide temperature range is a pumped loop radiator system. However, pumped loop systems are more vulnerable to single point failures. This vulnerability can be reduced by a combination of armoring, redundancy and loop-segmenting at the expense of additional mass. Further study is necessary to fully examine the advantages and disadvantages of pumped loop heat rejection for this application.

Power Conditioning, Control and Distribution

The power conditioning, control and distribution (PCC&D) approach for both the 100 and 550 kWe systems is presented in Figure 4. For both systems, an ac-ac converter is included to convert the alternator output to a suitable voltage for long distance transmission. Studies have shown that high voltage transmission is a favorable method for reducing the mass of the cabling for lunar base missions with multiple, distributed users (Gordon 1990). Long distance cabling might also be desirable when nuclear systems are utilized so as to provide safe separation distances between crew and power plant. The ac-ac converter consists of transformers, ancillary control and circuit protection in a thermally conductive enclosure with heat pipe radiator cooling. The output of the converter is assumed to be 5000 Vrms which is transmitted to a switching station located 1 km from the power plant. The transmission lines to and from the ac-ac converter were assumed as vacuum insulated, three wire, aluminum ac conductors. The Brayton alternators for the 100 kWe system would utilize a shared power conditioning and control system with the number of independent channels and output transmission lines matching the total number of rotating units. The 550 kWe system would employ an engine dedicated power conditioning and control system with redundant channels for each converter. The output of the converters would then be combined into a switchgear box with a single, redundant transmission line output.

100 KWe SYSTEM DESIGN:

The 100 kWe design has the reactor, power conversion units, and heat exchangers enclosed in a cylindrical protective shell which is supported by the lander structure. The deployed system is presented in Figure 5. A cross section of the internal components of that shell is shown in Figure 6 with the cylindrical man-rated shield cut away to display the reactor core. The range of temperatures for heat rejection required that a combination of mercury and water heat pipes be used for both designs. A schematic of the radiator design is presented in Figure 7. In an attempt to maximize the radiator's view of deep space, the final radiator design consists of four horizontal panels which extend radially from the common waste heat exchanger. Each of the panels is 4 m

TABLE 1.   Parameters Varied in Determining Design Point Selection.

Reactor Subsystem:
    Core L/D
    Fuel Pin Diameter
    Fuel Smear Density
    Gamma Shield Thickness
    Heat Source Heat Exchanger Effectiveness
    Heat Source Heat Exchanger $\Delta P/P$
    Heat Source Heat Exchanger Heat Capacity Ratio

Power Conversion Subsystem:
    Number of Engines
    Compressor Inlet Temperature
    Compressor Pressure Ratio
    Compressor Bleed Flow
    Compressor Inlet Pressure
    Recuperator Effectiveness
    Recuperator $\Delta P/P$
    Duct Diameters
    Rotating Speed
    He-Xe Molecular Weight

Heat Rejection Subsystem:
    Heat Transport Duct $\Delta P/P$
    Heat Transport Duct Length
    Fin Width
    Materials Selection
    Waste Heat Exchanger Effectiveness
    Waste Heat Exchanger $\Delta P/P$
    Number of Heat Pipes
    Heat Pipe Redundancy
    Heat Pipe Length (550 kW system only)

Power Conditioning, Control & Distribution Subsystem:
    Transformer Efficiency
    Alternator Output Conductor Temperature
    Alternator Output Conductor Diameter
    Alternator Output Conductor Efficiency
    Power Conditioning Output Conductor Temperature
    Power Conditioning Output Conductor Diameter
    Power Conditioning Output Conductor Efficiency

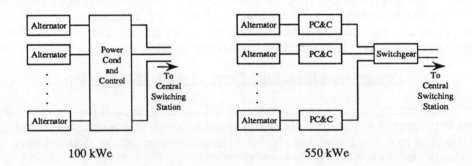

100 kWe   550 kWe

FIGURE 4.   Power Conditioning, Control and Distribution Approaches.

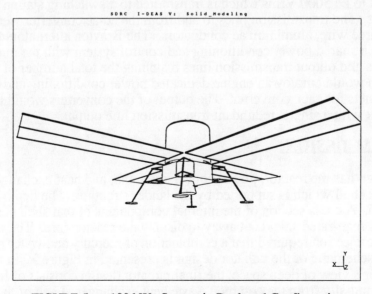

FIGURE 5.   100 kWe System in Deployed Configuration.

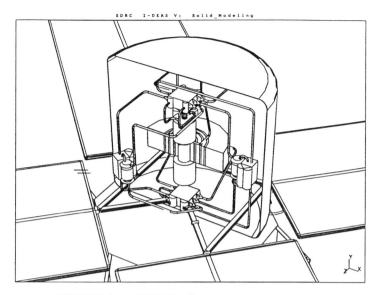

FIGURE 6.   100 kWe System Design Layout.

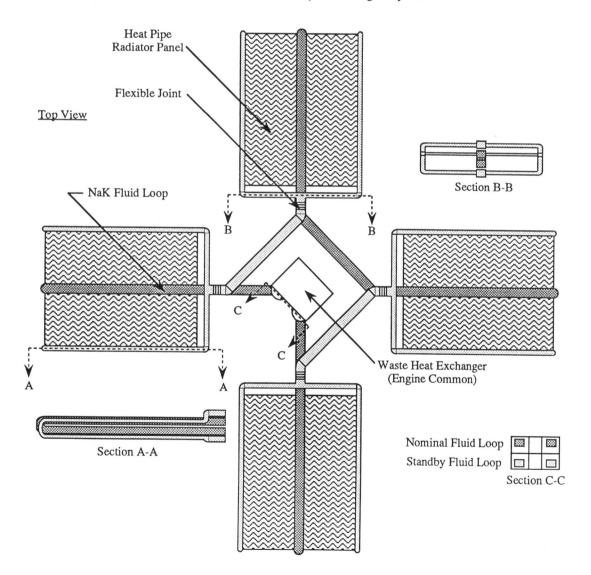

FIGURE 7.   100 kWe System Radiator Design.

872

wide. The heat transport duct lengths were found to be 16 m for the baseline case and 11 m for the advanced case. In order to minimize the effect of incident sunlight, optical solar reflectors are employed on the radiator which render a surface emissivity of 0.80 and a solar absoptivity of 0.08. Radiation scattering is avoided since the deployed radiator panels are within the zone protected by the circumferential shield. The system would be delivered on the lander with the radiator panels stowed upward as shown in Figure 8 and deployed by battery powered motors upon landing.

Figure 9 shows the effect of the number of engines on system reliability and mass. Component reliabilities were assumed as 0.99 for the reactor, 0.95 for each Brayton engine, 0.99 for the radiator, and 0.98 for the PCC&D. The inherent approximation resulting from applying reliability equations to first-of-a-kind systems such as the ones described in this report is acknowledged. In the context of this study, the particular values determined for system reliability were of minimal value. Rather, it was the trend which influenced the selection of the number of engines based on the point where increasing system mass was of little benefit in improving system reliability. Curves are provided for cases of both one and two spare engines. System reliability is highest for designs employing a minimal number of engines. These designs have a greater fraction of engine redundancy (ratio of spare engines to total engines). System mass is also highest for these designs since such a large portion of the power conversion capacity goes unused under normal circumstances. Lower system mass is achieved as the number of engines is increased and the fraction of engine redundancy is reduced. However, the mass savings become negligible for large numbers of engines. As exhibited in the baseline case, this trend eventually yields a minimum mass design after which the addition of more engines acts as a liability to both mass and reliability. In an attempt to satisfy both high system reliability and low system mass while keeping the number of engines to a minimum to simplify interface piping, a final design point of 3 engines including one spare was selected for both the baseline and advanced 100 kWe cases. The comparison between one and two redundant engine systems shows that for the single spare designs, low mass and high reliability can be achieved with much fewer total engines.

Design point performance and mass for the two 100 kWe cases is presented in Table 2. The engine design power coincides with the case of all engines operating in parallel to produce the 100 kWe net power output. In the event of a failure, the two remaining engines would operate off-design at approximately 50 kWe per unit to provide full power output. Figure 10 shows the trade-off of system mass and radiator area. The right most data point of each curve represents the global minimum mass design. The data points to the left are cases in which the mass has been minimized while constraining radiator area to be less than the global mass minimum. The design point was chosen to reflect a compromise between minimum mass and minimum radiator area. As indicated in the performance summary, the man-rated shield is the dominant mass item of the two systems comprising nearly 50% of the overall mass. A modest mass savings of 855 kg is realized for the higher turbine inlet temperature design conditions.

550 KWe SYSTEM DESIGN:

The 550 kWe system is depicted in Figure 11 showing a close-up of the reactor and power conversion system. The radiator panels are arranged in a vertical orientation and extend radially from the engine dedicated waste heat exchangers. As in the 100 kWe cases, mercury and water heat pipes are utilized for heat rejection. The radiator design schematic for this system is shown in Figure 12. The panels are tapered with three sections of varying height for the high temperature mercury, low temperature mercury, and water heat pipes. The total duct length of a radiator panel was determined to be 35 m. These panels would be delivered to the lunar surface in sections and assembled at the site. An insulating lunar surface blanket was employed to improve the spectral characteristics of the lunar soil and reduce the effective sink temperature to 225 K (Bien, Guentert, 1969). In order to minimize the probability of secondary loop meteoroid punctures, the radiator piping and heat pipe evaporator sections would be buried in the lunar regolith.

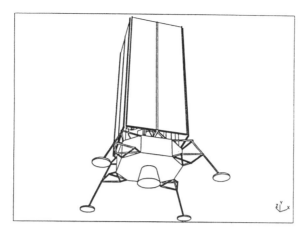

FIGURE 8.   Stowed Configuration for Landing.

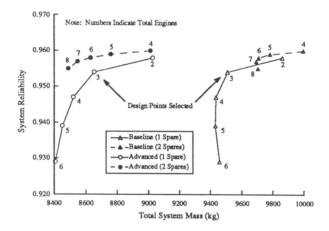

FIGURE 9.   100 kWe System Reliability and Mass Trade-off.

TABLE 2.   100 kWe Performance Summary.

	Baseline	Advanced
Turbine Inlet Temp, K	1144	1300
Compressor Inlet Temp, K	379	414
Compressor Inlet Pressure, kPa	137	157
Compressor Pressure Ratio	1.86	1.81
Recuperator Effectiveness	0.939	0.918
He-Xe Molecular Weight	22.8	25.8
Rotating Speed, rpm	58097	57950
Engine Design Power, kWe	35.6	35.6
Engine Efficiency	0.279	0.273
Net System Efficiency	0.261	0.256
Reactor Thermal Power, kWt	383	391
Total Main Radiator Area, m2	260	180
Reactor Subsystem, kg	1136	1143
Man-rated Shield Subsystem, kg	4193	4251
Brayton Subsystem, kg	1463	1082
Heat Reject Subsystem, kg	2219	1680
PCC&D Subsystem, kg	495	495
Total Mass, kg	9506	8651
System Specific Mass, kg/kWe	95	87

874

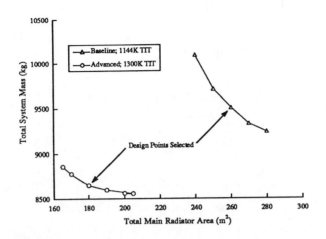

FIGURE 10.   100 kWe System Mass and Radiator Area Trade-off.

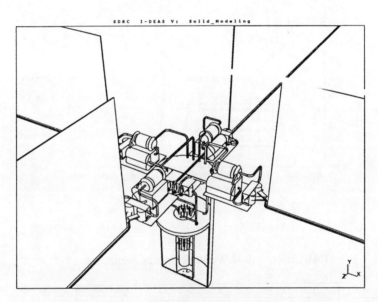

FIGURE 11.   550 kWe System Design Layout.

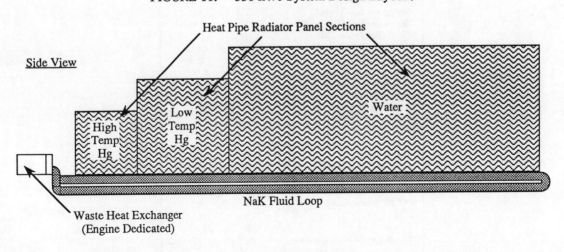

FIGURE 12.   550 kWe System Radiator Design.

875

The effect of the number of engines on mass and reliability for the 550 kWe system is provided in Figure 13. Component reliabilities were identical to those assumed for the 100 kWe cases. The effect of engine redundancy on system mass is more pronounced in this design since each engine has a dedicated radiator and PCC&D subsystem. For this application, a final design point of 4 engines including one spare was chosen to reflect the best compromise of system reliability and mass. As in the 100 kWe cases, there appears to be no advantage for multi-spare designs.

System performance and mass is presented in Table 3. The 196.2 kWe engine design power is based on the situation where one engine has failed and the remaining three must produce the rated 550 kWe system output. This approach was used in sizing the higher power system so that worst case conditions could be imposed on the heat rejection subsystem. Initially, the engines would be operated off-design at a lesser power level to achieve the rated system power. The trade-off of system mass and radiator area is presented in Figure 14. The global mass minimum design is the furthest right of the data points with area constrained mass minimum design points toward the left. By accepting a 2% mass penalty from the global mass minimum, almost 10% radiator area savings is realized for the final design point. The system specific mass represents a significant improvement over the two 100 kWe systems. This can be attributed to component economies of scale resulting from the higher power output and the utilization of in-situ materials to provide man-rated shielding.

CONCLUSION

Both of the designs presented in this paper offer distinct advantages for lunar surface power generation. The 100 kWe system provides a safe, reliable design which requires minimal manpower for installation and uses relatively near term technology. It is ideal for initial lunar base power requirements. The 550 kWe system is applicable when power requirements have increased to accommodate extensive in-situ resource utilization. Its design is consistent with the needs of an evolved lunar base: performance, safety, long life and operational flexibility. The potential advantages of the centralized power utility approach suggested in this report include redundancy, user growth accommodation, reduced development cost and simplified logistics. For nuclear reactor systems, an additional advantage of a central utility is that all of the power systems can be co-located in a single remote area for safety.

Acknowledgments

This work was performed at the NASA Lewis Research Center and accomplished through a combined effort of the Advanced Space Analysis Office and the Power Technology Division.

References

America at the Threshold - America's Space Exploration Initiative, Report of the Synthesis Group, U.S. Government Printing Office, Washington D.C., 20402, June, 1991.

Hickman, J.M. and H.S. Bloomfield (1989) "Comparison of Solar Photovoltaic and Nuclear Reactor Power Systems for a Human-Tended Lunar Observatory," NASA TM 102015.

NCRP (1989) *Guidance on Radiation Received in Space Activities*, Report No. 98, National Council on Radiation Protection and Measurements, Bethesda, MD, 20814, July, 1989.

Gordon, L.B. (1990) "Electrical Transmission on the Lunar Surface," Interim Report under NASA Grant # NASA-NAG-3-1055, Space Power Institute, Auburn University, April, 1990.

Bien, D.D. and D.G. Guentert (1969) "A Method for Reducing the Equivalent Sink Temperature of a Vertically Oriented Radiator on the Lunar Surface," NASA TM X-1729.

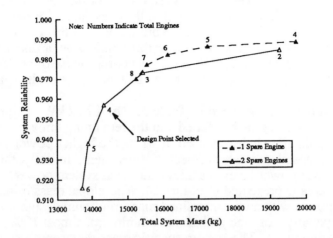

FIGURE 13.   550 kWe System Reliability and Mass Trade-off.

TABLE 3.   550 kWe Performance Summary.

Turbine Inlet Temp, K	1300
Compressor Inlet Temp, K	450
Compressor Inlet Pressure, kPa	508
Compressor Pressure Ratio	1.90
Recuperator Effectiveness	0.872
He-Xe Molecular Weight	28.2
Rotating Speed, rpm	47424
Engine Design Power, kWe	196.2
Engine Efficiency	0.222
Net System Efficiency	0.207
Reactor Thermal Power, kWt	2657
Total Main Radiator Area, m2	1100
Reactor Subsystem, kg	1658
Excavation Structure, kg	960
Brayton Subsystem, kg	2526
Heat Reject Subsystem, kg	6791
PCC&D Subsystem, kg	2402
Total Mass, kg	14337
System Specific Mass, kg/kWe	26

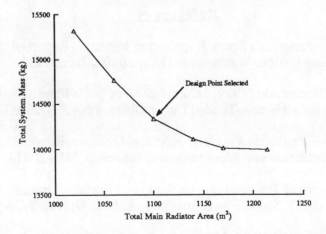

FIGURE 14.   550 kWe System Mass and Radiator Area Trade-off.

877

A LOW-ALPHA NUCLEAR ELECTRIC PROPULSION SYSTEM FOR LUNAR AND MARS MISSIONS

Edmund P. Coomes and Jeffery E. Dagle
Pacific Northwest Laboratory
P.O. Box 999 M/S K5-21
Richland, WA 99352
(509) 375-2549

Abstract

The advantages of using electric propulsion for propulsion are well-known in the aerospace community. The high specific impulse and, therefore, lower propellant requirements make it a very attractive propulsion option for the Space Exploration Initiative (SEI). Recent studies have shown that nuclear electric propulsion (NEP) is not only attractive for the transport of cargo but that fast piloted missions to Mars are possible as well, with alphas on the order of 7.5 kg/kW. An advanced NEP system with a specific power (alpha) of 2.5 kg/kW or less would significantly enhance the manned mission option of NEP by reducing the trip time even further. This paper describes an advanced system that combines the PEGASUS Drive with systems of the Rotating Multimegawatt Boiling Liquid Metal (RMBLR) power system that was developed as part of the DOE multimegawatt program and just recently declassified. In its original configuration, the PEGASUS Drive was a 10-MWe propulsion system. The RMBLR was a 20-MW electric power system. By combining the two, a second-generation PEGASUS Drive can be developed with an alpha less than 2.5 kg/kW. This paper will address the technology advancements incorporated into the PEGASUS Drive, the analysis of a fast piloted mission and an unmanned cargo transport Mars mission, and the integration of laser power beaming to provide surface power.

INTRODUCTION

The current goal of the American space program is the expansion of human presence into the solar system. The development of a permanent presence in space will require a continuous movement outward from Earth. The desire is to establish a permanent base on the moon and to land an American on Mars by the 50th Anniversary of the Apollo 11 lunar landing. The Space Synthesis Group, under Retired General Thomas Stafford, examined various architectures and technologies necessary to accomplish these goals, and a major concern was the logistic support required for both personnel and supplies. A global support infrastructure for power and propulsion must be developed to sustain this outward movement. An infrastructure based on an integrated energy approach is needed. Two areas that easily lend themselves to this are electric propulsion and planetary surface power. Incorporating the space transportation requirements into integrated planning and taking advantage of hardware commonality, a multimegawatt nuclear electric propulsion can meet the needs for both fast piloted missions and much slower cargo transport missions. In addition, if the NEP system included a long-life (7 to 10 years) power source, it looks very attractive as a dual function system that would double as the surface energy source when coupled with a beam-power transmission system (Bamberger 1990). This integrated energy approach to space power and space propulsion would reduce the number of new systems and technologies that need to be developed, provide a commonality of system hardware, and reduce the economic impact of SEI.

The high specific impulse and, therefore, lower propellant requirements make it a very attractive propulsion option for the Space Exploration Initiative (SEI). Recent studies (George, Hack and Dudzinski 1991) have shown that not only is electric propulsion attractive for the transport of cargo, but fast piloted missions to Mars, on the order of 400 days, are possible using near-term low power nuclear electric propulsion (NEP) with an alpha on the order of 7.5 kg/kW.

An advanced NEP system with an alpha of 2.5 kg/kW or less would significantly enhance the manned mission option of NEP by reducing the trip time even further.

The Rotating Multimegawatt Boiling Liquid Metal (RMBLR) power system (Johnson et al. 1987) was developed as part of the DOE multimegawatt program and just recently declassified. The RMBLR was a family of reactor concepts ranging in power output from 10 MWe to 500 MWe. All of these were direct Rankine cycle systems. The PEGASUS Drive (Coomes et al. 1987), in its original configuration, was a 10-MW nuclear electric propulsion system. By incorporating the technology advances of the RMBLR concept in reactor design, power conversion, and ultra-light fabric heat rejection systems, a second-generation PEGASUS Drive can be developed with an alpha less than 2.5 kg/kW. This propulsion system would allow a 400-MT Mars cargo vehicle, assembled and loaded in low Earth orbit (LEO), to deliver 193 MT of supplies and hardware to low Mars orbit 282 days after escaping Earth orbit.

Incorporating a laser power transmitter into the PEGASUS Drive and adding a laser power receiver to the Mars surface cargo manifest would make available 2 MW of electric power on the Mars surface. Upon arrival at Mars, the cargo transport vehicle would place its cargo into the desired parking orbit around Mars. The unloaded spacecraft would then proceed to Mars synchronous orbit above the desired landing sight. The laser transmitter would be activated and PEGASUS would be ready to beam energy to the surface to operate automated systems deployed from the cargo load. When the crew arrives later, this same system would provide the surface power needed to maintain the crew and support exploration activities. The availability of megawatt levels of electric power on the Mars surface would greatly enhance and even expand the mission options possible for Mars exploration.

MISSION ANALYSIS

Any number of mission profiles may be devised for a manned Mars expedition, depending on the assumptions made concerning desired objectives and available technology. The split-sprint Mars mission scenario allows optimum transport of personnel and materials, making this the mission of choice. An unmanned nuclear electric cargo vehicle, staged in LEO, is sent to Mars carrying all the necessary supplies and equipment needed to establish the Mars outpost. The manned mission, also staged in LEO, would follow later. Using a nuclear electric propulsion system with an alpha of 7.3 kg/kWe, a 303-MT spacecraft in LEO could transport astronauts to Mars, allow them to stay 30 days and then return all in 412 days (George, Hack and Dudzinski 1991), an outbound leg of 165 days, a stay in Mars orbit of 30 days, and a return leg of 217 days. The mass breakdown for a typical fast piloted Mars mission is presented in Table 1.

The transport time for the cargo spacecraft is not a critical issue but efficient mass transport is. The outbound leg for the cargo vehicle is 282 days to deliver 193 MT to Mars orbit. If the cargo vehicle is not returned to earth, several unique mission options (Coomes and Bamberger 1990) available for the system left in Mars orbit. Coupling the nuclear electric power system to a beam-power transmission system, electric power can be supplied from geosynchronous Mars orbit (GMO) to the surface to support a broad range of activities. Automated systems deployed from the cargo vehicle or left operating when the crew returns to Earth could convert raw materials on the Mars surface into usable form, which would then be available to the manned flights that follow. Finally, a backup return vehicle would be available, should the primary manned spacecraft become disabled. The crew could reconfigure the two spacecraft (already in orbit around Mars) and use the propulsion system of the cargo vehicle to propel them home. The mass breakdown for a Mars NEP cargo vehicle is presented in Table 2. The trip time includes 282 days for the outbound leg. With a 30-day stay time at Mars and reconfiguring the cargo vehicle as a piloted emergency return vehicle, the return trip would take 345 days. This is an abort or backup option offered by the split mission approach.

TABLE 1. Mars Piloted Vehicle Mass Summary.

SPACECRAFT COMPONENTS	MASS (kg)
Transit Habitat and ECCV	53,300
Power System	62,400
Propulsion System	5,800
Propellant	169,900
Propellant Tankage	16,100
Structure & Miscellaneous	4,500
Total Spacecraft Mass	303,000

TABLE 2. Mars Cargo Vehicle Mass Summary.

PROPULSION SYSTEM	MASS(kg)	POWER SYSTEM		MASS(kg)
MPD Thruster Assembly	1,160	Reactor System		3,620
Engine Control Assembly	80	UN Fuel	860	
High-Current Buss	1,360	Molybdenum	1,520	
Rectifier Assembly	330	Vessel & Reflector	1,240	
Transformer Assembly	3,190	Support Structure		350
AC Buss Components	110	Potassium Inventory		200
Thermal Control Systems	280	Shadow Shield		5,100
Subsystem Total	6,510	Turbines & Alternators		3,120
Contingency Mass (10%)	650	Fabric Heat Pipe Radiator		1,700
Structural Mass (20%)	1,300	Auxiliary Cooling		500
Propulsion System Total	8,460	Power System Total		14,590

SPACECRAFT COMPONENTS	MASS (kg)
Supplies, Lander, Rover, etc.	168,750
Power & Propulsion	23,050
Power Transmitter	25,000
Propellant (Outbound)	55,000
Propellant (Crew return)	66,000
Propellant Tankage	18,000
Navigation, Command, & Control	13,200
Structure & Miscellaneous	30,000
Total Spacecraft Mass	400,000

PROPULSION SYSTEM

Electric propulsion systems have not been seriously considered for use with large spacecraft because of the lack of a suitable electric power source to drive them. However, recent efforts to develop megawatt-class space power sources show such systems to be technologically feasible, and a multimegawatt lightweight nuclear electric propulsion system would enable missions of almost any conceivable duration and scope. A 10-MWe Mars cargo transport vehicle is a very viable system. A magnitoplasmadynamic (MPD) thruster coupled with a nuclear electric power plant provides the most attractive electric propulsion system. The propulsion system itself is simple, compact, and extremely rugged. Such a propulsion system is described in the following sections.

MPD Thruster System

The MPD thruster system is composed of thrusters, propellant tanks, power conditioning, and thermal control subsystems. The thruster assembly consists of seven MPD engines that are used one at a time. All seven engines are connected in parallel to a high-current buss, but each has a separate propellant valve and contactor for the cathode current feed. Performance development may change the electrode shape somewhat, but the overall dimensions of a multimegawatt thruster will remain about the same (King and Vondra 1984). Each engine is assumed to have a center-body cathode 3 cm in diameter and 10 cm long, with an anode about 12 cm inside diameter and of a comparable length. With the present understanding of cathode physics, thruster lifetime (which is limited by the cathode) is about 2,000 hours. The most massive part of the MPD engine is the anode heat removal system.

Electric Power Source

PEGASUS, the proposed power source for the MPD thruster system, is a 10-MWe boiling liquid-metal reactor power system. The system employs a direct rankine power cycle and is designed to meet the power requirements for a 10-MWe electric propulsion system and the operational needs of the power system and spacecraft itself. The power system is composed of five major subsystems or components: a cermet-fueled, boiling liquid-metal fast reactor; a shadow shield; three radial flow Ljungström derivative turbines each, driving a counterrotating superconducting alternator; a power conditioning subsystem; and a heat rejection/thermal control subsystem.

Reactor The reactor selected for the PEGASUS system is a fast reactor using a boiling alkali metal coolant and cermet fuel. A reactor system designed to operate in space (Coomes 1988) should take maximum advantage of the space environment rather than be a simple extrapolation of terrestrial reactor systems. The RMBLR (Johnson et al. 1987) does just that. The cermet fuel is composed of a refractory-metal alloy (Mo-7Re-3Hf) matrix with highly enriched uranium nitride (UN) as the fuel material. The UN/moly alloy cermet fuel was selected because of the high fissile density of UN (30% greater than UO_2), the high-temperature strength of the molybdenum alloy, and the ruggedness of cermet. The use of cermet fuel blocks with internal coolant flow channels, the use of UN in a matrix of refractory metal alloy of molybdenum, hafnium, and rhenium, and the arrangement of these fuel blocks provides unique characteristics (Barner et al. 1986) that are in direct contrast to typical clad fuel-pin core designs, even those using UN as the fuel. Also, the enclosed coolant channel and radial-inflow configuration eliminates crossflow instabilities that occur during boiling in open lattice cores.

Power Conversion Although the energy conversion system is the heart of any power system, it has received relatively little or no development attention. The belief is that sufficient experience and technology exist to design, build, test and produce turbomachinery for space-power systems. The only major issues needing resolution are the development of the high temperature materials and the development of bearings and seals to provide the desired system lifetimes. For terrestrial systems, this belief may be true. However, for space and even extraterrestrial applications, this is not the case. The absence of gravity, the lack of a stable mounting platform, and the hard vacuum that exists in space place special requirements on rotating machinery not normally considered on earth. The most important is the need for a system that will not destabilize the platform to which it is attached during startup/shutdown or load changes. Of secondary importance is the need for a high-power density; its importance depends on launch costs. These issues lead to a very different power conversion approach in the development of the RMBLR and in the power conversion system studies conducted by the Air Force Wright Research and Development Center (Giellis et al. 1990).

Because the largest alkali metal turbines ever developed were systems of less than 50 kWe output, the decision was made to review various turbine configurations to determine whether any one of the available basic designs had any intrinsic qualities that would make it more suitable for space applications. This effort led to the choice of the Ljungström turbine (Ljungström 1949) as the ideal turbine design for space applications. This design is unique in several respects. It has no stators; all blade rows rotate. Each set of blades in a stage is attached to blade rings which are attached by expansion rings to the two rotating disks. Staging occurs in the radial direction and results in a series of concentric blade rings attached to each disk. These disks rotate in opposite directions, providing the machine with two counterrotating shafts, each driving an alternator. This provides a natural means of balancing rotational torques that result during startup, shutdown, and load variations. By taking advantage of this counter rotation and modifying the standard, a single, very compact machine can be built. The flow through the machine is perpendicular to the axis of rotation and parallel to the acceleration forces acting on the blades. The blade speed is uniform along the entire length of each blade, resulting in maximum and average blade efficiency being the same. Because the flow vector and the acceleration vectors are parallel, this design is very tolerant of moist vapor. Typical exit qualities of axial flow machines are on the order of 88% or greater, while the Ljungström turbine should have the same erosion resistance with exit qualities as low as 70%. Through careful design to accommodate thermal expansion, it also has the capability of rapid startup.

The modified Ljungström turbine is coupled to a counterrotating superconducting alternator to provide the electrical output. Superconducting alternators are chosen to develop the electrical power because of their high power-to-weight ratio. Each superconducting alternator is expected to operate at 1500 Hz and 10 to 20 kVA with a continuous output power capability of 5 MWe (Dodge et al. 1987). The overall power conversion subsystem is expected to have a specific weight on the order of 0.5 kg/kW and an efficiency of 98%.

Heat Rejection and Thermal Control Heat rejection for the PEGASUS is accomplished by means of high- and low-temperature heat rejection subsystems. The high temperature subsystem handles waste heat rejection from the turbines. The low temperature subsystem takes care of waste heat from the alternator and other components requiring cooling. The high-temperature heat rejection system consists of three ceramic fabric heat-pipe radiators, associated pumps, piping, and structure. The radiator is sized to reject 21.5 MW of waste heat from the system during full-power operation.

The low-temperature heat rejection system consists of an auxiliary cooling system designed to reject waste heat produced within the alternator and other equipment operating at much lower temperatures. The auxiliary cooling system has its own working fluid, pumps, and radiator. This system utilizes helium as its working fluid and is composed of a Sterling cycle cryogenic cooler, an auxiliary chiller, a low-temperature radiator, and associated pumps and piping. The cryogenic cooler removes heat from the liquid helium and transfers this heat to the auxiliary coolant in the chiller. A closed loop system is used to pump this coolant through the auxiliary cooling radiator located around the perimeter of the main radiator. Before returning to the chiller, this coolant is used to cool pumps and other components.

CONCLUSIONS

Cargo transport has never been a time-critical activity. This makes a nuclear electric cargo vehicle the ideal spacecraft to carry the equipment and supplies necessary to sustain human expansion into the solar system. However, nuclear electric propulsion also has the ability to preform fast piloted missions with trip times on the order of 400 days with only a moderate alpha. Utilizing the advanced technologies developed under the Department of Energy's Multi-Megawatt Program, an upgraded PEGASUS Drive propulsion system, with an alpha less than 2.5, would significantly enhance the mission performance of NEP used for the Mars cargo transport and fast piloted missions. In addition, nuclear electric propulsion, when coupled with a beam-power transmission system, provides significant Mars surface mission enhancement, opening many new mission options not previously possible with chemical or nuclear thermal propulsion systems.

Acknowledgments

This work was performed at the Pacific Northwest Laboratory operated for the U.S. Department of Energy under Contract DE-AC06-76RLO 1830 by Battelle Memorial Institute.

References

Bamberger, J.A. (1990) "A New Method for Power Generation and Distribution in Outer Space," in Proceedings: 7th Symposium on Space Nuclear Power Systems, M. S. El-Genk and M. D. Hoover, eds., Vol. 1, pp. 387-392, Institute for Space Nuclear Power Studies Albuquerque, New Nexico.

Barner, J. O., E. P. Coomes, R. E. Williford, and L. A. Neimark (1986) "Cermet Fuels for Space Power Systems," in Proceedings: 3rd Symposium on Space Nuclear Power Systems, M. S. El-Genk and M. D. Hoover, eds., pp RF-10.1-10.4, Institute for Space Nuclear Power Studies, Albuquerque, New Nexico.

Coomes, E.P. and J.A. Bamberger (1990) "Applications and Space Mission Enhancements Made Possible with a Nuclear-Driven Beam-Power System," in Proceedings: 7th Symposium on Space Nuclear Power Systems, M.S. El-Genk and M.D. Hoover, eds., Vol. 1, pp 16-20, Institute for Space Nuclear Power Studies, Albuquerque, New Nexico.

Coomes, E.P. (1988) "A High Power Density Radial-In-Flow Reactor Split-Core Design for Space Power Systems," in Proceedings: 5th Symposium on Space Nuclear Power Systems, M. S. El-Genk and M. D. Hoover, eds., pp. 397-401, Institute for Space Nuclear Power Studies Albuquerque, New Nexico.

Coomes, E.P.,J.M. Cuta, B.J. Webb, and D.Q. King (1987) "The PEGASUS Drive: A Multimegawatt Nuclear Electric Propulsion System," in Space Nuclear Power Systems, M. S. El-Genk and M. D. Hoover, eds., Orbit Book Co., Malabar, Florida,Vol. V, pp. 365-373.

Dodge, R.E., E.P. Coomes, J.L. Kirtley, and S.J. McCabe (1987) "Design of a Superconducting Alternator for Space-Based Power Generation," in Proceedings: 4th Symposium on Space Nuclear Power Systems, M. S. El-Genk and M. D. Hoover, eds., Institute for Space Nuclear Power Studies, Albuquerque, New Nexico.

George, J. A., J. H. Hack, and L. D. Dudzinski (1991) "Fast Piloted Missions to Mars Using Near-Term, Low-Power Nuclear Electric Propulsion," presented to the Joint NEP Technology Panel, Cleveland, Ohio, July 1991.

Giellis, R., G. Hosford, K. Weber, W. Strohmayer, and H. Snyder (1990) "Megawatt Burst Power Thermal Management, Tasks 3&4 - System Assessments", WRDC-TR-90-2023, WRDC, Air Force Systems Command, Wright-Patterson AFB, Ohio, July 1990.

Johnson, B. M., E. P. Coomes, Z. I. Antoniak, B. J. Webb, W. W. Little, J. R. Friley, A. T. Luksic, and J. O. Barner (1987), The Rotating Multimegawatt Boiling Liquid Metal Reactor (RMBLR) Rev 2., PNL-D-675-DR, Pacific Northwest Laboratory, Richland, Washington, March 1987.

King, D. Q. and R. J. Vondra (1984) "Development Status and Projected Power Requirements of MPD Thrusters," presented at the Space Nuclear Propulsion Conference, Los Alamos, New Nexico, December 1984.

Ljungström, B. (1949) "The Development of the Ljungstrom Steam Turbine and Air Preheater", in Proceedings International Mechanical Engineering Conference, Vol. 160, pp. 262-265, April 1949.

CLOSED BRAYTON POWER CONVERSION SYSTEM DESIGN AND OPERATIONAL FLEXIBILITY

Thomas L. Ashe
William B. Harper, Jr.
Allied-Signal Aerospace Company
Garrett Fluid Systems Division
1300 West Warner Road
P.O.Box 22200 M/S 1207/4R
Tempe, AZ 85285-2200
[602] 893-4831

Abstract

The flexibility of the gas turbine engine is enhanced in closed Brayton cycle (CBC) power conversion systems for space power applications. The closed-loop configuration not only allows the use of the gas turbine (Brayton) heat engine in a vacuum environment, but also significantly increases the choices available to the system designer to optimize the power system design to provide the most desirable characteristics for a particular application. This increase in flexibility is a result of two factors, the ability to control working fluid characteristics, including both pressure level and molecular weight, and the use of heat exchangers to control and facilitate the movement of heat into, out of, and within the system. This paper reviews the system level characteristics of CBC systems, with emphasis on conversion systems applicable to space nuclear power applications, and the variations which can be achieved by altering working fluid and heat exchanger characteristics. The discussion includes the impact of working fluid pressure variability on both design and operational flexibility.

INTRODUCTION

Open-cycle gas turbine (Brayton) engines are tailored to specific applications by varying unit size, aerodynamic design (including both type - axial or radial - and stage count), rotor speed, pressure ratio and TIT (turbine inlet temperature). In addition, conversion efficiency can be improved by the incorporation of recuperation or regeneration. Operational variability is achieved by altering TIT (through fuel flow) or rotor speed. Additional flexibility in design and operation is achieved when the Brayton cycle engine is operated in a closed loop using an external heat source and heat sink. Closed Brayton cycle (CBC) systems can utilize, in addition to the all the variables applicable to open-cycle systems, variations in cycle gas type and molecular weight, CIT (compressor inlet temperature) and gas inventory (system pressure level) to achieve the operational characteristics required to provide the optimum system for a specific application. As a result, CBC systems offer the designer of space power systems an exceptional capability to tailor the design and operation of the power conversion system to the mission requirements. This paper examines some of the system variables which can be utilized by the designer to tailor the characteristics of the system, from both a design and operational standpoint, to specific operational requirements.

DESIGN FLEXIBILITY

Working Fluid Selection

Closed-loop systems remove the limitation imposed on open-cycle systems with regard to cycle gas selection. CBC equipment has been designed and built using a wide array of gasses and gas mixtures as the cycle working fluid. Utility scale equipment has been configured using air, helium, and helium-xenon mixtures as the cycle working fluid. Space and undersea CBC power systems have used pure noble gasses (helium, neon, argon, and krypton) as well as helium-xenon mixtures. Infinite variation in design working fluid molecular weight is possible with binary mixtures of noble gasses. Mixtures of xenon (molecular weight of 131) and helium (molecular weight of 4) can provide the full range of molecular weights between those of the constituent gasses. In addition, these mixtures are highly compatible with materials of construction and are chemically stable in the most hostile temperature and radiation environments.

The use of helium-xenon mixtures in CBC systems is particularly attractive from two standpoints. The infinitely variable molecular weight (from 4 to 131) allows selection of a molecular weight which balances the requirements of the mechanical, aerodynamic and heat-transfer components. In addition, helium-xenon mixtures result in lower working fluid Prandtl Numbers (Pr), providing heat transport conductance far exceeding that available from the pure noble gas of similar molecular weight and yielding lighter heat-transfer equipment (recuperators, heat rejection heat exchangers [HRHX], and heat source heat exchangers [HSHX]).

The noble gas mixtures generally considered for use as CBC working fluid fall in a range of Prandtl Numbers unusual for heat-transfer fluids. Most fluids, both liquid and gas, possess Prandtl Numbers greater than 0.65 and numerous heat-transfer correlations exist for these fluids. Commonly used gas-film heat-transfer correlations generally assume "Reynolds' Analogy" holds true. This analogy assumes Pr is near unity where the fundamental relationship between heat transfer and fluid friction is valid. Low Pr gas mixtures, such as the helium-xenon working fluids, may depart significantly from these correlations.

Liquid metals exhibit extremely low Prandtl Numbers (generally less than 0.02) due to their extremely high thermal conductivity. Extensive data has also been generated to support evaluation of liquid metal heat exchanger design. Until recent years, the area of fluids with Prandtl Numbers in the range of 0.1 to 0.6 was largely ignored.

A significant body of work, focused on CBC configurations, has recently been accomplished to improve understanding of heat-transfer film coefficients for gases in the 0.2 TO 0.5 range of Pr[2,3,4,5,6]. Result of this work has not been generalized, however, and is limited to application to similar geometries and range of Reynolds Number. It is clear that low Pr definitely yields much improved heat transfer, but results to date are configuration specific and not suitable for universal application.

As noted above, the use of helium-xenon mixtures to vary working fluid molecular weight is a valuable design tool. In general, lower power systems favor high molecular weight to keep rotating surface speeds low (windage loss, gas bearing loss) maintain low physical rotor speed (rotor dynamic considerations), and to yield higher design pressure levels consistent with gas bearing lubrication. For a fixed aerodynamic rotor configuration, a desired pressure ratio is achieved at a design physical speed which is inversely proportional to the square root of molecular weight. A broad trend exists, dependent on specific system design attributes, of decreasing optimum molecular weight with increasing power. This trend continues down to molecular weights of 35 to 40 for single-stage radial equipment. At lower molecular weights, aerodynamic tips speeds and rotor stress levels increase. Although CBC rotor stress levels are typically far below conventional aerospace design practice, the extreme life and reliability considerations of most space power applications place a lower practical limit on molecular weight $\geq$ 35 for single-stage radial CBC systems. Below this range of molecular weights, the tip speeds required to achieve the desired compressor pressure ratio (generally about 2:1) become excessive.

Multi-stage designs, more practical with axial aerodynamics since a lower pressure ratio per stage is needed to achieve the required cycle pressure ratio, allow consideration of lower molecular weight working fluids. Axial design history generally spans consideration of molecular weights in the 4 to 40 range. Gas-cooled reactor CBC studies in the multi-megawatt range have shown that the total system mass is insensitive to molecular weight in the range of 4 to 40. The major trade which is apparent is system complexity (high stage count at low molecular weight) versus larger heat-transfer equipment. Choice of pure helium working fluid may be dictated by specific heat source cooling requirements or experience data base (e.g., gas cooled reactors). With axial aerodynamic equipment, most systems optimize with a molecular weight in the 20 to 40 range, when molecular weight (e.g., helium) is not dictated by reactor cooling. Weight penalties are slight over the range of 4 to 40[1]. Recent work has shown that pure helium designs tend toward relatively short blades which accentuate tip clearance losses except at very high megawatt power levels, where blade size and hub-to-tip ratios increase and reduce clearance impact.

Figure 1 summarizes optimum molecular weight as a function of output power level for numerous CBC systems, both axial and radial, from the sub-kilowatt to multi-megawatt power range. This figure features a broad band of optimum molecular weight for radials to about 1 megawatt (molecular weight from 105 down to 37), a discontinuity with the transition to axial equipment at about 1 MW_e as it becomes mass and volume competitive with radial designs (molecular weight of 20 to 60), and a continuing decrease in molecular weight as power increases toward 30 megawatts. The design drivers and selection criteria are so diverse in this summary that only the broadest trend should be extracted (i.e., optimum molecular weight has downward trend with power). As indicated above, molecular weight selection for gas-cooled reactor cases may be totally dictated by the reactor cooling requirements (e.g., may dictate He).

In order to illustrate the impact on system design of cycle gas properties, optimization studies were conducted for a 2 kW_e isotope CBC system using krypton and two different helium-xenon gas mixtures. The optimization process considered both system mass and radiator area as the selection criteria. The process was conducted to systematically minimize the value of a parameter defined by system mass (kg) plus 15 times radiator area (m^2). The cases using helium-xenon as the cycle gas were run with molecular weight frozen at two different values, 60 and 83.8. The results of this study are shown in Table I.

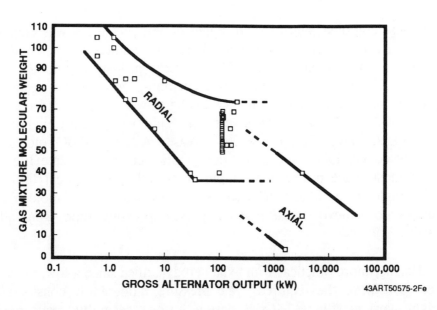

FIGURE 1. CBC Working Fluid Molecular Weight Trend with System Power Level.

The first comparison shown by the table is between systems using krypton working fluid (molecular weight 83.8, Pr=0.67) and helium-xenon working fluid (Pr=0.24) of the same molecular weight. The Prandtl Number exponents used in the calculations were taken from References [2] and [6] for the radiator heat-transfer passages and the recuperator passages, respectively. The krypton system shows a de-emphasis on heat transfer (lower recuperator effectiveness), slightly higher pressure drops (tougher heat transfer), and increased system mass. In order to compensate for diminished heat transfer, the optimized kryton system operates at a lower heat rejection temperature in order to recover the efficiency lost to lower recuperator effectiveness and higher recuperator delta P.

A comparison of the results for the two helium-xenon systems shows a mass penalty associated with the system using gas with a molecular weight of 60, although radiator area is slightly reduced. The lower molecular weight gas results in reduced system pressure levels in proportion to the square root of the molecular weight. Thus, the optimization process utilizes other design variables (e.g., aerodynamic design, rotor speed) to restore system pressure levels in order to optimize heat exchanger and ducting masses. The principal mechanism used to provide pressure recovery is reduced pressure ratio, which requires an increase in system pressure (inventory) to maintain mass flow. The increased volume flows associated with the lower molecular weight also cause an increase in radiator manifold pressure drop.   Thus, variation in molecular weight allows reduction in radiator area by increasing mass flow without increasing volume flow.

Radiator Area versus System Mass

The space power system designer is often required to define CBC system   configurations which provide balance between total system mass and required heat rejection radiator area. Radiator area is often limited by launch volume constraints for both deployable and non-deployable configurations. Radiator area limits may also be desired to limit spacecraft flexural oscillation when on station. In order to define reasonably balanced candidate cycles, an

Table 1. Comparison of Optimized Cycle Characteristics Using Three Candidate Working Fluids - 2 Kw Isotope Brayton Systems.

System Parameter		Krypton 83.8	He-Xe 83.8	He-Xe 60
System Mass	(kg)	253	234	242
Radiator Area	(m^2)	5.43	5.14	4.98
(Δ P/P)$_{radiator}$		0.0045	0.0040	0.0055
Compressor Diameter	(cm)	6.35	6.32	6.98
Compressor Inlet Temperature	(K)	367	372	383
Turbine Inlet Temperature	(K)	1117	1117	1117
Pressure Ratio		1.74	1.72	1.60
Compressor Diameter	(cm)	6.35	6.32	6.98
Compressor Discharge Pressure	(MPa)	0.359	0.372	0.324
Recuperator Efficiency		0.9175	0.9225	0.9325
(Δ P/P)$_{recuperator}$		0.0140	0.0120	0.0120
Recuperator Mass	(kg)	41.4	26.4	29.2
Optimization Parameter[*]		334	311	317

[*] Optimization Parameter for these cases = Mass (kg) + 15 x Area (m^2)

optimization parameter is defined which includes both mass and area, and may include conversion efficiency. Such an optimization parameter (Optimization Parameter = Mass (kg) + 15 x Area (m^2)) was used to define the cases presented in Table I. Figure 2 shows the lower bound of the attainable mass/area relationship for the reference 2 kW$_e$ isotope CBC system discussed previously. The mass/area lower bound is defined by five optimization cases which were completed using varying emphasis on the required radiator area (Optimization Parameter = Mass (kg) + C x Area (m^2); where C = 0, 5, 10, 15, 25). Mass and radiator area may be traded along or above this boundary. The broad range of area at near constant mass allows significant reduction in radiator area to be accomplished with minimal increase in mass. Inspection of Figure 2 reveals that a 20 percent reduction in radiator area may be achieved with only a 2.7 percent growth in system mass (C = 15 versus C = 0). Thus, the selected "optimum" for the 2 kW$_e$ isotope system would probably be represented by the target mark on Figure 2, rather than the minimum mass point. Most design point selections will occur to the left of the minimum mass point in order to make some reduction in area. Systems of higher conversion efficiency also exist above this minimum boundary. If efficiency is added to the optimization criteria, similarly shaped boundary loci can be generated for constant conversion efficiency.

System Mass versus Efficiency

In specific applications, conversion efficiency may or may not be an important cycle driver. CBC cycle designs can readily accommodate wide-ranging trades between system mass and conversion efficiency. The design flexibility of CBC systems is illustrated by the cycle design comparison of Table 2 for three 100 kW$_e$ systems designed for minimum mass at three different target efficiencies [7]. For this comparison, the heat source subassembly is not included in the total system mass.

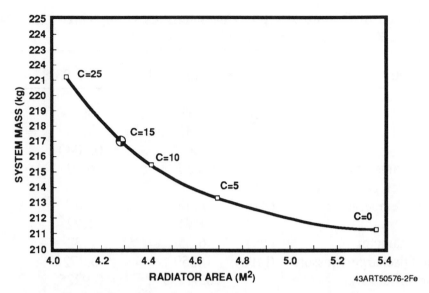

FIGURE 2. 2 kW Isotope CBC Mass - versus - Area Lower Bound.

Note that the efficiencies in Table 2 span nearly a 5:1 ratio from the minimum mass case to the high efficiency case. The CBC cycle shown in the first column (minimum mass) does not utilize a recuperator and yields a conversion efficiency of 0.112. A second system, of nearly the same mass but significantly higher efficiency, is shown in the second column. This system represents the minimum mass recuperated system and optimizes at a recuperator effectiveness of 0.708. The final cycle presented represents optimization for ultra-high conversion efficiency to the point of diminishing return relative to mass growth.

The addition of recuperation and the increase in efficiency are the notable differences between cases 1 and 2. Case 3, the high efficiency design, utilizes the combination of low CIT, high recuperator effectiveness, and low-heat exchanger and ducting pressure losses in order to enhance performance. The substantial gains in conversion efficiency, require significant mass increases for the radiator, heat exchangers, and integration structure (included in Miscellaneous).

Influence of Compressor and Turbine Inlet Temperatures

Changes to system mass and radiator area can also be effected by varying TIT and CIT. Due to the fourth-power relationship of the radiation heat rejection mechanism, radiator area and radiator mass are a strong function of CIT. CIT and TIT also influence the system by changing the cycle Carnot (and thereby conversion) efficiency. Increasing TIT and decreasing CIT increase conversion efficiency, thereby reducing both the heat input required and the amount of waste heat that must be rejected, thus reducing the size and mass of the heat engine, heat source and heat exchangers. These effects are illustrated in Figure 3, which shows additional data from the 2 kW_e isotope system study discussed earlier. These data are for systems optimized using an optimization parameter with a radiator area coefficient of 15 (Optimization Parameter = Mass [kg] + 15 times Area [m^2]). This data shows that radiator area is a strong function of CIT and a weak function of TIT. Although both CIT and TIT drive system efficiency, lowering CIT requires an increased radiator area (and thus mass). Therefore, as illustrated in Figure 3, system weight is little affected by variations in CIT, but a strongly affected by changes in TIT.

Table 2. Comparison of 100 Kw$_e$ Reactor/CBC Designs at Three Efficiency Levels.

Cycle Efficiency		0.112	0.211	0.506
Gross Alternator Output	(kW)	100	100	100
Working Fluid		XE/HE58	XE/HE58	XE/HE59
Turbine Inlet Temperature	(K)	1250	1250	1250
Rotor Speed	(krpm)	28	28	28
Thermal Input Required	(kW$_t$)	850	474	198
Compressor Inlet Temp	(K)	468	408	246
Compressor Pressure Ratio		2.08	2.09	1.91
Compressor Flowrate	(kg/s)	4.01	3.42	2.10
Compressor Disch Pressure	(MPa	1.03	1.03	0.72
Recuperator Effectiveness		0	0.708	0.968
System Total (Δ P/P)	(%)	4.3	6.6	3.2
Radiator Area	(m^2)	166	172	410
Radiator Mass	(kg)	580	600	1435
Combined Rotating Unit Mass	(kg)	183	178	165
Recuperator Mass	(kg)	0	110	714
HRHX Mass	(kg)	96	86	149
Duct Mass	(kg)	59	49	98
Miscellaneous Mass	(kg)	148	150	301
TOTAL SYSTEM MASS	(kg[*])	1066	1173	2861

[*] LESS HEAT SOURCE ASSEMBLY

OPERATIONAL FLEXIBILITY

With respect to operational flexibility, the CBC possesses one additional operational degree of freedom compared to the conventional open-cycle gas turbine powerplant. The open-cycle machine is constrained to operation at the prevailing atmospheric pressure level while the CBC pressure level (working fluid inventory) may be independently varied as a method of power control. For operation at fixed aerodynamic inlet conditions, the CBC power output is essentially proportional to the system pressure level. Part-load conversion efficiency can remain very high over large power turndown ratios. Non-linear second-order effects (e.g., parasitic drag loads, alternator efficiency variation, aerodynamic efficiency and fractional pressure loss variation with Reynolds Number) are the only causes for departure from a purely linear relationship between inventory and power output.

Turndown Capability

Figures 4, 5 and 6 illustrate CBC operational flexibility with test data obtained with the US Navy's CCPS-40 shaft power CBC [8] and the NASA BRU (Brayton Rotating Unit) CBC[9]. Figure 5 shows the CCPS-40's conversion efficiency with varying inventory for five sets of fixed coolant inlet (essentially fixed CIT) and TIT. Note that for any of the test conditions, conversion efficiency is near constant over a wide power turndown ratio. This is due to the fact that the aerodynamic matching of components, in corrected flow terms, is essentially unaffected by changes in inventory level (pressure level).

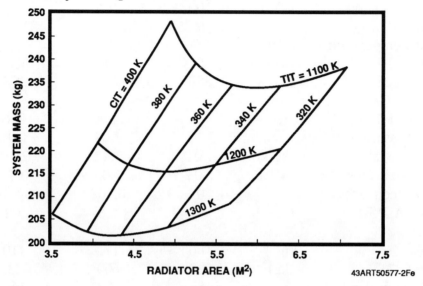

FIGURE 3. Mass and Area for 2 kW Isotope CBC System.

Figure 5 presents similar data and relates the NASA BRU CBC alternator output power to the system P_{cd} (compressor discharge pressure) for TIT values of 1144 and 978 K. Also indicated on Figure 5 (solid symbols) are two potential methods for power turndown; 1) constant TIT operation with variable working fluid inventory, and 2) constant inventory with variable TIT. Both examples accomplish the same 2:1 power turndown.

Both lines of data in Figure 5 were obtained with a CIT of 311 K. Note the high degree of linearity and large power turndown ratio in the system output power with varying P_{cd} at both TIT values. Applications requiring variable power can be readily accommodated by a CBC system equipped with a GMS (gas [inventory] management system). Adding or subtracting working fluid from the CBC at constant CIT and TIT results in proportional change in power output.

Figure 6 is a companion figure to Figure 5 showing the off-design conversion efficiency test data for the NASA BRU at the CIT/TIT conditions of the previous figure. Also indicated in Figure 6 is the conversion efficiency path which results from the control modes discussed above. Note the near constant efficiency for each fixed temperature set.

CBC power turndown can also be accomplished through TIT control as implied by Figure 6. Simply throttling TIT from 1144 k to 978 k at constant inventory is sufficient to achieve an approximate 2:1 power turndown as denoted by the dotted transition line on that figure. A slight reduction in P_{cd} can be observed as a result of decreased TIT at constant inventory. Since the CBC is a closed loop of constant volume, the decrease in hot-end temperatures results in an overall pressure decrease. In addition, the reduction in TIT will also induce an aerodynamic component rematch for flow continuity which results in a small reduction in compressor discharge pressure. The reduced pressure ratio has the effect of further reducing P_{cd}.

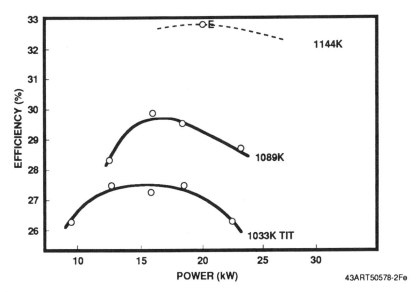

FIGURE 4. CCPS-40 Performance at Off Design Power and Turbine Inlet Temperature.

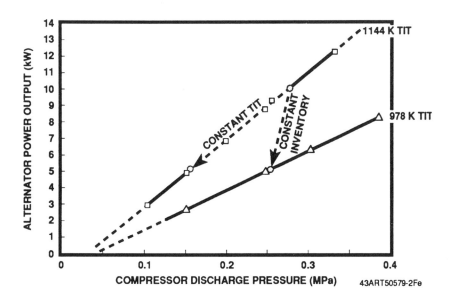

FIGURE 5. NASA BRU System Test Data at Two Turbine Inlet Temperature.

When TIT control is employed, some significant time may be required for the new TIT setpoint to be stabilized as the result of stored sensible heat in the hot components. During the transition, excess generated power is generally dissipated in a parasitic load bank also used for fast response engine speed control.

CONCLUSIONS

As a result of both the inherent design flexibility of the gas turbine heat engine and the additional freedom to select appropriate cycle parameters provided by the closed-cycle configuration, CBC systems provide the designer of space power systems with exceptional design and operational flexibility.

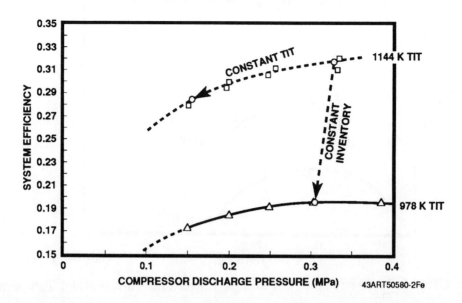

FIGURE 6. NASA BRU System Test Data at Two Turbine Inlet Temperatures.

Acknowledgments

The work reported in this paper, other than that extracted from the references, was conducted at the Allied-Signal Aerospace Company, Garrett Fluid Systems Division, under internal funding support.

References

[1]   D. J. Brandes, "High-Temperature Nuclear Closed Brayton Cycle Power Conversion System for the Space Exploration Initiative", Eighth Symposium on Space Nuclear Power Systems, Albuquerque, NM, 6-9 Jan 1991

[2]   D. M. McEligot, M. F. Taylor, and F. Durst, "Internal Forced Convection to Mixtures of Inert Gases", Int. J. Heat Mass Transfer, Vol 20, pp. 474-486, 28 May 1976

[3]   P. E. Pickett, M. F. Taylor, and D. M. McEligot, "Heated Turbulent Flow of Helium-Argon Mixtures in Tubes", Int. J. Heat Mass Transfer, Vol 22, pp. 705-719, 17 May 1977

[4]   M. F. Taylor, K. E. Bauer, and D. M. McEligot, "Internal Forced Convection to Low Prandtl Number Gas Mixtures", Interim Report for the Office of Naval Research, 30 June 1984

[5]   J. L. Mason, E. A. Mock, R. T. Caldwell, and A. Pietsch, "Monatomic Working Fluids VS Air for the Closed Brayton Cycle", 1985 Beijing International Gas Turbine Symposium and Exposition, 1-7 Sept 1985

[6]   A. V. von Arx, I. Ceyhan, "Laminar Heat Transfer for Low Prandtl Gases", Eighth Symposium on Space Nuclear Power Systems, Albuquerque, NM, 6-9 Jan 1991

[7]   T. L. Ashe, W. G. Baggenstoss, and R. Bons, "Nuclear Reactor Closed Brayton Cycle Power Conversion System Optimization Trends for Extra-Terrestrial Applications", 1990 IECEC

[8]   G. D. Duvall, "Operational Evaluation of a Closed Brayton Cycle Laboratory Engine", 11th IECEC, 1976

[9]   A. S. Valerino, R. P. Macosko, A. S. Asadoruian, T. P. Hecker, and R. Kruchowy, "Preliminary Performance of a Brayton-Cycle-Power-System Gas Loop Operating with Krypton Over a Turbine Inlet Temperature Range of 1200 °F to 1600 °F ", NASA TM X-52769, March 1970

FREE-PISTON STIRLING COMPONENT TEST POWER CONVERTER
TEST RESULTS OF THE INITIAL TEST PHASE

George R. Dochat
Mechanical Technology Incorporated
968 Albany-Shaker Road
Latham, NY 12110
(518) 785-2242

James E. Dudenhoefer
NASA-Lewis Research Center
21000 Brookpark Road
Cleveland, OH 44135
(216) 433-6140

Abstract

The National Aeronautics and Space Administration (NASA) - Lewis Research Center (LeRC) has the responsibility to develop power technologies that have the potential of satisfying anticipated future space mission power requirements. The Free-Piston Stirling Power Converter (FPSPC) is one of the many power technologies being evaluated and developed by NASA. FPSPCs have the potential to provide high reliability, long life, efficient operation; and they can be coupled with all potential heat sources, nuclear, radioisotope and solar, various heat input, heat rejection systems, and various power management and distribution systems. FPSPCs can compete favorably with alternative power conversion systems over a range of hundreds of watts to hundreds of kilowatts and to megawatts. Mechanical Technology Incorporated (MTI) is developing FPSPC technology under contract to NASA-LeRC and will demonstrate this technology in two full-scale power converters. The first of these, the Component Test Power Converter (CTPC), initiated testing in Spring 1991 to evaluate mechanical operation at space operating temperatures. This paper reviews the testing of the CTPC at MTI and the companion testing of the earlier technology engine, the Space Power Research Engine (SPRE) at NASA-LeRC.

BACKGROUND

In 1984, NASA awarded MTI a contract to demonstrate the Free-Piston Stirling Engine (FPSE) for space applications. The resulting Space Power Demonstrator Engine (SPDE) was operating 16 months after program initiation and had demonstrated the majority of its design goals by October 1986 (Dhar, M. et al.(1987), Dochat, G.R. (1991), Slaby, J. and D. Alger (1987), Slaby, J. (1987)). The SPDE was designed for 25 kW$_e$ and operated at a temperature ratio of 2.0, but its absolute temperature of 630 K at the heater head and ambient temperature at the cooler significantly reduced development time and cost.

The SPDE consisted of two engines sharing a common hot end and operating in an opposed configuration for dynamic balance. The SPDE was divided into two half engines, Space Power Research Engine (SPRE I and SPRE II), each providing 12.5 kW$_e$ power and testing was continued at both NASA-LeRC and MTI. Recently the second has been delivered to NASA.

Based on the success of the SPDE, NASA has continued the development of Stirling engine power for future space missions. NASA awarded a competitive multiyear Space Stirling Technology Program to MTI. The primary objective of this ongoing program is to develop Stirling technology for space power applications and to demonstrate that technology in full-scale hardware tests. The effort is focused on a mid power range (25 to 50 kW$_e$) Space Stirling Power Converter (SSPC) operating at a hot end temperature of 1050 K and a temperature ratio of 2.0 (see Table 1 for additional design goals). Figure 1 contains a timeline showing both the past and projected evolution of the SSPC. Note the development of the CTPC proceeds in parallel with continued technology development of the SPRE at NASA.

Following an initial technology assessment, NASA and MTI determined that a CTPC would be desirable to expedite development of critical technologies for the program. The CTPC would be designed for full space operating temperatures, but would be 12.5 kW$_e$ per cylinder.

The critical concern to be evaluated using the CTPC is proper mechanical operation of the cold end at 525 K. The FPSPC relies on close-clearance, noncontacting seals for good efficiency and good specific power. The CTPC will be the first power converter in which the cold end hardware must operate from ambient to 525 K and still maintain the necessary close clearances between the dynamic members and the stationary cylinders. Therefore, MTI is developing the CTPC in two phases. The first phase, cold end testing at 525K, will verify the mechanical design, internal bearing operation, and structural, mechanical, and electrical design of the linear alternator. The second phase will consist of full CTPC hot testing.

The first phase of the CTPC test program is nearing completion and test results to date are documented herein. In addition, testing of the SPRE at NASA-LeRC continues to provide technology understanding. Recent results of the SPRE also are contained herein.

CTPC DESIGN/HARDWARE

The design of the CTPC is shown in Figure 2. Complete layout of CTPC cold end hardware is shown in Figure 3. The CTPC is the first hardware designed for full expected space temperatures (i.e., T_H = 1050 K and T_C = 525 K). Figure 4 shows the assembled CTPC installed in the test cell. The SPRE is similar in design, however, the temperature capability is limited to 630 K heater head temperature and ambient cooler temperature.

SPRE TECHNOLOGY TESTING RESULTS

At present, both SPRE I and SPRE II are at NASA-LeRC for test and evaluation. SPRE I has been at NASA since May of 1987, while SPRE II was recently transferred to NASA in April of 1991. SPRE II had been modified by MTI to reduce the magnetic structure surrounding the linear alternator and had been tested to verify a linear alternator performance increase from 70% to 90% efficiency. NASA will integrate the SPRE II alternator end with the SPRE I displacer end to continue Stirling engine development. In the past year testing of the SPRE I at NASA-LeRC has focused on generating experimental data for validating performance codes. Two areas of particular interest are 1) the sensitivity of engine performance to the displacer seal clearance, and 2) the effects of varying piston centering port area. (Cairelli, J. et al. (1991))

The primary objective of the displacer seal clearance tests was to determine the sensitivity of the SPRE I pV power and efficiency to the displacer seal radial clearance. The results, shown in Figures 5 and 6 indicate pV power diminishes as the displacer seal clearance is increased, while efficiency remains nearly constant over the range investigated. The constant efficiency of SPRE I with varying displacer seal clearance is at odds with testing performed at MTI using a 1 kW and 3 kW test engine and which did show a significant efficiency drop with increasing displacer seal clearance. These results are being evaluated and compared to code to provide understanding of the loss mechanisms involved.

The primary objective of the piston centering port tests was to determine piston mid-stroke position sensitivity to available centering port area. Figure 7 shows the piston mid-stroke position variation with two centering ports and six centering ports as a function of piston amplitude and mean operating pressure. As expected, the greater number of ports resulted in less variation of mid-stroke position over the range of piston amplitude and less variation with engine mean pressure. The choice of the "right" amount of centering ports (i.e., centering port area) is dependent on the effect on power and efficiency (generally greater centerport area means additional centerport losses and hence less pV power and less efficiency), the sensitivity of mid-stroke position from engine start-up to full operating temperature (more port area usually implies less sensitivity), and engine stability.

PHASE I - CTPC TEST

The purpose of the cold end test is to verify mechanical operation, internal bearing operation, and alternator performance before integrating with the full CTPC. Because of fabrication delays of the CTPC hot heat exchangers, the cold end testing will be expanded to also evaluate internal engine performance. The cold end includes all dynamic components and internal hardware of the CTPC; it does not include the heat exchangers. The cold end hardware is installed with a dummy heater head and cooler heat exchanger. Cold end testing proceeds with motoring the linear alternator to provide motion of the piston and displacer. Start-up is performed at ambient temperature. Temperatures are increased by flowing the hot oil through the cooler heat exchanger and the facility cooling loop, which has a temperature capability up to 600 K.

TABLE 1. 1050 K SSPC Design Goals

> End of life power: **25 kW$_e$**
>
> Efficiency: >25%
>
> Life: 60,000 hr
>
> **Hot side interface:** heat pipe
>
> **Heater** temperature: 1050 K
>
> **Cooler** temperature: 525 K
>
> Vibration-casing peak-peak: <0.04 mm
>
> Bearings: internally pumped – working gas
>
> Specific mass: 6.0 kg/kW$_e$
>
> Frequency: 70 Hz
>
> Pressure: 15.0 MPa

P399

EVOLUTION OF HIGH-TEMPERATURE (1300 K) STIRLING SPACE POWER CONVERTER

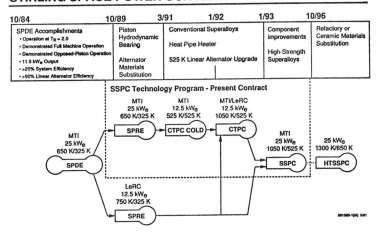

Mechanical Technology Incorporated   FIGURE 1. Stirling Technology Program Timeline

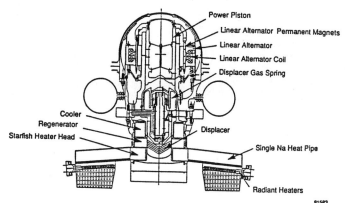

FIGURE 2. Design Layout of CTPC

896

FIGURE 3. "Cold End" Hardware

FIGURE 4. CTPC "Cold End" Installed in Test Cell

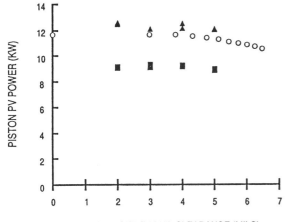

EFFECT OF THE DISPLACER SEAL CLEARANCE ON
PV POWER AT A TEMPERATURE RATIO OF 2.0 AND
MEAN PRESSURE OF 15 MPa

PISTON AMPLITUDE
■ 8 mm, SPRE DATA
▲ 10 mm, SPRE DATA
○ 9 mm, HFAST PRED.

FIGURE 5. Effect of the Displacer Seal Clearance on
PV Power at a Temperature Ratio of 2.0
and Mean Pressure of 15 MPa

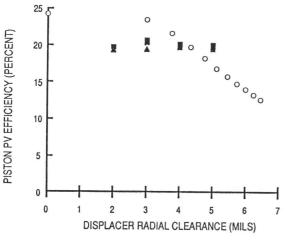

EFFECT OF THE DISPLACER SEAL CLEARANCE ON
PV EFFICIENCY AT A TEMPERATURE RATIO OF 2.0
AND MEAN PRESSURE OF 15 MPa

PISTON AMPLITUDE
■ 8 mm, SPRE DATA
▲ 10 mm, SPRE DATA
○ 9 mm, HFAST PRED.

FIGURE 6. Effect of the Displacer Seal Clearance on
PV Efficiency at a Temperature Ratio of
2.0 and Mean Pressure of 15 MPa

MEAN PRESSURE DEPENDENCE OF THE PISTON
MID-STROKE POSITION AT A TEMPERATURE RATIO OF 2.0

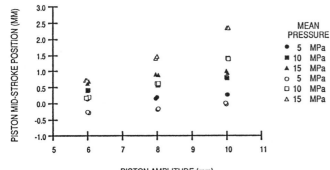

	MEAN PRESSURE	PORTS
●	5 MPa	6
■	10 MPa	6
▲	15 MPa	6
○	5 MPa	2
□	10 MPa	2
△	15 MPa	2

FIGURE 7. Mid-Stroke Position Variation as a Function of
Number of Centering Ports and Mean Pressure

898

Initial Ambient Checkout Test

The first test was a checkout of the initial assembly, the new test facility, and the start-up procedures. The cold end was started and motored for 15 min at ambient temperature. The system performed properly, with both the displacer and piston operating at 80% of design stroke. Post-test inspection of the hardware showed no evidence of mechanical rubs nor physical interferences.

Test at 338 K

Following post-test inspection for the initial test, the CTPC cold end was reassembled and motored at ambient temperature. Temperature through the cooler heat exchanger was slowly increased to 338 K. As planned, the CTPC operated for 1 hr at this temperature. The strokes of both the displacer and power piston were increased to 28 mm (design stroke). Post-test inspection of the hardware showed no evidence of mechanical rubs nor physical interferences.

Test at 390 K

The cold end of the CTPC was reassembled. The engine was motored at design strokes for the displacer and power piston of 28 mm. Temperature of the system was slowly raised toward the 423 K goal. At 390 K, the piston bearing pressure suddenly dropped to zero and the facility performed an automatic shutdown of the system. Visual inspection of the hardware showed that one of four carbon-graphite liners in the power piston cylinder slipped out of position, closing the bearing supply ports.

MTI has already received redesigned carbon-graphite liners, with improved tighter tolerances. No other damage was observed and the choice of bearing materials appears to be good. Operation to 390 K was excellent.

Initial Bearing Operation

The CTPC cold end was assembled for operation with both the power piston and displacer on internally pumped hydrostatic gas bearings. The test called for motoring the alternator for about 1 hr at the following operating conditions:

Mean Pressure:	150 Bar
Operating Frequency:	70 Hz
Piston and Displacer Amplitude:	14 mm

The CTPC cold end was motored at 70 Hz frequency and 150 Bar mean pressure with both the displacer and the power piston radially supported on the internally charged hydrostatic bearings. The supply voltage to the alternator was set to achieve piston stroke of 28 mm resulting in the displacer stroke of just over 20 mm. The gas bearings appeared to operate flawlessly. The test was continued at this operating condition for approximately 1 hr, with the temperatures of the displacer post and flange, and gas spring cylinder monitored. The temperatures were increased to 363 K. After about an hour of operation, the displacer mean position changed suddenly from +1.3 mm to -2 mm, then shifted back close to mid-stroke. The test was then stopped. Inspection of the hardware showed that the short liner of the displacer gas spring cylinder carbon-graphite sleeve had moved about 6 mm towards the compression space. There were marks on the sleeve indicating slight contact with the piston but with no evidence of any significant rub. This was the first time the displacer drive was tested with full engine pressure. Some analysis was performed on the gas spring cylinder liner. The results indicate that the pressure wave in the compression space was the cause of the slippage. As a fix, the cylinders will be modified to accept retaining rings which will restrain the liners from moving.

Test Results to Date

Testing to date has demonstrated the adequacy of the mechanical design (including the internal gas bearing design) of the CTPC at temperatures to 390 K. Testing has been limited because of loose fitting carbon-graphite liners and the resultant movement of these liners during operation. MTI has determined it is best to modify all hardware with carbon-graphite liners at this time. Modifications will include not only tighter interference fit carbon-graphite liners, but also retaining rings to physically prevent movement of the liners. The hardware will be available in October 1991, at which time testing will continue. The plan is to assemble the CTPC and test the cold end of the CTPC on internal bearings to 525 K. After mechanical operation is verified at temperature, the CTPC cold end will be used to validate performance (internal leakage losses, gas spring losses, etc.).

CONCLUSIONS

NASA/MTI continue to make progress in the development of free-piston Stirling technology for space applications. The cold end test of CTPC has demonstrated internal bearing operation of both the displacer and power piston, and mechanical operation up to 390 K. Testing is continuing toward operation at 525 K. The full CTPC will be the first free-piston Stirling power converter to operate at 1050 K and a temperature ratio of only 2.0. The CTPC is scheduled to begin testing in February 1992. Stirling technology being developed at NASA and other contractors (such as MTI) under NASA direction, will have application for space missions with power requirements from hundreds of watts to hundreds of kilowatts.

Acknowledgments

The authors wish to thank NASA for their continuing program support, Don Alger of Sverdrup Technology for technical guidance, and the strong technology development teams at NASA-Lewis Research Center and MTI for their dedication to success.

References

Cairelli, J., D. Swec, W. Wong, and T. Doeberling (1991) "Update on Results of SPRE Testing at NASA-Lewis," 26th IECEC.

Dochat, G. R. (1991) "Free-Piston Stirling Engine System Considerations of Various Space Power Applications," 8th Symposium on Space Nuclear Power Systems.

Slaby, J. (1988) "1988 Overview of Free-Piston Stirling Technology for Space Power at the NASA-LeRC," Cleveland, OH.

Slaby, J. and D. Alger (1987) "1987 Overview of Free-Piston Stirling Technology for Space Power Applications," NASA TM 89832, NASA-LeRC, Cleveland, OH.

Dhar, M., et al. (1987) "Design and Performance of a 25 kW$_e$ Free-Piston Stirling Space Power Demonstrator Engine," Intersociety Energy Conversion Engineering Conference, Philadelphia, PA.

COMBINED-BRAYTON CYCLE, SPACE NUCLEAR POWER SYSTEMS

Zéphyr P. Tilliette
Commissariat à l´Energie Atomique, DRN/DMT
Centre d´Etudes de Saclay, 91191, Gif-sur-Yvette, Cedex, France
(33) 1 69 08 32 87

Abstract

Because it is a widely recognized dynamic space conversion system, the Brayton cycle has been studied in France since several years, especially within the framework of a limited space program. A recuperated cycle of 20 to 30 kWe has been considered so far. However, possible applications could evolve and the need for an extended, diversified utilization of the Brayton cycle could appear. So, for Lunar or Mars bases which would accept large radiators and can benefit from a certain gravity level, combined cycle systems could be proposed. Following a reference to past works on space combined cycles, a possible association of a Brayton cycle with a thermoionic reactor is presented. The power level of a "Topaz-2" type space nuclear system can be boosted from 8 kWe to around 36 to 53 kWe, at the expense of a large radiator of course. Furthermore, combined Brayton-Rankine, organic (toluene) or steam, cycles can pave the way to a simpler gas-cooled, particle bed reactor concept. A particular arrangement of HeXe heater and boiler or steam generator in series is proposed. It makes it possible to lower the reactor inlet temperature, which is quite adequate for the use of light water as moderator. Oustanding net efficiencies of 25.8 to 27.6 per cent, given the reactor temperature profile, are obtained. Consequences on the reactor design are mentioned.

INTRODUCTION

Following two preliminary study phases carried out from 1983 to 1990, the French space nuclear program has entered a period of reconsideration and of lower activity (Carré et al. 1990 and Tilliette et al. 1990). Given the decrease in power needs for currently identified missions, the decision to develop a space nuclear power system in Europe is being held in abeyance (Luton 1990). However, future ambitious missions will be proposed and the advantages and the future necessity of nuclear energy are recognized for several of them. That is the reason why a conceptual, information work is being pursued.

These last years, emphasis was laid on small nuclear power systems in the range of 10 to 30 kWe. The leading project concerned a 930 K, NaK-cooled, fast spectrum reactor. Later, it appeared advisable to also investigate thermal spectrum reactor solutions, which could lead to dramatic advantages. Core concepts developed from ZrH-moderated to H_2O-moderated ones, which could seem surprising at first sight. In fact, the use of the Brayton cycle as the energy converter offers a means of effective moderator temperature conditioning. Through the combination of Brayton and Rankine cycles, a simpler H_2O-moderated gas-cooled reactor design can be proposed and higher power levels of 50 to 100 kWe can be reached. The link between the energy conversion and the nuclear heat source concept is strengthened in this way.

Before dealing with the association of combined Brayton-Rankine cycles with gas-cooled reactors, some examples of combined cycles previously studied for space applications and the possible addition of a bottoming Brayton cycle to a topping thermoionic reactor are successively presented in order to confirm this line of reflection.

EXAMPLES OF SPACE COMBINED CYCLES

The combination of a topping thermoelectric converter with a Stirling Cycle was mentioned several years ago as a means of better utilizing the heat of a space nuclear reactor. In

Figure 1a, thermoelectric cells are associated with organic Rankine units (Bland et al. 1986). The power level is here limited to a few kWe, because it relates to dynamic isotope power systems (DIPS) fitted with "General Purpose Heat Sources".

As an example of the more conventional combination of two dynamic cycles, two Rankine cycles in series, a mercury topping one and a toluene organic one, are shown in Figure 1b. A system efficiency of about 40 per cent is claimed for a mercury turbine inlet temperature of 1100 K and a condensing temperature of 375 K (Fox and Louis 1987). The proposed heat source is a solar receiver and the application could be a future manned space station.

The systems illustrated in Figure 1c and Figure 1d are more directly connected to the arrangement discussed in the last section, because both are Brayton-Rankine combined cycles. The concept of Figure 1c refers to aircraft space conditioning but it could apply to space converters, given the small size of its components (Friedman 1978). The cycle arrangement of Figure 1d studied for a space dynamic application can be taken as conventional inasmuch as the organic Rankine cycle is totally heated by the waste heat of the Brayton cycle (Massardo, 1990).

COMBINED THERMOIONIC-BRAYTON CYCLE SPACE POWER SYSTEMS

One of the most recent proposal for a space combined energy conversion concerns the association of a topping thermoionic reactor of the "Topaz-2" type with a bottoming Stirling cycle, intended to supply power to Lunar or Mars bases. As the diagram of Figure 2 shows, the Stirling cycle operates within the 900 to 400 K temperature range and uses a low temperature radiator much larger than the high temperature one of the thermoionic part (Nickitin 1990).

A first study of the addition of a bottoming Brayton cycle instead of a Stirling one to a thermoionic reactor has been carried out. The simplified system diagram and the approximate temperature conditions are shown in Figure 3 and Figure 4 respectively. Key data are given by Table 1. The superposition of a Brayton cycle to a 160 kWt "Topaz-2" thermoionic reactor makes it possible to increase the power level from 8 kWe to 36 kWe or 53 kWe according as the corresponding compressor inlet temperature is 373 K or 293 K. Such applications have to take into account radiators of a relatively large area, 83 m^2 and 164 m^2 respectively.

GAS-COOLED REACTOR, BRAYTON-RANKINE CYCLE POWER SYSTEM

The investigation of combined topping Brayton and bottoming Rankine cycles is driven by the search, not for a high efficiency at all costs, but for a convenient adaptation to a gas-cooled reactor. In the present case of a light water moderated, particle bed reactor, the reactor inlet temperature (RIT) has to be relatively low, typically around 500 K, in order to provide an adequate moderator temperature conditioning. Two means of conversion capable of achieving this aim are considered: a Brayton-toluene Rankine combined cycle shown in Figure 5 and the Brayton-steam combined cycle of Figure 6. The main feature of the cycle arrangement is the location of the Rankine cycle boiler or steam generator on the high pressure side of the HeXe circuit downstream of the recuperator (Tilliette 1990 and 1991). Concerning the cases of Figure 5 and Figure 6, there are differences in the degree of the gas cycle waste heat recovery, in the power distribution between the two cycles and in the RIT, 530 K and 475 K respectively, an attractive level indeed. The simplified toluene cycle does not need a recuperator.

Consequences for the reactor concept and design are undoubted. In case of a 10 to 30 kWe nuclear system and a single direct Brayton cycle, the HeXe pressure lies between 1 and 0.5 MPa, and the reactor vessel has not to withstand a higher light water pressure. So, a derivation D (Figure 6) of cold gas from the compressor outlet is used to keep the fuel assembly outside temperature below the H_2O boiling point. That results in some complexity and efficiency penalty. As far as a 50 to 100kWe combined cycle power system is concerned, the RIT is much lower, around 500 K, and the gas (HeXe) and H_2O pressures can be balanced at about 2.5/3 MPa. A se-

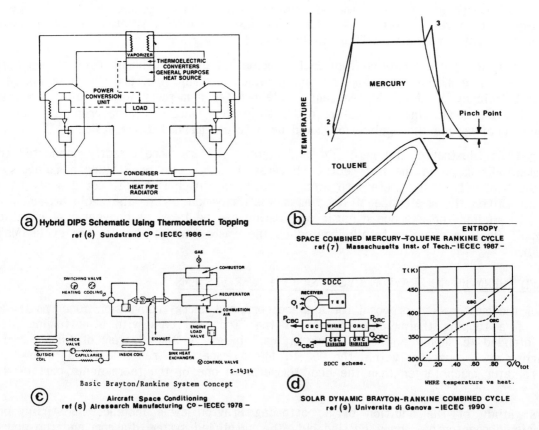

(a) **Hybrid DIPS Schematic Using Thermoelectric Topping**
ref (6) Sundstrand C⁰ –IECEC 1986 –

(b) **SPACE COMBINED MERCURY–TOLUENE RANKINE CYCLE**
ref (7) Massachusetts Inst. of Tech.– IECEC 1987 –

(c) **Basic Brayton/Rankine System Concept**
Aircraft Space Conditioning
ref (8) Airesearch Manufacturing C⁰ –IECEC 1978 –

(d) **SOLAR DYNAMIC BRAYTON-RANKINE COMBINED CYCLE**
ref (9) Universita di Genova –IECEC 1990 –

FIGURE 1. Examples of Combined Cycles Studied for Space Applications.

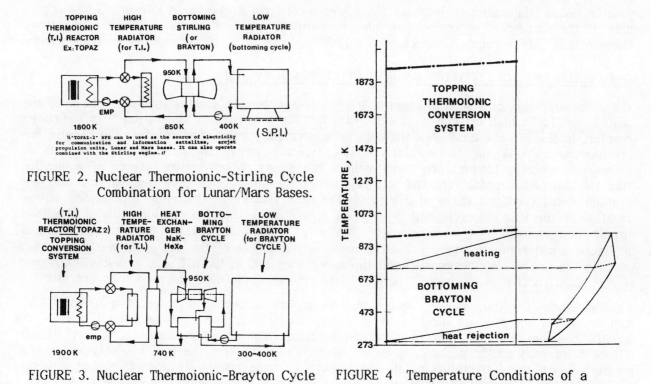

FIGURE 2. Nuclear Thermoionic-Stirling Cycle Combination for Lunar/Mars Bases.

FIGURE 3. Nuclear Thermoionic-Brayton Cycle Combined Power System for Lunar / Mars Bases.

FIGURE 4 Temperature Conditions of a Nuclear Thermoionic-Brayton Cycle Combined Space Power System.

903

TABLE 1. Key Data of Combined Nuclear Thermoionic-Brayton Cycle Space Power Systems.

TOPPING SYSTEM		THERMOIONIC (ex:~TOPAZ-2)	
BOTTOMING CYCLE		NO	BRAYTON
REACTOR THERMAL POWER kWt		←————— 160 —————→	
Thermoionic Conversion			
RADIATOR TEMPERATURE (mean) K		870	950
NET POWER OUTPUT kWe		8	7
NET EFFICIENCY per cent		5	4.4
RADIATOR AREA m²		6	5
Brayton Conversion			
HEAT INPUT kWt		←———150———→	
TURBINE INLET TEMPERATURE K		←— 950 —→	
COMPRESSOR INLET TEMPERATURE K		373	293
NET POWER OUTPUT kWe		29	46
NET EFFICIENCY per cent		19.3	30.7
RADIATOR AREA m²		83	164
Conversion Systems Combination			
NET POWER OUTPUT kWe		36	53
NET EFFICIENCY per cent		22.5	33.1

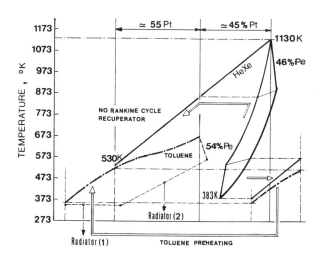

FIGURE 5. Brayton-Organic Toluene Rankine
    Combined Cycle
    for Space Nuclear Power Systems.

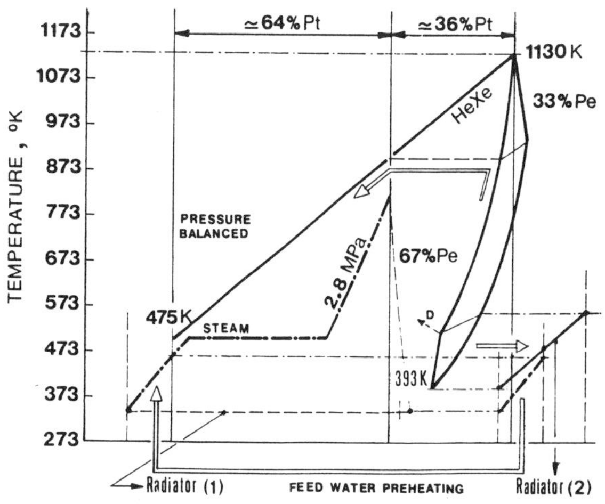

FIGURE 6. Brayton-Steam Rankine
    Combined Cycle
    for Space Nuclear Power Systems.

paration structure for the shunt cold gas is no longer required around the fuel assemblies. The corresponding particle bed reactor is shown in Figure 7. For simplification, it is depicted with seven fuel elements, but it could advantageously be made up of nineteen of them above a 50 kWe power level. Figure 8 shows how the reactor design can significantly benefit from a low RIT. Superfluous reactor structures are removed.

CONCLUSION

For space nuclear power systems, it would be hard to justify the use of combined exotic cycles which could lead to an increase in efficiency, but at the expense of great technological difficulties. The combination of two well developed systems, like the Brayton cycle, Rankine cycle or thermoionics, is not so beyond consideration. In case such an arrangement leads to satisfactory efficiency and heat rejection conditions, to new redundancy and stability approaches, and to an attractive reactor inlet and containment strucures temperature level, resulting in a coolant mass flow decrease favorable to a power output extrapolation, it is worth being mentioned, particularly for Lunar and Mars bases. Should an adequate combined cycle converter paves the way for a significant simplification of the reactor design, it surely deserves attention.

Acknowledgments

This work on space nuclear power systems is carried out for the French Government Agencies "Commissariat à l´Energie Atomique" (CEA) and "Centre National d´Etudes Spatiales" CNES).

References

Bland T.J. et al. (1986) "Integration Considerations of a Dynamic Power Conversion System for Spacecraft Applications," in Proc. 21st Intersociety Energy Conversion Engineering Conference, Paper n°869460, Vol.3, pp.2016-2019, held in San Diego, CA, August.1986.

Carré F.O. et al. (1990) "Update of the Erato Program and Conceptual Studies on LMFBR Derivative Space Power Systems," in Trans. 7th Symposium on Space Nuclear Power Systems, Conf-900109-Summs, held in Albuquerque, NM, 7-11 January 1990.

Fox A. and Louis J.F. (1987) "Binary Mercury/Organic Rankine Power Systems," in Proc. 22nd Intersociety Energy Conversion Engineering Conference, Paper n°879162, held in Philadelphia, PA, August, 1987.

Friedman D. (1978) "Light Commercial Brayon/Rankine Space Conditioning System," in Proc. 13rd Intersociety Energy Conversion Engineering Conference, August 1978.

Luton J.M. (1990) "Recherche et Technologie," in l´"Aéronautique et l´Astronautique", n°142,1990.

Massardo A. (1990) "High Efficiency Solar Dynamic Space Power Generation System," in Proc. 25th Intersociety Energy Conversion Engineering Conference, Paper n°900319, Vol.1, held in Reno, NV, August 1990.

Nickitin V.P. et al. (1991) "Topaz-2 Thermoionic Space Nuclear Power System and the Perspectives of its Development," in Proc. 8th Symposium on Space Nuclear Power Systems, Vol.2, pp. 631-635, held in Albuquerque, NM, 6-10 January 1991.

Tilliette Z.P. et al. (1988) "Adaptability of Brayton Conversion Systems to Fast, Epithermal and Thermal Spectrum Space Nuclear Reactors," in Proc. 23rd Intersociety Energy Conversion Engineering Conference, Paper n°889240, held in Denver, CO, 31 July-5 August 1988.

Tilliette Z.P. (1990) "Small Space Nuclear Reactors, Closed Brayton Cycle and Effective Moderators," in Proc. 25th Intersociety Energy Conversion Engineering Conference, Paper n°900234, Vol.1, pp. 115-120, held in Reno, NV, August 1990.

Tilliette Z.P. (1991) "Evolution in the Approach of Brayton Cycle-Space Nuclear Reactors," in Proc. 26th Intersociety Energy Conversion Engineering Conference, Paper n°910084, Vol.1, held in Boston, MA, 3-9 August 1991.

TABLE 2. Key Data of Combined Brayton- Rankine Cycle
Space Nuclear Power Systems.

TOPPING CYCLE		BRAYTON	
BOTTOMING CYCLE (Rankine)		TOLUENE	STEAM
NET POWER OUTPUT	kWe	75	
NET EFFICIENCY	per cent	25.8	27.6
TOTAL THERMAL POWER	kWt	291	272
Rankine Boiler	kWt	159	175
HeXe Cycle Heater	kWt	132	97
TOTAL RADIATOR AREA	m²	315	275
Condensation (1)	m²	172	262
Dry Gas (2)	m²	143	13
REACTOR INLET TEMPERATURE	K	530	475
HeXe Compressor Inlet Temperature	K	383	393
Rankine Turbine Exhaust Pressure	kPa	38	20

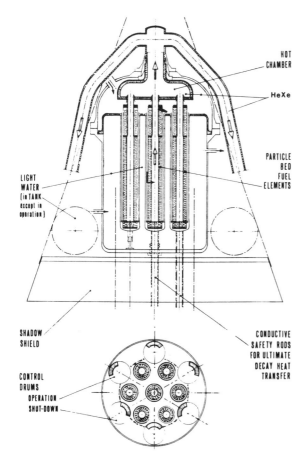

FIGURE 7. Space 50 to 100 kWe Combined Cycle
Light-Water Moderated
Particle Bed Reactor.

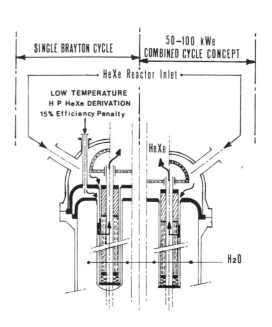

FIGURE 8. Space Combined Cycle
Light Water-Moderated
Particle Bed Reactor
Design Simplification.

A METHODOLOGY FOR THE RISK ANALYSIS
OF FISSION-REACTOR SPACE NUCLEAR SYSTEMS

Bart W. Bartram, Seshagiri R. Tammara, and
Abraham Weitzberg
HALLIBURTON NUS
Environmental Corporation
910 Clopper Road
Gaithersburg, MD 20877-0962
(301) 258-8666

Abstract

Fission-reactor space nuclear power systems have been used on past space missions and their use is planned on future missions where power and mission profile requirements cannot not be satisfied by other power system types. Fission-reactor space nuclear power systems pose a degree of risk due to the potential for radiation exposure under certain normal operating and postulated accident conditions. The level of risk will be a function of the system design, engineered safety features, system reliability, mission profile, accident environments, operating period, and orbital lifetime. This paper presents a methodology for the the risk analysis of these systems. The focus of this paper is on the long-term modeling of such radioactivity released into the environment and the calculation of radiological risk to the population.

INTRODUCTION

There are several possible scenarios for the release of radioactivity, dispersion through the environment, and subsequent population exposure resulting from the use of fission-reactor space nuclear systems and postulated accidents. Releases could occur at high and low altitudes, on the land surface, or underwater, and then be dispersed into various environmental media (air, soil, and water). Interactions with people through various exposure pathways would lead to inhalation, ingestion, and external radiation doses and associated health effects. Intact components could result in external radiation doses to people.

The evaluation of the radiological impact of postulated scenarios involving releases of activity into the environment includes the following steps:

1. Identification of the concentration, physico-chemical form, and containment of any radionuclides in the space nuclear system.

2. Identification of postulated accident or normal operation release modes, including the probability of release and the release location.

3. Source term definition, including the activity of each radionuclide released and the corresponding chemical form and particle size distribution, or the characteristics of any intact components.

4. Analysis of the time behavior and dispersion of the released activity to determine the concentrations in environmental media (air, soil, and water) as a function of time.

5. Analysis of the interaction between the environmental concentrations and people, leading to ingestion, inhalation, and external doses through each environmental exposure pathway.

6. The evaluation of the radiological impact on people in terms of the population doses received and the resulting health effects.

The radiation doses to the general population following a postulated release can be calculated in various levels of detail depending on the type of input data available, the geographic area affected by the release, and the time scale of interest. For time scales on the order of a few years and in which the released activity is restricted to a well-defined localized geographical area, high-resolution dose calculations can be performed to determine the maximum individual dose, the population-versus-dose distribution, and the number of persons receiving a dose above a specified level for each exposure pathway. In order to carry out such calculations, spatial distributions of the time-integrated activity in the air, in water, and on land relative to population and food supply distributions can be determined using refined atmospheric transport and dispersion models. This type of calculation scheme is incorporated into the RISK II and FSAR-EMERGE models developed by the NUS Corporation (NUS) designed primarily for evaluating Pu-238 systems, as documented in NUS (1989 and 1990) and DOE (1982).

The problem addressed in this paper is the long-term risk due to radioactivity released into the environment when the time scale of interest is greater than a few years, and the radionuclide spectrum is changing significantly with time, even during early times. These conditions apply in the case of fission-reactor systems. While the former methods use population wheels to describe localized distributions of population and other parameters, the methodology described in this paper uses average densities and incorporates more exposure pathways. It is also assumed that a linear dose response model relates radiological doses to associated health effects. The complexities of population distribution are traded for the complexities of the source terms inherent with fission-reactor systems. Conventional environmental dispersion modeling, as in RISK II, is required only if it is desired to calculate the dose-rate distribution over the population. Since the objective of the present method is to calculate population exposure only in terms of collective dose (person-rems) and apply the linear dose response model to determine health effects, a considerable simplification in the calculations results. In an air environment, for example, the critical points of contamination are people and food crops or animals. If it is assumed that the critical points of contamination are uniformly distributed throughout the environmental medium of interest, then the exposure is directly proportional to the total quantity of activity in the dispersing medium and independent of the distribution of the activity in that dispersing medium. This is a direct result of the linear dose response model and the Mean Value Theorem.

TECHNICAL APPROACH

Details of the general method described herein is presented in Bartram (1982). Key features of the method are illustrated in Figure 1 and outlined below.

Accident Scenario and Source Term Definition

The accident scenarios, probabilities, and source terms for each reference design and mission considered are evaluated based on available reference design data, the literature, and analysis. The specific activity of each radionuclide in the fuel at the time of initial release is evaluated using the ORIGEN code (ORNL 1976). The operating history of the RTG or reactor system is accounted for in arriving at the source term radionuclide spectrum at the time of released fuel in the environment. Specific activities are calculated for order-of-magnitude time steps from 10^2 seconds to 10^{11} seconds following release.

Environmental Radioactive Concentrations

Environmental radioactive concentrations are calculated as a function of time for the following type of releases:

- High altitude vapor releases,
- High altitude particulate releases,
- Low altitude (including ground level) releases, and
- Aquatic releases.

In order to account for the radiological impact over long time periods, the concentrations due to a given release are calculated in ten time steps identified previously. The evaluation of environmental concentrations is performed initially on a fuel mass basis due to the large number of radionuclides of interest. Mass concentration is then converted to radioactivity concentration of each radionuclide by applying the time dependent specific activity factors generated as described above.

In evaluating environmental transfer processes, consideration is given to interaction statistics. That is, for a given release scenario, the regional environment is described so that the interaction between the source term and the environment can be evaluated. Interaction statistics for a given release scenario would include information such as the distribution of land, fresh water, ocean, population, and food supplies in the region that is expected to be affected by the release. Information of this nature is presented in DOE (1982) for releases occurring in the Kennedy Space Center vicinity and on a worldwide basis.

- ### High-Altitude Vapor Releases

High altitude vapor releases (aerosols less than about 10 μm in diameter) result from partial or complete burnup due to reentry heating. The transport of such vapor is governed primarily by worldwide atmospheric circulation patterns. The atmosphere can be divided into three compartments: the mesosphere (above 4.5×10^4 m altitude), the troposphere (0- 10^4 m), and the stratosphere (10^4 - 4.5×10^4 m). Transport between these atmospheric compartments is described in an approximate manner by an exponential box model, analogous to radioactive decay, in which a residence half-life is assigned to each compartment. The residence half-lives are based on fallout data derived from nuclear weapons testing. Latitudinal variation of the ground level activity derived from the high altitude release at a given latitude is also based on nuclear weapons fallout data. For high altitude vapor releases at a given latitude, the model predicts the normalized ground plane and airborne concentrations as a function of latitude and time after release.

- ### High-Altitude Particulate Releases

High altitude particulate releases (aerosols greater than about 10 μm in diameter) can occur under the same conditions that result in high altitude vapor releases described above. The resulting particles will follow trajectories to the earth's surface that are determined by the mass, size, and shape of each particle, the release altitude, as well as the winds and density profile of the atmosphere. The trajectories of individual mass elements are determined based on the rate of change of latitude, longitude and altitude during local wind speed components, and the terminal fall velocity of the particles.

Using the release altitude and latitude as input, the model predicts the ground plane concentration as a function of affected area. Resuspension and weathering of ground plane deposition is taken into account.

- ### Low-Altitude Releases

Low-altitude atmospheric transport and dispersion of fuel within the troposphere can result from (1) ground-level releases following land impact or (2) resuspension of material previously deposited as a result of both high and low altitude releases.

Radiological doses to the general population due to low altitude releases must account for vanishingly small concentrations at great distances from the point of impact because such concentrations are considered important when a linear dose response model as used. A model is used in calculating population dose in which all released activity is accounted for in an average concentration calculation.

- Aquatic Releases

Aquatic transfer processes include the interaction of airborne dispersed activity and intact sources within water bodies. These interactions include fuel dissolution, transfers to sediment, and concurrent covering of intact sources by sedimentation and encrustation. These processes are characterized with an exponential box model with results evaluated parametrically as a function of time required to breach encapsulation, fuel dissolution rate, transfer rate of dissolved fuel to sediment, and removal rate of intact sources by sedimentation or encrustation. Average aquatic concentrations are calculated as a function of time for the ten discrete time steps mentioned previously. The average concentrations are calculated based on reference volumes required to produce annual fish/seafood and drinking water to persons within the affected area, accounting for the land/water fraction within the area.

- Environmental Concentration Outputs

The outputs of the environmental concentration models and interaction statistics described above include the following information for ten time steps centered from 10^2 to 10^{11} seconds after the accident:

- Normalized airborne concentration (g/m^3/g released),
- Normalized ground concentration (g/m^2/g released),
- Normalized vegetation concentration (g/m^2 leaf area/ g released),
- Normalized aquatic concentration (g/m^3/g released),
- Total grams of fuel released, and
- Number of persons affected.

These results serve as input to the radiation dose calculations described below.

Radiation Dose Calculations

Maximum individual and population doses are calculated for both dispersed fuel and intact sources as described below.

- Radiation Doses (Dispersed Fuel)

Radiation doses due to dispersed fuel are calculated in discrete time steps following the release using as inputs the results of the environmental concentration calculations described in the previous section. A computer code, TDOS, was developed by NUS to evaluate the doses in discrete time steps using the exposure pathway dose methodology presented in NRC Regulatory Guide 1.109 (Bartram 1984 and NRC 1977). For individual and population exposures during each time step, the resulting 50-year integrated dose commitments are evaluated for the following pathways:

- Inhalation,
- Cloud immersion,

910

- Ground Plane radiation,
- Vegetation ingestion,
- Meat ingestion,
- Milk ingestion,
- Water ingestion,
- Aquatic foods ingestion.

Resulting health effects are also calculated by TDOS. The TDOS code output includes the following:

- User supplied input parametrics;

- User supplied specific activities by nuclide and time;

- Dose factor library;

- Bio-transfer factors;

- Dose or health effects by nuclide, time, organ and pathway; and

- Environmental dose commitments and health effects for the first year, first 100 years, and the total for all time periods.

- Radiological Doses (Intact Sources)

Radiological doses due to external radiation from intact sources have been evaluated using standard methods. The intact sources were treated as point sources with radiation fields attenuated inversely as the square of the distance and by absorption. Buildup in the air media and interactions at the air/ground interface are accounted for in a limited manner.

Radiological Health Effects

Health effects resulting from the calculated doses are determined using a linear dose response model for low linear-energy-transfer (low-LET) radiation (gammas and betas) based on the NAS-NRC Committee on the Biological Effects of Ionizing Radiation (BEIR). The linear dose response model assumes the health effects, in terms of excess cancer fatalities, are proportional to the dose, even for vanishingly small doses. The health effects models from the BEIR Committee have been reviewed by the International Commission on Radiological Protection (ICRP). The individual probability of cancer fatality due to radiation exposure (risk coefficient) used in ICRP-26 was 10^{-2} Sv^{-1} (equivalent to 100 fatalities per million person-rem; Note: 1 Sievert = 1 Sv = 100 rem). This value is consistent with estimates in the 1972 BEIR-I Report and the 1977 UNSCEAR Report (NAS, 1980 and UNSCEAR, 1977) for low-LET (linear energy transfer) radiation such as gammas and betas. Based on results of more recent health effects studies contained in the 1988 UNSCEAR Report and 1990 BEIR-V Report, ICRP recommends the following nominal risk coefficients for the general population, including children (UNSCEAR 1988 and NAS 1990).

- 5×10^{-2} Sv^{-1} for low doses, and
- 10×10^{-2} Sv^{-1} for high doses.

In this case "low doses" are absorbed doses below 0.2 Gy (1 Gray - 1 Gy = 100 rads) and from higher absorbed doses when the dose rate is less than 0.05 Gy per minute. For gammas, this corresponds to 20 rem and 5 rem/minute, respectively.

The higher risk coefficient, 10 x 10^{-2} Sv^{-1}, results from interpreting data at high doses to extrapolate to lower doses. ICRP reduced this by a factor of 2 for low doses to allow for the reduced effect of cell death at low doses.

CONCLUSION

An overall approach for risk assessment of environmental releases of radioactivity has been described above is summarized in flow diagram form in Figure 1, and its implementation is described in Bartram (1981, 1982, and 1984). The method outlined above has been applied to a number of fission-reactor space nuclear power systems as used on actual or hypothetical missions, as detailed in Bartram (1984, 1988, and 1989).

Acknowledgments

This work was sponsored by the U.S. Department of Energy under Contract No. DE-A-C01-88NE32135.

References

Bartram, B. W., D. K. Dougherty, and S. R. Tammara (1981) *TDOS Computer Code User's Manual*, NUS-3851 NUS Corporation, Gaithersburg, MD.

Bartram, B. W. and D. K. Dougherty (1981) *A Long Term Radiological Risk Model for Plutonium-Fueled and Fission Reactor Space Nuclear Systems*, NUS-3845, NUS Corporation, Gaithersburg, MD.

Bartram, B. W., R. W. Englehart, S. R. Tammara, and A. Weitzberg (1984) *Comparative Risk Analysis of Selected Missions Utilizing Space Nuclear Electric Power Systems*, NUS-4083, NUS Corporation, Gaithersburg, MD.

Bartram, B. W. and A. Weitzberg (1988) *Radiological Risk Analysis of Potential SP-100 Space Mission Scenarios*, Prepared for NUS-5125, NUS Corporation, Gaithersburg, MD.

Bartram, B. W. and A. Weitzberg (1989) "Radiological Risk Assessment of Potential SP-100 Use in Space", Presented at the Sixth Symposium on Space Nuclear Power Systems held in Albuquerque, NM 6-12 January 1989.

ICRP (1977), *Recommendations of the ICRP*, ICRP Publication 26, International Commission on Radiological Protection, Pergammon Press, New York, NY.

ICRP (1990) 1990 *Recommendations of the International Commission on Radiological Protection*, ICRP Publication 60 International Commission on Radiological Protection, Pergammon Press, New York, NY.

ICRP (1979) *Limits for the Intake of Radionuclides by Workers*, ICRP-30 (1979 and additions thru 1988) International Commission on Radiological Protection, Pergammon Press, New York, NY.

INSRP (1990) *Safety Evaluation Report for the Ulysses Mission*, INSRP-90-01 thru 03 Interagency Nuclear Safety Review Panel.

NAS (1990) *Health Effects of Exposure to Low Levels of Ionizing Radiation*, BEIR-III Report, National Academy of Sciences, Committee on the Biological Effects of Ionizing Radiation, Washington, DC.

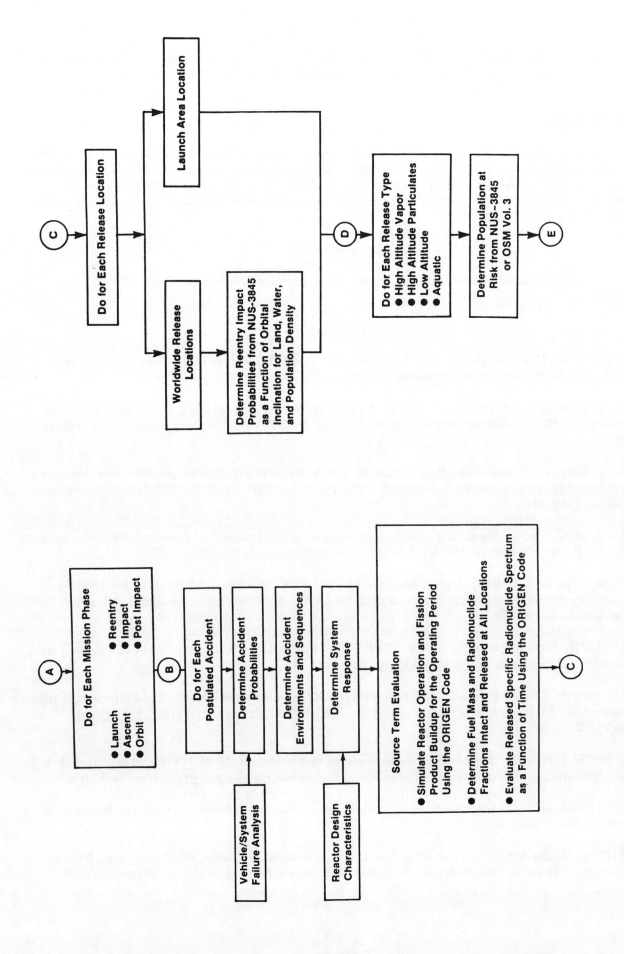

FIGURE 1. NUS Risk Assessment Method for Fission-Reactor Space Nuclear Systems (Page 1 of 2).

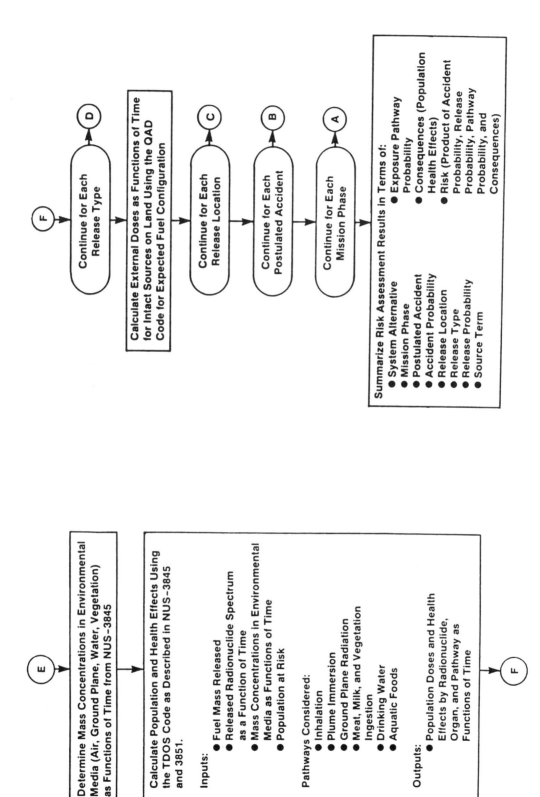

FIGURE 1. NUS Risk Assessment Method for Fission-Reactor Space Nuclear Systems (Page 2 of 2).

NAS (1988) *Health Risks of Radon and Other Internally Deposited Alpha Emitters*, BEIR-IV Report National Academy of Sciences, Committee on the Biological Effects of Ionizing Radiation

NAS (1990) *Health Effects of Exposure to Low Levels of Ionizing Radiation*, BEIR-V Report National Academy of Sciences, Committee on the Biological Effects of Ionizing Radiation, Washington, DC.

NAS (1987) *Recommendations on Limits of Exposure to Ionizing Radiation*, NCRP Report 91 National Council on Radiation Protection and Measurements, Washington, DC.

NRC (1977) *Calculation of Annual Doses to Man from Routine Releases of Reactor Effluents for the Purpose of Evaluating Compliance with 10CFR50*, USNRC Regulatory Guide 1.109, Rev. 1, U.S. Nuclear Regulatory Commission, Washington, DC.

NUS (1982) *Overall Safety Manual*, Prepared for U.S. Department of Energy (Updated for DOE through 1982), NUS Corporation, Gaithersburg, MD.

NUS (1989), *Final Safety Analysis Report for the Galileo Mission Volume III, Books 1 and 2, Nuclear Risk Analysis Document*, Prepared for U.S. Department of Energy, NUS Corporation, Gaithersburg, MD.

NUS (1990) *Final Safety Analysis Report for the Ulysses Mission, Volume III, Books 1 and 2, Nuclear Risk Analyses Document*, Prepared for U.S. Department of Energy, ULS-FSAR-004 and 005, NUS Corporation, Gaithersburg, MD.

ORNL 1976 *ORIGEN, Isotope Generation and Depletion Code-Matrix Exponential Method*, CCC-217 Oak Ridge National Laboratory, Oak Ridge, TN.

UNSCEAR (1988) *Sources, Effects and Risks of Ionizing Radiation*, E.88.IX.7 United Nations Scientific Committee on the Effects of Atomic Radiation, New York, NY.

UNSCEAR (1977) *Sources and Effects of Ionizing Radiation*, E.77.IX.I United Nations Scientific Committee on the Effects of Atomic Radiation, New York, NY.

THE STUDIES IN THE SUBSTANTIATION OF THE NUCLEAR SAFETY CONCEPTION OF NPS FOR MANNED MISSIONS

V.Ya.Pupko, F.P.Raskach, V.A.Lititsky, A.G.Shestyorkin, M.K.Ovcharenko, V.V.Volnistov.
Institute of Physics and Power Engineering, Obninsk, Kaluga region,USSR.

P.I.Bystrov, Yu.A.Sobolev, N.M.Lipovy, Research and Production Association "Energiya", Kaliningrad, Moscow region, USSR.

F.M.Arinkin,G.A.Batyrbekov, S.U.Talanov, Sh.Kh.Gizatulin Institute of Nuclear Physics, Kazahstan Acad. of Sci.

The paper to be presented at the 9th Scientific Symposium on Space NPSs, January 1992, Albuquerque, New Mexico, USA.

INTRODUCTION

The convincing evidences are currently available, that prove the lack of technically substantiated alternative to nuclear power and electric propulsion installations [1-7] for a number of space programs.

An application of NPS(NPRPS) as part of space craft (SC) implies the solution of radiation and nuclear safety control problems. In our view, the proper solution of these problems would be a decisive factor in NPS(NPRPS) utilization as part of SC. Therefore, giving primary prominence to this aspect in the development of national space programs in the USSR, USA and France using NPS as part of SC seems natural.

The concept and basic principles of safety control for the NPSs being currently designed in the USSR, USA and France are stated in detail in a number of papers [8-12] at the conference "Nuclear Power in Space" held in May 15-19, 1990 in Obninsk (USSR).

Bearing in mind an active interest in such programs, we find it important once again to pay attention on the determination of principles, criteria of nuclear safety control to be followed from the earliest stages of projects implementation. It is equally essential, that these principles were accepted and approved by the international public organization, e.g. by the Committee on the Utilization of Space for the UN organization and its subcommittees.

Among nuclear safety features of NPS(NPRPS) being devised, which represent the subject of further consideration within this report will be:
1. Space NPS(NPRPS) safety is to be substantiated as early as at the conceptual design stage;
2. No events related to a take-off area flight path, flight abortion or reentry with subsequent dynamic impacts can initiate an occurrence of critical geometry in a nuclear reactor core.
3. The reactor is to demonstrate a negative density prompt temperature and power reactivity coeffients.
4. The reactor is to be sustained subcritical when it immerses in water or other fluids.

5. The application of hostile chemically toxic material in the reactor should be minimized.

1. CALCULATIONAL AND EXPERIMENTAL STUDIES ON CRITICALITY ISSUES FOR THERMIONIC FAST REACTOR CONVERTERS (TRC) IN THE EMERGENCY SITUATIONS.

This section deals with the results of calculational-experimental studies on Keff and reactivity effects of thermionic fast reactor-converter (TRC) involved in Ref.[2] for particular emergency situations.
The TRC is characterized by:
1. Fissile material is highly enriched U-235 (or U-233 in the future).
2. Liquid-metal (entectic Na-K, Li-7) cooling; heat pipes cooler-radiator.
3. Refractory radiation-resistant structural materials (steels, niobium,tungsten).
4. This is a fast neutron TRC with a negative temperature effect.
The results stated below concerned the matters of:
- dependence of TRC Keff on the external surrounding by water or ingress of water inside the core (into the coolant loop);
- research into breeding properties of uranium-water lattices composed of various TRC core elements (fuel elements, fuel subassemblies);
- impact of resonance absorbers (Eu, Hf, Gd) on TRC Keff and uranium-water lattices.
The experiments were held at a critical facility [13] with FS-1-4.19 critical assemblies. The cores of 19 subassemblies in the form of hexahedions made of niobium. Inside its casing over the triangular lattice 37 thin-walled niobium tubes were arranged with a diameter 12.5*0.4mm. Cylindric fuel elements with fuel cores of highly enriched uranium dioxide (90% enrichment for U-235) were loaded inside the internal cavities of the above mentioned tubes. The height of the active section of fuel element was 800mm. The effective diameter of critical assembly cores were 300...600mm; the core volume - 96 lit.
The lateral berrilium reflector of critical assemblies with an effective thickness 150mm incorporated 12 control drums of rotating typ with boron carbide inserts absorbing neutrons. On the outside the critical assembly was surrounded by a water tank with a total thickness of water layer 400mm.
The calculational investigations were made by the Monte-Carlo Technique using the MMKFK code system [14] and the ARAMACO-C1 nuclear data set [15] following the technique described in Ref.[16]. This enabled the calculations of critical assemblies FS-1-4.19 (and appropriate TRC) to be performed with a subassembly configuration of the core and a heterogeneous representation of both the fuel elements in the subassemblies, and the control drums - in the lateral reflector.
Table 1 presents the measured and calculated values of Keff, reactivity effects and rotating-type individual control drum efficiently for one of the FS-1-4.19 critical assembly options with neutron-absorbing inserts of the control drums in the position out of the core (n=684 fuel elements) for the subsequent cases:
1. no water("dry" critical assembly)
2. critioal assembly is surrounded by the layer of water 400mm in thickness on the outside; no water in the core.

3. water in the core ("wet" critical assembly).
  Note, that the experimental value of Keff is obtained via extrapolation using the measured values of reactivity effects from alternative amounts of fuel element in subassemblies.
  Extrapolation of hydrogen reactivity effects to full-load core (684 fuel elements) from the critical state (440 fuel elements) of the critical assembly filled with water represents a problematic task. Owing to this the reliability of the calculational and experimental results presented in table 1 demands supplementary investigations.

Table 1

ΔKeff and Keff values for FS-1-4.19 critical assembly.

№ П/П	parameter			type of assembly		
				1	2	3
1	Keff	experiment		1.050 ±0.01	1.053 ±0.01	1.185 ±0.01
		calculation		1.056 ±0.006	1.058 ±0.006	1.132 ±0.006
2	effect for water, ΔKeff	experiment		0	0.30 ±0.02	13.5 ±0.4
		calculation		0	0.32 ±0.02	6.4 ±0.6
3	controller drum efficiency, βэфф	experiment		1.61 ±0.005	1.700 ±0.005	1.12 ±0.005

  Table 2 presents the results of Keff calculation for one of the FS-1-4.19 critical assembly options with the under loaded core (with and without water in the core); for these cases with water the Keff values were also calculated in applying thin coatings of gadolinium and hafnium with a layer thickness 0.1mm to the outer cans of fuel element tubes. This way of resonance absorber arrangement in the TRC core for positive-in-value hydrogen effect compensation was considered as an alternative to safety rods adoption.

918

Table 2.

Calculational values of Keff for FS-1-4.19 critical assembly using resonance absorbers.

№ П/П	condition of calculation	Keff values		ΔKeff	
		without water	with water	K^1ef	K^2ef
1	without absorber	1.04 ±0.005	1.09 ±0.005	4.4 ±0.007	
2	Gd	1.00 ±0.005	1.02 ±0.005	1.96 ±0.007	
3	Hf	1.02 ±0.005	1.025 ±0.005	0.0048 ±0.007	

It follows from Table 2 that the utilization of hafnium is preferable to compensate for the positive hydrogen effect.

The results of investigations for multiplication parameters (Kef) of uranium-water lattices formed of fuel elements (subassemblies) of the FS-1-4.19 critical assembly depending on the spacing pitch and amount of fuel elements(subassemblies) are shown in Figs.1 and 2. (In the experiments and calculations the water layer height was 120cm, the water lateral reflection of the side - over 100cm).

As it follows from fig.1 the maximum value of Keff is reached with a pace about 30mm; the results of calculation and experiment virtually agree within the margin of errors of calculation and experiment (relative error in the Keff values was as low as 1%).

The determination of suinable place for resonance absorber insertion in the core - into the fuel core or into the collector subassembly composition was of the direct practical interest. Fig.3 (for one of the FS-1-4.19 critical assembly options) shows the dependence of Keff on the concentration of gadolinium introduced into the composition of fuel core (curve 1) or into the composition of fuel collector subassembly (curve 2). As can be seen in this figure, the introduction of resonance absorber into the composition of collector subassembly is preferable.

Fig.4 shows the calculational dependences of Keff on the pace for uranium-water lattices composed of 631 fuel elements of the FS-1-4.19 critical assembly; the calculations were made for the following cases:
- no resonance absorber in the niobium tube of fuel element (curve 1);
- europium with thickness of coating - 0.075mm (curve 2) on the niobium tube;
- gadolinium with thickness of coating - 0.012mm (curve 3) on the niobium tube;

The dependence of K_{ef} on the spacing of fuel elements lattice

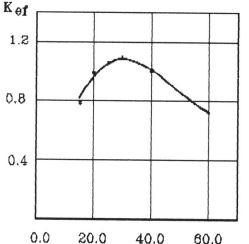

spacing of fuel elements lattice, mm

— — calculation; *— experiment

Fig. 1

The dependence of K_{ef} on the amount of fuel elements for spacing of lattice 13.5 mm.

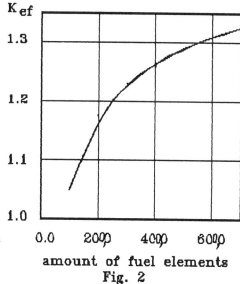

amount of fuel elements

Fig. 2

The dependences of K_{ef} FS-1-4.19 on the concentration Gd.

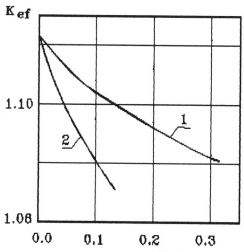

Gd concentration, 10^{20} nucl/cm^3

1 — Gd introduced into the composition of fuel core;
2 — Gd introduced into the collector subassembly.

Fig. 3

The dependences of K_{ef} on the spacing for uranium-water latties composed of 631 fuel elements of the FS-1-4.19 critical assembly.

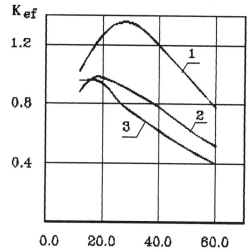

spacing of fuel elements lattice, mm

1 — no resonance absorber; 2 — Eu in the Nb-tube of fuel elements;
3 — Gd in the Nb-tube of fuel elements.

Fig. 4

920

As it follows from this figure, an introduction of resonance absorber about 1mm in thickness into the composition of collector subassembly niobium tube may provide for the subcriticality of the set of 631 fuel elements within the range of spacings 12.5 to 60mm. There is an optimum of the resonance absorbers insertion effect depending on their arrangement over the core volume .(The resonance absorbers should be arranged at a distance 2/3 of core radius from the core centre.)

CONCLUSION

The resultsof investigations presented in the paper show, that the introduction of resonance absorber (Hf, Gd or Eu) into the composition of core TFE collector subassemblies for the TRC class considered may provide for the TRC subcriticality of virtually all accidents related to TRC break-up (under impacts) and its subsequent ingress into water. In other words, an internal safety of the TRC class considered may be ensured. An approach described above gives, firstly, an essential advantage in the sense of reliability as compared with the use of safety rods set in the TRC design, and, secondly, in this case the design of the inherent TRC is significantly simplified.

REFERENCES

1. Gryaznov G.M. et al. Thermionic Reactors-Converters for Space NPS. Atomnaya Energiya, 1989, v.66, N6, 374.
2. Yu.P.Semenov, P.I.Bystrov, V.Ya.Pupko et al. Fast Neutron Thermionic Reactor-Converters for High-Power Space NPS. Presentation at the 8th National Symposium on Space NPS, January 6-9, 1991, Albuquerque, New Mexico, USA.
3. Yu.P.Semenov, P.I.Bystrov et al. Nuclear Power Propulsion Unit on a Basis of Thermoionic Nuclear Power Supply System for Mars Manned Mission. Anniversary Specialist Conference on Nuclear Power Engineering in Space, Obninsk, May 15-19, 1990.
4. Vincent C. Truscello. The SP-100 Power System. Anniversary Specialist Conference on Nuclear Power Engineering in Space, Obninsk, May 15-19, 1990.
5. Earl Wahlquist. U.S. Space Reactor Programs. Anniversary Specialist Conference on Nuclear Power Engineering in Space, Obninsk, May 15-19, 1990.
6. Claude Poher. Space Missions studies with Nuclear Generators.Anniversary Specialist Conference on Nuclear Power Engineering in Space, Obninsk, May 15-19, 1990.
7. Frank Carre et al. Current Trends in the Erato Study Program on Space Nuclear Brayton Systems. Anniversary Specialist Conference on Nuclear Power Engineering in Space, Obninsk, May 15-19, 1990.
8. Neil W.Brown et al. U.S.A. Approach to Space Nuclear Power Safety.Anniversary Specialist Conference on Nuclear Power Engineering in Space, Obninsk, May 15-19, 1990.
9. Alan J. Willoughby. Nuclear Radiation Issues for Human Exploration of Space. Anniversary Specialist Conference on Nuclear Power Engineering in Space, Obninsk, May 15-19, 1990.
10. N.V.Ganov et al. Criteria, Basic Principles and Engineering Measures to Ensure SNPS Nuclear Safety. Anniversary Specialist Conference on Nuclear Power Engineering in Space, Obninsk, May 15-19, 1990.

11. Robert G. Lange et al. Safety Review Process for Space
    Nuclear Power Sources. Anniversary Specialist Conference on
    Nuclear Power Engineering in Space, Obninsk, May 15-19, 1990.
    Anniversary Specialist Conference on
    Nuclear Power Engineering in Space, Obninsk, May 15-19, 1990.
12. James Lee.Technology Requirements for the Disposal of Space
    Nuclear Power Sources and Implications for Space Debris
    Management Strategies. Anniversary Specialist Conference on
    Nuclear Power Engineering in Space, Obninsk, May 15-19, 1990.
13. V.Ya.Pupko et al. Critical Assembly for Studies of Space
    Purpose Reactors Physical Parameters. Anniversary Specialist
    Conference on Nuclear Power Engineering in Space, Obninsk, May
    15-19, 1990. nuclear
14. A.D. Frank-Kamenetsky.Abstract of Program Complex
    MMKFK for Reactors Calculations via Monte-Carlo method.
    Vaprosy   Atomnoy   Nauki   i   Techniki.   Serics:
    Fisica i Technika Yadernih Reactorov, 1981,   8(21), p 16.
15. M.N.Nikolaev et al. System of Nuclear Data Provision
    ARAMACO-C1 IPPE report №7012, 1984
16. Polevoy V.B. Peculiarities of Neutron Physics Calculation of
    a Fast Reactor with Moderating Reflector. Anniversary
    Specialist Conference on Nuclear Power Engineering in
    Space, Obninsk, May 15-19, 1990.

NUCLEAR THERMAL ROCKET ENTRY HEATING
and THERMAL RESPONSE PRELIMINARY ANALYSIS

Leonard W. Connell, Donald L. Potter,
C. Channy Wong, and Marc W. Kniskern
Sandia National Laboratories
Albuquerque, NM 87185
(505) 846-0459

Abstract

This preliminary study analyzes the atmospheric entry of a solid core nuclear thermal rocket (NTR) engine under several entry accident scenarios. Results of the analysis showed that, without external thermal protection, an aluminum pressure vessel will fail at altitudes ranging from 20 to 65 km, depending on the scenario, thus releasing the core materials. The graphitic based core materials will undergo only partial ablation, with the percent mass loss depending on the geometry of the fuel form. A thermal protection system was sized to prevent pressure vessel aerothermal failure and was found to increase the mass of the NTR by approximately 15 percent.

INTRODUCTION

The NASA/DOE/DoD Joint Nuclear Thermal Rocket Safety Policy Working Group has recently recommended broad policy guidelines for the entry safety of a Nuclear Thermal Rocket (NTR) (Marshall 1991). These guidelines allow for two distinct design options: (1) intact entry, or (2) complete dispersal upon entry (complete breakup and ablation of the NTR such that all core materials are dispersed as an aerosol). The objective of this study was to develop a preliminary understanding of the atmospheric entry behavior of an NTR under several entry scenarios and to investigate the feasibility of achieving these design options. Since this is not a risk analysis, the probability of occurrence of these scenarios and the resultant hazards was not evaluated. Specifically, the study focused on three issues:

1. The altitude regime in which a thermally unprotected pressure vessel will fail, thus releasing core materials;
2. The extent of fuel element ablation mass loss; and
3. The mass of a pressure vessel heatshield required to ensure intact-to-impact entry of the pressure vessel.

Since the entry response of an NTR is somewhat dependent on the specific design, the results presented cannot be generalized to all designs. They are pertinent only to those designs that have similar mass, material, and geometry properties as the one chosen for this study. However, the procedures and calculations described here provide a framework for entry analysis of all classes of NTRs.

NTR DESIGN CHOSEN

The Small Nuclear Rocket Engine (SNRE) (Durham 1972) was chosen for analysis. The mass and material properties of the SNRE were well defined, as shown in Table 1. The SNRE is based on NERVA technology but is scaled down to one-fourth the mass and thrust of NERVA. Its mass and size are therefore representative of a Particle Bed Reactor (PBR) engine of the 444 kN (100 klbf) class. Three distinct fuel forms were analyzed in order to obtain an understanding of the sensitivity of fuel ablation to fuel geometry: PBR particles (0.05 cm dia, 350 mg), Pellet Bed Reactor (PeBR) pellets (1.0 cm dia, 5.0 g), and NERVA prismatic rods (1.91 cm width x 88.9 cm length, 767 g) (NASA 1990). The elements were modeled as pure graphite for the thermal response calculations.

ACCIDENT SCENARIOS

Three entry scenarios were analyzed: Orbital decay, Mars Boost Misfire maximum velocity (MBMV), and Mars Boost Misfire maximum entry angle (MBMγ). Entry conditions for each scenario are: V = 7.8 km/s, γ = -0.5 deg. (Orbital Decay), V = 11.1, γ = -12.4 (MBMV), and V = 6.0, γ = -35 (MBMγ). The MBM scenarios are caused by a misalignment of the NTR thrust resulting in the Mars transfer delta-V being misaligned to the orbital velocity. Rather than being placed on a transfer orbit to Mars, the spacecraft is placed on a trajectory that intersects the earth's atmosphere with entry velocity V and entry angle γ.

PRE-BREAKUP AERODYNAMICS

TABLE 1. NTR Materials and Masses.

Component	Material	Mass (kg)
Pressure Vessel	Al 7050-T73	150
Reflector	Beryllium	570
Internal Shield	ZrH	240
Core Support Plate	Al 2219-T87	37
Core	Graphite	868
Nozzle (Regen)	Inconel 625	177
Nozzle (Aft Skirt)	Wound Graphite	41
TPA	Stainless Steel	41
Thrust Structure	Inconel 718	30
Propellant Lines	Al 219-T6	16
Valves	Stainless Steel	210
Electronics		160
Total		2540

An aerodynamic stability analysis was performed to determine if the NTR would be stable in the "nose first" (nozzle aft) configuration. The analysis consisted of determining the location of the center of pressure (cp) relative to the center of gravity (cg) and is termed a static stability analysis. If the cp is located aft of the cg then any disturbance that pitches the body axis about the cg away from the velocity vector, thus producing an angle of attack, would be resisted by a pitching moment which would realign the system back to a zero angle of attack configuration. Location of the cp requires computation of the flowfield about the body. The NTR was modeled as a combined sphere-cone-cylinder-flare (Figure 1). This approximation takes into account the dominant physical characteristics of the vehicle. Results indicated that for moderate angles of attack, the cp is located aft of the cg which implies that the "nose forward", zero angle of attack configuration exhibits static stability.

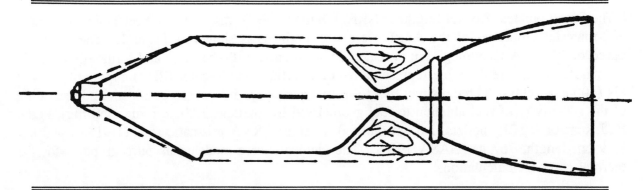

FIGURE 1. NTR Aerodynamic Model.

If the NTR entered attached to a propellent tank which was subsequently torn away by aerodynamic forces, an initial angular momentum could be imparted to the NTR resulting in a random tumble and spin entry. A dynamic stability analysis would be needed to investigate the ability of the NTR to damp out these random gyrations and achieve a stable flight profile. Since a dynamic stability analysis was beyond the level of effort of this study, random tumble and spin entry, although still a possibility, was not considered. The Trajectory Simulation and Analysis Program (TSAP) (Outka 1990) was used to compute stable, zero angle of attack entry trajectories of the NTR. Figure 2 presents the trajectory for the MBMV scenario. Trajectories were also obtained for the other two scenarios using the same techniques.

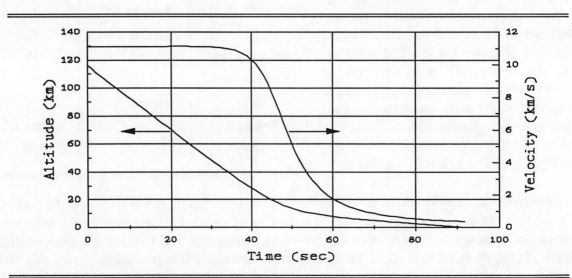

FIGURE 2. MBM Maximum Velocity Trajectory.

PRE-BREAKUP AEROTHERMAL ANALYSIS

The Hypersonic Integral Boundary Layer Analysis of Reentry Geometries (HIBLARG) code (Polansky 1990), was used to predict cold-wall heat transfer rates to the reactor structure. Correction for the hotwall effects on the heating parameters are made during the thermal response calculations. The Charring Materials Ablation (CMA) code (Blackwell and Kaestner 1970) was used to evaluate the reactor's thermal response. Thermal response calculations were performed for three key components: the inconel thrust structure at the front of the NTR, the aluminum pressure vessel both with and without a carbon-phenolic heatshield, and the carbon-carbon nozzle aft skirt.

The highest heating rates occur for the MBMV scenario, peaking at approximately 13.7 kW/cm^2 (12,000 Btu/ft^2-sec) on the thrust structure. However, the orbital decay trajectory leads to the highest integral heat input. For comparative purposes the integrated stagnation point cold wall heat flux for the three scenarios are 400 kJ/cm^2 (Orbital Decay), 250 kJ/cm^2 (MBMV) and 40 kJ/cm^2 (MBMγ). Aerothermal failure was defined for the metal components as the altitude at which melt temperature is reached. For all scenarios, the inconel thrust structure failed first followed quickly by the aluminum pressure vessel. The carbon-carbon nozzle did not show complete ablation of the cross section for any of the scenarios. Figure 3 depicts the stagnation point heating and pressure vessel wall temperature for the orbital decay scenario. Melt through of the pressure vessel occurs at 65 km compared to 45 km and 20 km for MBMV and MBMγ.

POST-BREAKUP AERODYNAMICS

Reactor core materials will be released after aerothermal failure of the NTR pressure vessel. The hypersonic ballistic coefficients, beta, of the fuel forms are approximately .4 (PBR particle), 8 (NERVA rods), and 15 (PeBR pellets) (units are lbf/ft^2) which are relatively small values in comparison with that of the NTR (1000 lbf/ft^2). Beta is a measure of the ability of an object to maintain its linear momentum. Based on the low beta of the PBR particles, we expect them to rapidly decelerate, while the PeBR pellets will be the most persistent in maintaining entry velocity. This is demonstrated in Figure 4.

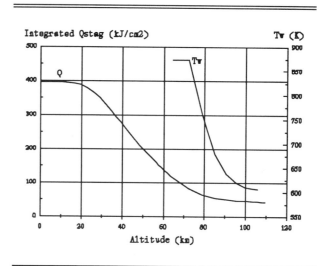

FIGURE 3. NTR orbital Decay Heating.

POST-BREAKUP AEROTHERMAL ANALYSIS

Empirical heating distribution correlations (Klett 1964) are incorporated in the HANDI code (Potter 1989) which is used to compute the heating around the fuel elements. Since the entry modes for the spherical and cylindrical elements were determined to be random tumbling and spinning and side on spinning respectively, the heating distributions were integrated and averaged over the appropriate surface area prior to use by the thermal response code.

Figure 5 presents integrated heating inputs for the orbital decay scenario and Table 2 illustrates the CMA computed percent mass loss of each element relative to its initial mass. For a given fuel element type, the percent mass loss between entry scenarios correlates with integrated heating. Correlation of mass loss mass (ΔM/Mo) within a scenario between two different element types is given by the following relation

$$\left(\frac{\Delta M}{M_0}\right)_2 = \frac{Q_2}{Q_1} \frac{(A/V)_2}{(A/V)_1} \left(\frac{\Delta M}{M_0}\right)_1$$

where Q is the integrated heating and A/V is the area to volume ratio for the fuel elements (120 for the PBR, 6 for the PeBR, and 2.1 for NERVA, in units of cm^{-1}). The ablation results presented in Table 2 indicate that complete dispersal does not occur for any of the scenarios, thus failing to satisfy safety design option 2.

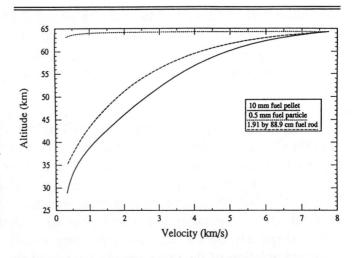

FIGURE 4. Fuel Trajectory, Orbital Decay.

TABLE 2. Fuel Element % Mass Loss.

Scenario	PBR	PeBR	NERVA
Orbital Decay	38.0	38.0	6.0
MBM Max V	66.0	29.0	2.3
MBM Max Gamma	53.0	6.4	0.8

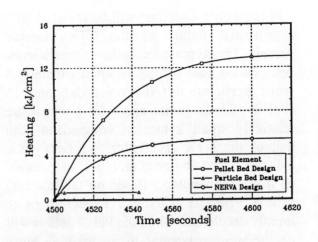

FIGURE 5. Fuel Heating: Orbital Decay.

PRESSURE VESSEL HEATSHIELD SIZING

Parametric CMA calculations were performed for all three entry scenarios to determine the necessary heatshield thickness to satisfy safety design option 1 (intact entry). The heatshield design requirement is to maintain the aluminum pressure vessel temperature below 550 K (1000 R). At this temperature, the aluminum has approximately 40% of its initial strength, sufficient to maintain structural integrity. As expected from the integrated heating results, the orbital decay was the design limiting trajectory. A heatshield thickness of 2.5 cm kept the aluminum pressure vessel below 550 K. Total mass of this shield is approximately 375 kg or 15 percent of the mass of the NTR.

CONCLUSIONS

This paper has examined the NTR entry problem with preliminary calculations of entry breakup and ablation. For the scenarios considered, complete dispersal of the NTR did not occur. Furthermore, the additional mass needed to thermally protect the NTR and prevent aerodynamic breakup is not large, representing roughly 15 percent of the unprotected NTR mass. Therefore, of the two entry safety designs, intact entry appears to represent a more practical design approach. Further work is needed which examines the sensitivity of the results to NTR mass, materials, geometry, and to random tumble and spin entry.

Acknowledgments

The authors wish to thank James Hipp for his advice on entry safety issues and his calculations of the entry scenario state vectors. This work was conducted at Sandia National Laboratories, supported by the U.S. Dept. of Energy under contract number DE-AC04-76DP00789.

References

Blackwell, B. F. and P. C. Kaestner (1970) *Operation Instructions for Charring Materials Ablation Code*, SC-DR-70-140, Sandia National Laboratories, Albuquerque, NM, March 1970.

Durham, F. P. (1972) *Nuclear Engine Definition Study, Preliminary Report Volume 1*, LA-5044-MS, Los Alamos Scientific Laboratory, Los Alamos, NM, September 1972.

Klett, R. D. (1964) *Drag Coefficients and Heating Ratios for Right Circular Cylinders in Free-Molecular and Continuum Flow from Mach 10 to 30*, SC-RR-64-2141, Sandia National Laboratories, Albuquerque, NM, December 1964.

Marshall, A. C. and J. C. Sawyer Jr (1992) "Nuclear Safety Policy Working Group Recommendations for SEI Nuclear Propulsion Safety," in *Proc. Ninth Symposium on Space Nuclear Power Systems*, CONF-920104, M. S. El-Genk and M. D. Hoover, eds., American Institute of Physics, New York, 1992.

NASA (1990) *Nuclear Thermal Propulsion, A Joint NASA/DOE/DOD Workshop*, NASA Lewis Research Center Advanced Space Analysis Office, Cleveland, Ohio, July 10-12 1990.

Outka, D. E. (1990) *User's Manual for the Trajectory Simulation and Analysis Program*, SAND88-3158, Sandia National Laboratories, Albuquerque, NM, July 1990.

Polansky, G. F. (1990) *Hypersonic Integral Boundary Layer Analysis of Reentry Geometries (HIBLARG) Code Description and Users Manual Version 2.0*, SAND89-0552, Sandia National Laboratories, Albuquerque, NM, March 1990.

Potter, D. L. (1989) *Approximate Heating Analysis Methods for Appended Bodies*, Internal Memorandom, Sandia National Laboratories, Albuquerque, NM, 13 October 1989.

AGING EFFECTS OF U.S. SPACE NUCLEAR SYSTEMS, CURRENTLY IN ORBIT

Bart W. Bartram and Seshagiri R. Tammara
HALLIBURTON NUS
Environmental Corporation
910 Clopper Road
Gaithersburg, MD 20877-0962
(301) 258-8666

Abstract

There are currently nine U.S. space nuclear systems in orbit, include eight with plutonium-238 fueled radioisotope thermoelectric generators and one fission-reactor system. Due to the inventories of radioactive materials onboard these spacecraft, the potential exists for radioactive releases during or following reentry. Projected orbital lifetimes, aging of containment materials, and radioactive decay will determine the degree of potential hazard associated with spacecraft reentry. These factors also affect consideration of possibly retrieving these spacecraft prior to reentry in order to minimize such hazards. This paper reviews the present U.S. space nuclear systems in orbit and summarizes the results of studies related to the aging effects of containment materials; describes changes in radionuclide inventory and external radiation field due to radioactive decay; and estimates the potential radioactive releases during or following reentry.

INTRODUCTION

The U.S. currently has nine spacecraft in orbit that incorporate nuclear systems for spacecraft electrical power, (see Table 1). These spacecraft, launched during the period 1961 to 1976, reflect changes in design philosophy and improvements with respect to development of space nuclear power systems. Eight spacecraft have radioisotope thermoelectric generators (RTGs) that use plutonium-238 fuel in various forms as a source of heat, including the metallic form in those systems designed for reentry burn-up (SNAP-3 and SNAP 9A); PuO_2 molybdenum cermet fuel (Transit Triad RTG), PuO_2 microspheres (SNAP-19), and pressed PuO_2 (MHW-RTG) for those systems designed for intact reentry. One spacecraft, SNAPSHOT, incorporated a fission-reactor system, SNAP-10A, that was inadvertently shut-down by a non-nuclear mechanical failure after 43 days of operation in orbit. The orbital lifetimes of these spacecraft are generally long, ranging from 150 to 1 million years, with Transit Triad and Transit 4A having the shortest orbital lifetimes of 150 and 570 years, respectively.

Due to the inventories of radioactive materials onboard these spacecraft, the potential exists for on-orbit radioactive releases due to containment disruption from debris in orbit; during reentry as a result of reentry burnup; or following reentry due to land impact on rock surfaces. Projected orbital lifetimes, aging of containment materials, and radioactive decay will determine the degree of potential hazards. These factors also affect consideration of possibly retrieving these spacecraft prior to reentry in order to minimize such hazards. The feasibility of retrieving nuclear power systems in space has been reviewed by Pyatt (1982). This paper summarizes a study of existing U.S. space nuclear systems and factors affecting considerations of retrievability, including the effects of aging on fuel containment materials over the projected orbit lifetimes, changes in radionuclide inventory and external radiation field due to radioactive decay, and estimates the potential radioactive releases during or following reentry (Bartram 1982).

TABLE 1.   Characteristics of U.S. Space Nuclear Systems in Earth Orbit as of Launch Date .

Spacecraft	Launch Date	Period (min)	Inc. (deg)	Apogee (km)	Perigee (km)	Power Source	Inventory of Pu-238 at Launch (curies)	Expected Orbital Lifetime (years)	Disposal Mode
Transit 4A	6/29/61	103.7	66.8	993	881	SNAP-3	1,700	570	Burnup
Transit 4B	11/15/61	105.7	32.4	1,103	953	SNAP-3	1,700	1,200	Burnup
Transit 5BN1	9/28/63	107.3	89.8	1,135	1,072	SNAP-9A	16,300	1,900	Burnup
Transit 5BN2	12/05/63	107.0	89.9	1,119	1,067	SNAP-9A	16,300	1,800	Burnup
SNAPSHOT	4/08/65	111.6	90.0	1,313	1,282	SNAP-10A	Reactor	3,000-5,000	Burnup
Nimbus III	4/14/69	107.3	99.7	1,135	1.074	SNAP-19	36,000	3,600	Intact Reentry
"Transit" (TRIAD-01-0X)	9/02/72	100.6	90.1	838	742	RTG	28,000	150	Intact Reentry
LES-8	3/14/76	1440.0	25.0	35,900	35,900	MHW-RTG	151,000	10^6	Intact Reentry
LES-9	3/14/76	1440	25.0	35,900	35,900	MHW-RTG	151,000	10^6	Intact Reentry

TECHNICAL APPROACH

Each of the U.S. space nuclear systems presently in orbit and identified in Table 1 were evaluated in terms of the following:

- Radioisotope inventory versus time following launch;

- External radiation field versus time following launch;

- Effects of aging on fuel containment materials over the projected lifetime of the system, and condition of the space nuclear system at the time of reentry;

- Potential radioactive releases during and following reentry.

Radionuclide Inventory

The initial radionuclide fuel mix in the Pu-238 fueled systems will be transformed in time as the initial radionuclide decay and daughter radionuclide accumulate. In the case of SNAP-10A, a fission reactor system, fission products were generated during operation followed by fission product decay after shutdown. The radioisotopic mix and, therefore, the radiological hazard of each system will vary over time. The radionuclide inventory versus time for each system was calculated using the ORIGEN code (ORNL, 1975). Inputs to the code include the initial curie activity of each radionuclide in a given fuel type and the output includes curie activity versus time for all parent and daughter radionuclide, the gamma photon production rate and energy spectrum versus time, and the neutron production rate and energy spectrum versus time. Neutron production due to plutonium mix spontaneous fissioning and (a, n) reactions was taken into account. ORIGEN was also used to simulate SNAP-10A operation and fission product buildup prior to shutdown.

External Radiation Field

The external radiation field dose rates due to a given system will change with time due to radionuclide decay and buildup, resulting in varying gamma and neutron energy spectra with time. The external dose rates from a given system was calculated using an NUS-modified version of the QAD-P5 code, derived from the QAD code originally developed by Los Alamos National Laboratory (Malenfant 1967, ORNL 1970; and Huang 1973).

QAD is a point kernel code system designed to calculate fast neutron and gamma-ray penetration of various shield configurations. Estimates are made of gamma-ray flux, dose rate, and energy deposition at specified detector locations, accounting for buildup within the source using Capo's parameter definitions and attenuation by shielding materials. The fast neutron dose obtained using a modified Albert-Wetton kernel was found to agree most closely with measured values.

Input data for the code consists of (1) a description of the source distribution and intensity by a number of point isotopic sources, (2) a mathematical representation of the physical geometry with quadratic surfaces, and (3) a tabulation of attenuation coefficients applicable to the shielding materials, buildup factors, and conversion factors.

In applying the code, equivalent cylindrical geometries were assumed for all systems considered. Thickness and material type were specified for each layer of the system. Gamma photon and neutron production rates and energy spectra as a function of time, derived from the ORIGEN runs described above, were input into successive QAD runs to determine the external dose rates versus time at specified distances in the axial and radial directions of each system. Measured gamma and neutron dose rates from actual systems were then used to calibrate the calculated values at time t=0, assuming that the measured values represented fresh fuel. Calculated values for subsequent time periods were scaled accordingly. The results indicated that the calculated and measured values at time t=0 were in good agreement, being within a factor of two at most.

Aging of Materials

The significance of a number of materials degradation mechanisms that may occur during the expected orbital lifetimes of each system was evaluated. The postulated degradation could influence a decision in the future to retrieve one or more of the systems while still in orbit, by use of an appropriately designed Space-Shuttle subsystem. In turn, a reevaluation of the conditions to be expected at reentry, interpreted in the light of the projected technological developments and political climate, may reinforce the conclusions of the original safety analysis reports, acknowledging burnup or reentry, or may warrant a later exercise of the retrieval alternative.

The following mechanisms were reviewed, where applicable:

- Radionuclide fuel/containment interactions,
- Phase changes or other thermal effects,
- Containment material fatigue,
- Pressure effects due to gaseous decay products,
- Interaction of containment materials,
- Radiation damage,
- Space irradiation,
- Evaporation sublimation, and
- Meteoroid damage.

The cumulative effect of the relevant degradation mechanisms was assessed and reentry source terms estimated accordingly.

RESULTS

Detailed results of the evaluation described above for each system considered are presented in Bartram (1982). Key features and highlights of the results are summarized below.

Radionuclide Inventory

The resulting total curie activity versus time for all systems considered is summarized in Figure 1. The radioactivity versus time of the plutonium-fueled systems is essentially dominated by Pu-238 with the other radionuclides contributing in only a minor way, except after approximately 300 years. A typical breakdown of the contribution of each radionuclide versus time for the Transit Triad RTG is presented in Figure 2. The SNAP-10A fission reactor system was shutdown after 43 days of operation in orbit after being launched in 1965. The resulting fission activity dropped rapidly within one year following shutdown and present radioactivity levels are dominated by Cs-137 and Sr-90.

Total production of helium and radon gas versus time have also been estimated for each spacecraft based on the ORIGEN output by converting time integrated alpha activity into a helium inventory, and activity of all radon isotopes into radon gas inventory. The results are presented for the Transit Triad RTG are presented in Figure 3 as an example.

External Radiation Field Versus Time

The external radiation field dose rates versus time for each system resulting from variations in gamma and neutron energy spectra versus time were calculated. In general the external dose rates from the plutonium fueled systems are dominated by spontaneous fission neutrons and (α, n) reaction neutrons from Pu-238 during approximately the first 300 years, with dose rates ranging on the order of 0.01 to 0.09 mR/h at 1 m. After approximately 300 years, radon isotopes and their short-lived daughters begin to buildup resulting from plutonium isotope decay. Gamma dose rates in turn begin the monotonicly increase after approximately 300 years and dominate thereafter. An example of these results is presented for the Transit Triad RTG in Figure 4. External dose rates from the SNAP-10A fission reactor system were quite high immediately after shutdown (approximately 60 rem/h at 1 m due primarily to gamma photons), but fell off rapidly after that. Present dose rates are on the order of 0.8 mR/h and are continuing to drop rapidly.

Aging of Materials

A summary of the evaluation of material aging effects for each system is presented in Table 2. The results, supported by the details in Bartram (1982), reveals several instances of sufficient concern or sufficient uncertainty to warrant individual study in depth, if retrieval missions are to be mounted for technical or tactical reasons.

Reentry Source Terms

A given space nuclear system in earth orbit will reenter the earth's atmosphere following orbital decay. Reentry scenarios that could release fuel include 1) reentry burnup, 2) land impact, and 3) ocean impact. The release mode probability and source term (percent inventory released) for each system were evaluated based on the evaluation of material aging effects. The results assume that reentry and associated release modes occur at the end of predicted orbit life. The expectation of source term released, representing the product of release probability and source term at the time of release, is presented in Table 3. Due to the early reentry time of Transit Triad compared to the other systems, it has the highest expectation source of 5180 curies (Ci), including 26.5 Ci to land surfaces, and 5150, Ci to ocean surfaces.

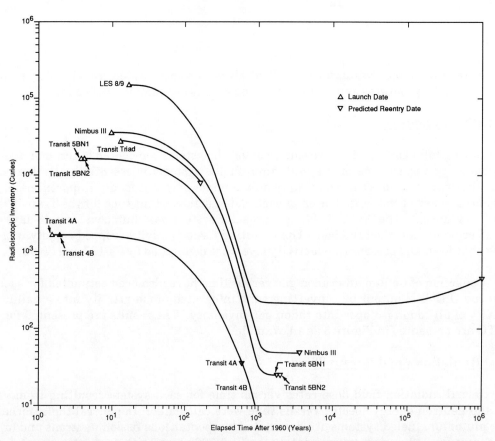

Figure 1. Total Radionuclide Inventory Versus Time for U.S. Space Nuclear Systems in Orbit.

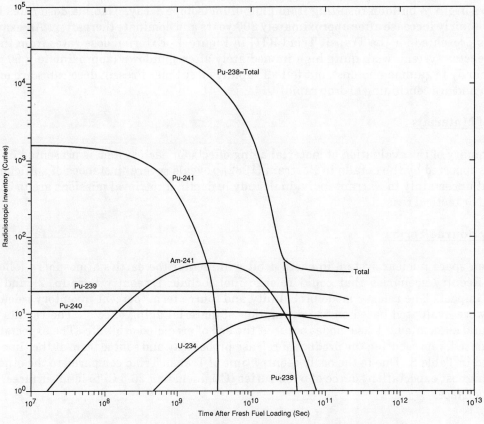

Figure 2. Transit Triad RTG Radionuclide Inventory Versus Time.

933

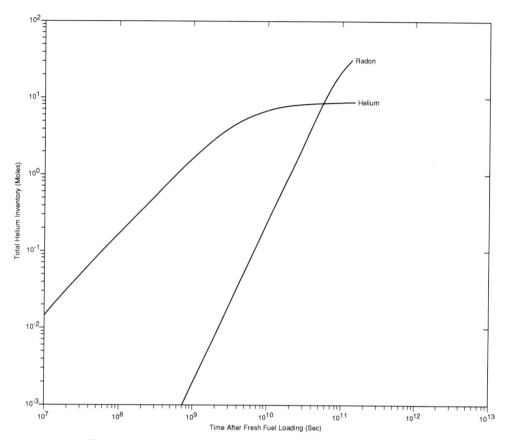

Figure 3. Transit Triad RTG Helium and Radon Gas Production Versus Time .

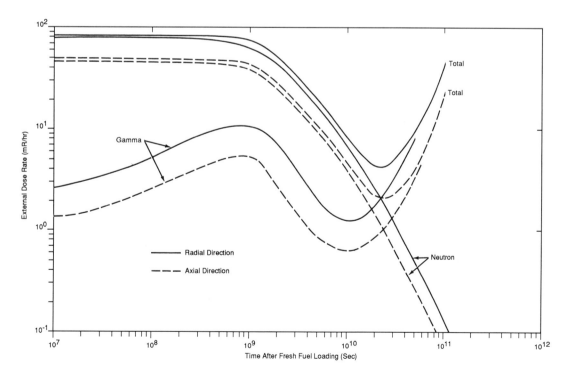

Figure 4. Transit Triad RTG External Dose Rates Versus Time at 1m Distance .

934

TABLE 2.   Summary of Material Aging Effects of U.S. Space Nuclear Systems in Orbit.

Orbiting System	Fuel: Containment Interaction	Phase Changes	Pressure Effects	Interaction of Containment Materials	Radiation Damage - Internal	Space Irradiation Effects	Evaporation - Sublimation	Meteoroid Damage
Transit 4A/4B	FF LP SMD	N	PCR	N	N	N	N	MD
Transit TRIAD-01X	FF,S LP SMD	SD	VPP PCR	SD	MD	N	VPP	MD
Transit 5BN1/5BN2	FF LP SMD	N	SD	MD	N	N	N	MD
Snapshot-10A	N	N	N	N	N	N	N	MD
Nimbus III	FF	N	VPP	N	N	N	N	MD
LES 8/9	FF LD	N	VPP	N	N	N	VPP	SD

Legend:

FF	Fuel Fragmentation	S	Swelling
LP	Liner Penetration	LD	Liner Degradation
SMD	Strength Member Degradation	MD	Minor Degradation
PCR	Probable Capsule Rupture	SD	Significant Damage
VPP	Vent Plugging Probable	N	Negligible

TABLE 3.   Expectation of Reentry Source Terms for U.S. Space Nuclear Systems in Orbit.

	Expected Radioactive Source Terms at Reentry (Curies)					
	High Altitude Burnup					
System	Vapor Release	Particulate Release Over Land	Particulate Release Over Ocean	Land Impact	Ocean Impact	Total
Transit 4A (SNAP-3)	3.65E+01	-	-	-	-	3.65E+01
Transit 4B (SNAP-3)	4.20E−01	-	-	-	-	4.20E−01
Transit Triad (RTG)	-	-	-	2.65E+01	5.15E+03	5.18E+03
Transit 5BN1 (SNAP-9A)	2.50E+01	-	-	-	-	2.50E+01
Transit 5BN2 (SNAP-9A)	2.48E+01	-	-	-	-	2.48E+01
Snapshot (SNAP-10A)	2.65E−02	-	-	-	-	2.65E−02
Nimbus III (SNAP-19)	-	-	-	1.63E−01	3.17E+01	3.19E+01
LES 8/9 (MHW-RTG)	-	-	-	1.20	3.24E+02	3.25E+02

CONCLUSIONS

The results of this study indicate that aging effects of containment materials of present U.S. space nuclear systems will result in degradation to the extent that the releases during and following reentry will be likely. The Transit Triad RTG will be the first of these systems to reenter in approximately the year 2122 (150 years after launch in 1972) and will have the highest expectation of release of 5180 Ci.

Acknowledgement

This work was sponsored by the U.S. Department of Energy under Contract No. DE-AC01-80ET32079.

References

Bartram, B. W., R. Huang, S. R. Tammara, and N. R. Thielke (1982) *Aging Effects of U.S. Space Nuclear Systems in Orbit*, NUS-3812 NUS Corporation, Gaithersburg, MD.

Pyatt, D. W., *The Feasibility of Retrieving Nuclear Heat Sources from Orbit with the Space Shuttle*, NUS-3495 NUS Corporation, Gaithersburg, MD.

ORNL (1975) *ORIGEN Isotope Generation and Depletion Code, Matrix Exponential Method*, CCC-217 (1975), Oak Ridge National Laboratory, Oak Ridge, TN.

Malenfant, R. E.(1967) *QAD: A Series of Point Kernel General Purpose Shielding Programs*, LA-3573, Los Alamos National Laboratory, Oak Ridge, TN.

ORNL (1970) *QAD, Point Kernel General Purpose Shielding Codes: QAD IV, QAD P-5, QAD-B, QAD-V, QAD-INT, QAD-HD, QAD-5K, QAD-BR, and QAD-P5A*, ORNL-4181 Oak Ridge National Laboratory, Oak Ridge, TN.

Huang, R. (1973) *QAD-SQ: NUS Modified QAD-P5*, NUS-TM-EC-16 NUS Corporation, Gaithersburg, MD.

NUCLEAR THERMAL PROPULSION
ENGINE SYSTEM DESIGN ANALYSIS CODE DEVELOPMENT

Dennis G. Pelaccio and Christine M. Scheil
Science Applications International Corporation
21151 Western Avenue
Torrance, CA 90501
(213) 781-8690

Lyman J. Petrosky and Joseph F. Ivanenok III
Westinghouse Electric Corporation
Advanced Energy Systems
Waltz Mill Site
Madison, PA 15663
(412) 722-5110

Abstract

A Nuclear Thermal Propulsion (NTP) Engine System Design Analysis Code has recently been developed to characterize key NTP engine system design features. Such a versatile, standalone NTP system performance and engine design code is required to support ongoing and future engine system and vehicle design efforts associated with proposed Space Exploration Initiative (SEI) missions of interest. Key areas of interest in the engine system modeling effort were the reactor, shielding, and inclusion of an engine multi-redundant propellant pump feed system design option. A solid-core nuclear thermal reactor and internal shielding code model was developed to estimate the reactor's thermal-hydraulic and physical parameters based on a prescribed thermal output which was integrated into a state-of-the-art engine system design model. The reactor code module has the capability to model graphite, composite, or carbide fuels. Key output from the model consists of reactor parameters such as thermal power, pressure drop, thermal profile, and heat generation in cooled structures (reflector, shield, and core supports), as well as the engine system parameters such as weight, dimensions, pressures, temperatures, mass flows, and performance. The model's overall analysis methodology and its key assumptions and capabilities are summarized in this paper.

INTRODUCTION

An accurate, standalone, preliminary NTP engine system design analysis tool is required to support current and future SEI propulsion and vehicle design studies. Currently available NTP engine design models are those developed during the NERVA program in the 1960s and early 1970s and are highly unique to that design (Plebach, McDougall, Spencer, and Weaver 1965) or are modifications of current liquid propulsion system design models. To date, NTP engine-based liquid design models lack integrated design of key NTP engine design features, such as in the areas of reactor, shielding, multi-propellant capability, and multi-redundant pump feed fuel systems. Additionally, since the SEI effort is in the initial development stage, a robust, verified NTP analysis design tool could be of great use to the community.

A recent effort was undertaken that developed an accurate, versatile NTP engine system design analysis program (tool) to support ongoing and future engine system and stage design study efforts. In this effort, Science Applications International Corporation's (SAIC) NTP version of the Expanded Liquid Engine Simulation (ELES) was modified extensively to include Westinghouse Electric Corporation's near-term solid-core reactor design model. The ELES program has extensive capability to conduct preliminary system design analysis of liquid rocket systems and vehicles. The program is modular in nature and is versatile in terms of modeling state-of-the-art component and system options (Taylor 1984). The Westinghouse reactor design model, which was integrated in the NTP ELES model, is based on the near-term solid-core ENABLER NTR reactor design concept (Livingston and Pierce 1991).

This program is now capable of accurately modeling (characterizing) a complete near-term solid-core NTP engine system in great detail, for a number of design options, in an efficient manner. The following discussion summarizes the overall analysis methodology, key assumptions, and capabilities associated with the NTP ELES program.

ENGINE SYSTEM MODEL

The ELES model was developed by Aerojet in the early 1980s to conduct preliminary system design analysis of liquid rocket systems and vehicles. The code uses both mechanistic and empirical models of subsystems and components. ELES has been verified by modeling the Delta 2nd stage, the Transtar (Titan 3rd stage), the Centaur/RL-10, and the Space Shuttle Main Engine, and has since been used to model a wide variety of different engines and vehicle systems.

Over the past year, ELES has been modified to analyze nuclear thermal rocket engine designs. Westinghouse's ENABLER solid core reactor design code has been integrated with it to perform a detailed design analysis of both the reactor and key engine subsystems. This new code, called NTP ELES, designs the reactor, turbomachinery, tankage, nozzle, lines, and valves in terms of both weight and performance/operating characteristics. A schematic of a baseline NTP expander cycle engine system which NTP ELES models is shown in Figure 1.

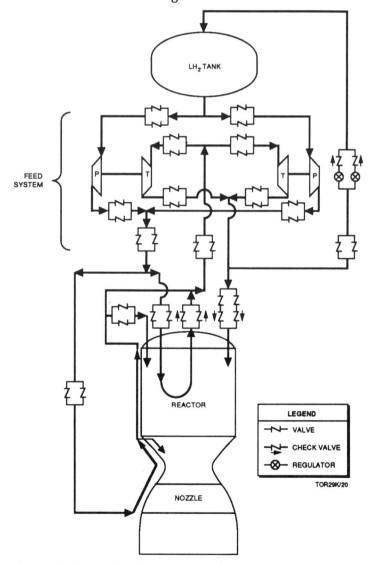

FIGURE 1.   Baseline NTP Expander Cycle Engine System Schematic Modeled by NTP ELES.

The NTP ELES user has a wide variety of design options from which to choose. The most pertinent of these are summarized in Table 1. The code can analyze both expander and gas generator (GG) cycles using hydrogen as propellant (LO_2 is used as needed for the GG cycle). NTP ELES can evaluate both steady-state and off-design engine performance by use of a user-defined turbomachinery option. It also handles a multiple-leg propellant feed system. The user selects the nozzle configuration and cooling methods, along with the materials of construction of all key components. A wide range of tankage configurations is available.

TABLE 1. Key NTP ELES Features and Options.

SYSTEM	FEATURES/OPTIONS
Cycle Type	Expander or gas generator
Nozzle	Area ratio, cooling method, number of nozzle sections and their construction types, materials, shape, thickness, gimballing, translating
Reactor	Fuel type, support configuration, operating temperature and pressure
Turbomachinery	Boost pumps, number of identical feed systems, ELES-designed pumps and turbines or user-defined, materials
Lines, Valves	Materials, pressure drops
Tankage	Tandem or nonconventional, materials, pressurization method, acquisition device

A typical NTP ELES run takes from 1 to 5 minutes on a VAX mainframe computer system depending on the complexity of the engine design. The important user inputs include the thrust level, reactor pressure and temperature, nozzle exit area ratio, and propellant flow paths. The extensive output consists of weights and sizes of components, performance of the reactor, turbomachinery, and the overall engine as well as temperatures, pressures, and mass flowrates at key engine cycle path locations.

The NTP ELES program logic is shown in Figure 2. Several routines are called more than once to improve accuracy. The first pass through the reactor routine uses an estimated nozzle heat load; the actual heat load is calculated later and compared with this initial estimate, and the reactor routine is rerun with this new value if they are not within 10% of each other. This ensures that the reactor is sized within an accuracy of 1%.

REACTOR SYSTEM MODEL

The ENABLER class of NTP engine systems (Livingston and Pierce 1991) is based on the nuclear rocket technology developed in the Rover/NERVA programs (WANL 1972). The ENABLER reactor designs incorporate NERVA-type fuel elements that are 19 mm (0.75 inch) hexagonal extrusions of graphite-based fuel with a 19-coolant channel array within the element. The fuel elements are fabricated from the (U,Zr) C-Graphite composite material developed late in the Rover/NERVA program, which exhibits improved corrosion resistance and allows higher operating temperatures (Lyon 1973, Taub 1975). Zirconium-hydride moderator is placed in the core support elements (demonstrated in the Pewee reactor) to increase the neutronic reactivity and thereby decrease the required uranium fuel loading.

The Rover/NERVA database provides numerous reference designs for reactors and engines in the size range of 65 kN to greater than 1000 kN (15 Klbf to >250 Klbf) thrust range. Detailed data is available on the breakdown of actual reactor system component masses. The R-1 engine design shown in Figure 3 was selected as the baseline for the reactor model. The nominal core dimensions are 96 cm (38 inch) diameter by 132 cm (52 inch) long. Numerous components surround the core to satisfy structural and neutronic requirements. The major components are the core barrel, reflector, pressure vessel, core support plate, and top shields.

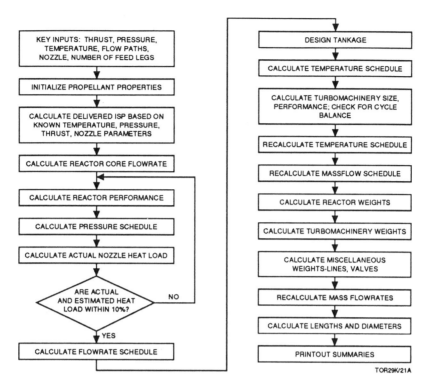

FIGURE 2.   NTP ELES Analysis Logic.

The reactor code is divided into two modules. The first determines the thermal hydraulics of the core reactor peripheral components. The required core power level is determined from the specified engine flow and chamber temperature. The thermal and pressure profile is calculated for the peak channel of the peak element in the core. This determines the core pressure drop. The core power level and the average heat generation of a fuel element determine the total number of fuel elements and support elements which yields the size of the core. At this point the heat generation rates for the core peripheral regions are also calculated. After completion of these calculations, control returns to the NTP ELES engine code for determination of the cycle balances.

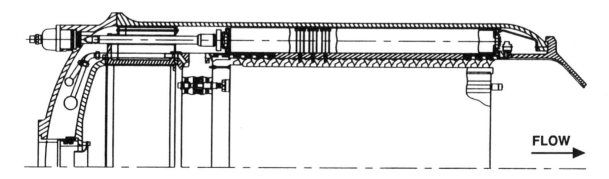

FIGURE 3.   Layout Drawing of the R-1 Reactor.

The second reactor module determines the reactor masses. The mass model is divided into 53 regions in an R-Z model as shown in Figure 4. Each region contains one, or at most a few, components. The masses of all the components and their constituent parts within a region have been tallied and converted into a pseudodensity for each region. The dimensions of the regions are based on the core size

940

determined above, with appropriate dimensional dependency algorithms. The pseudodensity is applied to each region to yield the mass schedule of the reactor for everything out to and including the pressure vessel. Thrust structure, turbopumps, and nozzle masses are not calculated in this module; the ELES code determines the balance of engine masses.

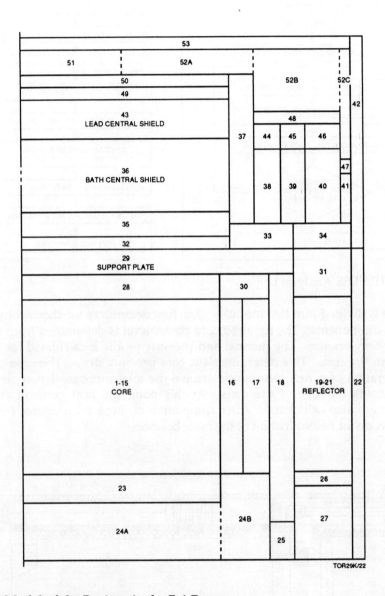

FIGURE 4.   R-Z Model of the Regions in the R-1 Reactor.

The primary input to the reactor modules is the engine flow and chamber conditions. The reactor code modules allows one of three fuel materials to be selected: graphite (UC_2 beads in graphite), composite ((U,Zr)C-Graphite), or carbide ((U,Zr)C). Each fuel type exhibits different properties with regard to mass density, power density, and temperature limit. The fuel-to-support ratio within the core may be set to one of three patterns: 2:1, 3:1, or 6:1. The user has control over the power distribution in the core and peripheral components.

The code assumes that the same basic design will be used at every size level within the specified code domain. This provides the basis for calculating the size of the core periphery. It also assumes that the user has specified an attainable combination of input criteria. For example, the code does not verify core criticality and control span. This cannot be accomplished until core neutronics is integrated into the code. Similarly, power distribution in the peripheral regions is based on external data sources such as test measurements.

CONCLUSIONS

A near-term solid-core NTP engine system preliminary design analysis tool is now available to the SEI community to support in the assessment of promising NTP engine and vehicle systems. This engine system preliminary design model characterizes NTP engine system performance, weight and size, and key operating parameters in detail for the overall system and its associated subsystems. Development of an analysis capability is considered to be one of many key first steps required to support NTP development. Because of the modular nature of the NTP ELES program, it has great potential for further upgrades in its design/technology options and analysis capabilities. It is envisioned that NTP ELES could be integrated as a key element into an advanced NTP engine system design workstation.

Acknowledgments

The work discussed in this paper is a study performed for the Nuclear Project Office of NASA Lewis Research Center under contract NAS3-25809. The authors would like to acknowledge the support of Dr. Stanley Borowski of NASA Lewis Research Center, Cleveland, OH, and Mr. Michael Stancati of SAIC's Schaumburg, IL, office. Additionally, the authors would like to acknowledge Mr. Jim Mangus, Ms. Julie Livingston and Mr. Bill Pierce (retired) of the Advanced Energy Systems Division of Westinghouse Electric Corporation in Madison, PA for their support.

References

Plebuch, R. K., J. R. McDougall, R. B. Spencer, and K. R. Wener (1965) "Volume IV: Detailed Technical Report, Nuclear Rocket Engine Analysis," TRW Report No. 8423-6008-RL000, TRW Space Technology Laboratories, Redondo Beach, CA, March 1965.

Taylor, C. E (1984) "Expanded Liquid Engine Simulation Computer Program – Technical Information Manual," Aerojet Report No. ELES-1984, Aerojet TechSystems Company, Sacramento, CA, August 1984.

Livingston, J. M., B. L. Pierce (1991) "The ENABLER – Based on Proven NERVA Technology," Proceedings of the Eighth Symposium on Space Nuclear Power Systems, Vol. 2, pp. 598-602, January 1991.

WANL (1972) "Technical Summary Report of NERVA Program," TNR-230, Westinghouse Astronuclear Laboratory (currently Westinghouse AES), July 1972.

Taub, J. M. (1975) "A Review of Fuel Element Development for Nuclear Rocket Engines," Los Alamos National Laboratory, LA-5931, June 1975.

Lyon, L. L. (1973) "Performance of (U,Zr)C-Graphite (Composite) and of (U,Zr)C (Carbide) Fuel Elements in the Nuclear Furnace 1 Test Reactor," Los Alamos National Laboratory, LA-5398-MS, September 1973.

COMPUTATIONAL STUDY OF NONEQUILIBRIUM HYDROGEN FLOW IN A LOW PRESSURE NUCLEAR THERMAL ROCKET NOZZLE

Dana A. Knoll
Idaho National Engineering Laboratory
EG&G Idaho, Inc.
P.O. Box 1625
Idaho Falls, ID 83415
(208) 526-8862

Abstract

The low pressure nuclear thermal rocket (LPNTR) concept offers increased specific impulse over high pressure concepts by utilizing hydrogen dissociation and recombination. To study the nozzle flow of the partially dissociated hydrogen requires a computational tool capable of modeling chemical nonequilibrium, and possible thermal nonequilibrium. We are currently developing such a tool, and report here the results from our effort to date. These results show that chemical nonequilibrium dominates down stream of the nozzle throat and that thermal nonequilibrium could also have a significant effect on vehicle I_{sp}

INTRODUCTION

There has been a revived interest in nuclear thermal propulsion as a result of the Space Exploration Initiative. This interest has prompted increased investigation of various nuclear thermal propulsion concepts. The concept which may be able to offer the most significant increase in specific impulse (I_{sp}) is the low pressure radial flow reactor concept (Leyse et al. 1990). This concept utilizes the dissociation of molecular hydrogen at high temperatures and low pressures. Since I_{sp} is inversely proportional to the square root of molecular weight, dissociation will increase I_{sp}. Recombination of atomic hydrogen is an exothermic reaction, and thus recombination in the rocket nozzle may further increase I_{sp} if the recombination energy can be converted to fluid kinetic energy. To study this problem computationally we must recognize the important time scales. If the characteristic times required to achieve chemical and thermal equilibrium in the rocket nozzle are comparable to the residence time of a molecule in the nozzle, then this problem must be studied with thermochemical nonequilibrium computational fluid dynamics (CFD). By thermal equilibrium, we refer to the distribution of the internal energy of the hydrogen molecules. The internal energy is distributed in translational, rotational, vibrational, and electronic states, and will have a most probable, Boltzmann, distribution at a given temperature (Vincenti and Kruger 1965). For a system of molecules in chemical and thermal nonequilibrium, thermal equilibrium will be achieved faster than chemical equilibrium (Vincenti and Kruger 1965). This means that for a given reacting flow system, if chemical equilibrium is a good approximation then so is thermal equilibrium. But, if the chemistry is not in equilibrium then thermal nonequilibrium may also exist. The reason that we are concerned with these issues is that only translational energy will be converted to fluid kinetic energy by the rocket nozzle. Thus, energy used to populate other internal modes, such as vibration, will not be useful for propulsive purposes. Davis et al. (1991) have performed one-dimensional chemical equilibrium computations and then estimated the nonequilibrium results by applying kinetic loss terms to the equilibrium results. Their results show a significant difference between computed equilibrium results and estimated nonequilibrium results.

In this paper we will use a chemical nonequilibrium CFD code to perform a more accurate study of the effects of dissociation-recombination on the nozzle performance of a low pressure nuclear thermal rocket. Our code, LAVA1D, is a variable area one-dimensional version of LAVA (Ramshaw and Chang 1991). We will also illustrate the possible effects of thermal nonequilibrium. The extension of LAVA1D to full thermochemical nonequilibrium is currently underway.

NONEQUILIBRIUM HYDROGEN NOZZLE FLOW

The two dissociation-recombination reactions that occur in a pure hydrogen flow are ;

$$3[H] \leftrightarrow [H] + [H_2] \tag{1}$$

and

$$2[H] + [H_2] \leftrightarrow 2[H_2] \tag{2}$$

The initial product of three body recombination is an excited hydrogen molecule. The internal energy of this molecule will have a nonequilibrium distribution with much of the recombination energy used to populate upper vibrational and rotational states (Orel 1987 and Schwenke 1990). Through molecular collisions the internal energy distribution of the excited molecule will eventually relax to a Boltzmann distribution with a large fraction of the energy in the translational mode. A critical issue is how fast this relaxation occurs in comparison to the residence time in the nozzle.

The low pressure concept increases H mole fraction because recombination is a three body process. The rate coefficients are functions only of temperature, so for a fixed temperature and decreasing pressure and density the backward reactions will increase and the forward reactions will decrease. This will cause an increase in the equilibrium mole fraction of H as can be seen in Figure 1. The data in this figure were computed using the equilibrium constant given by Cohen and Westberg (1983). The rapid drop in temperature experienced by the hydrogen flow in the diverging section of a rocket nozzle will promote recombination. However, the associated rapid drop in pressure will promote dissociation. To study this process we must couple fluid dynamics and nonequilibrium chemistry, but this may not be sufficient to model the flow since equilibrium is assumed for thermal energy (internal energy states). The importance of the thermal equilibrium assumption for nozzle performance can be assessed by looking at two bounding cases. First, it must be recognized that only translational internal energy is converted to fluid kinetic energy in a rocket nozzle. The best case for propulsion is thermal equilibrium. In this situation all of the recombination energy is immediately equilibrated among the translational, rotational and vibrational levels of the recombined hydrogen molecule. This will give the maximum amount of translational energy. In the worst case, all of the recombination energy is used to populate the upper vibrational states of the recombined hydrogen molecule and the energy exchange rate is not fast enough, compared to the residence time, to relax the thermal energy distribution to equilibrium. In this case we assume that none of the recombination energy can be converted into fluid kinetic energy, which should represent the worst case scenario.

GOVERNING EQUATIONS AND NUMERICAL METHOD

The fluid equations in LAVA1D consist of momentum and internal energy equations for the two-component fluid mixture and a continuity equation for each of the fluid components, H and H_2. The two continuity equations are coupled through finite rate dissociation-recombination chemistry, and the state relations are those of an ideal gas with temperature-dependent specific heats. Our quasi one-dimensional model is

species continuity (i = H , H_2)

$$A \frac{\partial \rho_i}{\partial t} + \frac{\partial \rho_i A u}{\partial x} = A \dot{\rho}_i^c \tag{3}$$

momentum

$$A \frac{\partial \rho u}{\partial t} + \frac{\partial \rho A u^2}{\partial x} = -A \frac{\partial P}{\partial x} \tag{4}$$

internal energy

$$A \frac{\partial \rho e}{\partial t} + \frac{\partial \rho A u e}{\partial x} = -P \frac{\partial A u}{\partial x} + A \dot{Q}_c \tag{5}$$

with

944

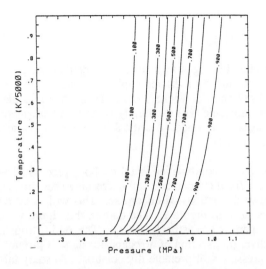

FIGURE 1. Equilibrium H Mole Fraction.

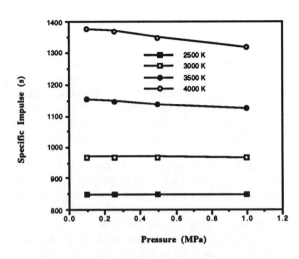

FIGURE 2. Specific Impulse for Thermal Equilibrium.

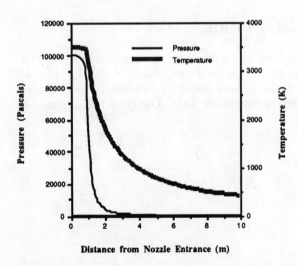

FIGURE 3. Nozzle Temperature and Pressure for Core
Exit 0.1 MPa and 3500 K.

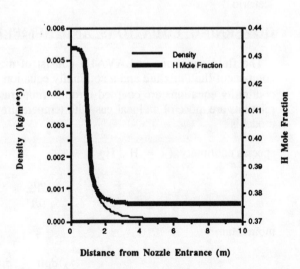

FIGURE 4. Nozzle Density and H Mole Fraction for
Core Exit 0.1 MPa and 3500 K.

945

$$\rho = \rho_H + \rho_{H_2} \tag{6}$$

A, the cross sectional area, is a function of position. The terms $\dot{\rho}_i^c$ and $\dot{Q}_c$ are the rates of change of ρ_i and ρe respectively, due to chemical reactions. The state relation for pressure is

$$P = R_g T \sum_i \frac{\rho_i}{Mw_i} \tag{7}$$

where R_g is the universal gas constant and Mw_i is the molecular weight of species i. The state relation between internal energy and temperature is

$$\rho e = \sum_i \rho_i e_i(T) \tag{8}$$

where $e_i(T)$ is the specific internal energy of species i at temperature T, tabulations of which are available. The forward and backward rate coefficients, k_{fr} and k_{br}, are of the general Arrhenius form

$$k = \alpha \, T^\beta \exp\left(\frac{-\varepsilon}{T}\right) \tag{9}$$

In LAVA1D, Equations (3) through (5) are solved using a fully explicit finite difference method. A staggered grid is used for velocities, and the convective terms are differenced with a weighted average of donor cell and interpolated donor cell differencing (Harlow and Amsden 1975). Given initial conditions LAVA1D integrates the governing equations in time out to a steady solution.

SELECTED RESULTS

Our computational study is in the early stages and only limited results are available at this time. Results will be shown for chemical nonequilibrium and thermal equilibrium. We will also show results which illustrate the possible effect of thermal nonequilibrium. The sensitivity of these results to uncertainties in the reaction rates are also investigated. Computations have been made at core exit temperatures of 2500, 3000, 3500, and 4000 K, and core exit pressures of 0.1, 0.25, 0.5, and 1 MPa (0.1 MPa $\approx$ 1 atm). We have used a generic conical nozzle with a 30 degree half angle in the converging section and a 15 degree half angle in the diverging section. The area ratio is 100, and the total length is 10 meters with the throat being 1 meter from the core exit. At the core exit, which is our inlet boundary condition, pressure, temperature, and equilibrium mole fractions are specified. The inlet velocity is allowed to develop self-consistently such that choking occurs at the nozzle throat. The reaction rates we have used for Equations (1) and (2) are from Cohen and Westberg (1983) and are given in Table 1.

TABLE 1. Constants for Hydrogen Dissociation-Recombination Reaction Rates.

Reaction	α	β	ε
k_{f1}	3.2e+3 $\frac{m^6}{mol^2 \, sec}$	0.0	0.0 K
k_{b1}	2.71e+10 $\frac{m^3}{mol \, sec}$	- 0.1	5.235e+4 K
k_{f2}	1.0e+5 $\frac{m^6}{mol^2 \, sec}$	- 0.6	0.0 K
k_{b2}	8.47e+11 $\frac{m^3}{mol \, sec}$	- 0.7	5.235e+4 K

Figure 2 shows I_{sp} as a function of temperature and pressure assuming thermal equilibrium. For this series of runs, I_{sp} is a stronger function of temperature than pressure. Figures 3 through 5 isolate the results of the run with core exit conditions of 3500 K and 0.1 MPa. There is a steep gradient in all variables in the throat region. These sharp gradients, and a millisecond scale residence time, are the reasons that nonequilibrium chemistry CFD must be used. A fluid element is accelerating through rapidly changing temperature and pressure profiles and the molecules in that fluid element do not suffer an adequate number of collisions to achieve chemical equilibrium at the local temperature and pressure. Figure 6 shows the computed H mole fraction and the equilibrium H mole fraction using the local temperature and pressure from Figure 3, for a few points around the throat. This plot is intended to demonstrate that the chemistry is clearly not in equilibrium.

Since the chemistry does not reach equilibrium, we must be concerned with the possibility that thermal equilibrium is not achieved either. That is, we must be concerned with the distribution of internal energy of the newly recombined molecules. To illustrate the possible effects of thermal nonequilibrium we have repeated the runs of Figure 2 with $\dot{Q}_c$ in Equation (5) set equal to zero. This is intended to model a situation where all of the recombination energy is in the upper vibrational levels of the recombined H_2 and does not contribute to the fluid kinetic energy. We consider this to be a worst case scenario for propulsion. The results of these computations can be seen in Figure 7. Comparing Figure 7 with Figure 2, we see that the loss of the recombination energy is detrimental to I_{sp} as expected. This problem becomes more significant as we move to higher temperatures and lower pressures, where thermal nonequilibrium may reduce I_{sp} on the order of 10 percent. It is also important to note that for a core exit of 3500 K and 0.1 MPa roughly 10 percent of the total internal energy is in the vibrational mode. It is possible that some of this energy will be frozen in the vibrational state due to thermal nonequilibrium in the throat region.

As with all chemical reaction rates, our dissociation-recombination rates have some uncertainty associated with them. To understand the effect this uncertainty may have on our results, we have made computations for three additional sets of rates for a core exit temperature of 3500 K and our previous range of core exit pressures. To obtain our three new sets of rates we have multiplied the constant α in Equation (9) by 0.5, 2.0, and 5.0 for all of the reactions. This shifts the reaction rates without changing the equilibrium concentrations. The activation temperature, ε, was not varied, as it simply corresponds to the known dissociation energy. The results of the computations can be seen in Figure 8. The sensitivity of I_{sp} to the rates increases for decreasing pressure and increasing temperature. This is because the equilibrium mole fraction of H entering the nozzle is larger and thus there is a greater degree of nonequilibrium beyond the nozzle throat. However, a factor of 5 increase in the rates increases I_{sp} by only 4 percent, so our results for this nozzle geometry are not very sensitive to the reaction rates.

CONCLUSIONS

This study has shown that the time scale for chemical equilibrium is longer then the residence time of the molecules in the nozzle for a low pressure NTR. This is due to the rapid expansion which causes a decrease in density, and thus a decrease in collision frequency. This nonequilibrium situation may be improved by optimizing the nozzle geometry, especially in the throat region.

We have also shown that the effect of thermal nonequilibrium could be significant on I_{sp}. Since only translational energy can be converted to fluid kinetic energy we want to maximize the translational energy. This is not the situation in the initial internal energy distribution of a newly recombined hydrogen molecule. We are currently working on an enhanced physical model to analyze the problem. The simplest enhancement would be the addition of a conventional vibrational energy equation (Anderson 1982) which would have terms representing convection, relaxation towards equilibrium, and a source term from recombination. A more satisfactory alternative is to model the individual vibrationally excited states of the hydrogen molecule as separate species with their own continuity equations and associated source and sink terms due to collisional transitions. This method has been used by the chemical laser community (Ramshaw and Dukowicz 1979). It will require reaction rates to model collisional transitions among the vibrationally excited states.

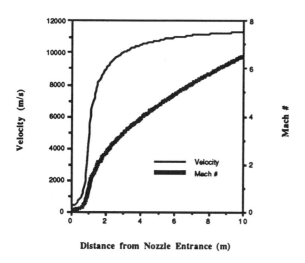

FIGURE 5. Nozzle Velocity and Mach Number for
Core Exit 0.1 MPa and 3500 K.

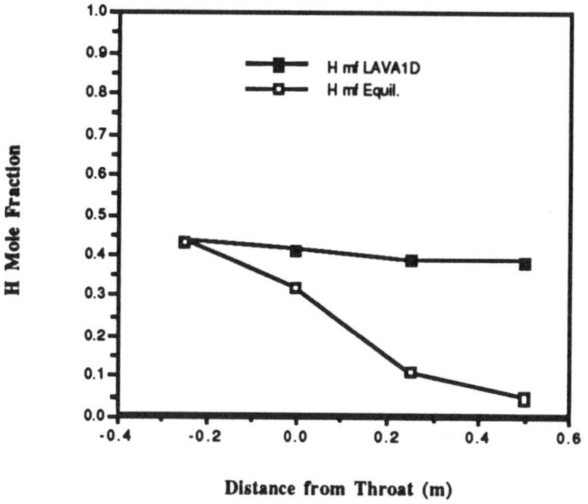

FIGURE 6. Chemical Nonequilibrium near the Throat
for Core Exit 0.1 MPa and 3500 K.

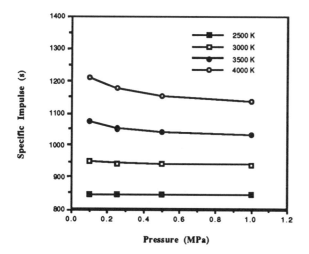

FIGURE 7. Specific Impulse with the Assumption
of Thermal Nonequilibrium.

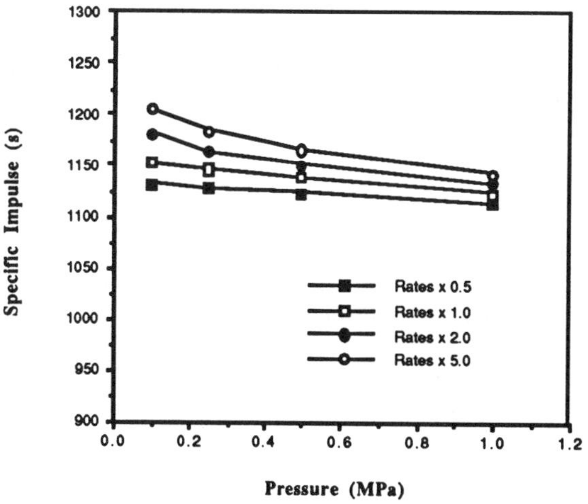

FIGURE 8. Specific Impulse for Various Sets of
Reaction Rates.

Our results did not show much sensitivity to changes in the reaction rates. However this is for a single nozzle geometry and we may well see more sensitivity to reaction rates in a nozzle geometry optimized for recombination.

Finally, our enhanced version of LAVA1D, which will include thermal nonequilibrium, will still only be quasi one-dimensional. It is necessary to move to a two-dimensional description in order to study the effects of the boundary layer and wall recombination in a self-consistent manner. We will then be in a position to computationally optimize the nozzle geometry for a low pressure NTR.

Acknowledgments

The author wishes to express his appreciation to Dr. J. D. Ramshaw and Dr. C. H. Chang of INEL for invaluable technical discussions and their timely assistance in producing LAVA1D. The author also wishes to express his appreciation to Professor Norman Roderick of the University of New Mexico for useful discussions. This work was performed under the auspices of the U.S. Department of Energy, contract number DE-AC07-76-IDO1570, and was partially supported by the INEL Long Term Research Initiative in Computational Mechanics.

References

Anderson, John D. Jr. (1982) *MODERN COMPRESSIBLE FLOW with Historical Perspectives*, McGraw-Hill Inc., New York.

Cohen, N. and K.R. Westberg (1983) "Chemical Kinetic Data Sheets For High-Temperature Reactions", *J. Phys. Chem. Ref. Data*, 12(3): 531.

Davis, J. Wesley, Joesph C. Mills, James F. Glass, and Wen-Hsiung Tu (1991) "Advanced NTR Options", *in Proc. Eighth Symposium on Space Nuclear Power Systems*, CONF-910116, M.S. El-Genk and M.D. Hoover, eds., American Institute of Physics, New York, 1991.

Harlow , F. H., and A. A. Amsden (1975) "Numerical Calculation of Multiphase Fluif Flow", *J. Comp. Phys.*, 17: 19.

Leyse, C. F., J. H. Ramsthaler, and B. G. Schnitzler (1990) "Potential for Nuclear Propulsion", Informal Report Prepared for USDOE, Contract No. DE-AC07- 76ID01570, Idaho National Engineering Laboratory.

Orel A.E. (1987) "Nascent vibrational /rotational distribution produced by hydrogen atom recombination", *J. Chem. Phys.*, 87(1): 314 .

Ramshaw, J. D., and C. H. Chang "Computational Fluid Dynamics Modeling of Multicomponent Thermal Plasmas", *Plasma Chem. Plasma Process.*, Submitted.

Ramshaw, J. D. and J. K. Dukowicz (1979) "APACHE: A Generalized-Mesh Eulerian Computer Code for Multicomponent Chemically Reactive Fluid Flow", Los Alamos Scientific Laboratory Report LA-7427 .

Schwenke, David W. (1990) "A theoretical prediction of hydrogen molecule dissociation-recombination rates including an accurate treatment of internal state nonequilibrium effects", *J Chem. Phys.*, 92(12): 7267.

Vincenti, Walter G., and Charles H. Kruger Jr. (1965) *Introduction to Physical Gas Dynamics*, John Wiley and Sons Inc, New York.

RESTARTABLE NUCLEAR THERMAL PROPULSION CONSIDERATIONS

Charlton Dunn
Space Systems Engineering
Rocketdyne Division
Rockwell International Corporation

6633 Canoga Avenue
Canoga Park, CA 91303
(818) 718-3424

Abstract

Nuclear thermal propulsion (NTP) offers significant improvements in rocket engine specific impulse and, therefore, reductions in vehicle gross weight and/or mission duration compared to vehicles employing chemical propulsion. Restarting NTP systems, however, for use in multiple maneuvers, requires that decay power be removed from the NTP to preserve it for reuse. Open and closed–cycle methods of decay power removal are compared for a Rover–type reactor. Use of closed–cycle decay power removal commencing 3 h after shutdown for approximately 4 d reduces consumed propellant for decay power removal to approximately 50% of that for open–cycle decay power removal. Closed–cycle decay power removal also eliminates the approximate 4–d thrust tail–off from open–cycle cooling which may be inconvenient, or even useless, with regard to mission performance.

INTRODUCTION

Restartable NTP systems require a method for removing exponentially decaying power due to fission products from the shut–down reactor. This must be done while maintaining components within temperature limits. To achieve this, the reactor coolant exit temperature must be reduced as the flow rate is reduced. This causes performance penalties since the specific impulse is reduced. If an engine is to be restarted several times, it may benefit from a closed–cycle system with a radiator to dump the fission decay power to space. The radiator may be incorporated with the radiation–cooled nozzle skirt to save weight. Several days after reactor shutdown, active cooling can be terminated and the decay power is then radiated to space from the pressure vessel and nozzle. Since piloted Mars missions being considered will require from 3 to 12 restarts when engines are reused, the cooldown systems required for an NTP system are a vital part of the engine design. In cases where the open cycle cooling thrust cannot be used propulsively, the penalty is more severe, and the thrust must also be negated.

RESULTS AND DISCUSSION

The decay power as a function of time after shutdown following 30 min of full power operation at 5,500 Mw is shown as a function of full power in Figure 1. For isothermal (2700 K) aftercooling, the flow rate fraction is the same as the power fraction. After approximately 6 min, the flow rate can be provided by tank pressure. After approximately 4 d, when the power has dropped to 30 kw, flow can be stopped with the power being radiated to space.

However, isothermal aftercooling at 2700 K is not practical. At full power, the liquid hydrogen from the pump maintains the thrust–chamber, pressure vessel, and reflector regions of the engine in the range of a few hundred degrees Kelvin. It is not practical to consider designing these regions of the engine to allow temperatures as high as 2700 K. A value of 800 K was selected as a reasonably difficult limit, considering that control drums, bearings, and actuators will reach this temperature. The cool–down

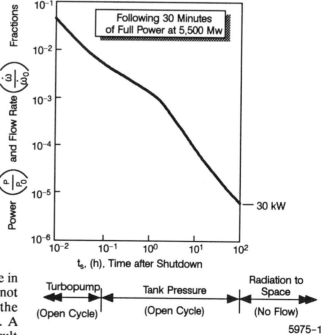

FIGURE 1. Power and Isothermal Flow Rate Fractions.

flow rate – chamber temperature schedule is shown in Figure 2. For the 800 K, temperature–limited case, the flow rate versus temperature difference was taken as linear from the full–power temperature of 2700 K to the initial space radiation temperature of 800 K when active cooling is terminated.

The temperature–limited flow rate as a fraction of mainstage flow rate is presented in Figure 3. The flow rate fraction in this case is greater than the power fraction due to imposition of the cooldown flow rate – chamber temperature ramp of Figure 2. As noted on Figure 3, after approximately 3 h, the power level has dropped to 2 Mw, the cooldown flow rate is being supplied by tank pressure, and this condition provides a reasonable opportunity to employ a closed–cycle system using a radiator to eliminate decay power and provide a substantial savings of propellant.

The aftercooling propellant requirement as a fraction of mainstage propellant usage during the 30 min of full power operation is shown in Figure 4 for isothermal and temperature–limited cases. As shown, the temperature–limited case consumes approximately three times the propellant used in the hypothetical isothermal case. Also, after 3 h of cooldown, at the potential closed–cycle initiation, approximately half of the aftercooling propellant for the temperature–limited case has been expelled.

The aftercooling average–specific–impulse, chamber–temperature, and impulse as fractions of the mainstage values are shown in Figure 5 for the temperature–limited case. As indicated, the thermal transient is primarily during the first 36 s after shutdown. Conditions during this time are somewhat affected by the shutdown excess and core thermal capacitance; however, these don't have a significant effect on the overall transient. As shown, for the 800 K final chamber temperature, the average cooldown specific impulse is finally approximately half the mainstage value. For the initiation of closed–cycle cooldown operation, 3 h after shutdown, the thermal transient is essentially complete, and any continued open–cycle cooling is essentially isothermal at 800 K. Open–cycle cooldown impulse is, finally, approximately 6% of the mainstage impulse.

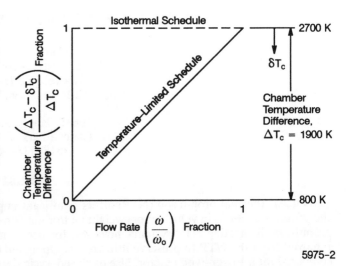

FIGURE 2.   Flow Rate–Chamber Temperature Schedule.

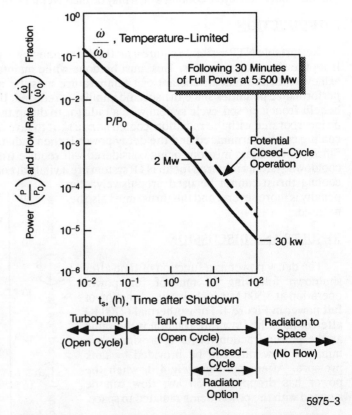

FIGURE 3.   Power and Temperature–Limited Flow Rate Fractions.

The overall parameters for the 1.1 MN (750 kLb) thrust nuclear thermal rocket (NTR) used in this evaluation are shown in Figure 6. The radiation–cooled nozzle skirt, from an area ratio of 150 to 500:1, has potential for incorporation of the radiator for a closed–cycle aftercooling system. The skirt is approximately 13 m long with an area of approximately 240 m^2. The fuel element and support configuration for the core of this Rover/NERVA derivative NTR is shown in Figure 7. The tie–tubes form the basis for extraction of core decay power. These two concentric tubes provide a circuit within a cluster of fuel elements and the two sets of tubes are separately manifolded above the core inlet to provide a source of turbine power during mainstage operation. The closed–cycle

951

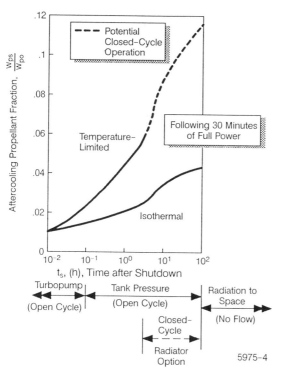

FIGURE 4.   Isothermal and Temperature-Limited Aftercooling Propellant Fraction.

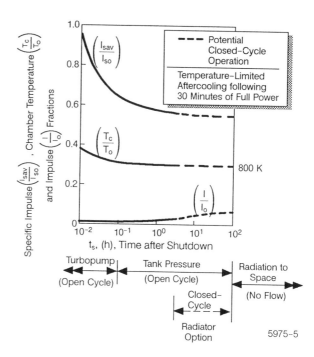

FIGURE 5.   Aftercooling Average Specific Impulse, Chamber Temperature, and Impulse Fractions.

- Radiation-cooled skirt
  - 150 to 500:1
  - Stowed for boost
- Expander cycle
- Turbopump-out capability

Flow rate	123 kg/s
P_c	7 MPa
T_c	2700 K
I_{sp}	8983 $\frac{N-s}{kg}$
Reactor power	5,500 Mw
T/P power (total of 3)	22 Mw
Weight (w/o shield)	17,000 kg
Engine length	25 m
Nozzle exit diameter	7.6 m

5975-6

FIGURE 6.   1.1 MN Thrust NTR

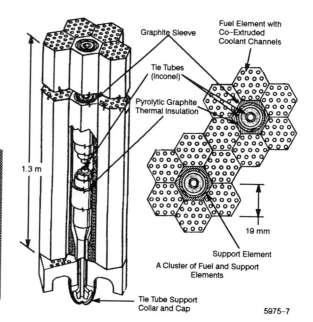

FIGURE 7.   Schematic of Fuel Elements, Support Elements, and Hot-End Support Hardware.

aftercooling system accesses the inlet and outlet manifolds of the tie–tube cooling system to remove decay power. A schematic for the NTR is shown in Figure 8. Three flow circuits are indicated; namely, (1) the mainstage circuit where the turbine is powered by the tie–tube cooling circuit, (2) the open–cycle aftercooling circuit where turbopump or tank pressures are used to provide coolant flow, and (3) the closed–cycle aftercooling circuit where the tie–tube system is used to extract core decay power and transfer it to a Rankine loop. The Rankine loop provides pumping power, and the radiation–cooled nozzle skirt of the NTR can be considered for incorporation of the radiator function of the Rankine loop. This should provide for packaging and support of the radiator while increasing the rigidity of the radiation– cooled skirt and resulting in a net system weight savings.

The results are summarized in Table 1. Open–cycle, temperature–limited aftercooling requires 26,000 kg of hydrogen propellant. This is three times the hypothetical isothermal case, and represents 12% of mainstage propellant or 150% of the engine weight. Assuming that the 4 d of thrust tail–off can be productively used in the mission profile, the open–cycle propellant weight penalty is 11,700 kg for the tem– perature–limited case compared to the hypo– thetical isothermal case. This penalty can be cut by 6,400 kg to 5,300 kg by going to closed–cycle aftercooling 3 h after shutdown. The 6,400 kg per restart represents a significant potential weight savings by employing a closed–cycle aftercooling system when multiple restarts are required.

CONCLUSIONS

- Isothermal (2700 K) aftercooling is unrealistic

- Open–cycle, temperature–limited (800 K) aftercooling requires 12% of main stage propellant

- For an engine requiring multiple restarts, a closed–cycle aftercooling system should be incorporated, which reduces open– cycle aftercooling requirements from 12% to 6% of mainstage propellant

- The high–expansion–ratio, radiation– cooled, nozzle extension (area of 240 m^2) of the NTP system should be considered for multiple use as the radiator for the closed–cycle aftercooling system

Acknowledgments

This evaluation was performed at the Rocketdyne Division of Rockwell International Corporation under internal funding support of Space Systems Engineering. Also, this is to recognize Margaret Clarke, Estella Langford, Donald Savitt, and Noel Thorp for their efforts in preparation of this manuscript.

- **Nomenclature** -

I_{sav}:	Aftercooling average specific impulse	δT_c:	The difference between mainstage chamber temperature (2700 K) and chamber temperature during aftercooling for the temperature–limited case
I_{so}:	Mainstage specific impulse, $8,983 \frac{N-s}{kg}$		
I:	Aftercooling engine impulse		
I_o:	Mainstage engine impulse, 1,980 MN – s	t_s:	Time after reactor shutdown
P:	Decay power after shutdown	W_e:	Engine weight, 17,000 kg
P_o:	Mainstage power, 5,500 Mw	W_{po}:	Weight of propellant expelled during mainstage operation, 221, 400 kg
T_c:	Aftercooling chamber temperature	W_{ps}:	Weight of propellant expelled during aftercooling
T_o:	Mainstage chamber temperature, 2700 K	$\dot{w}$:	Flow rate during aftercooling
ΔT_c:	The difference, 1900 K, between mainstage chamber temperature (2700 K) and the final chamber temperature (800 K) at termination of aftercooling flow for the temperature–limited case.	$\dot{w}_o$:	Mainstage flow rate, 123 kg/s

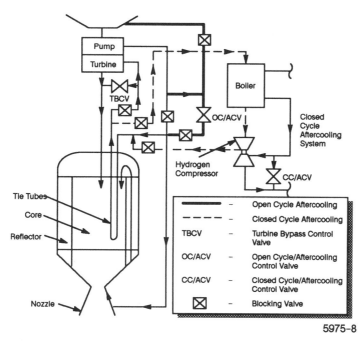

5975-8

FIGURE 8.   Tie Tube Turbine Power and Decay Power
            Removal Source.

TABLE 1.   Restartable 1.1 MN–Thrust NTP System

Aftercooling Requirements Following 30 Min of Full Power.	Open Cycle		Open/Closed Cycle
	Isothermal	Temperature–Limited	Temperature–Limited
Chamber Temperature	2700 K	2700 ➔ 800 K	2700 ➔ 800 K
Open–Cycle Aftercooling Time	100 h	100 h	3 h
Aftercool Propellant			
Weight	9,600 kg	26,000 kg	12,000 kg
% Mainstage Propellant	4%	12%	6%
% Engine Weight	56%	150%	70%
Avg. Aftercool Specific Impulse/ Mainstage Specific Impulse	100%	55%	56%
After cool Impulse			
% Mainstage Impulse	4%	6%	3%
Aftercool Propellant Penalty[a]			
Weight	0	11,700 kg	5,300 kg
% Mainstage Propellant	0	5%	3%
% Engine Weight	0	68%	32%
Closed–Cycle Aftercooling System	N/A	N/A	
Peak Power Level			2 Mw
Operating Time per cooldown			97 h
Nozzle Skirt Area[b]			240 m²

[a]Assuming that the long duration aftercooling thrust can be used propulsively.
[b]Potential use of the high–expansion–ratio, radiation–cooled, nozzle extension
   for the radiator of the closed–cycle aftercooling system.

D635-0211

954

PASSIVE DECAY HEAT REMOVAL IN THE PeBR CONCEPT
FOR NUCLEAR THERMAL PROPULSION

Nicholas J. Morley and Mohamed S. El-Genk
Institute for Space Nuclear Power Studies
Chemical and Nuclear Engineering Department
The University of New Mexico
Albuquerque, NM 87131
(505) 277-3321/5442

Abstract

Passive decay heat removal enhances the safety of a nuclear propelled spacecraft and could increase its I_{sp} by up to 4 percent. The large height-to-diameter ratio (>1.5) of the PeBR provides an effective means for passive removal of decay heat from the reactor core after firing. A multi-dimensional transient heat conduction/radiation model of the PeBR core and surrounding structures is developed to investigate the potential for passive decay heat removal. The effective conductance of the core region includes both conduction and radiation contributions and the thermophysical properties of the fuel and core structure materials are taken to be temperature dependent. Results indicate that passive decay heat removal, following a short period of active cooling (600 - 1000 seconds), would maintain the PeBR core safely coolable with the peak fuel temperature well below 3400 K.

INTRODUCTION

In the 1960s and 1970s the NERVA and Rover programs set out to develop rockets engines powered by solid core nuclear reactors. Hydrogen, having the lowest molecular weight, has been used as the propellant of choice for these engines. Solid core nuclear rockets promise a 2-3 times increase in performance level over chemical rocket, with a projected increase to 10 times with the development of gaseous core reactors. This increased performance would literally cut months off a round trip to Mars and save thousands of tonnes of materials that would otherwise be needed in Low Earth Orbit (LEO). While these earlier programs had been technical successes, changing objectives and budgetary constrains brought them to an end in 1973.

Recently, there has been a renewed interest in nuclear propulsion to support the Space Exploration Initiative (SEI) for manned missions to Mars. The Exploration Technology Program (ETP) of SEI is aimed at developing a broad range of technologies to enable future space exploration missions. Nuclear propulsion is one of eight major technology areas in the ETP. The technology development in the nuclear propulsion area will address multiple approaches to applying space nuclear power systems to manned missions to Mars. Three activities under investigation by National Aeronautic and Space Administration (NASA) are: (a) solid core and gaseous nuclear thermal propulsion (NTP) reactor concepts capable of long life and multiple starts, (b) nuclear electric propulsion (NEP) systems, including nuclear reactor, advanced low mass radiators, and power management systems technologies, and (c) high powered long life electric thrusters for piloted missions to Mars (Bennett 1990).

The Pellet Bed Reactor (PeBR), originally conceived at the University of New Mexico's Institute for Space Nuclear Power Studies (ISNPS), has unique design and safety features which make it suitable for both thermal and electric propulsion missions (El-Genk et al. 1990a, 1990b, 1991a and 1991b). The PeBR concept draws on established pebble bed and high temperature gas cooled terrestrial reactors and NERVA technology bases. The PeBR is a fast-flux, annular core reactor that is cooled by hydrogen gas and fueled with (U-Nb)C microspheres (~ 500 μm in diameter) dispersed in ZrC spherical fuel pellets (~1 cm in diameter).

The most recent design of the fuel microspheres for the PeBR has been a TRISO type fuel, where the fuel kernel is coated by two layers of low density and high density graphite, respectively, and an outer layer of ZrC carbide (El-Genk et al. 1991a and 1991b). Although this fuel particle design ensures full retention of fission products, it limits the maximum operating temperature of the fuel because of the high carbon content at the interface between the fuel kernel and the carbon coating. The stability of the of the high melting point of the carbide fuel kernel (> 3400 K) is sensitive to carbon concentration, where a small deviation from the optimum stoichiometry would significantly reduce the melting temperature of the fuel below 3000 K (Matthews et al. 1991). Such low fuel melting temperature would lower the maximum propellant temperature to about 2700 K and the Isp to less than 820 Seconds. To alleviate this problem a new fuel microsphere design is developed in which the (U-Na)C fuel kernel is coated by a thin layer of NbC (~ 20 μm) followed by a layer of ZrC carbide (~ 15 μm). Additionally, the graphite matrix in the fuel pellets is replaced with ZrC for better compatibility with hydrogen coolant and higher thermal conductivity for passive decay heat removal from the PeBR core. Figures 1 and 2 present cross-sectional views of the new fuel microsphere and fuel pellet for the PeBR concept for nuclear thermal propulsion. For nuclear electric propulsion applications, since the reactor core will be cooled with helium gas instead of hydrogen, the original fuel pellet and microsphere designs will be retained (El-Genk 1991 a and 1991b); (U-Zr)C fuel kernel in a TRISO type microspheres (~ 500 μm in dia.) which are dispersed in a graphite matrix formed into spherical pellets ~ 1.0 cm in diameter.

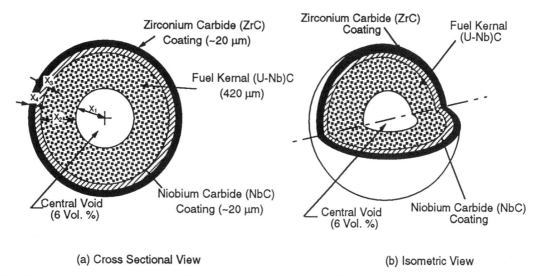

(a) Cross Sectional View   (b) Isometric View

FIGURE 1. Microsphere Design for the Pellet Bed Reactor for Nuclear Thermal Propulsion.

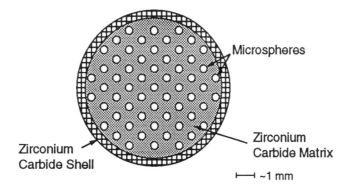

FIGURE 2. Cross Sectional View of the Fuel Pellet Design for Pellet Bed Reactor for Nuclear Thermal Propulsion.

The proposed choices of materials for the PeBR fuel pellet for NTP applications should enable reactor operation at an exit propellant temperature in excess of 3000 K and Isp in excess of 1000 seconds. A further increase in propellant temperature can be realized by lowering the maximum fuel temperature in the fuel microspheres via the formation a of central void in the fuel kernel during fabrication. The formation of a central void (~ 6 vol.%) in the fuel kernel not only reduces the stresses in the NbC and ZrC coatings but also lowers the maximum fuel temperature in the microspheres. This fuel microsphere design concept is first introduced by Matthews et al. (1991).

In the PeBR core the fuel pellets are randomly stacked between inner (hot) and outer (cold) porous frits. During full power operation the propellant (hydrogen gas) enters the core radially though the outer frit at about 120-200 K and exits at the hot frit at about 3000 K to the 18 cm diameter central channel (see Figure 3 and 4). The propellant in the central channel then exits the reactor through the rocket expansion nozzle. The fuel pellets in the PeBR core are self supported, which eliminates the need for internal core structure, simplifies the core design and reduces the size and mass of the reactor.

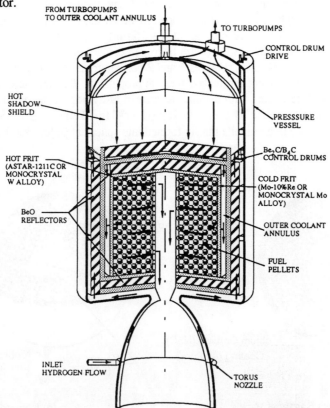

FIGURE 3. A layout of the Pellet Bed Reactor Nuclear for a Nuclear Thermal Rocket.

A safety and operational concern in solid core NTP reactors is the need to cool off the reactor core after each firing operation. Active cooling of the core had been considered for NERVA reactors using discrete pulses of hydrogen through the core over a period of 24 hours (Misra et al. 1971 and Buden 1970). The penalty of using active cooling, however, is that additional propellant mass is needed to flush the reactor after firing, which could reduce the Isp by as much as 4% (El-Genk et al. 1991a). Therefore, a partial or a total passive removal of the decay heat is preferable, since it would enhance not only the operation, but also the safety of the spacecraft.

OBJECTIVES

The objective of this work is to investigate the potential for partial or total passive removal of decay heat in a PeBR nuclear thermal rocket. A multi-dimensional transient heat conduction/radiation model of the PeBR core and surrounding structures is developed and used to perform a parametric analysis of passive cooling of PeBR after firing. The analysis investigates the effects of core dimension, peak thermal power and firing time, on the coolability of the reactor core by passive means.

GOVERNING EQUATIONS

Post reactor shutdown decay heat in generated in the core due to the radioactive decay of fission products in the fuel. The decay heat power, g(r,z,t), depends on the operation power and operation time of the reactor as well as the the time after reactor shutdown. The axial and radial variations of the decay power in the PeBR core follow those in the reactor during full power operation. The radial and axial power distributions in the core are taken as cosine and Bessel functions, respectively, with a peak-to-average power ration of 1.18 in both directions. The average value for g(r,z,t) is calculated based on the reactor operational power level, the reactor operating time, and the time since

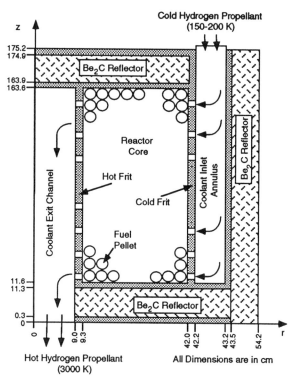

FIGURE 4. Cross-sectional Layout of the Pellet Bed Reactor with Dimensions (not to scale).

reactor shutdown, by the following equation (El-Wakil, 1978):

$$g(r,z,t) = g_o(r,z)\left[0.1(\tau_s + 10)^{-0.2} - 0.087(\tau_s + 2\times10^7)^{-0.2}\right] - \left[0.1(\tau_s + \tau_o + 10)^{-0.2} - 0.087\left(\tau_s + \tau_o + 2\times10^7\right)^{-0.2}\right], \quad (1)$$

where g_o is the operational power per unit volume (W/cm^2), τ_s is the time after shutdown (s), and τ_o is the reactor operating time, (s). The decay heat removal from the reactor core is governed by conduction through the fuel pellets and radiation in the interstitial voids. While both conduction and radiation contributions depends on the fuel pellet temperature, the latter also depends on the optical properties of the ZrC surface (Chen and Churchill 1963). The effective thermal conductivity of the PeBR core, k_{eff}, is taken as the sum of the conduction and radiation components, k_c and k_r, respectively:

$$k_{eff} = k_c + k_r . \quad (2)$$

In Equation (2) the radiation contribution, k_r, is given as:

$$k_r = 8\ \sigma\ \varepsilon\ r_p\ T^3, \quad (3)$$

where, σ is Stefan-Boltzmann constant (5.6669×10^{-8} W/m$^2\cdot$K^4), ε is the surface emissivity of the ZrC pellet surface, which is taken to be dependent on pellet surface temperature, T, in the reactor core (Touloukian, ed. 1967). The conduction component, k_c, in Equation (2) is the sum of two contributions: conduction through the solid fuel pellets, k_s, and conduction through the gas, k_g, that fills the interstitial voids (Chen and Churchill 1963):

$$k_c = \left((1\text{-}\gamma)\left(k_s^{-1} + k_g^{-1}\right)^{-1}\right) + \gamma\ k_g, \quad (4)$$

where γ is the void fraction in the PeBR core, k_g is the thermal conductivity of the hydrogen gas, and k_s is the effective conductivity of the fuel pellets. For vacuum between pellets k_g equals 0 and k_c equals $(1\text{-}\gamma)k$.

958

Because the PeBR core is composed of a randomly packed fuel spheres, determining the effective thermal conductivity of the solid fuel pellet, k_s, is not straight forward. This effective conductivity is calculated using the following expression for the conductance through a packed bed of spheres in a vacuum (Chan and Tein 1973):

$$k_s = S_p \, k_f \left(\frac{1-\mu^2}{E} \, P \right)^{1/3} .$$

(5)

In Equation (5), μ, is the Poisson's ratio for ZrC, E, is Young's modulus of elasticity for the ZrC, P is the interfacial pressure at the contact points between neighboring pellets in the core, and Sp is a geometry factor which depends solely on the packing pattern of the pellets in the core (Chan and Tein 1973). For the random packing condition expected in the PeBR core Sp has a value of 1.85.

The effective thermal conductivity of the fuel pellet material, k_f, is determined assuming a homogeneous compositions using the following conductivity correlation for distributed spheres in a ZrC medium:

$$k_f = k_{ZrC} \left\{ \frac{2 + (k_{micro}/k_{ZrC}) - 2\gamma_{micro}[1 - (k_{micro}/k_{ZrC})]}{2 + (k_{micro}/k_{ZrC}) + \gamma_{micro}[1 - (k_{micro}/k_{ZrC})]} \right\} ,$$

(6a)

where γ_{micro} is the volume fraction of microspheres in the pellet. The effective conductivity of the microsphere, k_{micro}, is determined using a heterogeneous configuration of (U-Nb)C fuel, NbC , and ZrC with appropriate volume fractions , γ_2, γ_3, and γ_4 and equivalent thicknesses, x_2, x_3, and x_4, respectively (see Figure 1). This yields the following equation:

$$k_{micro} = \left[\left\{ (x_2/r_p)^2 \left(2 \, k_{fk}\gamma_2 \right)^{-1} \right\} + \left\{ (x_3/r_p)^2 \left(k_{NbC}\gamma_3 \right)^{-1} \right\} + \left\{ (x_4/r_p)^2 \left(k_{ZrC}\gamma_4 \right)^{-1} \right\} \right]^{-1} .$$

(6b)

In Equation (6) r_p is the fuel pellet radius, and the subscripts micro, fk, NbC, and ZrC denote microsphere, fuel kernel, niobium carbide, and zirconium carbide, respectively.

The transient heat transfer equations in the PeBR core and in the reflector regions are solved simultaneously, using the above heat transfer correlations, to determine the axial and radial temperature distributions in the core and the in reflector regions as functions of time after reactor shutdown. The heat transfer in the annulus separating the core region from the reflector (see Figure 4) and at the outer surface of the reflector is by radiation.

The transient 2-D heat conduction equations in the core and reflector regions are:

(a) Reactor Core Region ($R_{in} < r < R_{core}$):

$$\frac{1}{r} \frac{\partial}{\partial r} \left(r \, k_{eff} \frac{\partial T_{core}}{\partial r} \right) + \frac{\partial}{\partial z} \left(k_{eff} \frac{\partial T_{core}}{\partial z} \right) + g \, (r,z,t) = \rho_{eff} \, C_{eff} \frac{\partial T_{core}}{\partial t} ,$$

(7)

where ρ_{eff}, and C_{eff} are the effective density and heat capacity of the PeBR core, which are given as:

$$\rho_{eff} = (1-\gamma_c) \, \rho_p ,$$

(8a)

where,

$$\rho_p = \gamma_2 \, \rho_{fk} + \gamma_3 \, \rho_{NbC} + \left(\gamma_4 + (1-\gamma_{micro}) \right) \rho_{ZrC},$$

(8b)

and,

$$C_{eff} = C_p \left(1 - \gamma_c \right) \rho_p / \rho_{ef},$$

(8c)

where

$$C_p = \upsilon_1 \, C_{fk} + \upsilon_2 \, C_{NbC} + \upsilon_3 \, C_{ZrC} .$$

(8d)

In Equation (8) ρ_p is the average density and and C_p is the average heat capacity of the fuel pellets, while, υ_1, υ_2, υ_3 are the mass fractions of the fuel kernel, NbC, and ZrC in the pellet.

(b) Reflector Region (($R_{core} + \delta g$)$< r < R_{out}$):

$$\frac{1}{r}\frac{\partial}{\partial r}\left(r\, k_{ref}\frac{\partial T_{ref}}{\partial r}\right) + \frac{\partial}{\partial z}\left(k_{ref}\frac{\partial T_{ref}}{\partial z}\right) = \rho_{ref}\, C_{ref}\frac{\partial T_{ref}}{\partial t}, \tag{9}$$

Equations (7) and (9) are solved numerically subject to the following initial and boundary conditions:

$$T_{core}\,(r,z,0) = f_1\,(r,z),\quad \text{and}\quad T_{ref}\,(r,z,0) = f_2\,(r,z), \tag{10a-b}$$

$$\frac{\partial T_{core}}{\partial r} = 0\ @\ r = R_{in}\,,\qquad k_{eff}\frac{\partial T_{core}}{\partial r} = q_r\big(R_{core},\,z,t\big)\ @\ r = R_{core}, \tag{10c-d}$$

$$\frac{\partial T_{core}}{\partial z} = 0\ @\ z = 0\,,\qquad \frac{\partial T_{core}}{\partial z} = 0\ @\ z = H\,, \tag{10e-f}$$

$$k_{ref}\frac{\partial T_{ref}}{\partial r} = q_r\big(R_{core+\delta g},\,z,t\big)\ @\ r = R_{core+\delta g}\,,\qquad k_{ref}\frac{\partial T_{ref}}{\partial r} = q_r\big(R_{out},\,z,t\big)\ @\ r = R_{out}\,, \tag{10g-h}$$

$$\frac{\partial T_{ref}}{\partial z} = 0\ @\ z = 0\,,\ \text{and}\ \frac{\partial T_{ref}}{\partial z} = 0\ @\ z = H\,. \tag{10i-j}$$

The values of the functions f_1 and f_2 in Equations (10a-b) are the temperature distributions in the core and reflector regions before scram. However, in this analysis, since these distributions are not available a priori, linear distributions is assumed for conservative considerations

METHODOLOGY

In order to solve Equations (7 and 9) with finite difference the reactor core is discritized into the computational mesh shown in Figure 5, with the radial index, i, ranging from 1 to IMAX and the axial index, k, ranging from 1 to KMAX. Each cell node is located at the geometric center of the control volume. The position of the nodes is given by the following equations:

$$r\,(i)\ = R_{IB} + (i-i_{IB}+0.5)\,\Delta r(i)\quad\text{and}\quad z\,(k)\ = (k-0.5)\,\Delta z\,, \tag{11a-b}$$

where R_{IB} is the material regions inner boundary and i_{IB} is the value of i in the first cell of the material zone. The global node number, m, as shown in Figure 5 is calculated by:

$$m = i +(k-1)\cdot IMAX\,. \tag{12}$$

For a given cell, m, the global node number of the adjacent cells is written as a function m(n), where n is the local node number and ranges from 1 to 4, as shown in Figure 5. Finally, the time domain is divided into small steps, Δt, such that:

$$t = p(\Delta t)\,. \tag{13}$$

Using Equations (11), (12), and (13) the temperature and decay heat volumetric generation terms at any location (r,z) and time (t) are represented by:

$$T(r,z,t) = T_m^{\,p}\quad\text{and}\quad g(r,z,t) = g_m^{\,p}\,. \tag{14a-b}$$

In order to discretize Equations (7 and 9), a weighted average (semi-implicit method) of the node temperatures is taken between time step p and p+1. For the semi-implicit scheme used, Equations (7 and 9) may now be written for

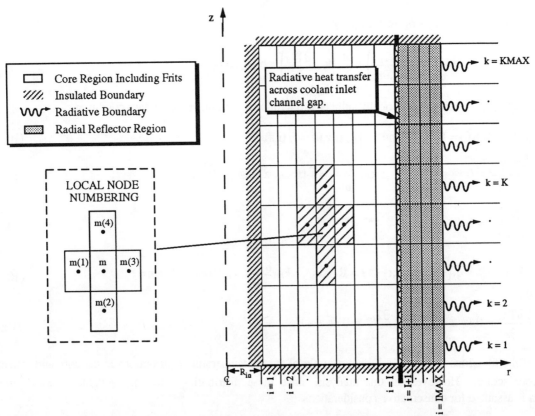

FIGURE 5. The r-z Mesh Layout for the Numerical Solution in the PeBR Decay Heat Model

each node, m, in the following form:

$$V_m\left[(1-\eta)g_m^p + \eta g_m^{p+1}\right] - (1-\eta)\sum_{n=1}^{4}\frac{T_m^p - T_{m(n)}^p}{R_{m,n}^p} - \eta\sum_{n=1}^{4}\frac{T_m^{p+1} - T_{m(n)}^{p+1}}{R_{m,n}^{p+1}} = \left(V_m\rho_m C_m\right)\frac{T_m^{p+1} - T_m^p}{\Delta t},\tag{15a}$$

or by rearranging,

$$\left(\frac{V_m\rho_m C_m}{\Delta t} - \eta\sum_{n=1}^{4}\frac{1}{R_{m,n}^{p+1}}\right)T_m^{p+1} - \sum_{n=1}^{4}\left[\left(\eta\frac{1}{R_{m,n}^{p+1}}\right)T_{m(n)}^{p+1}\right] = \frac{V_m\rho_m C_m}{\Delta t}T_m^p + V_m\overline{g}_m - (1-\eta)\sum_{n=1}^{4}\frac{T_m^p - T_{m(n)}^p}{R_{m,n}^p},\tag{15b}$$

where

$$R_{m,1} = \frac{\left(\Delta r_m / k_{eff,m} + \Delta r_{m(1)} / k_{eff,m(1)}\right)}{\left[(r_m - \Delta r_m)\,\Delta z\right]}, \qquad R_{m,3} = \frac{\left(\Delta r_m / k_{eff,m} + \Delta r_{m(3)} / k_{eff,m(3)}\right)}{\left[(r_m + \Delta r_m)\,\Delta z\right]},\tag{16a-b}$$

$$R_{m,2} = \frac{\left(\Delta z_m / k_{eff,m} + \Delta z_{m(2)} / k_{eff,m(2)}\right)}{\left(r_m\Delta r\right)}, \text{ and } \quad R_{m,4} = \frac{\left(\Delta z_m / k_{eff,m} + \Delta z_{m(4)} / k_{eff,m(4)}\right)}{\left(r_m\Delta r\right)}.\tag{16c-d}$$

In Equations (15a and 15b) η is the degree of implicitness, where $\eta = 1$ implies the numerical scheme is fully implicit and $\eta = 0$ implies the scheme is fully explicit. Equation (15) may now be represented in a simplified form (letting the left hand side equal f_m, the first term in parentheses on the left hand side equal d_m, and the second term in parentheses on the left hand side equal $c_{m,n}$) by:

$$d_m\,T_m^{p+1} - \sum_{n=1}^{4}\left[c_{m,n}\,T_{m(n)}^{p+1}\right] = f_m\tag{17}$$

For a given time step, p+1, the radiative heat flux at the outer reflector boundary is given as:

$$q_{out}^{p+1}(R_{out}, z) = A_{out} \sigma \varepsilon F \left[\left(T_w^{p+1} \right)^4 - T_{amb}^4 \right] \qquad (18)$$

This equation is linearized by assuming that the outer wall temperature, T_w, does not change significantly between time steps p and p+1:

$$q_{out}^{p+1}(R_{out}, z) = A_{out} h_r \left(T_w^{p+1} - T_{amb} \right) \qquad (19)$$

where,

$$h_r = \sigma \varepsilon F \left(\left(T_w^p \right)^2 - T_{amb}^2 \right) \left(T_w^p + T_{amb} \right) \qquad (20)$$

Similarly, the radiative heat transfer in the annulus separating the core from the reflector, as given by Equations (10d) and (10g), are linearized. The remainder of the boundary conditions in Equation (10) are easily reproduced in there respective discritized forms by applying Equations (11) - (14) and are note rewritten here.

The temperature in each cell is determined by applying the appropriate boundary conditions to Equation (17) and combining all terms into the following matrix:

$$
\begin{bmatrix}
d_1 & -c_{1,3} & 0 & .. & -c_{4,1} & .. & 0 \\
-c_{1,2} & d_2 & -c_{3,2} & 0 & .. & .. & 0 \\
.. & .. & .. & .. & .. & .. & -c_{4,mmax-imax} \\
- & .. & .. & - & .. & .. & . \\
-c_{2,imax} & 0 & .. & .. & .. & .. & 0 \\
0 & .. & .. & .. & -c_{1,mmax-1} & d_{mmax-1} & -c_{3,max-1} \\
0 & .. & -c_{2,mmax} & 0 & 0 & -c_{1,mmax} & d_{mmax}
\end{bmatrix}
\begin{bmatrix}
T_1^{p+1} \\
T_2^{p+1} \\
. \\
. \\
. \\
T_{mmax-1}^{p+1} \\
T_{mmax}^{p+1}
\end{bmatrix}
=
\begin{bmatrix}
f_2 \\
f_2 \\
. \\
. \\
. \\
f_{mmax} \\
f_{mmax}
\end{bmatrix}
\qquad (21)
$$

Equation (21) is positive definite and, therefore, T is solved using a banded Gaussian elimination matrix solver.

RESULTS AND DISCUSSION

In this section the results on the effect of fuel pellet diameter and temperature of the effective thermal conductivity in the PeBR core after shutdown are presented. Also, the potential for partial and total passive removal of decay heat in the PeBR after firing operations is investigated and the results presented and discussed.

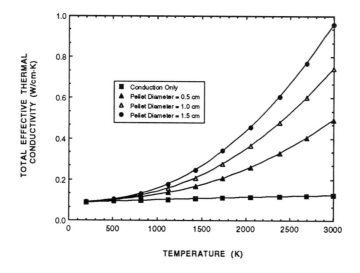

FIGURE 6. Effective Conductivity for the PeBR Core as a Function of Fuel Pellet Size.

962

TABLE 1.   Base Case Parameters for Decay Heat Removal in a Pellet Bed Reactor

Coolant Exit Channel Radius	9.0 cm
Core Diameter	40 cm
Active Core Height	158 cm
Cold Frit Thickness	0.3 cm
Coolant Inlet Channel Thickness	1.0 cm
Reflector Thickness	11 cm
Fuel Pellet Diameter	0.9 cm
Young's Modulus	2×10^5 MPa
Poisson's Ratio	0.2
Ambient Temperature	200 K
Operating Power	1000 MW
Start Time for Passive Cooling	1000 s

Effective Conductivity in the PeBR After Shutdown

During full power operation the fission heat generated in the reactor core is removed by convection to the high velocity flow of hydrogen through the core. After reactor shutdown and in the absent of active cooling, however, the heat transfer in the core is by conduction and radiation at high temperatures. The contributions of these two modes of heat transfer in the PeBR core after shutdown are quantified and presented in Figure 6. As shown in Figure 6, at temperatures below 500 K, the effective conductivity of the PeBR core is solely by conduction, and is independent of the pellet size. At higher temperatures, however, the radiation contribution becomes important and dominates at temperatures higher than 1000 K. At such temperatures, the effective bed conductivity increases significantly with pellet size. For example, at a bed temperature of 3000 K, increasing the pellet diameter from 1 cm to 1.5 cm, causes the effective bed conductivity to increase by approximately 40% (from 0.70 to 0.95 W/cm·K). These results demonstrate that decay heat removal in the PeBR will be controlled mostly by thermal radiation, indicating that ZrC coating is an excellent choice not only because of its compatibility with hydrogen, but also due to its good optical properties. The latter will significantly enhance the PeBR safety by improving the potential for passive decay heat removal in the PeBR. Results in Figure 6 also show that the effective conductivity of the PeBR after shutdown increases with increasing bed temperature; a unique self-regulating feature whereas the effective thermal conductivity is highest in the hottest part of the reactor core and the lowest in the cooler part of the core.

Passive Decay Heat Removal Analysis

The base case parameters of the PeBR used in the present analysis are listed in Table 1 and the results are presented in Figures 7 and 8. Note that the base case calls for active cooling of the PeBR for only 1000 seconds after firing, followed by passive removal of the decay heat. Figure 7 shows the calculated radial temperature distribution in the PeBR core and reflector regions at different times after shutdown. These results are for active cooling of 1000 s instantaneously after reactor shutdown, followed by active cooling. As this Figure shows, a decay heat removal strategy that involves active cooling of 1000 s, followed by passive removal of the decay heat will maintain the PeBR core safely coolable; whereas the maximum fuel temperature is kept below 3150 K. As delineated in Figure 7, the maximum core temperature after reactor shutdown always occurs at the hot frit and the temperature in the core decreases with radial distance from the center of the core. The decay heat arriving at the outer radius of the core is transported by radiation through the coolant annulus to the reflector structure, then it is removed from the reflector structure by radiation into space. Therefore, the rate of decay heat removal from the PeBR core is controlled not only by the conduction and radiation properties in the core, but also by the surface area and optical properties of the reflector outer surface. The effect of the latter is delineated in Figure 8.

After the termination of active cooling of the core, decay heat is removed from the center region of the core and absorbed in the outer core region, causing the maximum fuel temperature in the former to decrease and in the latter to increase. However, since decay heat generation rate is initially higher than the decay heat removal by radiation at

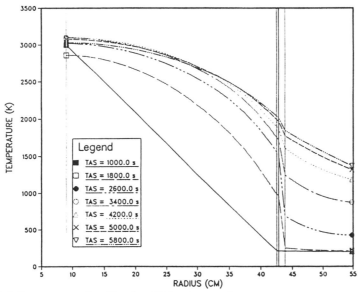

FIGURE 7. Radial Temperature Profiles for 1000 s Active Cooling.

the outer surface of the reflector, because of its low surface temperature, the fuel temperature at the center of the PeBR core begins to rise. Eventually, this fuel temperature reaches a maximum when the decay heat generation rate becomes equal to the the removal rate at the surface of the reflector. Beyond this point, the fuel temperature continues to decrease with time. As shown in Figure 7, the outer reflector surface temperature continues to rise with time after shutdown; hence increasing the rate of heat rejection which varies with the surface temperature raised to the fourth power.

The results in Figure 8 show that with initial active cooling of 1000 s, following a reactor shutdown (see Table 1), the maximum fuel temperature initially declines after active cooling is terminated, then begins to rise reaching a peak of 3100 K, 4000 s after shutdown. When the active cooling time is reduced from 1000 s to 600 s, the maximum fuel temperature in the core increased, but the time after shutdown when this temperature occurs decreased. As shown in Figure 8, with only 600 s of active cooling, the maximum core temperature peaked at 3400 K at 3000 s after shutdown. The results in Figure 8 indicate that complete passive cooling of the PeBR core is not attainable, after operating at a thermal power of 1000 MW, since the fuel temperature rises precipitously after reactor shutdown, exceeding 3500 K within 200 s. The effect of core height on the post operational coolability of the PeBR core is also investigated. The maximum temperatures in the core with 1000s of active cooling, for a core height of 158 and 168 cm are compared in Figure 8. Increasing the height of the core enhances its coolability, since it not only reduces the power density of the reactor but also increases the radiative surface area of the reflector. As Figure 8 shows increasing the core height from 158 cm to 168 cm reduces the peak fuel temperature after firing from approximately 3100 K to as low as 3035 K.

CONCLUSIONS

Preliminary results indicate that using only passive decay heat removal from a PeBR thermal rocket after operating at a thermal power of 1000 MW, could cause the temperature of the fuel pellet at the center of the core to overheat beyond 3500 K. However, using an active for only a short period (600 to 1000 s), following shutdown, it will be possible to safely remove the decay heat from the PeBR by passive means. Active cooling of the core for only 1000 seconds after firing, followed by passive decay heat removal could maintain the peak fuel temperature at approximately 3100 K. Increasing the core height, increases the decay heat removal capability by radiation at the outer surface of the reflector, hence it would be possible to reduce the duration of active cooling required after reactor shutdown below 600 s, without causing the core temperature to increase beyond 3400 K. Results show that decay heat removal in the PeBR is controlled mostly by thermal radiation, indicating that ZrC coating is an excellent choice not only because of its compatibility with hydrogen, but also due to its good optical properties. Results also show the effective conductivity of the PeBR after shutdown increases with increasing bed temperature; a unique self-regulating feature, whereas the effective thermal conductivity is highest in the hottest part of the reactor core and the lowest in the cooler part of the core.

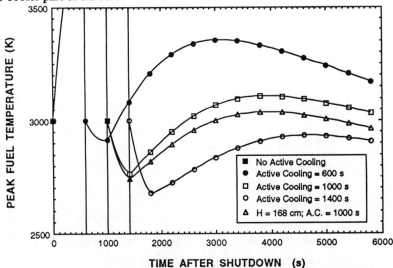

FIGURE 8. Radial Temperature Profiles in the PeBR for 1000 s Active Cooling.

Acknowledgments

This research is funded by the University of New Mexico's Institute for Space Nuclear Power Studies.

References

Bennett, G. (1990) "Nuclear Thermal Propulsion Program Overview," *NASA Nuclear Thermal Propulsion Workshop*, Cleveland, OH, 10-12 July 1990, NASA CP-10079.

Buden, D. (1970) "Operational Characteristics of Nuclear Rockets," *J. Spacecraft*, 7(7):832-836.

Chan, C. K. and C. L.Tien (1973) "Conductance of Packed Spheres in Vacuum," *Journal of Heat Transfer*, August 1973, pp.302-306.

Chen, J. C. and S. W. Churchill (1963) "Radiant Heat Transfer in Packed Beds," *A.I.Ch.E. Journal*, 9(1): 35-41.

El-Genk, M. S., A. G. Parlos, J. M. McGhee, S. Lapin, D. Buden, and J. Mims (1990a) "System Design Optimization for Multimegawatt Space Nuclear Power Applications," *J. Propulsion and Power*, 6(2): 194-202.

El-Genk, M. S., N. J. Morley, and V. E. Haloulakos (1990b) "Pellet Bed Reactor for Nuclear Propelled Vehicles," *NASA Nuclear Thermal Propulsion Workshop*, Cleveland, OH, 10-12 July 1990, NASA CP-10079.

El-Genk, M. S., N. J. Morley, and V. E. Haloulakos (1991a) "Pellet Bed Reactor for Nuclear Propelled Vehicles," in *Proc. Eighth Symposium on Space Nuclear Power Systems*, Albuquerque, NM, M. S. El-Genk and M. D. Hoover, eds., Symposium conference Proceedings No. 217, American Institute of Physics, New York, pp. 607-611 (January 6-10, 1991).

El-Genk, M. S., J. Y. Yang, and N. J. Morley (1991b)" Thermal-Hydraulic Analysis of Pellet Bed Reactor," AIAA/NASA/OAI Conference on Advanced SEI Technologies, Cleveland, OH, Paper No. AIAA-91-3510, September 4-6, 1991.

El-Wakil, M.M. (1978) *Nuclear Heat Transport, American Nuclear Society*, La Grange Park, Illinois,

1978.

Matthews, R. B., H. T. Blair, K. M. Chidester, K. V. Davidson, W. A. Stark, Jr, and E. K. Storms (1991)"Carbide Fuels for Nuclear Thermal Propulsion," AIAA/NASA/OAI Conference on Advanced SEI Technologies, Cleveland, OH, Paper No. AIAA-91-3455, September 4-6, 1991.

Misra, B., J. H. Altsiemer, and G. D. Hart (1971) "In-Flight Coolant Management Considerations for the Nerva Reactor Cooldown," *J. Nuclear Technology*, 12(11): 298-306.

Ozisik, M. N. (1980) *Heat Conduction*, John Wiley and Sons, New York, 1980.

Touloukian, Y. S. (ed.) (1967), Thermophysical Properties of High Temperature Solid Materials, The Macmillan Company, NY, Vol. 5, pp. 263-293.

Vortmeyer, D. (1978) "Radiation in Packed Solids," in *Proc. Sixth Int. Heat Transfer Conference*, Vol. 6, Toronto, Canada, 7-11 August 1978, pp.525-539.

--Nomenclature--

English

A	:	Flow area (cm^2)
C	:	Heat Capacity (J/kg)
c	:	Off-diagonal matrix coeff.
d	:	Diagonal matrix coeff.
E	:	Young's Modulus (MPa)
e	:	Internal energy (J/kg)
F	:	View Factor
f	:	Source vector coefficients
g	:	Volumetric heat generation rate ($J/m^3 \cdot s$)
H	:	Core height (cm)
h	:	Radiative heat transfer coefficient ($W/cm^2 \cdot K$)
IMAX	:	Number of radial divisions
KMAX	:	Number of axial divisions
m	:	Global node number
MMAX	:	Total number of nodes
n	:	Local adjacent cell number
P	:	Pressure (Pa)
p	:	Time step
q	:	Total surface heat transfer (W)
r	:	Radial coordinate (cm)
S	:	Geometry factor
T	:	Absolute temperature (K)
t	:	Time (s)
V	:	Volume (cm^3)
x	:	Fuel kernel layer thickness (cm)
z	:	Axial coordinate (cm)

Greek

δ	:	Thickness (cm)
γ	:	Void Fraction
ε	:	Emissivity
μ	:	Poisson's Ratio
ρ	:	Density (g/cm^3)
σ	:	Stefan-Boltzmann constant
τ	:	Time
u	:	Mass fraction

Subscripts

amb	:	Ambient
c	:	Conduction
core	:	Core
eff	:	Effective
f	:	Fuel pellet
fk	:	Fuel kernel
g	:	Gap
IB	:	Inner Boundary
imax	:	Number of radial divisions
m	:	Node m
micro	:	microsphere
mmax	:	Total number of nodes
o	:	Full power operation
out	:	Outer edge of the reflector
p	:	Pellet
r	:	Radiation
ref	:	Reflector
s	:	After shutdown
w	:	Wall

HUMAN ROUND TRIP TO MARS: SIX MONTHS AND RADIATION-SAFE

Otto W. Lazareth, Eldon Schmidt, Hans Ludewig, and James R. Powell
Reactor Systems Division
Brookhaven National Laboratory
Building 701
Upton, NY 11973
(516) 282-3448, 282-5078, 282-2624, 282-2440

Abstract

We describe a different type of round trip to Mars, using a combination of spacecraft. Compared to typical proposals, this flight is relatively fast and relatively safe from biological radiation dosage. Our study is concerned with the trip from Earth orbit to Mars orbit. Four spacecraft are required for the round trip. The crew spends most of their time on board a comparatively large, well shielded spacecraft (LC) which is in free (non-powered) orbit about the sun. The crew travels from Earth orbit to the LC while on board a comparatively small, powered spacecraft (SC). At Mars, the procedure is reversed and the crew returns on a second LC. In addition, a cargo craft, with no crew, is sent to Mars prior to the crew leaving Earth orbit. The trip time is about six months and the radiation dose equivalent is within guidelines recommended by the National Commission on Radiation Protection and Measurements.

INTRODUCTION

The Radiation Environment

Outside the Earth's atmosphere, the radiation environment is quite severe. There are five sources of this radiation.

1. Galactic Cosmic Rays (GCR) are highly energetic particles which originate from outside the Solar System. They consist of 98% protons and heavier ions and 2% electrons and positrons. Their energies range from about 100 MeV to 10 GeV per nucleon. The radiation dose-equivalent (DE) from these particles is about .45 Sv/y (45 rem/y) to a human being with 2 g/cm^2 of Al shielding (NCRP Report No. 98, 1989).

2. Solar Event Particles (SEP) are energetic particles which originate from Solar Particle Events (SPE) of our Sun. They also are atomic nuclei, but their kinetic energies range from about one to 100 MeV per nucleon. The DE from these particles varies drastically according to the strength of the SPE. The frequency of occurrence of anomolously large SPEs ranges from several per year to one per decade.

3. Continuous solar cosmic rays (SCRs) are the relatively low energy particles given off by our Sun at a constant rate. Their energies range up to about two MeV per nucleon.

4. Trapped radiation consists primarily of protons and electrons confined to the "Van Allen Belts." They contribute a significant DE to astronauts only when a spacecraft passes through them.

5. If a nuclear engine is used, the radiation from the engine will be another source of DE. However, because of the distance from the crew to the engine (about 100 meters), and the large amount of structure and liquid hydrogen (LH) propellant between them, the radiation effects of the nuclear engine are expected to be small.

The National Council on Radiation Protection and Measurements (NCRP) has set up a guideline of .50 Sv (50 rem) for humans in outer space (NCRP Report No. 98, 1989).

Other Effects

There are two other effects which have quite serious implications for space travel. One is the psychological effect of loneliness during an extended (months) journey through space with very limited human interaction. The second effect is the physiological one of weightlessness for an extended period of time.

These factors, together with the danger inherent in the new technology of space travel, cause a voyage to Mars to be a dangerous enterprise. However, the adverse consequences of these effects will be reduced by a shortening of the trip time. This reduction is the main purpose of this study.

DESCRIPTION OF THE SYSTEM

Simplified Description

We propose an interplanetary transportation system which will provide a round trip for humans travelling from Earth to Mars and back. Compared to typical proposals, the journey will be relatively fast and relatively safe from radiation dosage.

The basic concept consists of four space vehicles and their trajectories. First, a heavily shielded, large spacecraft (LC#1), which usually is unmanned, is put into a free (non-powered) heliocentric orbit which closely passes Earth and Mars, in that order. Second, but during the same time frame, a second large spacecraft (LC#2), which will be used for the return trip, is put into an orbit which closely passes Mars and Earth in turn. Third, but also during the same time period, an unmanned cargo vessel, containing LH and other supplies, arrives at Mars before any humans depart from Earth. Fourth, a comparatively small spacecraft (SC) carrying the crew is launched from Earth orbit, travels to a rendezvous point with LC#1 and couples to it. The SC carries the minimal amount of supplies and only enough fuel to rendezvous with LC#1. Therefore, it is light and quickly can reach high speeds. The crew transfers to the relative comfort and safety of the LC. When they reach Mars, the SC decouples from LC#1 and orbits Mars, while LC#1 continues in its orbit. The crew spends 30 days on Mars, during which time the SC is re-supplied with materials from the cargo ship which was previously sent to Mars. The crew then leaves Mars in the SC and proceeds to a rendezvous point with LC#2 and couples to it. The crew transfers to it and travels back to the vicinity of the Earth. Then the crew transfers back to the SC and returns to the Earth orbit.

A More Detailed Description

Many options on this scenario are possible. However, to express it simply, we'll discuss one possible voyage, using a specific set of options. Then, we'll note some of the options.

Large Spacecraft #1 (LC1)

LC#1, without a crew, is put into a heliocentric orbit which will pass close to the Earth, and secondly, also will pass close to Mars (see Figure 1). The LC is powered by three Particle Bed Reactors (PBRs) (two nuclear electric reactors and one nuclear thermal reactor). It carries extra LH and most materials required by the crew. It is stocked by robots and its engines are automatic but can be overridden either by instructions from the crew (when they are on board) or by radio signals from mission control. It will be put into orbit over a period of months, or years, during which time, no humans will be on board. Propulsion will be supplied by very efficient but low thrust nuclear electric rocket engines.

The LC is heavily shielded and therefore does not have storm shelters. (A storm shelter is a small, heavily shielded volume in which the crew would be located in the event of an SPE.) The external shielding consists of the wall of the vessel, the surrounding store of supplies and the filled LH tanks. (Hydrogen is the most effective shielding per unit mass.) After the LC reaches Mars, the LH will be pumped to the SC, where it will be used as propellant.

Small Spacecraft (SC)

An SC, carrying the crew, is powered by two PBRs. This craft is envisioned as a scaled-down version of the Grumman Aerospace Corporation (GAC) Split/Sprint Mars Vehicle Design (Venetoklis 1991). That design is particularly good and has many laudable features.

Since the SC requires supplies and LH for a trip of about one day, it is relatively light, and therefore, relatively fast.

The SC is not as heavily shielded as the LC and therefore has a storm shelter. The crew will travel in the SC for about one day to intersect with the LC's orbit. The SC will couple to the LC and the crew will transfer to it and remain in it until the vicinity of Mars is reached. During the trip, the SC will be resupplied with cargo and LH from the LC. While at Mars, the SC will again be resupplied with cargo and LH, this time from the cargo vessel which previously was sent to Mars.

After spending 30 days on Mars, the crew, on board the SC, leaves the planet and meets up with LC#2. As before, the crew transfers to the LC, then returns to the vicinity of the Earth while resupplying the SC. Then, they again board the SC and return to Earth orbit. This is shown, conceptually, in Figure 2.

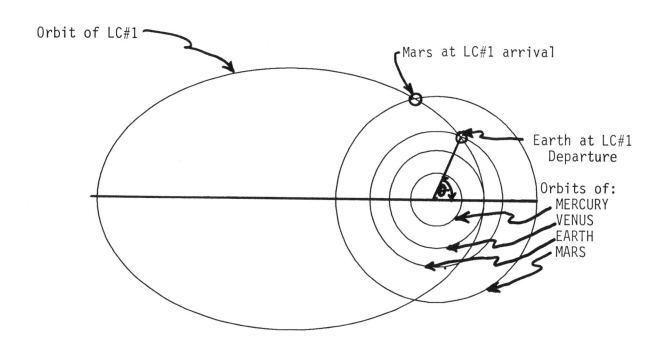

FIGURE 1. Orbits of the planets and LC#1.
Outbound trip (Earth to Mars)

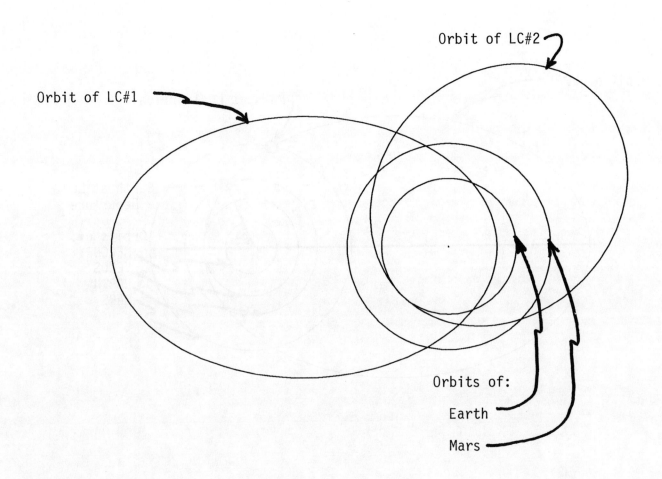

FIGURE 2. Orbits of Earth, Mars, LC#1 and LC#2
Inbound trip (Mars to Earth)

Radiation

Of the five sources of radiation dose, we can neglect the effect of both the trapped radiation and the nuclear engine. The continuous SCR will be stopped by less than 1 mm of Al and, therefore, also need not be considered. This leaves the GCR and SPE radiation.

The NCRP has established a guideline of a dose-equivalent (DE) 0.50 Sv/y (50 rem/y) for astronauts. Therefore, for one year:

Total Allowable DE = DE (GCR) + DE (SPE), where DE (GCR) and DE (SPE) are the radiation DEs from the GCRs and the SPEs, respectively, in units of Sv.

The GCR DE is constant (in time) for a particular thickness of shielding. The SPE DE varies not only with the amount of shielding but varies dramatically with the intensity and duration of the event. As a standard, we'll use the anomalously large SPE of August 1972, which was the largest SPE recorded up to 1989. Using the data calculated by Letaw (1989), our reference shielding is 29 g/cm^2. This correspond to 8.9 cm of Al plus 5.0 g/cm^2 self-shielding for the blood forming organs. For this amount of shielding, we have:

.50 (total) = .30 (GCR) + .20 N (SPE), where N is the number of SPEs, each of which contribute as much DE as the August 1972 event. This equation has the solution: N= 1. That is, with 29 g/cm^2 of shielding, the astronauts can tolerate, indefinitely, the yearly equivalent of the August 1972 SPE. The LH required for one leg of the journey will provide about that much shielding.

Cycler

While the main objective of manned flights is to protect the passengers from radiation exposure with quick trips, that of unmanned cargo flights is economy. Low power, propellant expenditure, and lighter vehicles with reduced radiation shielding are important considerations. The operation would be very similar to that of a freight train following a regular schedule. Cargo units (CARS) would be dropped off into and picked up from local satellite orbits during flybys of Earth and Mars. This is essentially the operation of a cycler (Friedlander, 1986 and Aldrin, 1989).

Cycling transfer orbits have been proposed since the 1960s. The one we propose would follow the synodic period of Earth to Mars opposition. The vehicle would thus be in position to transfer cargo at each window opportunity. On Earth flyby the vehicle would ΔV into the next cycler orbit. We illustrate the actual orbits in Figure 3 using an idealized solar system for simplicity. We assumed the orbits for Mars and Earth to be circular and coplanar. The orbital parameters of an ellipse are:

1. Semi-major axis, a, in astronomical units (AU),
2. Eccentricity, ϵ, and
3. Period, Y, in years. Their values are given in Table 1.

TABLE 1. Orbital Parameters

Orbit	Semi-Major Axis a(AU)	Eccentricity ϵ	Period Y (years)
Earth	1.00	0.0	1.000
Mars	1.52	0.0	1.874
Vehicle	1.60	0.393	2.025

The synotic period for outbound windows is 2.144 y. Note that the vehicle overlaps its orbit by 0.119 y to hit the window with Earth. At that point it ΔVs into the next cycler orbit. The turn is 14.5° which rotates the vehicular orbital periapsis by 51.9°. In Figure 3 the elapsed times for the first three encounters with Earth and the first two encounters with Mars are given in years. The first Earth encounter is time 0.0. The trip time from Earth to Mars would be 0.399 y (146 d) and from Mars to Earth, 1.63 y (637 d).

The Earth flyby is used to rotate the orbit. The geocentric closing velocity (V∞) is 6.6 km/s. The maximum turn angle is about 75° with 84.3° needed for the proper periapsis rotation. In the heliocentric system this corresponds to ΔVs of about 7.6 km/s out of 8.9 km/s needed. The additional 1.3+ km/s ΔV would require power. The exact value would depend on how close to Earth the vehicle passes. At Mars the vehicle would not execute a flyby and stay well away from the Martian surface. The V∞ here is 9.8 km/s.

The total ΔV per trip is 1.3+ km/s at Earth flyby for the vehicle plus 6.6 km/s and 9.8 km/s at Earth and Mars respectively for local transfers with satellite orbits. This certainly qualifies for the low power and propellant expenditure desired. The low local relative velocities are due partially to low angle approaches to the planetary orbits. In the actual solar system allowing for the eccentricities and inclinations of the true planetary orbits along with other perturbations, the vehicle orbits would require additional adjustments to properly hit the windows. However, such a system still qualifies for low power/propellant transfer orbits for cargo shipments.

Though designed for cargo, a heavily shielded passenger "car" could be added for emergency passenger return if other vehicles failed. This would be a trip of 637 days unless a second cycling vehicle for the returning Mars to Earth window is used with the 146-day trip time.

Orbits

A computer program, ELLIPSE, was written to calculate parameters for, and information about, particular elliptical orbits. As input data for a particular orbit, the program requires the orbital period and the perihelion distance (the minimum distance to the focus of the ellipse, which is the Sun).

For ease of calculation, all orbits were assumed to be co-planar. Circles were used for the orbits of the planets and non-perturbed ellipses were used for the orbits of the spacecraft.

As a test case, the Hohmann Transfer was run and the results compared with those given in a standard astrodynamics textbook (Bates 1971). See Table 2.

TABLE 2. Comparison of Accepted Values for the Hohmann Transfer Parameters To Values Predicted By The ELLIPSE Model

Parameter	This Paper	Accepted Value[a]
Angle at Earth Departure (deg)	.11	0.00
Angle (θ) at Mars Arrival (deg)	179.83	180.00
Elapsed Time (d)	258.4	258.9
Speed of Earth (km/s)	29.79	29.78
Speed of Spacecraft (km/s)	32.73	32.73

[a]From Bates 1971, p 365.

The angle, (θ), is the angle between the major axis of the orbit and a line connecting the Sun and the planet. See Figure 1.

For this paper, a reasonable orbit was chosen but not optimized. For this orbit, the period is 5.0 y and the perihelion distance is .72 AU (the orbital radius of Venus). The following results were obtained:

Parameters calculated for an Earth to Mars transfer with a period of 5.0 y and a perihelion of .72 AU:

Angle (θ) at Earth Departure	69.21 deg
Angle (θ) at Mars Arrival	102.93 deg
Elapsed Time	46.22 d
ΔV	18.99 + 19.53 = 38.52 km/s

The outbound trip time (elapsed time from Earth to Mars) is quite reasonable at 46.2 d, without optimization. For a smaller perihelion, the trip time would be less but the LC (although uninhabited except between Earth and Mars) would pass closer to the Sun, thereby increasing its residual radioactivity.

For the return (inbound) trip, we calculate a requirement of 127 d (see Figure 2).

The total trip time is:

Total time = A + B + C + D + E + F + G

where A is the SC trip from Earth orbit to LC#1, accelerating only (1 d).

B is the LC#1 trip from Earth locale to Mars locale (46 d).

974

CYCLIC VEHICLE

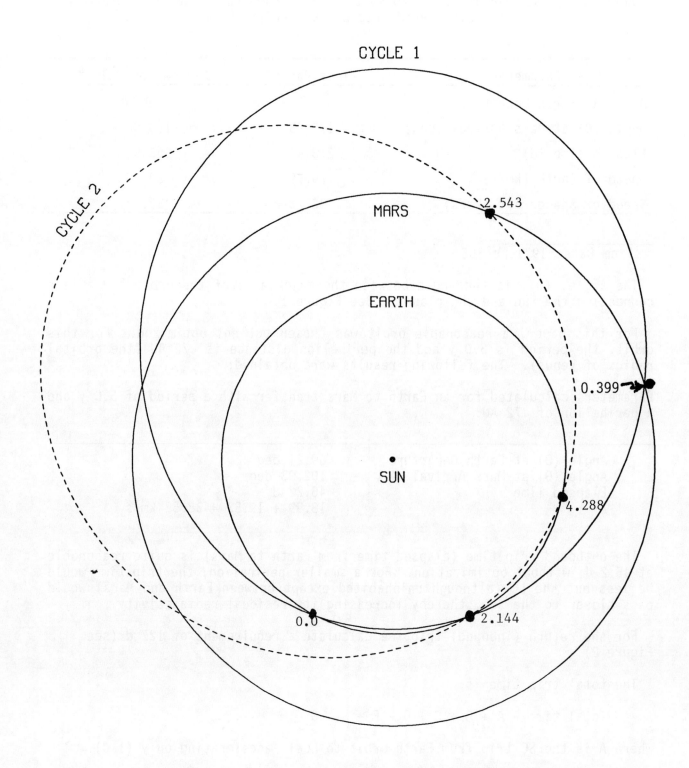

FIGURE 3. First Two Orbits of Cyclic Transfer Vehicle.
Points are Elapsed Time in Years Starting with
an Earth Encounter at 0.0.

C is the SC trip from LC#1 to Mars, braking only (1 d).

D is the 30 day stay on Mars (30 d).

E is the SC trip from Mars orbit to LC#2, accelerating only (1 d).

F is the LC#2 trip from Mars locale to Earth locale (127 d).

G is the SC trip from LC#2 to Earth orbit, braking only (1 d).

```
Outbound time = A + B + C = 1 + 46 + 1 =  48 d
Stay on Mars  = D                       =  30 d
Inbound time  = E + F + G               = 129 d
                                          _____
Total time = 207 d.                        207 d
```

This trip was not optimized. The orbits of Earth and Mars were assumed to be circles. In reality, Mars' orbit is significantly elliptical and the minimum distance between the two planets is less than that between two circles having the same center. The calculated travel time is 177 d (48 + 129 d). If the trip is optimized, we estimate that this can be reduced by about 15%, to a value of about 153 d. The optimized total trip time is then 183 d (153 + 30 d), or about six months.

Options

An overriding principle in choosing any of various options is safety. Redundancy is always desirable for system elements so that a failed component can be bypassed and its function performed by another component. The balancing traits of redundancy are system complexity and cost.

Using this principle, the LC is designed with three engines, with their capability such that the mission can be accomplished with any two of them. Similarly, the SC is designed with two engines and its mission can be performed with either one alone.

An even more desirable option is to use two SCs, each with a complete crew, and each able to hold both crews (in case of the failure of one SC) for the limited required time.

DISCUSSION

Complexity. The LC/SC System is complex in that two rendezvous of the spacecraft are required. However, similar capabilities must be developed for any large scale space program - with or without human crews.

Funding. The LC/SC System also requires a large amount of funding. However, some of that funding can be amortized by using the LCs for transporting cargo with the Cycler concept.

In general, a manned round trip to Mars is an enormous undertaking and must be sufficiently funded. Several spacecraft and several crews should be used. Even if a continuous program of spaceflight to Mars is not envisioned, a single craft is not a viable option. It would be better to not attempt the journey than to undertake it with insufficient support.

CONCLUSIONS

The concept of an LC/SC transportation system between Earth and Mars is feasible and results in:

1. A tolerable, long-term radiation DE absorbed by the crew.

2. A relatively fast round trip to Mars for the human crew.

3. A continuous, cyclical program of moving cargo to and from Mars economically.

4. A possible method of transportation to the outer planets.

References

Aldrin, Buzz, "The Mars Transit System", Air and Space, October/November 1990.

Bate, Roger R., and others (1973) "Fundamentals of Astrodynamics."

Friedlander, Alan L., and others, "Circulating Transportation Orbits Between Earth and Mars," Science Applications International Corp., Report No. 86-2009.

Letaw, J. R., Silberberg, R., and Tsao, C. H. (1989), "Radiation Hazards on Space Missions Outside the Magnetosphere," Adv. Space Res., Vol. 9, No. 10, pp. (10)285 - (10)291.

National Council on Radiation Protection and Measurements (1989) Report No. 98, "Guidance on Radiation Received in Space Activities."

Venetoklis, P., Pallmer, R., and Gustafson, E., Grumman Aerospace Corporation, "Fast Missions To Mars With A Particle Bed Reactor Propulsion System,"AIAA/NASA/OAI Conference on Advanced Technologies, September 4-6, 1991, Cleveland, OH

Acknowledgments

One of the authors (O.W.L.) is grateful for discussions with Harold C. Berry and Arthur D. S. Harris of Brookhaven National Laboratory, New York. Also, this study is the result of concerns of trip time expressed by Franklin Chang-Diaz of NASA-Houston.

Work performed under the auspices of the U.S. Department of Energy under Contract No. DE-AC02-76CH00012.

Glossary

AU	Astronomical unit (the distance from the Sun to Earth)
CARS	Cargo units of the cycler
DE	Dose-equivalent
GAC	Grumman Aerospace Corporation
GCR	Galactic cosmic rays
LC	Large spacecraft
LH	Liquid hydrogen
NCRP	National Council on Radiation Protection and Measurements
PBR	Particle bed reactor
SC	Small spacecraft
SCR	Solar Cosmic Rays
SEP	Solar event particles (which originate from Solar Particle Events [SPEs]
SPE	Solar particle events

Definitions

1. Radioactivity is the disintegration of nuclei together with the emission of particles of radiation. The S.I. unit of the rate of radioactive decay is the Becquerel (or Curie in atomic units).

2. Absorbed dose is the radiation energy absorbed per unit mass of absorber material. The unit is Gray (or Rad).

3. Dose-equivalent (DE) is the product of the quality factor and the dose. The quality factor is a measure of the effects of different types of radiation. The DE, therefore, is a measure of the biological damage caused by various sources of radioactivity. The unit is Sievert (or rem).

PRACTICAL NUCLEAR THERMAL ROCKET STAGES SUPPORTING HUMAN EXPLORATION OF THE MOON AND MARS

Robert M. Zubrin
Martin Marietta Astronautics
PO Box 179
Denver, CO 80201
(303)-977-1925

Abstract

This paper discusses the use of practical nuclear thermal rocket (NTR) stages to support the human exploration of the Moon and Mars. The NTR engines employed are small (44 kN) and have modest performance (850 s Isp, T/W= 3), so that they can be developed on a short time scale with a constrained development budget. The spacecraft configurations employed are simple stages that can be launched fully integrated with their payloads, rather than the gargantuan assemblies of trusses, engines, and liquid hydrogen filled drop tanks that have frequently been visualized as the appropriate form for the NTR spaceships of the future. Mission profiles using such simple stages are presented which accomplish all Space Exploration Initiative goals. It is shown that such near-term technology NTRs are fully adequate to meet the requirements of such missions, and that the sensitivity of mission performance to advancing NTR technology beyond this level is low. It is also shown that the more demanding mission profiles frequently cited as the basis for advanced technology NTRs operating at high thrust levels lack rationale. It is further shown that, in addition to being easier to develop, the small thrust NTR engine is more versatile than the large thrust NTR, being applicable across a much wider range of potential missions. It is concluded that the current NASA/DOE/DOD effort to develop an NTR should focus on the development of such a small engine on a short time schedule, and that NASA Space Exploration Initiative mission studies should be redirected towards mission modes that can be carried out by such practical stages.

INTRODUCTION

Nuclear thermal rocketry has been identified as an enabling technology for manned Mars missions (Stafford et al. 1991). Yet the requirements that have been proposed for such engines, in terms of thrust, thrust/weight and specific impulse, as well as the technology requirements for the gigantic NTR spaceships envisioned, in terms of required launch capability, on-orbit assembly, checkout and cryogenic fluid management, are so formidable that they relegate Mars missions based upon such concepts to the distant future. Thus, so conceived, NTR becomes not an enabling technology, but a disabling technology for the Mars mission. Furthermore, by acting as an expensive and perpetual toll-gate blocking the initiation of the manned Mars exploration, NTR actually sabotages itself, for so long as its chosen application remains several decades in the future there will never be a strong incentive to put the NTR development program on a schedule that will produce real results. A vicious circle is thus set up in which the advanced NTR and the Mars mega-mission application each act to prevent the other from ever coming to be. Clearly what is needed is to break this circle, and the way to do it is to put the NTR program on a fast track to produce a useful, albeit not optimum engine, that can meet a host of mission needs that already exist, and will continue to exist regardless of the political fortunes of the manned Lunar and Mars program. Once such an engine exists, it in-turn will greatly help the manned Lunar and Mars mission by providing real, flight demonstrated hardware that can be incorporated into practical, cost-effective mission plans in which payloads and trajectories are chosen to accommodate the limitations of real (although not optimal) launch and space propulsion systems. In this paper we shall examine the application of such near-term NTR and spacecraft technology to Lunar and Mars missions and show that it is fully capable of meeting all the requirements of the Space Exploration Initiative.

ASSUMED TECHNOLOGY BASELINE

The technology baseline assumed in this paper is NTR engines with a specific impulse of 850 s, a thrust of 44 kN (10 klbf) and a thrust to weight (T/W) of 3. The rationale for such a baseline is that the 2500 K propellant temperature required to produce 850 s Isp is achievable with materials proven in the NERVA program, so that an extensive program of fuel development will be unnecessary. The reason for the modest thrust level is to enable the engine to be tested at 50% power within existing ground facilities at the Idaho National Engineering Lab (Buden 1991 and Parsley et al. 1991), after which a useful engine with a thrust of 22 kN will have been demonstrated. The engine's full design thrust of 44 kN can then be demonstrated in an escalating series of flights with smaller payloads. The modest T/W of 3 is chosen so as not to force the pushing of unproven technology or design concepts, and to allow mass for sufficient subsystem (turbo-pump, for example) redundancy to assure adequate engine reliability.

It is also assumed that a launch vehicle is available with the capability of lifting 150 tonnes to low Earth orbit (LEO) and a 10 m x 30 m fairing. Such a vehicle has been recommended by the Stafford Commission (Stafford et al. 1991) as the baseline for the re-initation of manned Lunar exploration. The Stafford commission also recommended that such a vehicle later be upgraded to a 250 tonne to LEO booster, however in the belief that funding constraints may preclude such a development, no such augmented booster has been assumed. No orbital infrastructure for service or construction of space transfer vehicles is assumed. This is logical, since both the Stafford and Augustine commissions recommended against designing the Space Station Freedom (SSF) as a construction or servicing node for space transfer vehicles, the current SSF design does not include such capability, and no alternative facilities have been seriously proposed. Indeed, with the well known problems that NASA has been experiencing to get SSF funded, the possibility of building a second space station for space transfer vehicle support can be considered very remote. Furthermore the technologies required for such on-orbit assembly and servicing of space transfer vehicles do not exist, and thus their inclusion within the mission plan would thus relegate the mission to the distant future. Furthermore, since on-orbit assembly and servicing increases mission costs while reducing safety and reliability due to the extreme difficultly in verifying quality control of any construction or servicing done under orbital conditions, such scenarios are to be avoided even when they do become technologically feasible.

The chosen mission mode is thus to simply integrate the NTR stage with its payload within the fairing of the heavy lift booster, and launch the assembly from the ground. This, in fact, is the way every mission ever conducted beyond LEO, manned or unmanned, has been launched.

Rather than employ rubber tanks customized for each mission, a single propellant tank size has been chosen for use on all missions. The tank has a diameter of 9.5 m, with a cylindrical barrel section of 10 m, root 2 hemi half domes at each end, for a total tank length of 16.8 m. It contains a total of 75 tonnes of liquid hydrogen at normal boiling point, and has a mass, including a covering of 0.03 m of multi-layer insulation, of 7.5 tonnes. We call a stage composed of such a propellant tank with a group of from one to six 44-kN NTR engines with an Isp of 850 s a Practical NTR Stage (PNTRS).

LUNAR MISSION APPLICATION

Let us first consider the case of Lunar missions. While transportation architectures incorporating Lunar Orbit Rendezvous (LOR) have been popular even since Apollo, they are very undesirable for support of a manned Lunar base. The reason for this is that if the surface stay time is long, so is the orbit time. A LOR (or Mars Orbit Rendezvous) architecture is therefore left in a predicament of whether to leave someone in the mothership during the extended surface stay, exposed to cosmic rays and the rigors of zero-gravity conditions and accomplishing nothing; or leave the mothership unmanned for an extended period and have the returning crew trust to fate that it will be ship-shape when they return. If it isn't, their predicament may be hopeless. The alternative to LOR and MOR architectures are those that employ Direct Return to Earth from the planets' surface. Such architectures offer great safety advantages for both Lunar and Mars base support transportation systems. In a Direct Return (DR) architecture, if anything is wrong with the return transportation, the astronauts have immediately available to them the safe haven afforded by the surface base and all its supporting infrastructure. Furthermore, the Earth return launch windows possible in DR architectures are much wider than those of LOR or MOR architectures, thus greatly reducing the probability that a launch delay will prevent a timely return flight home.

Direct Return is possible to do on a Lunar mission with all terrestrial produced propellents with a minor (and transitory) mass hit compared to a LOR architecture provided either NTR or cryogenic chemical propulsion is used. However once Lunar LOX becomes available, the DR mission architecture gains a permanent mass advantage (lower initial mass in LEO (IMLEO)) compared to the LOR option. The DR option has an added advantage in total system simplicity, as the hardware required to deliver cargo and manned missions to the Lunar surface is identical, a feature

980

which offers significant cost savings. Taken together, these operational, mass, and cost advantages speak decisively in favor of DR as the basis of a Lunar base support architecture, and therefore DR shall form the basis of the transportation study presented here. Direct return from the surface of Mars is also possible, but absolutely demands that indigenous propellants be used. It is, however, advantageous, and we will discuss a Mars DR architecture below.

Let us consider a Lunar transportation architecture based upon single launches of a 150 tonne to LEO class heavy lift vehicle (HLV), with no on-orbit assembly. All flights deliver payloads all the way to the Lunar surface. Depending upon whether the mission is unmanned or manned, the payload delivered to the surface can be either a large amount of cargo or a small amount of cargo plus a wet Earth Return Vehicle. We consider 3 options for LEO to the Lunar surface delivery:

1) Cryogenic chemical propulsion with a specific impulse of 465 s and a stage dry mass fraction of 0.12. The transfer vehicle is composed of two stages, with the first stage dropped after insertion in Low Lunar Orbit (LLO).

2) NTR propulsion using a PNTRS to deliver the payload from LEO to LLO, after which it is delivered to the Lunar surface by a cryogenic lander, as in Option 1. In the trades shown, the NTR stage is expended in LLO; at a cost of about 1 tonne in landed payload, sufficient propellant can be retained after Lunar orbit capture to allow the PNTRS to propel itself out of LLO into interplanetary space for final disposal.

3) NTR propulsion using a PNTRS is used to deliver the payload from LEO to LLO, after which the NTR executes a 1.7 km/s burn to bring the payload to an approximate dead halt about 2 km above the Lunar surface. The NTR is then turned off, and the NTR and the payload separate, and are each landed on the moon at separate locations using small sets of storable bipropellant engines with a specific impulse of 320 s. Since the ideal delta-V to land on the Moon from a dead halt at an altitude of 2 km is only 0.08 km/s (0.3 km/s was used in the calculation to allow for hover and landing margin,) this alternative effectively uses the NTR's high specific impulse to effect the nearly 6 km/s delta-V between LEO and the Lunar surface.

A comparison of the payload delivery capability of these options is given in Table 1. The PNTRS in the cases shown utilizes a pod of 2 44 kN engines. Three perigee burns are required for trans-lunar injection; all other NTR maneuvers are done with a single burn.

It can be seen that the proper use of near-term practical NTR stages for Lunar missions allows a 89% increase of useful cargo delivered to the Lunar surface for each HLV flight compared to cryogenic chemical propulsion, and a doubling of the transportation efficiency ratio (payload delivered/hardware expended.)

TABLE 1. Lunar Surface Delivery Capability off of 150 Tonne to LEO HLV.

Option	Surface Payload	Hardware Expended	Payload/ Expended
1. Cryo	29.4 tonnes	12.9 tonnes	2.28
2. NTR/Cryo	45.6	13.3	3.42
3. NTR	55.5	11.5	4.83

It may be objected that the landing of a used NTR on either the Moon or Mars poses an environmental hazard due to the inventory of long term radionuclides that are created when the engine is fired. In fact, this objection is unfounded, as the United States has already landed several RTGs on both the Moon and Mars, each of which contained over two orders of magnitude greater long term radiological inventory than a used NTR.

There are many options for achieving trans-lunar injection that trade off number of engines against the required number and duration of the perigee kicks. In Table 2 we see the results of a study using a 4th order Runge-Kutta code which simulates orbit transfer trajectories by directly integrating the primitive equations of motion of a rocket vehicle operating in a planetary gravitational field. The case examined was for various sets of 44-kN NTR engines to hurl 100 tonnes of mass (including the PNTRS tank, propellant for further maneuvers, and actual payload) out of a 300-km LEO orbit onto tran-lunar injection (TLI) with a C3 of 3.8 km^2/s^2. The trajectory options examined included either a single continuous burn, or perigee kicks with thrust arcs firing either 30 or 60 degrees around the periapsis of the orbit (engines firing from a true anomaly of -30 to +30 degrees around periapsis.)

TABLE 2. Perigee Kick Options for TLI of 100 tonne Mass with Sets of 44 kN NTR Engines.

	Number of Engines																
	1			2			3			4			5			6		
	Mass	#kicks	hrs	Mass	#kicks	hrs	Mass	#kicks	hrs	Mass	#kicks	hrs	Mass	#kicks	hrs	Mass	#kicks	hrs
Contin. Burn	191	1	4.7	182	1	2.1	172	1	1.2	169	1	0.8	168	1	0.6	168	1	0.5
60 deg. kicks	157	6	73.9	159	3	10.5	161	2	4.2	164	2	6.7	166	2	20.1	168	1	0.5
30 deg. kicks	153	12	221.4	156	6	34.0	158	4	15.3	160	4	202.1	161	3	16.3	164	3	81.0

In Table 2, mass is the initial mass in LEO of each mission, in tonnes, #kicks is the number of perigee kicks required, and "hrs" is the number of hours from the initiation of the series of burns to escape onto TLI. The actual burn time of the engines is the same as this for the continuous burn options. In the case of the perigee kick options, the required engine burn time may be estimated from the fact that each 30 degree perigee kick requires about 15 minutes of burn time; each 60 degree kick requires about 30 minutes of burn time. It can be seen that while going from a continuous burn strategy to one involving 60 degree kicks results in a significant lowering of mission mass, going from 60 degree kicks to 30 degree kicks doubles the required number of kicks with only marginal mass benefits. The time required to escape is not a monotone function of thrust because some kick series involve a very large elliptical orbit prior to the last perigee kick.

MANNED MARS MISSION APPLICATION

Mars Mission Strategy

Vehicle configurations for accomplishing manned Mars missions with two alternative strategies, Mars orbit rendezvous (MOR) and direct return (DR), are depicted in fig.1. Both mission plans have an initial mass in LEO of 300 tonnes and are launchable all-up with two launches of a 150 tonne to LEO class heavy lift booster.

In Figure 1a we show a conventional MOR conjunction mission. In this case two payloads are each launched to Mars on medium energy Type 1 trajectories using a PNTRS off of separate boosters. The PNTRSs are expended after trans Mars injection (TMI.) One of the payloads, which flies out unmanned, is a Mars surface habitat plus equipment and a wet Mars Ascent Vehicle (MAV) all positioned on a lander. The other payload, which flies out manned, is a Mars Transfer Vehicle (MTV) habitat, a Earth Crew Capture Vehicle (ECCV), and a cryogenic hydrogen/oxygen trans Earth injection (TEI) stage. Both payloads perform a low energy aerocapture into a 250 km x 1 sol elliptical orbit about Mars, after which they rendezvous and the crew of the MTV transfer to the MAV for descent to the surface. After a 1.5 year surface stay, the crew ascends in the MAV and rendezvous with the MTV. The hydrogen/oxygen stage is then used to send the MTV onto trans-Earth injection. Upon arrival at Earth, the crew bails out in the ECCV to perform an Apollo type direct entry, and the remainder of the spacecraft is expended.

In figure 1b we show a version of the Mars Direct (Zubrin and Baker 1990 and Zubrin et al. 1991) mission concept using a PNTRS and low energy aerocapture. (Aerocapture is highly advantageous for Mars missions incorporating large surface payloads, since all payloads destined for the Mars surface must carry an aeroshield in any case.) In this mission a PNTRS is used to throw a payload lifted to LEO by a 150 tonne ETO class HLV on direct trans-Mars injection, after which the payload aerocaptures and lands on Mars, while the PNTRS is expended into interplanetary space. The payload consists of an unfueled two-stage Earth Return Vehicle (ERV) driven by methane/oxygen engines, a propellant processing plant (PPP), and about 10 tonnes of liquid hydrogen which subsequent to landing is reacted with the CO_2 martian atmosphere to produce the methane and oxygen required by the ERV. (184 tonnes of CH_4/LOX are produced; 166 for the ERV and 18 to enable long range ground operations high-powered internal combustion engine driven rovers.) Power to drive this process can be provided by a 100-kWe power reactor landed with payload. Two years (the next Earth-Mars launch window) after the ERV has landed and fueled itself, a second

982

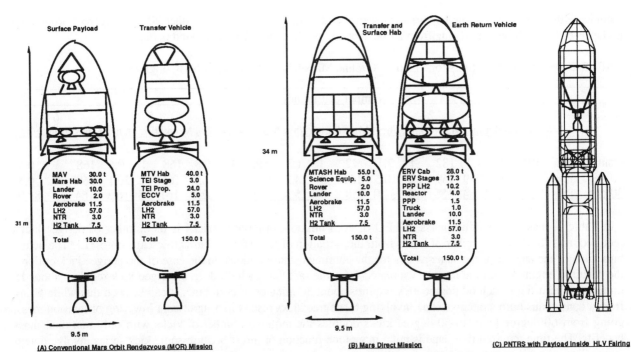

Figure 1. Manned Mars missions using Practical NTR Stages. No on-orbit assembly is required.

HLV is launched sending to Mars a crew of 6 astronauts in a Mars Transfer and Surface Hab (MTASH), which lands in the immediate vicinity of the ERV. The crew will remain on the Martian surface for 1.5 years, exploring widely with the aid of combustion engine driven ground vehicles, and then execute a direct return to Earth in the ERV. Simultaneous with the MTASH launch, a second ERV payload is also sent out to Mars where it aerocaptures into orbit. If anything is wrong with the ERV launched earlier, this second ERV can be landed near the MTASH and used as a backup, otherwise it can be landed elsewhere to open up a new potential landing site for the next MTASH which will be launched 2 years later. Thus every two years two HLVs are launched, an average of one per year, to sustain a continuous and very robust program of manned Mars exploration.

While the Mars Direct option requires the in-situ manufacture of propellant, the processes required (a simple Sabatier reactor to react the hydrogen with Martian CO_2 to produce CH_4 and H_2O, after which the CH_4 is stored as fuel, while the H_2O is electrolysed to produce oxygen propellant and hydrogen feedstock for the Sabatier reaction) are 19th century technology. On the other hand, the Mars Direct approach allows the use the the preferable DR mission architecture, much greater surface mobility by enabling the use of combustion engine driven ground vehicles, and a much greater useful surface payload. In the case of the conventional mission plan shown in figure 1a the useful surface payload is 36 tonnes, about the same as that delivered by the reference mission in the NASA 90 Day Report (Cohen et al. 1989). The PNTRS Mars Direct mission described above has, between the ERV and MTASH landings, a useful surface payload of 111 tonnes, over 3 times as much as the conventional MOR mission.

By reducing mission mass to reasonable levels and eliminating on-orbit assembly or servicing, these two mission plans provide two alternative routes to a manned Mars mission using near-term technology. The question arises, however, how much could these missions be enhanced if the NTR technology level were raised beyond the modest level assumed for the PNTRS. Since the initial mass in LEO of these missions is fixed at 300 tonnes by the lift capacity of the launch vehicle, the relevant measure of transportation technology effectiveness is the useful payload delivered to the Martian surface. In Table 3. we show a comparison of useful payload delivered to the Martian surface for both mission plans assuming either cryogenic hydrogen/oxygen, PNTRS, or various levels of advanced NTR .

In Table 3 the entries marked with an asterisk represent mission plans using 3 perigee kicks for trans Mars injection; all other entries represent mission plans where TMI is accomplished with a single burn. The thrust levels chosen have been optimized for each option. The thrust levels of 176 or 88 kN selected for the PNTRS (850 s Isp) missions represent pods of either four or two 44 kN (10 klbf) engines, depending upon whether a single burn or 3 perigee kicks are employed for TMI.

TABLE 3. Useful Payload Delivered to Martian Surface in 300 tonne IMLEO Manned Mars Missions.

TMI Propulsion			Mission Plan	
Isp	T/W	Thrust	Conventional MOR	Mars Direct
465	50	343 kN	24 tonnes	75 tonnes
850	3	176 / 88*	33 / 36*	102 / 111*
900	6	235 / 88*	36 / 37*	111 / 116*
950	10	294	38	117
1000	40	294	40	124

It can be seen in Table 3 that the largest mission benefit, a tripling of capability, results from dropping the conventional MOR plan in favor of the Mars Direct plan; and this should clearly be done since the in-situ propellant production technology required for the Mars Direct plan can be developed for a cost about 2 orders of magnitude lower than any of the NTR options. Having adopted the Mars Direct plan, a 48% improvement in mission capability can be achieved by shifting from chemical propulsion to PNTRS technology for TMI. Advancing from PNTRS technology to the best medium-term technology NTR option, (950 s Isp, T/W=10) however, results in a mere 5% further improvement, while even invoking the performance levels of a super advanced NTR (1000 s Isp, T/W=40) only increases the mission capability by 12% over that of the PNTRS. The fact that the benefits offered by advanced NTR over near-term NTR in an optimized Mars mission plan are so marginal makes the wisdom of stretching the NTR development program to achieve such difficult levels of engine performance highly questionable.

Fast Missions and Good Missions

The Mars mission plans that we recommend use Type 1 conjunction class (slightly above minimum energy) orbits for interplanetary transfer. The use of such "medium energy" orbits require Mars mission round trip times of roughly 2.6 years. It has been sometimes remarked that such long missions create excessive mission risk due to the deleterious effects of radiation in space and zero gravity. For this reason it has been argued that opposition class (high energy) missions are necessary, as these can reduce the round trip mission time to about 1.6 years.

The opposition class mission has numerous disadvantages. In the first place, its higher energy trajectories increase the total mission delta-V to the point where the total initial mass of the mission is driven up by about a factor of 2 compared to conjunction class missions. The larger amounts of propellant utilized necessitate a correspondingly greater total number of engine-burn-minutes, with the probability of an engine failure increasing in direct proportion. The higher energy orbits also entail higher energy aerocaptures maneuvers, with increased probability of skip out from the planetary atmosphere, and deceleration g loads increased to about 8 to 10 g's compared to the 3 to 4 g's common on conjunction class mission aerocapture maneuvers. The high g loads of opposition class mission aerocapture could pose an unacceptable hazard to crews weakened by 1.5 years in a low gravity environment, navigation and control margins are narrowed by the higher entry velocity, and the violence of the maneuver could create thermal and mechanical problems for the spacecraft as well.

The opposition mission reduces total round trip mission time primarily by reducing the stay time at Mars down to a minimal 0.1 years, with interplanetary round trip transit time remaining at about 1.5 years. This tends to give the mission a very inefficient, if not slightly absurd, character, as the time available to accomplish Mars exploration is reduced to nil. The situation is somewhat analogous to a family which decides to fly to Hawaii for Christmas vacation, taking a 6 hour flight out to the island, 20 minutes to taxi up to the airport terminal, exiting for a 20 minute sortie around the airport, and then returning to the aircraft for a 20 minute wait on the runway followed by another 6 hour flight home.

Finally the opposition class mission must spend part of its flight in a swing into the inner solar system to a distance of about 0.65 astronomical units from the Sun. At this distance, the radiation dose experienced from a solar flare would be 2.4 times that felt at Earth's distance, and 5.5 times that felt by a spacecraft in orbit about Mars. This is very important, because the effect of high sudden doses of radiation are non-linear, and a single 200 rem dose experienced by an opposition mission crew as they flew within the orbit of Venus would be far more dangerous (severe radiation sickness would result) than 5 doses of 40 rems delivered over a 1.5 year period to a conjunction class mission crew hanging in orbit about Mars (no observable symptoms would be expected). In addition, the doubling of heat loads during the opposition class mission transit through the inner solar system would create a major thermal design problem, and result in the catastrophic failure of the mission should the cooling system fail.

The opposition mission is a dead end. It drives mission mass up to the point where the repeated missions required for a sustained human presence on Mars would be prohibitively expensive, and it is incapable of supporting any significant program of surface exploration. The question is, is the taking on of the burden of all the additional risks, cost, and inefficiency entailed by the opposition class mission really necessary? It is not. As we shall see, the proper design of a conjunction class mission can eliminate the zero gravity and radiation exposure rationales that have repeatedly been cited as the basis for choosing the opposition class option.

While it is the case that the canonical minimum energy (Hohmann transfer) flight time to Mars is 0.707 years (258 days) each way, assuming circular planetary orbits, the fact of the matter is that for the real orbits of Earth and Mars, trajectories can be found that are marginally more than minimum energy which reduce one way transit times to as little as 0.345 years (126 days). If we assume a trans-Mars injection C_3 of 15 km/s^2, modest aerocapture C_3 limits at both Earth and Mars of 25 km/s^2, and a trans-Earth injection C_3 limit of 10 km/s^2, then a set of type 1 "medium energy" trajectories can be defined with a canonical Earth-Mars transit time (each way) of 180 days, and stay times at Mars of about 1.5 years. A comparison of such a mission plan with the alternative minimum energy conjunction or opposition plans is shown in Table 4.

TABLE 4. Flight Times and Stay Times of Mars Missions.

	Minimum Energy	Medium Energy	Opposition
Total Transit	1.4 years	1.0 years	1.5 years
Mars Staytime	1.2 years	1.5 years	0.1 years

If the interplanetary transit is to be done in zero gravity, then the reduced transit time of the conjunction class mission reduces zero g exposure to a corresponding degree. On the other hand, if artificial gravity is desired, then the reduced delta-V requirement of the conjunction class mission makes the mass penalty associated with rigid artificial gravity systems that much easier to tolerate. If tethered artificial gravity systems are anticipated, then the use of conjunction class trajectories allows the reduction of trans-Mars payload mass to the point where useful levels of artificial gravity can be generated by tethering off the burnt-out TMI stage. Thus the counterweight at the end of the tether is not mission critical, and if tether dynamics were to go awry, the tether could be dropped without loss of the mission. Such an artificial gravity option is open to both of the PNTRS Mars mission plans presented here. This could not be the case with an opposition mission tether system, which due to increased TMI payload mass would require mission critical payload elements to be placed at both ends of the rotating tether.

The question at hand is, how do these mission profiles effect the radiation dose the crew may be expected to receive in the course of a manned Mars mission, and what is the magnitude of the health hazard that such a dose represents?

It has been shown (Clark and Mason 1990, Simonson et al. 1990; Letaw et al. 1987) that the Mars surface doses are much less than those experienced in 1 AU interplanetary space. This is because the Martian atmosphere, while only 20 g/cm^2 thick in the vertical direction, actually provides the equivalent of 65 g/cm^2 of shielding when rays impacting the astronauts after an oblique transversal through the atmosphere are averaged in. In addition, the surface of Mars itself blocks out 50% of all cosmic rays. Finally, with a distance from the Sun averaging 1.52 AU, the dose from a solar flare experienced in near Mars space would be 43% of that at 1 AU.

Using the data presented in the cited references, we can estimate the "worst case" dose to the expedition crew on each alternative mission plan. We assume that a solar flare equal to the average of the three worst recorded cases (2/56, 11/60, and 8/72) occurs at a rate of 1 per year during solar max, and at a rate of 0.2 per year during solar min. We estimate the average distance from the Sun during a conjunction class orbit at 1.3 AU, and during an opposition class orbit as 1.2 AU. We assume that 25% of the time of each solar flare is spent unsheltered (5 gm/cm^2 of shielding), and 75% within shelter (35 g/cm^2 of shielding.) We assume that the crew of the spacecraft sleep in the storm shelter and thus spend 30% of their galactic cosmic ray (GCR) exposure time within shelter, and 70% out of shelter. Since on the surface of Mars, the entire habitat could be sheltered we assign 70% of the Mars surface GCR exposure to the sheltered category, and 30% to the unsheltered. The calculated doses for the three mission plans (opposition, Minimum energy, and Medium energy) are given in Table 5.

TABLE 5 . Radiation Dose Experienced on Mars Missions.

	Min. Energy	Medium Energy	Opposition
Transit Doses			
GCR (Solar Min)	63.0 rem	45.0 rem	67.5 rem
GCR (Solar Max)	25.9	18.5	27.8
Sol. Flare (S. Min)	2.5	1.8	3.2
Sol. Flare (S. Max)	12.7	9.1	16.0
Mars Doses			
GCR (Solar Min)	11.4 rem	14.3 rem	1.0 rem
GCR (Solar Max)	5.5	6.9	0.5
Sol. Flare (S. Min)	1.1	1.3	0.1
Sol. Flare (S. Max)	5.4	6.8	0.5
Total Dose			
Solar Min	78.0 rem	62.4 rem	71.8 rem
Solar Max	49.5	41.3	44.8

To place these doses in perspective, it may be noted that every 60 rem of radiation (received over an extended period, such as the doses given above) adds 1% of extra risk of a fatal cancer at some point later in life to a 35 year old woman, while 80 rem adds 1% of extra risk of fatal cancer to a 35 year old man.

It can be seen in Table 5. that the medium energy Mars Direct mission, with its slightly accelerated conjunction class orbits combined with a long surface stay, actually offers an average mission radiation dose somewhat less than the opposition class mission. As stated above, however, the freak chance of a single large catastrophic dose is much higher on the opposition class mission due to its close-in pass to the Sun. There is thus no radiation-dosage rationale for choosing the opposition mission plan over Mars Direct.

In summary, we find that neither zero-gravity nor radiation exposure concerns offer any countervailing advantages to offset the inefficiency, risk, complexity, and high cost of the opposition class mission. We therefore recommend that the opposition mission plan be dropped from the NASA baseline.

The other mission option that is sometimes identified as leading to a requirement for very high performing NTR is to fly the mission on a very fast (90 day transit) conjunction class trajectory. In this case, the total mission time is about 2.6 years, while the round trip transit time is reduced to 0.5 years. Such trajectories are superior to opposition trajectories in that the inner solar system is avoided and plenty of time is available at Mars for useful exploration. However the mass of such missions tend to be comparable to opposition class (two to three times medium energy) with all the complications identified above, as well as all the safety disadvantages associated with greater engine burn times and high energy planetary capture maneuvers. Because the transit time is less, the mission round trip radiation doses can be 20 to 30 rem less than that of the medium energy transfer. Such a dose, however, represents only a 0.25 to 0.5% risk of cancer, and is very small compared to all the other risks associated with the Mars mission. While it might be argued that any risk reduction is worthwhile, the increased risks associated with high energy trajectories, more complex on-orbit assembly (both risky itself as an activity, and a creator of mission risk due to difficulty of quality control), and the lesser degree of redundancy of critical systems forced by the much tighter constraints on high energy mission dry mass, dwarf the benefit of radiation risk reduction. Thus on grounds of safety, as well as cost, the high energy, very fast transit, conjunction mission is to be rejected.

Does Low Thrust Constrain the Available Mission Options?

It might be argued that while the low mass mission plans described above might be attractive, a situation might arrive in which mission planners were awash with money and therefore opted for a gargantuan mission instead. Wouldn't the small size of the PNTR preclude such a choice? The answer is that it would not.

Let us consider three cases: In case A we throw a 300-tonne payload on trans-Mars injection with a high energy C_3 of 30 km^2/s^2, as would be done in the piloted outbound leg of a split sprint-mission. In case B we throw a 400 tonne payload onto TMI with a medium energy C_3 of 15 km^2/s^2, as could be done in either a piloted or cargo mission. In case C we throw a 500-tonne payload onto TMI with a C_3 of 9 km^2/s^2, as would be done in a outbound cargo flight of a very large split-sprint mission. (We limit our analysis to trans-Mars injection since that is the portion of the mission where thrust will be critical. Thrust requirements for Mars orbital capture (MOC) are typically a factor of 6 less than those for TMI, due to Mars' 3/8 g, reduced spacecraft mass, and reduced ΔV of MOC compared to TMI.) In the big-bucks universe where all of this is occurring, we shall assume that customized tanks are available (10% mass fraction) in any size required.

We examine three cases. In the first instance the NTR propulsion is a pod of of 1 to 6 PNTR (44 kN, T/W=3, Isp=850 s) engines which are used in a series of perigee kicks until TMI is achieved. In the second case, the NTR unit is used to raise the payload from LEO to a 1-day elliptical orbit, after which the payload stages off onto TMI using pod of 7 augmented RL-10 engines with a thrust of 154 kN (35 klbf) each and an Isp of 465 s. In the third case, high thrust, 330 kN (75 klbf) NTR units with a specific impulse of 850 s and a T/W of 5 are used in a pod to achieve TMI with minimum mass after a series of perigee kicks.

TABLE 6. Very Large Manned Mars Missions Done with Small Engines

Propulsion Option	Case A 300 t C_3=30				Case B 400 t C_3=15				Case C 500 t C_3=9			
	Mass/	LH2	/#Kks/	hrs	Mass/	LH2	/#Kks/	hrs	Mass/	LH2	/Kks	/hrs
1 44 kN PNTR	715	376	24	213	785	348	27	471	899	361	31	639
2 44 kN PNTRs	667	331	12	259	745	311	13	135	877	340	15	129
4 44 kN PNTRs	623	288	6	85	738	302	7	145	863	324	8	120
6 44 kN PNTRs	621	284	4	23	723	286	5	196	874	332	5	22
1 44 kN PNTR /7 RL-10s	692	207	21	91	785	235	24	110	916	275	28	130
2 44 kN PNTRs/7 RL-10s	695	209	11	49	787	236	12	49	918	275	14	58
4 44 kN PNTRs/7 RL-10s	700	210	6	28	790	235	6	20	922	276	7	25
6 44 kN PNTRs/7 RL-10s	702	209	4	14	794	237	4	11	927	277	5	18
1 330 kN NTR	623	287	3	10	718	282	4	47	866	327	4	15
2 330 kN NTRs	611	270	2	7	727	285	2	5	874	328	2	4
3 330 kN NTRs	637	288	1	1	742	292	2	23	881	329	2	8

In Table 6, "Mass" is mission initial mass in LEO in tonnes, "LH2" is the amount of hydrogen required for trans-Mars injection in tonnes, "#Kks" is the number of perigee kicks required for TMI, and "hrs" is the number of hours between initiation of LEO departure and completion of the TMI maneuver. All perigee kicks are 60 degrees.

It can be seen that if grouped in pods of 4 or 6 engines, the 44 kN PNTR can accomplish even these very demanding missions with the same IMLEO as the 330 kN NTR. If podding engines is not allowed, the PNTR takes significant gravity losses, and these would increase if the TMI payloads were increased beyond the (already very large) levels analyzed here. However it can be seen that the IMLEO of the PNTR/RL-10 option is flat, regardless of the number of NTRs employed, and this would remain true regardless of how much the TMI payload were increased, or the NTR engine thrust dropped. The IMLEO of the PNTR/RL-10 options are only 6 to 11% more than that of the fully optimized 330 kN NTR options, and this modest mass disadvantage will be further reduced by the fact that the 44 kN engine will be better for the MOC and TEI burns, where the 330 kN engine is over-powered and over-weight.

The bottom line is that no matter how ambitious the manned Mars mission strategy, there is no need for an NTR engine with a thrust greater than 44 kN.

NON-SEI APPLICATIONS

There are many other potential applications for NTR propulsion besides the Space Exploration Initiative (SEI) manned Lunar and Mars missions. Such missions include DoD and civil missions to near Earth space (for example geosynchronous orbit, or GEO), and unmanned science missions to Mars and the outer solar system. A large number (hundreds) of such missions are currently planned for the next two decades, and more are certain to be planned for the years following, regardless of the fate of any one particular program. Because there are so many such missions it is very worthwhile to examine how effective an NTR engine is in accomplishing them. Indeed, in the long run, the performance of an engine on these missions may be much more relevant in quantifying its potential benefits to the nation's overall space program than its utility on SEI missions alone.

In Table 7, we show the benefits of NTR for 3 unmanned missions; TMI, GEO, and trans-Jupiter injection (TJI.) The importance of the GEO mission is self evident, as over half of all civil space missions and numerous military missions are flown to this destination. Mars is certain to be a continuing target of unmanned probes, regardless of the fate of SEI, and extensive planning is ongoing worldwide for such missions as Mars Rover, Mars Rover Sample Return, MESUR, and Mars Aeronomy Observer, among others, all of which could benefit by enhanced payload capabilities. Finally TJI is important since all missions to the outer solar system would use a gravity assist at Jupiter, and so the performance on this mission is generic to all exploration probes to Jupiter and beyond.

Trip time is not a driver on such unmanned missions, and so all missions analyzed here are flown on minimum energy trajectories. The initial mass in LEO of all missions is set at 20 tonnes, since this is the launch capability of both the Space Shuttle and the Titan IV.

TABLE 7. Payload Delivery Capability for Unmanned Missions with 20 Tonne IMLEO

| Mission | ΔV (km/s) | Cryo (465 s) | NTR 850 s | | NTR 950 s | |
			44 kN T/W=3	330 kN T/W=5	44 kN T/W=6	330 kN T/W=10
Trans-Mars	3.9	7.35 tonnes	10.27 tonnes	4.47 tonnes	11.72	9.07 tonnes
GEO	4.3	6.56	9.63	4.33	11.11	8.46
Trans-Jupiter	6.6	3.17	6.46	1.16	8.08	5.43

In Table 7. cryo stage and NTR tank dry mass fractions were taken as 0.1. It can be seen that the 330 kN/850 s NTR is inferior to cryogenic chemical propulsion on these missions, and while the 950 s/330 kN NTR is superior to chemical propulsion, it is inferior to the 44 kN/850 s NTR in every case, despite its vastly more advanced technology. The little 44 kN/850 s NTR significantly enhances all missions considered, increasing deliverable payload by 50 to 100% over the cryogenic chemical state of the art. The 950 s/44 kN NTR is certainly best, but whether the 14 to 25% improvement in capability it offers over the 850 s/44 kN model is worth the more difficult and risky development effort it entails is questionable.

CONCLUSIONS

We conclude that the rapid development of a modest thrust (44 kN) NTR engine with modest performance specifications is the optimal path for the nation's NTR development program. Such an engine can accomplish all SEI missions with only marginal mission performance disadvantage compared to NTR engines based on much more advanced and difficult technology and requiring more elaborate and expensive test facilities. The small engine is also capable of significantly enhancing a broad spectrum of unmanned missions for which large engines are inferior or useless. By developing such a small engine quickly, the benefits of NTR will be available much sooner in the SEI program, allowing major cost savings in the Lunar base build up, and precluding the possibility of NTR becoming a roadblock for the manned Mars mission. Such a small engine incorporating near-term technology can be developed at much quicker and at lower cost than a large high technology engine, thus reducing both cost and programmatic risk. In addition, by providing a quick route to a practical, flyable NTR, the rapid development of a small low technology engine will provide early NTR flight operations experience, and essential precondition to the prudent eventual development of larger, more advanced and more complex devices.

Acknowledgements

The author wishes to acknowledge many useful conversations with Jack Ramsthaler of the Idaho National Engineering Lab and Tal Sulmeisters of Martin Marietta on the subjects of the utility, testing requirements, and potential development schedule of small nuclear thermal rocket engines.

References

Buden, D.(1991) Private Communication, Idaho National Engineering Laboratory, Idaho Falls, ID, September 1991,

Cohen, A. et al, (1989) Report of the 90 Day Study on Human Exploration of the Moon and Mars, National Aeronautics and Space Administration Report, Washington, DC, November 1989.

Clark, B., and L. Mason, (1990) "The Radiation Show-stopper to Mars Missions: A Solution," presented to the AIAA Space Programs and Technologies Conference, Huntsville, AL, 28 September, 1990.

Letaw, J., R. Silberberg, and C. Tsao, (1987) "Radiation Hazards on Space Missions," Nature, 330(24):709-710.

Parsley, R., S. Peery, S. Earley, R. Zubrin, J. Ramsthaler, and J. Ivanenok, (1991) "A Low Thrust Near Term Nuclear Thermal Rocket Concept," AIAA 91-3352, AIAA/SAE/ASME 27th Joint Propulsion Conference, Sacramento, CA, June 1991.

Simonson, L., J. Nealy, L. Townsend, and J. Wilson, (1990) Radiation Exposure for Manned Mars Surface Missions, NASA TP-2979

Stafford, T. et al. (1991) America at the Threshold: Report of the Synthesis Group on America's Space Exploration Initiative, U.S. Government Printing Office, Washington DC.

Zubrin, R. and D. Baker, (1990) "Humans to Mars in 1999," Aerospace America, Vol. 28, No. 8, August 1990.

Zubrin, R., D. Baker, and O. Gwynne, "Mars Direct: A Simple, Robust, and Cost effective Architecture for the Space Exploration Initiative," AIAA 91-0326, 29th Aerospace Sciences Meeting, Reno, Nevada, 7-10 January, 1991.

MODELING THE FILM CONDENSATE FLUID DYNAMICS AND HEAT TRANSFER WITHIN THE BUBBLE MEMBRANE RADIATOR

Keith A. Pauley
Pacific Northwest Laboratory
MS: K6-47
P.O. Box 999
Richland, WA 99352
(509) 376-5779

John Thornborrow
NASA Johnson Space Center
MS: EC-2
NASA Loop Road 1
Houston, TX 77058
(713) 483-9130

Abstract

An analytical model of the fluid dynamics and heat transfer characteristics of the condensate within the rotating Bubble Membrane Radiator is developed. The steady-state, three-dimensional heat transfer and flow equations were reduced to a set of third-order ordinary differential equations by employing similarity transformation techniques. These equations are then solved for the radial, axial, and angular flow distributions in the film condensate. Pressure, temperature, heat transfer, film thickness and mass flow rate distributions are also calculated. The analytical model is the basis of the SCRABLE code which is used both as a zero-g design tool and a ground-test bed analyzer.

INTRODUCTION

The Bubble Membrane Radiator (BMR) is a member of a unique class of spacecraft thermal management devices which combine low-mass and low-launch volume with superior thermal performance and survivability. The BMR is an enclosed two-phase radiator which rotates about a central axis and consists of nine major components (See Figure 1): the attachment boom, rotation platform, central rotating shaft, central spray nozzle, thin film radiating surface, fluid collection troughs, pitot pumps, return piping and structure, and main feed/return lines. The common characteristic of this class of radiators, referred to as 'fabric radiators,' is the utilization of ultralite fabric (UF) composite materials.

The goals in the development of the BMR have been the reduction of cost-to-orbit, as well as increased systems efficiency and thermodynamic performance for space-based thermal management systems. To accomplish these criteria, a heat rejection system must transport significant quantities of thermal energy from the cycle working fluid and reject this energy directly to space. The BMR is a hybrid radiator design which incorporates the high surface heat fluxes of conventional heat pipes and the low system masses associated with liquid droplet radiators.

This paper documents work performed during the second phase of development of the BMR to predict the thermal hydraulic behavior of the condensate within the BMR under microgravity and 1 G conditions, for a variety of configurations. The specific phenomena of interest were condensate flow velocity, pressure gradients, temperature profiles, mass flow rates, condensation flow rates, condensate film thickness, heat transfer coefficients, and laminar-to-turbulent transition.

The Steady State Condensation with Rotational Acceleration Boundary Layer Examiner (SCRABLE) was written to determine the thermal hydraulic behavior of the BMR. The SCRABLE code analyzes both spherical and plate geometries with and without rotation.

The theory used in the SCRABLE code draws upon work in describing film condensation on spherical and rotating surfaces.

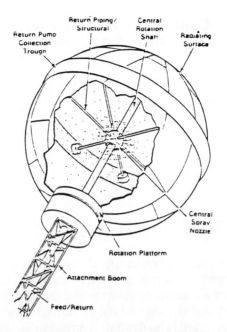

FIGURE 1. Spherical Bubble Membrane Radiator Design.

Film Condensation on a Non-Rotating Sphere

The heat and mass transfer phenomena associated with laminar film condensation on a spherical surface have been studied previously by Yang (1973). Yang developed a similarity transformation for the mass, momentum, and energy conservation equations as outlined below.

The coordinate system, shown in Figure 2, was selected to simplify the derivation of the conservation equations. The x-direction is arc length and the y-direction is the nominal radial distance from the surface. The origin is set at the upper flow stagnation point.

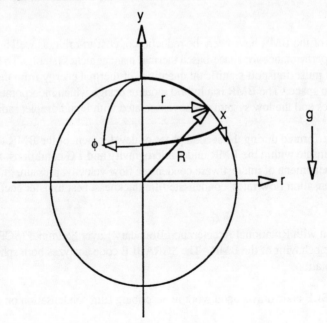

FIGURE 2.   Spherical Coordination System for Modeling of
             the Bubble Membrane Radiator.

The sphere is assumed to be immersed in an infinite body of pure vapor at its saturation temperature, called t_{sat}. The surface of the sphere is maintained at a uniform temperature, t_w. For laminar film condensation, a stream function representation using the Sparrow and Gregg (1959) similarity transformation of the steady, nondissipative, axisymmetric, constant-property flow is as follows:

$$f''' + 2ff'' - f'^2 + 1 = 0 \qquad \text{and} \qquad (1)$$

$$\Theta'' + 2Prf\Theta' = 0 \qquad (2)$$

The boundary conditions for these equations are:

$$
\begin{array}{ll}
\underline{\text{at } \eta = 0} & \underline{\text{at } \eta = 1} \qquad\qquad (3)\\
f = 0 & f'' = 0 \\
f' = 0 & \Theta = 0 \\
\Theta = 1 &
\end{array}
$$

The similarity transformation uses the dimensionless variable η, defined as

$$\eta = y\, C^{1/4}\, G/R, \qquad (4)$$

to derive the velocity function f and the dimensionless temperature Θ. The transformation equations for f and Θ are

$$\psi = \nu\, C^{1/4}\, Hf(\eta) \qquad \text{and} \qquad (5)$$
$$\Theta = (t-t_{sat})/(t_w-t_{sat}) \qquad\qquad\quad (6)$$

Film Condensation on a Rotating Disk

By creating an artificial body-force in the presence of film condensation, the heat transfer rate from the vapor to the condensing surface may be increased. Sparrow and Gregg (1959) explored the analysis of the flow and thermal behavior of laminar-film condensation on a horizontal plate. The system under consideration, shown in Figure 3, is a cooled rotating disk at a uniform temperature, t_w, located in an infinite body of pure saturated vapor. A film is assumed to form on the disk surface and will metriculate radially outward under the influence of the centrifugal force field.

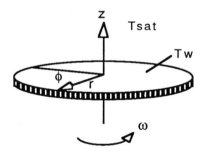

FIGURE 3. Rotational Coordinate System for a Flat Plate Geometry.

With some insight from Von Karman (Schlichting 1955), a similarity transformation may be made using the following new variables:

$$\eta = \left(\frac{\omega}{\nu}\right)^{1/2} z \qquad (7)$$

$$F(\eta) = \frac{V_r}{r\omega} \qquad (8)$$

$$H(\eta) = \frac{V_z}{(\omega \nu)^{1/2}} \qquad (9)$$

$$P(\eta) = \frac{p}{\mu\omega} \qquad (10)$$

and

$$\Theta(\eta) = \frac{T_{sat}-T}{T_{sat}-T_w} \qquad (11)$$

Two assumptions are made concerning directional dependence. First, there is angular symmetry such that $\partial/\partial f = 0$. Second, the shape of the velocity profiles in the r and f directions is constant in the radial direction. This implies simple stretching of a fundamental shaping function at various values of r. This yields:

$$H''' = HH'' - (H')^2/2 + 2G^2 \qquad \text{and} \qquad (12)$$

$$G'' = HG' - H'G \qquad (13)$$

The transformed boundary conditions are

at h = 0   at h = h$_\delta$

H = 0   H" = 0
H' = 0   G' = 0   (14)
G = 1 Θ = 0
Θ = 1

Rotational Superposition

To model condensation of the rotating sphere, the profiles of the spherical non-rotating and rotating flat plate boundary layer velocities were superpositioned (see Figure 4). The relationship between the x and r coordinates in Figure 4 is found to be

$$r = R \sin(x/R) \qquad (15)$$

This yields an effective radius for use in the boundary conditions for the rotational flat plate solution. Both the flat plate and non-rotating spherical models are solved and the velocity and temperature profiles are combined by superposition at each value of x.

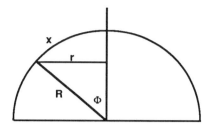

FIGURE 4. Superposition coordinate system.

OPERATION OF THE SCRABLE CODE

A typical set of results from the SCRABLE code are presented for the conditions found in Table 1.

TABLE 1.   Sample Conditions input to SCRABLE Code

Input Parameter	Parametric Values
Working Fluid	Water
Inlet Vapor Temperature	600.0 K
Heat Rejection Capability	10.0 kW
Surface Emissivity	0.90
View Factor	1.00
Space Environment Temperature	2.00 K
Radiator Radius	1.91 m
Rotational Speed	10.00 rpm

Figures 5 and 6 illustrate the predicted flow distributions over the condensate film thickness at two points on the radius. The dimensionless pressure and temperature profiles over the film thickness are shown in Figure 7.

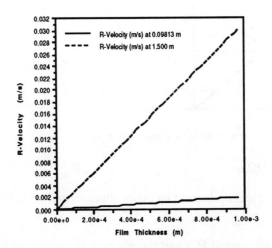

FIGURE 5. R-directional Velocity Component over the Film Thickness at Two Radial Locations.

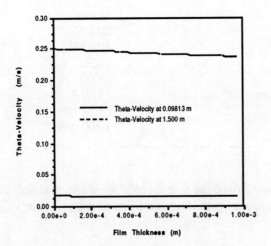

FIGURE 6. Theta-direction Velocity Component over the Film Thickness at Two Radial Locations.

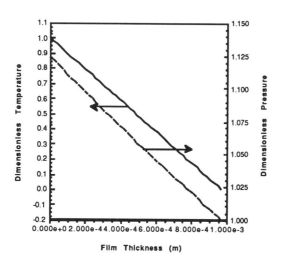

FIGURE 7. Dimensionless Temperature and Pressure over the Film Thickness.

The mean heat transfer characteristics of the film condensate are explained in Table 2. These values represent the average properties over the effective arc length from the stagnation point at the apex to the equator of rotation.

TABLE 2.   Heat Transfer Characteristics of Film Condensate from the Sample Model

Output Parameter	Parametric Values
Film Thickness	0.000969 m
Nusselt Number	0.0014
Heat Transfer Coefficient	7140.238 W/m^2-K
Required Mass Flow Rate	0.007814 kg/s
Limiting Mass Flow Rate	0.716204 kg/s
Limiting Rejection Power	916.5808 kW

Due to the assumption made in the derivation of the governing equations, results from the SCRABLE code are valid only for steady, nondissipative laminar flow. Transition to turbulence will increase the heat transfer characteristics of the flow and further improve the capabilities of the BMR design. However, further modelling will be necessary for this condition.

The Pacific Northwest Laboratory is working with Oregon State University in analyzing the BMR Ground Test Bed and validating the performance of the SCRABLE code.

Acknowledgments

Support for this work from the NASA Johnson Space Center through DE-ACO6-76RLO 1830 is gratefully acknowledged.

References

Schlichting, H. (1955) <u>Boundary Layer Theory</u>. McGraw-Hill Book Co., New York, NY.

Sparrow, E.M. and J. L. Gregg, (1959) "A Theory of Rotating Condensation," <u>Journal of Heat Transfer</u>. Vol. 113-120.

Yang, J.W. (1973) "Laminar Film Condensation on a Sphere." <u>Journal of Heat Transfer</u>. Vol.   174-178.

-- **Nomenclature** --

<u>English</u>

x: arc length in Fig. 2

y: outward normal direction in Fig. 2

r: distance from axis of rotation in Fig. 2.

R: distance from center of sphere in Fig. 2

g: gravitational acceleration

C: flow variable = $g(\rho - \rho_v) R^3/\rho v^2$

f: dependent dimensionless flow variable in Eq. 1.

t_w: wall temperature

t_{sat}:   saturation temperature

G: r- direction flow variable for plate development

H: ϕ- direction flow variable for plate development

F: z- direction flow variable for plate development

J: x- direction flow variable for spherical development

I: y -direction flow variable for spherical development

P: dimensionless pressure

<u>Greek</u>

η: Dimensionless independent variable defined in Eq. 4 and 7.

ψ: stream function defined in Eq. 5.

θ: dimensionless temperature

ρ: density of condensate

ρ_v: density of vapor

ϕ: angular coordinate

ω: rotational velocity

NUMERICAL PREDICTION OF AN AXISYMMETRIC TURBULENT MIXING LAYER USING TWO TURBULENCE MODELS

Richard W. Johnson
Idaho National Engineering Laboratory
EG&G Idaho, Inc.
P.O. Box 1625
Idaho Falls, ID 83415-2414
(208) 526-0955
FTS 583-0955

Abstract

Nuclear power, once considered and then rejected (in the U. S.) for application to space vehicle propulsion, is being reconsidered for powering space rockets, especially for interplanetary travel. The gas core reactor, a high risk, high payoff nuclear engine concept, is one that was considered in the 1960s and 70s. As envisioned then, the gas core reactor would consist of a heavy, slow moving core of fissioning uranium vapor surrounded by a fast moving outer stream of hydrogen propellant. Satisfactory operation of such a configuration would require stable nuclear reaction kinetics to occur simultaneously with a stable, coflowing, probably turbulent fluid system having a dense inner stream and a light outer stream. The present study examines the behavior of two turbulence models in numerically simulating an idealized version of the above coflowing fluid system. The two models are the standard $k\sim\varepsilon$ model and a thin shear algebraic stress model (ASM). The idealized flow system can be described as an axisymmetric mixing layer of constant density. Predictions for the radial distribution of the mean streamwise velocity and shear stress for several axial stations are compared with experiment. Results for the $k\sim\varepsilon$ predictions are broadly satisfactory while those for the ASM are distinctly poorer.

INTRODUCTION

One of the nuclear propulsion concepts for space vehicles that was investigated in the 1960s and early 1970s was the high potential gas core reactor. With its extremely high operating temperatures, the gas core reactor was to combine the dual attractions of theoretically high specific impulse and large hydrogen propellant flow rates leading to high thrust levels. With the current resurgence of interest in nuclear propulsion, a fresh look is being taken at the gas core reactor concept.

One of the configurations investigated in early times for the implementation of the gas core engine was the open cycle coaxial flow design wherein the light hydrogen propellant would flow annularly at a high rate past an inner core of slow moving, fissioning uranium vapor. The heavy uranium core would heat the hydrogen through convection and radiation heat transport. A number of experimental investigations were pursued during the 1960s and 70s to determine and investigate the critical operating parameters of the coaxial gas core configuration. Two of the critical aspects of the design concept are the hydrodynamic stability and the stability of the nuclear kinetics. That is, can sustained nuclear fission be achieved while hydrodynamically confining the uranium gas such that the heavy gas leaves the reactor vessel at a very low rate?

One of the early experimental investigations, undertaken to investigate turbulent mixing of coaxial streams, examined the dynamics of a slow inner stream with a faster outer stream surrounding it (Zawacki and Weinstein 1968). Data were taken for various ratios of mean outer bulk velocity to mean inner bulk velocity and for homogeneous and heterogeneous combinations of fluids. Because the boundary layer of the outer stream had time to transition to turbulence and because some of the inner streams were also turbulent, the coaxial flow can be described as an axisymmetric turbulent mixing layer.

The objectives of the present study are to simulate one of the homogeneous turbulent axisymmetric mixing layer flows of Zawacki and Weinstein (1968) using two turbulence models to determine the suitability of each for this flow and to assess the need for further model and experiment development. The two turbulence models are the standard $k\sim\varepsilon$ model and a thin shear algebraic stress model (ASM). Figure 1 provides a schematic of the flow geometry and conditions for the present axisymmetric mixing layer.

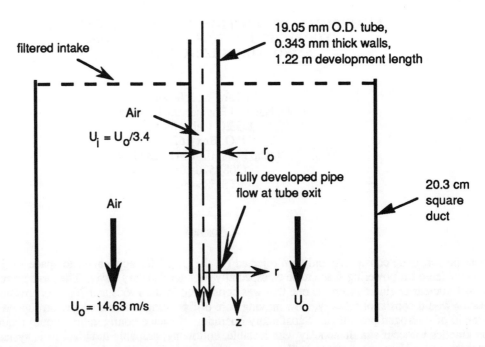

FIGURE 1.   Schematic of the Geometry and Conditions for the Axisymmetric Mixing Layer of the Present Study. U_O and U_i are Bulk Mean Velocities.

An early numerical simulation of the data of Zawacki and Weinstein (1968) for the same case investigated herein was reported by Zelazny et al (1973). They derived a highly empirical, zero equation eddy viscosity model for the turbulent shear stress based on data for axisymmetric jets, wakes and coflowing streams. They initiated their calculations downstream of the initial point of mixing where the data are complete enough to apply their model. Their calculations for two additional downstream stations show quite good agreement. However, because the turbulence model is so highly empirical, it would seem imprudent to apply the model to flows much different than the simple flows on which it is based.

A number of studies have been performed for the related flows of planar mixing layers and wakes. The level of turbulence models employed for simulating such flows ranges from simple mixing length models to sophisticated second moment closures. The study of Launder et al (1973) compares the results for a mixing length model and one and two equation turbulence models for the simulation of a plane mixing layer. Predictions for the transverse distributions of the mean streamwise velocity and the turbulent shear stresses are compared with experiment at two streamwise stations. The results for the $k\sim\varepsilon$ model are found to be very good for the mean velocity, while the other models show excessive diffusion. However, the $k\sim\varepsilon$ results for the shear stress, found to be good for the earlier streamwise station, are off by more than 100% for the station farther downstream. It is concluded by Launder et al that the shear stress data are suspect.

The conference report by Rodi (1986) describes a large number of turbulence modeling studies presented at the EUROMECH Colloquium 180 held in Karlsruhe, FRG, July 4-6, 1984. Summaries of studies relating to mixing layers and wakes by Radespiel, Schiestel, and Jones are presented therein. Radespiel employs a three equation $k\sim\varepsilon\sim\gamma$ turbulence model to simulate a plane mixing layer; γ, the intermittency factor, is modeled with a transport equation similar to those for k and ε. Schiestel uses a multiscale algebraic stress model (ASM) where the energy spectrum is divided into two ranges with equations for turbulence quantities provided for each range. He gives results for a plane wake. Jones reports computations for a plane mixing layer with a full second moment closure. In each of these three cases, the spreading rates for the mixing layer or wake are underpredicted, although improvements over results for the standard $k\sim\varepsilon$ model are reported for the first two studies.

Biringen (1978) reports results for simulating an axisymmetric wake flow for a body of revolution using a three equation model of turbulence. The three turbulence parameters for which transport equations are solved are the turbulent kinetic energy and shear stress, and an integral length scale. Results for one axial station are presented for the wake flow. Agreement for the radial distribution of the mean streamwise velocity is shown to be quite good, although the distributions for turbulent kinetic energy and shear stress are less satisfactory.

Further examples of second moment closures applied to mixing layers and wakes are provided by Launder et al (1975) and Lewellen et al (1974). Both report good results for mean velocities and Reynolds stresses, although complete agreement is not achieved.

It is further reported, for instance in Rodi (1986), that spreading rates for round jets are overpredicted while those for planar wakes are underpredicted, using the same turbulence model. It therefore seems prudent to test models for axisymmetric geometry even if they have been found to work well for planar flows. The present study examines the numerical simulations for the axisymmetric mixing layer (with thick boundary layers at the beginning of the mixing region) for the two equation $k\sim\varepsilon$ model and the anisotropic thin shear ASM.

The describing equations for two dimensional, incompressible, turbulent flow and the two turbulence models used for closure for the present study are detailed in the following section.

DESCRIBING EQUATIONS

The Mean Fluid Flow Equations

The Reynolds averaged continuity and Navier-Stokes equations for two dimensions are employed for the present incompressible, quasi-steady, isothermal, axisymmetric problem of coflowing streams. The mean velocities are represented by U and V in the x- and r-directions, respectively. The equations in cylindrical coordinates are

Continuity

$$\frac{\partial U}{\partial x} + \frac{\partial V}{\partial r} + \frac{V}{r} = 0 , \tag{1}$$

Momentum

$$U\frac{\partial U}{\partial x} + V\frac{\partial U}{\partial r} = -\frac{1}{\rho}\frac{\partial P}{\partial x} + \frac{\partial}{\partial x}\left[2v\frac{\partial U}{\partial x} - \overline{u^2}\right] + \frac{1}{r}\frac{\partial}{\partial r}r\left[v\left(\frac{\partial U}{\partial r} + \frac{\partial V}{\partial x}\right) - \overline{uv}\right] , \text{ and} \tag{2}$$

$$U\frac{\partial V}{\partial x} + V\frac{\partial V}{\partial r} = -\frac{1}{\rho}\frac{\partial P}{\partial r} + \frac{\partial}{\partial x}\left[v\left(\frac{\partial U}{\partial r} + \frac{\partial V}{\partial x}\right) - \overline{uv}\right] + \frac{1}{r}\frac{\partial}{\partial r}r\left[2v\frac{\partial V}{\partial r} - \overline{v^2}\right] - 2v\frac{V}{r^2} . \tag{3}$$

The quantities $\overline{u^2}$, $\overline{v^2}$, and $\overline{uv}$ are Reynolds stresses. The kinematic viscosity is represented by v. The Reynolds stresses must be modeled in order to close the system. The two turbulence models employed for closure in the present work are detailed in the following section.

Turbulence Models

The first turbulence model used is the standard $k\sim\varepsilon$ model (Jones and Launder 1972 and Launder and Spalding 1974). The $k\sim\varepsilon$ model employs the following expressions for the Reynolds stresses:

$$-\overline{u^2} = 2v_t\frac{\partial U}{\partial x} - \frac{2}{3}k , \qquad -\overline{v^2} = 2v_t\frac{\partial V}{\partial r} - \frac{2}{3}k , and \qquad -\overline{uv} = v_t\left(\frac{\partial U}{\partial r} + \frac{\partial V}{\partial x}\right) . \tag{4}$$

The kinematic turbulent viscosity v_t is modeled as $v_t = C_\mu k^2/\varepsilon$ where $k \equiv 0.5\left(\overline{u^2} + \overline{v^2} + \overline{w^2}\right)$ is the turbulent kinetic energy and ε is its rate of dissipation. The turbulence quantities k and ε are determined from modeled transport equations given in cylindrical coordinates as

$$U\frac{\partial k}{\partial x} + V\frac{\partial k}{\partial r} = \frac{\partial}{\partial x}\left(\frac{v + v_t}{\sigma_k}\right)\frac{\partial k}{\partial x} + \frac{1}{r}\frac{\partial}{\partial r}r\left(\frac{v + v_t}{\sigma_k}\right)\frac{\partial k}{\partial r} + P - \varepsilon \text{ , and} \tag{5}$$

$$U\frac{\partial \varepsilon}{\partial x} + V\frac{\partial \varepsilon}{\partial r} = \frac{\partial}{\partial x}\left(\frac{v + v_t}{\sigma_\varepsilon}\right)\frac{\partial \varepsilon}{\partial x} + \frac{1}{r}\frac{\partial}{\partial r}r\left(\frac{v + v_t}{\sigma_\varepsilon}\right)\frac{\partial \varepsilon}{\partial r} + C_{\varepsilon 1}\frac{\varepsilon}{k}P - C_{\varepsilon 2}\frac{\varepsilon^2}{k}, \tag{6}$$

where the production of turbulent kinetic energy P in cylindrical coordinates is given by

$$P = v_t\left[2\left(\frac{\partial U}{\partial x}\right)^2 + 2\left(\frac{\partial V}{\partial r}\right)^2 + 2\left(\frac{\partial U}{\partial r}\frac{\partial V}{\partial x}\right) + \left(\frac{\partial U}{\partial r}\right)^2 + \left(\frac{\partial V}{\partial x}\right)^2 + 2\frac{V^2}{r^2}\right]. \tag{7}$$

The modeling coefficients are given the usual optimized values, $C_\mu = 0.09$, $\sigma_k = 1.0$, $\sigma_\varepsilon = 1.22$, $C_{\varepsilon 1} = 1.44$, and $C_{\varepsilon 2} = 1.92$ (Launder and Spalding 1974 and Johnson 1984).

Inlet values for U, k, ε, and v_t were obtained from numerical simulations of pipe and annular flows. The wall treatment used for these precursory simulations was the standard wall function approach detailed in Launder and Spalding (1974). The inlet conditions for the annular part of the inlet plane were obtained by matching annular flow computations with the datapoint which lies in the annular boundary layer at $z/r_o = 0$ given in Zawacki and Weinstein (1968). The inlet conditions for the pipe portion of the flow are fully developed turbulent pipe flow. For the far radial boundary condition, both free and no slip prescriptions are used as will be discussed later. The standard wall functions are used in the coflowing domain at the outer boundary for the no slip case. The flow data are specified as having zero gradient at the outlet of the computational domain.

The second turbulence model employed in the present study is a thin shear form of the algebraic stress model (ASM) of Rodi (1972). The model employed by the ASM for the Reynolds stresses is given as

$$\overline{u_i u_j} = \frac{k(1 - C_2)}{\varepsilon(C_1 - 1 + P/\varepsilon)}\left(P_{ij} - \frac{2}{3}\delta_{ij}P\right) + \frac{2}{3}\delta_{ij}k. \tag{8}$$

P_{ij} and P are the production tensor for the Reynolds stresses and the production of k, respectively. For the present thin shear flow version of the ASM, the production P is given as

$$P = -\left[\overline{uv}\frac{\partial U}{\partial r} + \overline{w^2}\frac{V}{r}\right]. \tag{9}$$

The four components of the production tensor that are required for the present ASM are approximated as follows:

$$P_{12} = -\overline{v^2}\frac{\partial U}{\partial r}, \quad P_{11} = -2\overline{uv}\frac{\partial U}{\partial r} \approx 2P, \text{ and } P_{22} = P_{33} = 0. \tag{10}$$

Employing these relations along with Equation (8), the individual Reynolds stresses are modeled as:

$$-\overline{uv} = \frac{2}{3}\frac{k^2}{\varepsilon}(1 - C_2)\frac{(C_1 - 1 + C_2 P/\varepsilon)}{(C_1 - 1 + P/\varepsilon)^2}\frac{\partial U}{\partial r}, \tag{11}$$

$$-\overline{v^2} = -\overline{w^2} = -\frac{2}{3}k\frac{(C_1 - 1 + C_2 P/\varepsilon)}{(C_1 - 1 + P/\varepsilon)}, \text{ and} \tag{12}$$

$$-\overline{u^2} = -\frac{2}{3}k\frac{(C_1 - 1 + 1.5(2 - C_2)P/\varepsilon)}{(C_1 - 1 + P/\varepsilon)}. \tag{13}$$

Transport equations are again required for k and ε. The transport equations used for the present thin shear ASM for k and ε are

$$U\frac{\partial k}{\partial x} + V\frac{\partial k}{\partial r} = \frac{\partial}{\partial x}\left(\nu + C_{k1}\frac{k}{\varepsilon}\overline{u^2}\right)\frac{\partial k}{\partial x} + \frac{1}{r}\frac{\partial}{\partial r}\,r\left(\nu + C_{k1}\frac{k}{\varepsilon}\overline{v^2}\right)\frac{\partial k}{\partial r} + P - \varepsilon, \text{ and} \tag{14}$$

$$U\frac{\partial \varepsilon}{\partial x} + V\frac{\partial \varepsilon}{\partial r} = \frac{\partial}{\partial x}\left(\nu + C_{\varepsilon}\frac{k}{\varepsilon}\overline{u^2}\right)\frac{\partial \varepsilon}{\partial x} + \frac{1}{r}\frac{\partial}{\partial r}\left(\nu + C_{\varepsilon}\frac{k}{\varepsilon}\overline{v^2}\right)\frac{\partial \varepsilon}{\partial r} + C_{\varepsilon1}\frac{\varepsilon}{k}P - C_{\varepsilon2}\frac{\varepsilon^2}{k}, \tag{15}$$

where P is given by Equation (9). The model coefficients for the ASM are $C_1 = 1.8$, $C_2 = 0.6$ (Gibson and Launder 1978), $C_{k1} = 0.22$ (Launder and Morse 1979), $C_\mu = 0.09$, $C_{\varepsilon1} = 1.44$, $C_{\varepsilon2} = 1.92$ (Launder et al 1973), and $C_\varepsilon = 0.17$ (Launder 1980).

The inlet conditions used for the ASM computations are computed from precursory calculations similar to the $k\sim\varepsilon$ case except that the ASM is used. The standard wall corrections for the ASM are employed for these calculations (Johnson 1984). The same boundary conditions are used for the ASM simulations as for the $k\sim\varepsilon$ model predictions.

COMPUTING DETAILS

The above describing equations are solved using a steady, implicit finite volume code based on the staggered mesh SIMPLE algorithm of Patankar and Spalding (1972). The code belongs to the TEACH family (Gosman and Ideriah 1976) of finite volume codes. The iterative procedure of SIMPLE is executed until totaled mass and momentum residuals are each less than $1.0 \times 10^{-5}\%$. The grid used for free slip calculations is 60×52 cells while that for the no slip calculations is 66×81. The cells expand in length in the axial direction and radially beyond the inner tube location toward the outer radius. The cross-sectional area of the cylindrical computational domain is configured to have the same perimetral length as the square-sectioned test section of the experimental apparatus of Zawacki and Weinstein (1968). The mesh is especially fine near the inner and outer walls of the inner tube. The HYBRID scheme (upwinding for cell Reynolds number $\text{Re}_\Delta > 2$, centered differencing for $\text{Re}_\Delta < 2$) is used in the calculations. Typical run time for a simulation is 250 seconds on the INEL CRAY XMP/216.

Validation computations were made to compare $k\sim\varepsilon$ results for the same plane mixing layer computed by Launder et al (1973). Computed profiles of mean velocity and shear stress (not shown) were found to be very close to the calculations of Launder et al (1973).

RESULTS AND DISCUSSION

Experimental data are provided by Zawacki and Weinstein (1968) for several different ratios of outer to inner bulk mean velocities for the axisymmetric coflowing stream problem. While the same velocity is used for the outer stream in all cases, the bulk velocity of the inner stream is varied to obtain the different velocity ratios. Of these several experiments, only two involve inner streams that are fully turbulent: $U_0/U_i = 1.0$ and 3.4. To meet the objective of simulating an axisymmetric turbulent mixing layer, the velocity ratio of $U_0/U_i = 3.4$ was chosen. For this ratio, the Reynolds number of the inner stream is $\text{Re} = U_i D/\nu = 5040$.

The experimental data for the radial variation of the mean streamwise velocity are given in Zawacki and Weinstein (1968) for several axial stations along the axisymmetric mixing layer. Unfortunately, the inlet data are incomplete because only mean velocities are given. As mentioned earlier, the inlet data for the turbulence quantities are obtained from simulations made earlier for pipe and annular flows. Figure 2 presents results for the $k\sim\varepsilon$ model, using a free slip outer boundary, plotted against the data for the radial variation of mean streamwise velocity for various axial locations. The mean velocity is normalized by the bulk mean outer velocity $U_0 = 14.6304$ m/sec (48 ft/sec) and the axial location is nondimensionalized using the inner radius of the inner tube $r_0 = 9.1821$ mm (0.3615 in). The predictions, though not labeled to avoid clutter, sequentially follow the data as z/r_0 increases.

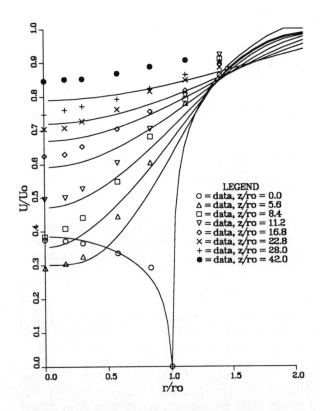

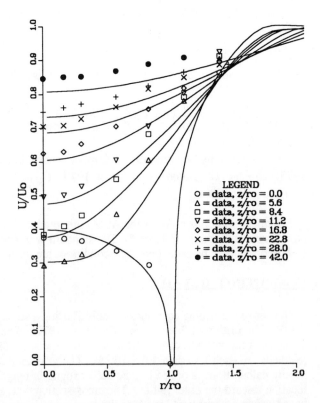

FIGURE 2. Mean Axial Velocity; —— $k{\sim}\varepsilon$ Predictions (Free Slip) ; Symbols, Data of Zawacki and Weinstein (1968).

FIGURE 3. Mean Axial Velocity; —— $k{\sim}\varepsilon$ Predictions (No Slip) ; Symbols, Data of Zawacki and Weinstein (1968).

As shown in Figure 2, the inlet data for the mean velocity are quite well predicted. The acceleration of the centerline velocity is somewhat underpredicted, but not more than by about 8%. This is quite reasonable as it can be expected that experimental uncertainty may reach 15%. The radial profiles at the various locations are also underpredicted, the largest discrepancy being about 13%.

Because the data are virtually all underpredicted beyond the inlet plane, it may be surmised that the overall mass flow rate is off. The growing boundary layer at the outer radial boundary would tend to accelerate the interior fluid which could account for some of the discrepancy. To estimate this effect, the cylindrical domain is extended to a radius such that the perimeter matches that of the experimental working section. Hence, the growing outer boundary layer causes about the same mass flow deficit as occurred in the experiment. A no slip condition is used to allow the boundary layer growth. Figure 3 illustrates the results of accounting for the outer boundary layer. It should be noted that the radial mesh spacing has been refined by an additional 50%. The results shown in Figure 3 indicate that some of the discrepancy of Figure 2 is indeed the result of an acceleration of the interior flow due to a growing boundary layer. The bulk of this effect is assumed to be accounted for accurately as the outer boundary layer is predicted to grow from about 6.5 mm (1/4 in.) to about 19 mm (3/4 in.), the same values reported by Zawacki and Weinstein (1968). The maximum underprediction for Figure 3, however, is still 16%. The above radial refinement of the grid as well as additional runs where the axial grid spacing was refined by a factor of three indicate that the results for the finer 66 x 81 grid are virtually grid independent.

Predictions for the thin shear ASM are given in Figure 4. The finer grid, used for the results of Figure 3, is used again. Inlet conditions are provided from thin shear ASM calculations of pipe and annular flows. The mean streamwise centerline velocity accelerates faster for the ASM than for the $k{\sim}\varepsilon$ model. However, the underprediction of the mean velocity in the developing mixing layer is somewhat greater for the ASM than for $k{\sim}\varepsilon$. One has to

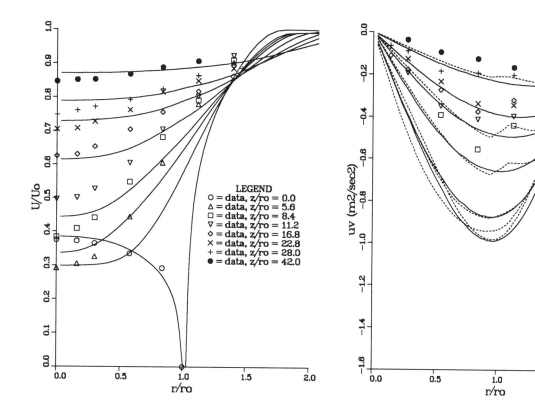

FIGURE 4.   Mean Axial Velocity; ——— ASM
Predictions (No Slip) ; Symbols, Data
of Zawacki and Weinstein (1968).

FIGURE 5.   Turbulent Shear Stress; ——— k~ε,
- - - - ASM Predictions; Symbols,
Data of Zawacki and Weinstein (1968).

conclude that the standard k~ε results are better than those for the ASM because the former are quantitatively closer to the data for the earlier stations (z/r_0 = 5.6, 8.4, and 11.2) (being qualitatively similar) and are qualitatively better for the outer stations. (Although the ASM results are closer to the data for $z/r_0 \geq 22.8$, the mean gradient $\partial U/\partial r$ is shallower than for the data or the k~ε results.)

The behavior of the ASM can be explained in part by comparing the two forms for the turbulent shear stress, Equations (4) and (11). While both formulae have an explicit dependence on the factor k^2/ε, the ASM formula also contains an explicit dependence on the factor P/ε. When the production is high, as it is for low z/r_0, the net effect is for the 'turbulent viscosity' (the coefficient of $\partial U/\partial r$) to diminish, thereby reducing the mixing effect. For the far downstream region, however, where production of turbulent kinetic energy has decreased substantially, the net effect of the factor P/ε is to increase the viscosity, enhancing the turbulent mixing and promoting the development of the core region. Unfortunately, these effects, which are due to the nature of the ASM, are shown to be undesirable for the present flow.

Figure 5 compares predictions for the turbulent shear stress for the two models with the data for $z/r_0 \geq 8.4$. The calculations show magnitudes of roughly a factor of two greater than the data! This might lead one to conclude that the turbulent mixing rate is drastically overpredicted. However, additional computations have been made for the k~ε model beginning at z/r_0 = 8.4 where shear stress data are reported (the same approach used by Zelazny et al 1973). The results for the mean streamwise velocities and shear stresses are plotted in Figures 6 and 7. While the values calculated for the shear stresses are clearly much closer to the reported data than shown in Figure 5, the results for the mean velocities are clearly poorer as the centerline velocity is more slowly accelerated than indicated for the k~ε results (Figure 2). Of course, the mean velocities depend on the gradients of shear stress, not its magnitude. Inspection of the shear stress predictions in Figure 5 shows that the radial gradient of the shear stress is represented

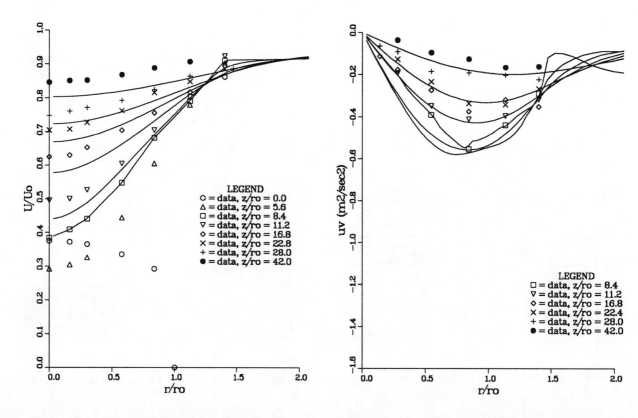

FIGURE 6.   Mean Axial Velocity, Inlet at $z/r_0 =$ 8.4 (No Slip); —— $k{\sim}\varepsilon$; Symbols, Data of Zawacki and Weinstein (1968).

FIGURE 7.   Turbulent Shear Stress, Inlet at $z/r_0 =$ 8.4 (No Slip); —— $k{\sim}\varepsilon$; Symbols, Data of Zawacki and Weinstein (1968).

fairly well. One concludes that the $k{\sim}\varepsilon$ model turns in a respectable performance for this constant density case, and that it should be useful as a modeling tool for similar flows. Further testing of the $k{\sim}\varepsilon$ model is required, however, for the mixing of fluids of much different densities and at much higher temperatures to be able to assess its value for use as a modeling tool in connection with the gas core reactor. However, it would be desirable to obtain new data for the turbulent axisymmetric mixing layer, documenting completely the initial mixing plane.

CONCLUSIONS

Numerical simulations are obtained for each of two turbulence models ($k{\sim}\varepsilon$ and a thin shear ASM) for the incompressible, quasi-steady, turbulent, axisymmetric mixing layer. The conclusions of the present study can be summarized as follows: (1) the $k{\sim}\varepsilon$ model simulations are broadly satisfactory, (2) the thin shear ASM simulations are generally inferior to the $k{\sim}\varepsilon$ predictions, (3) a significant portion of the errors in the predictions is likely due to inaccurate inlet conditions as the experimental inlet data are incomplete, (4) the available data are suspect because fairly good results are obtained for the mean velocities while results for the turbulent shear stresses differ greatly from experiment, and (5) additional data wherein initial mixing plane information is completely documented are desirable.

Acknowledgments

This work was performed at the Idaho National Engineering Laboratory, Idaho Falls, ID, under the auspices of the U.S. Dept. of Energy (DOE), Idaho Operations Office, under DOE Contract No. DE-AC07-76ID01570.

References

Biringen, S. (1978) "Calculation of Axisymmetric Jets and Wakes with a Three-Equation Model of Turbulence," *J. Fluid Mechanics*, 86(4): 745-759.

Gibson, M. M. and Launder, B. E. (1978) "Ground Effects on Pressure Fluctuations in the Atmospheric Boundary Layer," *J. Fluid Mechanics*, 86(3): 491-511.

Gosman, A. D. and Ideriah, F. J. K. (1976) *TEACH-2E: A General Computer Program for Two-Dimensional Turbulent, Recirculatory Flows*, Dept. Mech. Eng., Imperial College, London.

Johnson, R. W. (1984) "Turbulent Convecting Flow in a Square Duct with a 180° Bend; An Experimental and Numerical Study," Ph. D. Thesis, Faculty of Technology, University of Manchester, England.

Jones, W. P. and Launder, B. E. (1972) "The Prediction of Laminarization with a 2-Equation Model of Turbulence," *Int. j. Heat Mass Transfer*, (15): 301-314.

Launder, B. E. (1980) *Turbulence Transport Models for Numerical Computation of Fluid Flow*, course ME 213, Dept. Mech. Eng. University of California, Davis, CA.

Launder, B. E. and Morse, A. P. (1979) "Numerical Prediction of Axisymmetric Free Shear Flows with a Reynolds Stress Closure," *Turbulent Shear Flows - 1*, Springer-Verlag, Berlin, p. 279.

Launder, B. E., Morse, A. P., Spalding, D. B. and Rodi, W. (1973) "Prediction of Free Shear Flows - A Comparison of the Performance of Six Turbulence Models," *Proceedings of the 1972 Langley Free Shear Flows Conference*, NASA SP321, held in Hampton, VA, July 20-21, 1972, pp. 361-425.

Launder, B. E., Reece, G. J. and Rodi, W. (1975) "Progress in the development of a Reynolds-Stress Turbulence Closure," *J. Fluid Mechanics*, 68(3): 537-566.

Launder, B. E. and Spalding, D. B. (1974) "The Numerical Computation of Turbulent Flows," *Comp. Meth. Applied Mech. Eng.*, 3: 269-289.

Lewellen, W. S. , Teske, M. and Donaldson, C. duP. (1974) "Application of Turbulence Model Equations to Axisymmetric Wakes," *AIAA J.*, 12(5): 620-625.

Patankar, S. V. and Spalding, D. B. (1972) "A Calculation Procedure for Heat, Mass, and Momentum Transfer in Three Dimensional Parabolic Flows," *Int. j. Heat Mass Transfer*, 15: 1787-1805.

Rodi, W. (1972) "The Prediction of Free Boundary Layers by Use of a 2-Equation Model of Turbulence," Ph. D. Thesis, Faculty of Engineering, University of London, England.

Rodi, W. (1986) "Turbulence Modeling for Incompressible Flows," *PCH PhysicoChemical Hydrodynamics*, 7(5/6): 297-324.

Zawacki, T. S. and Weinstein, H. (1968) "Experimental Investigation of Turbulence in the Mixing Region Between Coaxial Streams," NASA CR-959, February, 1968.

Zelazny, S. W., Morgenthaler, J. H. and Herendeen, D. L (1973) "Shear Stress and Turbulence Intensity Models for Coflowing Axisymmetric Streams," *AIAA J.*, 11(8): 1165-1173.

MODELING THE BEHAVIOR OF A TWO-PHASE FLOW APPARATUS IN MICROGRAVITY

Eric W. Baker and Ronald F. Tuttle
Air Force Institute of Technology
Department of Engineering Physics
AFIT/ENP
Wright-Patterson AFB, OH 45433
(513) 255-4498

Abstract

There are many unknown parameters in two-phase flow in microgravity environment. The database is incomplete and therefore correlations are unknown. This has prompted theoretical and experimental work in the area. A Phillips Laboratory program is currently exploring this area. The Phillips Laboratory experiment is a closed loop rankine cycle with a boiler, condenser/subcooler, accumulator and a pump. The work reported herein attempts to model the Phillips Laboratory Apparatus using a thermal-hydraulic software modeling system called Sim-Tool developed by Mainstream Engineering. This work also explores the limitations of software modeling a microgravity environment. Results of this modeling effort indicate that Sim-Tool needs further development in order to correctly predict two-phase flow in a microgravity environment.

INTRODUCTION

Future space systems such as a NASA space station or a DoD space-based radar will require power in the multi-kilowatt and multi-megawatt ranges. These high power systems may require the use of two-phase fluids for electrical power production. Any two-phase power system operating in a microgravity environment will require thermal-hydraulic design information to correctly model the behavior of the system. Information on system pressure drops, heat transfer coefficients, and flow regimes are critical for design, but such data for two-phase flowing a microgravity environment are currently very limited.

Present two-phase flow information is primarily obtained from empirical or semi-empirical equations developed from drop tower experiments or aircraft flying a parabolic trajectory. These equations based on experimental data gathered in a 1-g environment or less. At present there is no significant reduced gravity data upon which to base any reduced gravity relationships (Antoniak 1978).

As a gas and liquid mixture flow through a pipe, the pattern of the two phases may take on different patterns, or regimes as they are called. The flow regimes are categorized into four distinct flows: dispersed flow, slug flow, stratified flow and annular flow.

Because the two-phase reduced gravity, thermal-hydraulic parameters are dependent on flow regime, it is necessary to fully understand how parameters vary in each type of flow. At the present time there is still a clear lack of understanding when each flow regime dominates, theoretical and experimental research in this area continues. Experimentation in reduced gravity can be accomplished with a drop tower test, airplane parabolic flights, rocket tests, shuttle flights, or possible space station testing (Antoniak 1978).

CURRENT EFFORTS

Best et al. (1986) attempted to map the flow regime against the Martinelli parameter verses a dimensionless force parameter to predict the flow regime present. The experimental data from this work was provided by KC-135 parabolic flights to simulate a reduced gravity environment. Experimental data was insufficient to verify the theoretically developed regime transition lines completely and more work was recommended.

Chen and Downing (1988) researched pressure drops associated with two-phase flow in reduced gravity. Crowley (1987) attempted to verify a unified, mechanistic method for determining two-phase flow regimes. This work verified other data in that the pressure drop is higher in a reduced gravity environment than in a 1-g environment.

PHILLIPS LABORATORY EXPERIMENT

Presently at the Phillips Laboratory, there is ongoing research in two-phase reduced gravity flow. An airplane has been used and rocket reduced gravity tests are planned in order to predict flow regimes, pressure drops and heat transfer coefficients. Testing has been done in a 1-g environments and aboard a KC-135 to simulate reduced gravity. The experiment will also be done with a rocket flight to yield longer times of microgravity. The ground and airplane tests have been completed. The rocket test is scheduled for summer of 1991.

PHILLIPS LABORATORY APPARATUS

The two-phase experiment is to demonstrate the operation of a two-phase pumped loop thermal system in microgravity, comparing data obtained in ground, airplane and in rocket tests. This data will be used to draw correlations for pressure drops, heat transfer coefficients, and other critical parameters.

The primary coolant loop is a two-phase pumped system using pressurized refrigerant-11 as the working fluid. The fluid is heated in a boiler section, then passing through an insulated (adiabatic) part of the system, then through a four part condenser/subcooler section. The secondary side fluid is water at atmospheric pressure. After the subcooler, there is a non-ideal accumulator, and then the flow returns to the pump. The hear exchanger is a tube-in-tube counter-flow device.

SOFTWARE MODELING

The purpose of the using software to model the Phillips Laboratory experiment is to add confidence in the pre-flight rocket test and afterwards to evaluate the results. Software has the advantage of being able to run an "experiment" many times as compared to the one-time rocket test which will destroy the Phillips Laboratory apparatus.

Sim-Tool Simulation Software was chosen for the task of evaluating the two-phase system. Sim-Tool is a tool that links software described components together to form thermal-hydraulic loops.

The approach on this system was to model the piping, boiler, condenser/subcooler, accumulator and pump all under the guidelines of the experiment. The first step was to link a simple system together for three reasons: to ensure the software is working, to gain experience with the component modules of the simulation, and to have a simple model to evaluate certain parameters of the system (such as gravity). The second step was to model the Phillips Laboratory experiment. The last step was to evaluate the results and the software.

SOFTWARE MODEL OF THE PHILLIPS LABORATORY APPARATUS

Sim-Tool uses a list of components as a model. These components have parameters, inputs, and outputs that each need to be specified. The model designed for this two-phase experiment had a source, lengths of pipes, square and round elbows of different arc lengths, boiler, heat exchanger and a source as an outlet. Sim-Tool has a series of pre-processors evaluate the model for different aspects. For example, these checks ensure that all components of a subsystem are of the same series (two-phase, single-phase, or ideal gasses) or that the nodes of a model have been connected properly with the correct number of inlets and outlets and that each node has been used properly throughout the system. Each subsystem working fluid, pressure, and the mass flow rate must be specified. The elevation for each node also needs to be input.

A Sim-Tool model consists of the following parts: a Master Control section, a Constants section, System Data section, Nodal Data section, and a Monitor output section. The Master Control section sets global parameters such as gravity; this section also sets whether the problem is transient or steady-state and the number of iterations to convergence as well as the time steps and which time steps get output. The Constants section is just that, a list of constants that are used in the model. This is optional, but if the constants are defined in one section, they act as variable in the main program and can be easily edited and run again for different numbers instead of hunting down numbers within the program. The System Data section is where all the components are listed. This section also specifies the working fluid. The Nodal Data section specifies the convergence criteria for the degrees of freedom for each node, states the initial values for all the nodes, and marks the nodal information desired from any or all of the nodes to be sent to the output file. The Monitor section controls what is sent to the monitor during the simulation. This section is also optional but important in order to track the progress of the simulation as it is running. This section sets whether the title and subtitle are printed to the screen, the time and frequency with which information is sent to the screen, also the format of the output specified in the Nodal Data section.

The first step mentioned above, was to model a simple system. This system was a source supplying water to a pipe inlet, the pipe, and a sink at the pipe outlet. This model was used to ensure a working Sim-Tool simulation program. This was key in the approach because the simulation program was not working. It took two sets of back-up programs to finally get Sim-Tool up and running. This model was then verified by matching the pressure drop through the pipe to the analytic result. This model was used parametricly to evaluate changing the system gravity. Raising the outlet of the pipe and varying the gravity to 0-g brought the pressure drop to the same value as the case where both inlet and outlet are at the same elevation. Then the fluid was changed from single-phase to two-phase and repeated. The 0-g elevated outlet case was the same as 1-g level case. This data confirmed that the simulation was correctly varying gravity for the potential energy head loss. The results of this model are tabulated in Table 1.

Table 1. Comparison between Single-Phase and Two-Phase Pressure Drops for Water with and within Gravity.

g (m/s2)	Inlet Elev (m)	Outlet Elev (m)	Single-Phase		Two-Phase	
			Inlet Pr(MPa)	Outlet Pr(MPa)	Inlet Pr(MPa)	Outlet Pr(MPa
9.80	0.0	0.0	1.0	0.884	10.0	7.279
9.80	0.0	100.0	1.0	0.036	10.0	7.179
0.098	0.0	100.0	1.0	0.876	10.0	7.278
0.00	0.0	100.0	1.0	0.884	10.0	7.279

The second step was to model the Phillips Laboratory apparatus as accurately as Sim-Tool allowed. This turned out to be less accurate than expected, but those limitations will be discussed in the next section. The start of the model was a source in order were: a pipe, a square 90o turn, a pipe, a square 90o turn, a pipe, a pipe with power added, a pipe, a 180o elbow, a pipe, a 180o elbow, a 180o elbow, a cross-flow shell and tube heat exchanger, a pipe, a sink at 295 kPa and 21o C. The secondary side was modeled as a source at 101.3 kPa and 37o C and a sink with a constant flow rate. Figure 1 is a schematic of the model. This is not the same order as the actual experiment, but is equivalent in turns and lengths for pressure drops and in heat exchanger area for heat transfer. The sink conditions will not affect the results because these conditions do not have any effect upstream in the system.

LIMITATIONS

In assessing the model performance and Sim-Tool's applicability to modeling the Phillips Laboratory apparatus, there are some drawbacks with this method and therefore the results should be put in that light. There are four main areas of concern with the Sim-Tool model, they are listed below.

1) Gravity in two-phase flow is not accounted for properly. It is essentially handled as a 1-g environment with head losses according to the gravity in the environment. Table 1 results highlight this case.
2) The heat exchanger (Sim-Tool component 4851) has at least one software "bug" in it which Mainstream Engineering was trying to correct as of this paper. The heat exchanger also treats the heat transfer as if in a 1-g environment even when the gravity is set lower in the model.
3) The Constant Pressure and Volume Tanks (Sim-Tool components 4883 and 4884, respectively) are not in version 3.11, so it was impossible to accurately model the accumulator of the Phillips Laboratory apparatus.
4) The MCSTR variable in Sim-Tool components represents the heat capacity. It is not used by the two-phase fluid model components, making it impossible to model this phenomenon.

These limitations reduced the effectiveness of Sim-Tool to correctly model two-phase reduced gravity experiment. Although the model has limitations, some improvements can be added within the constraints of the software.

MODIFYING THE OPEN LOOP MODEL

The Sim-Tool model previously presented is an open loop model. The part of the experiment from the square quarter turn after the heat exchanger to the system pump was left out in order to attempt to model the rest of the system accurately without these components effecting the model, because of unknown inputs and parameters of these components. Two additional Sim-Tool models were prepared. These models have closed loop systems. One has a constant pressure tank as the accumulator and one has a constant volume tank as the accumulator. These models presently will not run in Sim-Tool version 3.11 because of the limitation presented above. These models are presented for ease of adapting the apparatus into Sim-Tool format if these components become available. The pump in these two models is a centrifugal pump with estimated inputs and parameters. All inputs and parameters should be cross-checked with actual values to add accuracy to the model. Another possible addition to the model is the addition of plumbing fittings and valves (components 4006 through 4009) if this data is known. This will add accuracy to the system pressure drops.

MODEL RESULTS

The results (Baker 1991) of the Sim-Tool simulation for the low power/low primary flow and high power/high primary flow rates are presented in Table 2.

Table 2. Key System Parameters calculated by Sim-Tool.

Key System Parameters	Low Power	High Power
Temperature of Primary Fluid out of Pump (°C)	21.85	21.85
Pressure of Primary Fluid out of Pump (kPa)	300.0	300.0
Quality of Primary Fluid out of Pump	-1.0	-1.0
Primary System Mass Flow Rate (kg/s)	0.3035	0.437
Boiler Input Power (W)	310.0	450.0
Temperature of Primary out of the Boiler (°C)	58.36	58.36
Pressure of Primary out of the Boiler (kPa)	299.9	299.8
Quality of the Primary out of the Boiler	0.388	0.394
Temperature of Primary into Heat Exchanger (°C)	58.32	58.27
Pressure of Primary into Heat Exchanger (kPa)	299.7	299.3
Temperature of Primary at Exit of Model (°C)	54.74	47.12
Pressure of Primary at Exit of Model (kPa)	299.6	299.6
Quality of Primary at Exit of Model	-1.0	-1.0
Temperature of Secondary into Heat Exchanger (°C)	37.0	37.0
Pressure of Secondary into Heat Exchanger (kPa)	101.3	101.3
Secondary System Mass Flow Rate (kg/s)	0.300	0.300
Secondary Temperature out of Heat Exchanger (°C)	48.77	48.77
Pressure of Secondary out of Heat Exchanger (kPa)	101.3	101.3

Note: Quality of -1.0 is subcooled

CONCLUSION

Due to the current constraints of Sim-Tool software, results of the model are questionable if used to predict pressure drop and heat transfer for a two-phase flow in a microgravity environment. Sim-Tool excludes the pressure drop modeling that this simulation was trying to determine, so any software model will not accurately predict the Phillips Laboratory experiment. However, Sim-Tool is suited to model the 1-g tests that have been performed. Further development of Sim-Tool to include microgravity correlation needs to be encouraged to allow this capable software to perform the desired modeling.

Acknowledgments

This work was supported by the Phillips Laboratory and the was done at the Air Force Institute of Technology in partial fulfillment of the Master of Science Degree in Nuclear Engineering.

References

Antoniak, Z. I. (1988) "Two-Phase Alkali-Metal Experiments in Reduced Gravity," J.SPACECRAFT, 25(2):146-155.

Baker, E. W. (1991) "Modeling the Behavior of a Two-Phase Flow Apparatus in Microgravity," Report GNE92M/NENG-605-05/31/91, Air Force Institute of Technology, Wright-Patterson AFB, OH 45433.

Best, F., L. J. Kachnik, J. M. Cuta, T. E. Michener, Z. I. Antoniak, J. M. Bates, and W. J. Krotiuk (1986) "Reduced-Gravity Two-Phase Flow Experiments in NASA KC-135," in Trans. Fifth Symposium on Space Nuclear Power System, CONF-880122-- Summs., held in Albuquerque, NM, 11-14 January 1986.

Chen, I. Y. and S. R. Downing (1988) "A Reduced Gravity Flight Experiment: Observed Flow Regimes and Pressure Drops of Vapor and Liquid Glow in Adiabatic Piping," AICHE J. Symposium Series, 84(263):203-216.

Crowley, C. J. (1987) "Unified Flow Regime Predictions at Earth Gravity and Microgravity," Technical Note TN-483, CREARE Inc., Hanover, NH 03755

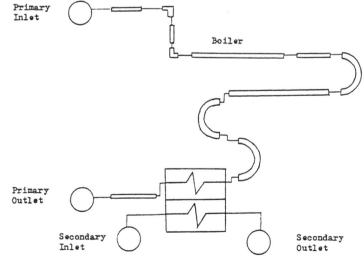

FIGURE 1. Schematic of Sim-Tool Model of Apparatus.

"TITAM" THERMIONIC INTEGRATED TRANSIENT ANALYSIS MODEL: LOAD-FOLLOWING OF A SINGLE-CELL TFE

Mohamed S. El-Genk, Huimin Xue, Chris Murray and Shobhik Chaudhuri
Institute for Space Nuclear Power Studies
Department of Chemical and Nuclear Engineering
University of New Mexico
Albuquerque, NM 87131
(505) 277-5635

Abstract

TITAM, a dynamic model that simulates transient and steady-state operation of a fully integrated single-cell thermionic fuel element (TFE), has been developed. Operation modes investigated include transient response to a step input in reactivity, effects of changing the Cs pressure and/or the width of the interelectrode gap on the load electric power, overall conversion efficiency, and the load-following characteristics of a TFE. Results show that although the nuclear reactor is always load following, a TFE is only partially load-following. Results also show that in a thermionic (TI) nuclear power system, which employs single-cell TFEs having a large interelectrode gap, it is desirable to conserve Cs by lowering its vapor pressure at the begining-of-life, since increasing the Cs pressure insignificantly affects the load electric power. However, should fuel swelling, after operating for an extended period of time, reduce the width of the interelectrode gap, both the conversion efficiency and the load electric power will decrease. In this case a partial compensation in the load electric power may be achieved by increasing the fission power and/or the Cs vapor pressure in the interelectrode gap of the TFEs.

INTRODUCTION

Design requirements of space nuclear power systems include low specific mass, no single point failure, high temperature and radiation survivability, low heat rejection radiator area and high reliability. TI power systems possess many attributes that make them attractive candidates for meeting space power needs ranging from a few tens to hundreds of kilowatts. The TI nuclear power systems have relatively high conversion efficiency (8%-10%) and high radiator temperature (>900 K). The former lowers the reactor thermal power, and hence the reactor and shield mass, while the latter results in a small radiator area and low specific mass. In addition, static TI conversion promotes high reliability due to the lack of moving parts, and redundancy, since a failure of a single TFE will have little effect on the performance of other TFEs in the reactor core.

An important issue related to the development of TI space nuclear power systems is understanding the effects of various operation and design parameters, such as changes in reactor thermal power due to reactivity insertion, changes in the width of interelectrode gap due to fuel swelling, changes in Cesium (Cs) vapor pressure in the interelectrode gap, and changes in the load demand on the steady-state and transient performance of the system. In order to quantify some of these effects, an integrated model, which simulates both transient and steady-state operation of a fully integrated TFE, needs to be developed and experimentally verified. Such a model should directly couple the TI converter to both the electric load and the nuclear fuel region and incorporate reactor kinetics with the appropriate reactivity feedback effects.

Existing models are limited to the analysis of steady-state operation of either a TI diode (Rasor and Associates Inc., 1990) or of a TFE (Pawlowski, Klien and McVey 1991, and Peters and Jekel, 1992). The input to the diode model (Rasor and Associates Inc., 1990) include the emitter and collector temperatures and the emission current density. The model of Pawlowski et al. (1991) basically couples the TI diode model of Rasor and Associates Inc. (1990) to a two-dimensional, steady-state thermal-hydraulic model of TFEs. Because this model is not directly coupled to the electric load and reactor kinetics, the load current and fission power must be specified in the input. In addition, neither model (Rasor and Associates Inc., 1990 and Pawlowski, Klien and McVey 1991) includes the unignited mode of emission and, hence, are suited only for high temperature operation.

OBJECTIVES

The objective of the research is to develop a dynamic model that simulates both steady-state and transient operation of a fully integrated single-cell TFE. Transient modes of operation, which are of interest, include start up, changes in load demand, changes in Cs vapor pressure and/or the width of interelectrode gap due to fuel swelling, and changes in control drums/rods position. Transient analysis is performed to investigate the effects of a step function input of reactivity on the TFE efficiency, load electric power and converter temperatures. The load-following characteristics of a single-cell TFE and the effects of Cs vapor pressure and width of the interelectrode gap are also investigated. The operation regimes, in which a TFE is inherently load-following, are identified.

MODEL DESCRIPTION

The developed Thermionic Integrated Transient Analysis Model, TITAM, consists of four interactively coupled models: (a) nuclear reactor six-group point-kinetics model, (b) lumped transient thermal model of integrated nuclear fuel element and TI converter, (c) thermal-hydraulics model of the liquid metal coolant, and (d) thermionic emission model (Rasor, 1982). The work functions of the emitter and the collector are determined as a function of the electrode-Cs reservoir temperature ratio using the correlations which are developed based on the data of Houston (1962). The input to the model include the dimensions, materials and reactivity feedback coefficients of the different regions of the TFE, cesium reservoir temperature, load electric resistance, either the fission power or the value and ramp rate of reactivity input to the reactor core, and the coolant inlet temperature and mass flow rate. The output of the model includes electric power to the load, load current and voltage, conversion efficiency, and temperatures in the different regions of the TFE. In the TFE model, the dimensions of the electric leads could be optimized for minimum electric losses.

The present TFE model can be used as an effective tool in the design optimization of a single-cell or multi-cell TFE and for simulating the transient and steady-state operation of in-core TI conversion elements. An example of potential in-core applications of such a model is in determining the design parameters for maintaining a constant emitter temperature in the TFEs, despite the radial and axial variations in fission power in the reactor core. Other applications include: (a) determining the operation regime where the system is inherently load-following, and (b) developing an operation strategy to optimize the utilization of Cs. As indicated earlier, this model extends the capability of the TI conversion model of El-Genk et al. (1991) by incorporating the TI unignited mode of operation, which makes it suitable for both high temperature and low temperature operations. The model also couples the TFE thermal model to a six-group point-kinetics model with reactivity feedback effects for simulating the transient response of a nuclear reactor core. The TFE model is detailed elsewhere (El-Genk et al., 1991), therefore will not be repeated here. Only the reactor kinetics model and the governing equations for the TI unignited mode of operation are presented in the following sections.

Six Group Point-Kinetics Model

The six-group point-kinetics model calculates the transient fission thermal power following an external reactivity insertion and incorporates reactivity feedback due to Doppler and temperature effects. The reactor point kinetics equations are:

$$\frac{dP_{in}(t)}{dt} = \frac{\rho(t) - \bar{\beta}}{\Lambda} P_{in}(t) + \sum_{i=1}^{6} \lambda_i C_i^*(t),$$

and,

$$\frac{dC_i^*(t)}{dt} = \frac{\beta_i P_{in}(t)}{\Lambda} - \lambda_i C_i^*(t) \qquad (i=1,2,..,6).$$

$$(1)$$

$$(2)$$

The total reactivity input to the reactor core is given as:

$$\rho(t) = \rho_{ex}(t) + \rho_f^D(t) + \sum_i \rho_i^E(t).$$

$$(3)$$

The second and third terms in Equation (3) are the sum of the reactivity feedback due to fuel material Doppler effect, $\rho_f^D(t)$, and temperature effects, $\rho_i^E(t)$, in different regions of the TFE, which can be expressed as:

$$\rho_f^D(t) = \alpha_f^D \ln\left(\overline{T_f(t)/T_f(0)}\right),$$

and,   (4)

$$\rho_i^E(t) = \rho_i^E \overline{(T_i(t) - T_i(0))}.$$ (5)

TI Unignited Mode of Operation

In the unignited mode of operation, surface ionization is the dominant mechanism for maintaining neutral plasma in the inter-electrode gap of the converter. In this mode of operation, volume ionization is absent. The plasma potential can be expressed as:

$$\phi_n = 0.5\left[h\left(T_e/T_R\right) + V_i\right], \text{ where } h = 0.76 \text{ eV}, V_i = 3.89 \text{ eV},$$ (6)

and the electron emission current is:

$$J_n = AT_e^2 \exp\left(\frac{-\phi_n}{k_B T_e}\right).$$ (7)

This electron current is partially neutralized by the ion emission current given by the Saha-Langmuir equation (Langmuir 1932):

$$J_i = e\,\mu\left[1 + 2\exp\left(\frac{V_i - \phi_n}{k_B T_e}\right)\right]^{-1},$$ (8)

where the arrival rate, μ, is given as:

$$\mu = D \exp(-h/kT_R).$$ (9)

In equation (9), D is taken equal to 10^{27} atoms/cm^{2}sec (Rasor, 1982). The unignited mode of operation has three regions of interest: (a) retarding region, (b) transition point, and (c) plasma saturation region. At the transition point (J", V"$_{out}$), which corresponds to the condition of zero electric field at the collector (no collector sheath), the current and voltage are given by:

$$J'' = J_n \left(\frac{3\delta}{4\lambda} + 1\right)^{-1}, \text{ and } V''_{out} = \phi_n - \phi_c.$$ (10)

In the retarding region ($V_{out} > V''_{out}$) the output current is given as:

$$J = J_n\left[\left(3\delta/4\lambda\right) + \exp\left((V''_{out} - V_{out})/k_B T_e\right)\right]^{-1}$$ (11)

In the plasma saturation region ($V_{out} < V''_{out}$), the current is limited by the ions reflected from the collector sheath and is given by:

$$J = 2J'' - J_n\left[\left(3\delta/4\lambda\right) + \exp\left((V''_{out} - V_{out})/k_B T_e\right)\right]^{-1}.$$ (12)

Equations (6) through (12) are solved iteratively to determine the current-voltage characteristics of the TI converter in the unignited mode.

RESULTS AND DISCUSSIONS

The steady-state and transient analyses presented herein are for optimized electric leads and for the base case design and operation parameters listed in Table 1. The following subsections present the results of the transient response of a single TFE to both a step function increase and a step function decrease in reactivity. Presented are the calculated changes in fission power, load electric power, overall conversion efficiency, and temperatures in the different regions. The TFE overall conversion efficiency is defined herein as *the ratio of the load electric power to the actual fission power generated in the TFE during the transient.* At steady-state, the fission power is equal to the thermal power supplied to the emitter. However, during a transient increase or decrease in fission power, the corresponding thermal power to the emitter is, respectively, lower or higher, than the fission power.

Transient Analysis of a Single Cell TFE

Figures 1a and 1b present the changes in the total reactivity, fission power, and temperatures of the fuel, emitter and collector, following a $0.5 positive step increase in reactivity. Initially, the steady-state fission thermal power is 1,000 W (see Figure 1b). As these figures show, the reactor reaches steady-state condition about 250 seconds after the initiation of the transient. Figure 2b shows the fission power and the temperatures of the TEF increasing initially very rapidly due to the step increase in reactivity. The increase in the amount of negative reactivity

TABLE 1. Base Case Parameters.

Parameter	Value	Parameter	Value
DIMENSIONS		OPERATING CONDITIONS	
TI Cell Height (cm)	8.532	Coolant Inlet Temperature (K)	850
Fuel Gap Size (mm)	0.01	Coolant Mass Flow Rate (g/s)	55.0
Inter-Electrode Gap (mm)	0.25	Nominal Load Resistance (mΩ)	2.5
Emitter Thickness (cm)	0.192	Initial Steady-State	
Collector Thickness (cm)	0.15	Fission Power (W_{th})	1000
Insulation Thickness (cm	0.1	Nominal Cs Reservoir	
Cladding Thickness (cm)	0.15	Temperature (K)	567
Coolant Channel Width (cm)	1.0	TEMPERATURE REACTIVITY	
Radius of Fuel Region (cm)	1.205	COEFFICIENTS	
MATERIALS		Fuel Material	-1.2×10^{-5}
Emitter Material	W	Emitter & Collector	-6.1×10^{-6}
Collector Material	Mo		
Insulation Material	Al_2O_3	Insulator	-8.7×10^{-6}
Cladding Material	Stainless-Steel	Cladding	-1.0×10^{-5}
Coolant	NaK(78%)	Coolant	-2.0×10^{-6}
Fuel Material	UO_2	Fuel Doppler Effect	-0.8×10^{-6}

feedback, due to the increase in temperatures, slows the rate of fission power increase. Eventually, the fission power as well as the temperatures in the different regions of the TFE peak, as the negative reactivity equals the input reactivity (see Figure 1a). Beyond this point, the total reactivity becomes negative, forcing the fission power and temperatures of the TFE to decrease. Ultimately, steady-state is reached as the net reactivity in the reactor core becomes equal to zero.

Figures 2a and 2b present the results of a negative step input of reactivity of $0.5. A comparison of the results in Figures 1 and 2 indicate that following a positive and a negative reactivity input of $0.5 the fission power changes

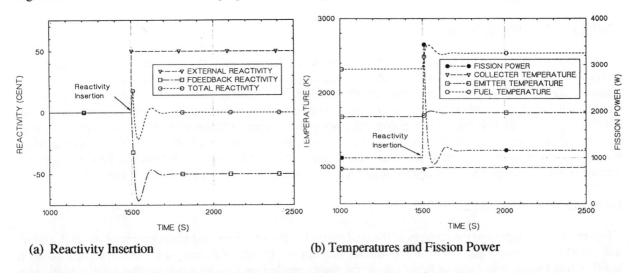

(a) Reactivity Insertion   (b) Temperatures and Fission Power

FIGURE 1. Transient Response to a $0.5 Reactivity Insertion.

initially by as much as +350% and -40%, respectively. This dramatic difference in response is also indicated not only by the change in the emitter temperature, but also by the changes in the load electric power and conversion efficiency of the TFE (see Figure 3a and 3b). Another major difference in the response of the TFE to positive and negative reactivity inputs of equal magnitude is that the second peaking in the fission thermal power for the former (1,150 W) is much larger than that for the latter (1,005 W).

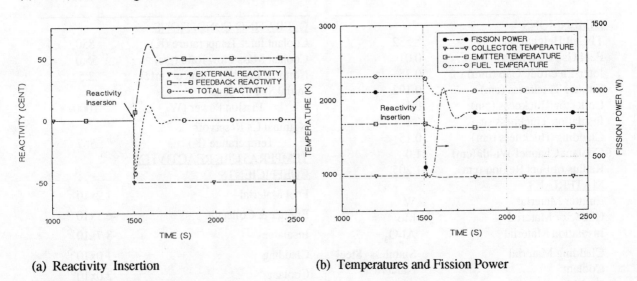

(a) Reactivity Insertion   (b) Temperatures and Fission Power

FIGURE 2. Transient Response of a Single-Cell TFE to a -$0.5 Reactivity Insertion.

In the case of a positive reactivity input of $0.5, because the initial rise in fission power is accompanied by a much slower rise in load electric power (see Figures 3a), the conversion efficiency of the TEF initially drops precipitously from about 10% to a minimum of 3% (see Figure 3b). The minimum efficiency corresponds to the peak fission power occurring after the initiation of the transient. Beyond this point, the TFE efficiency begins to rise as the fission power decreases and the load electric power continues to increase, reaching a higher steady-state value of approximately 11 percent; a net increase of one percent. The similar, but opposite responses in the load electric power and TFE conversion efficiency are shown following a negative reactivity input of $ 0.5 (see Figures 3a and 3b).

It is worth noting that the transient response of the TI conversion efficiency is different from that described above for the TFE overall conversion efficiency. While the TFE efficiency initially decreases (or increases) following a positive (or negative) step input in reactivity, the TI efficiency increases (or decreases) slowly with time. At steady-state the TI efficiency equals that of the TFE as the fission power becomes equal to the thermal power input to the emitter. At the beginning of the transient, the change in fission power mainly changes the thermal energy storage in the fuel, hence its temperature. The thermal input to the emitter changes very slowly with time because

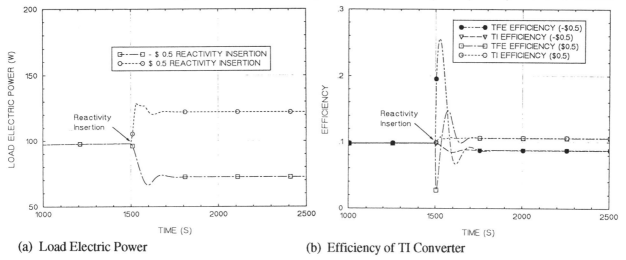

(a) Load Electric Power   (b) Efficiency of TI Converter

FIGURE 3. Transient Response of a Single-Cell TFE to a ± $0.5 Reactivity Insertion.

of the low thermal diffusivity of the UO_2 fuel and the presence of an open gap between the fuel pellets and the emitter.

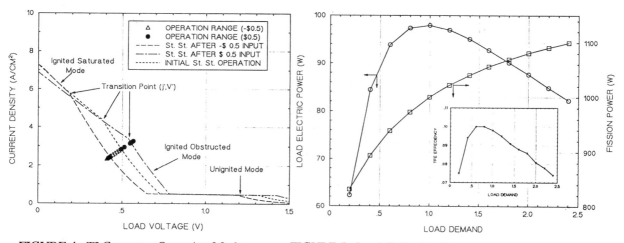

FIGURE 4. TI Converter Operation Modes.   FIGURE 5. Load-Following Response

Figure 4 presents the operating condition before and after the transients on a plot of the J-V characteristics of the TI converter. It can be seen that initially, the TI converter operates in the ignited obstructed mode. As shown in Figure 4, a positive step input in reactivity of $0.5 moves the operation conditions of the TI converter closer to the transition point (J', V') between the ignited saturated mode and the ignited obstructed mode, resulting in a higher conversion efficiency and higher load electric power (see Figures 3b and 4). Conversely, a negative step input in reactivity of $0.5 moves the operation conditions of the TI converter away from the transition point (J', V'), hence lowering both the TI conversion efficiency and the load electric power.

<u>Load-Following Characteristics</u>

For an initial steady-state fission power of 1,000 W, Figure 5 presents the load-following characteristics of the TFE. As this Figure shows, the nuclear fission power is always load-following. An increase in the load electric demand increases electron cooling of the emitter, therefore lowering its temperature and that of the fuel region. Such decrease in temperatures introduces a positive temperature reactivity causing the fission power to increase commensurate with the load demand. Conversely, a decrease in load demand increases the fuel and emitter temperatures, hence lowering fission power due to the negative temperature feedback reactivity effect.

While the fission power is always load-following, the integrated TFE is only partially load-following. As demonstrated in Figure 5, the load electric power for the base case TFE increases with increased load demand; it peaks at 98 W for a 100% load demand. A further increase in load demand is met with a decline in the load electric power. Therefore, the TFE being analyzed (corresponding load resistance is 2.5 mΩ) is load-following below 100 % load demand and is non load-following at higher load demands. It should be noted that changing the design and operating conditions of the TFE would shift the peak load electric power to the right or to the left of the point of 100% load demand in Figure 5.

Results indicate that when operating in the load-following half of Figure 5, a 50 percent reduction in load demand causes both the fission power and the load electric power to decrease by only 9.2%, from 1,000 W to 920 W and from 98 W to 88.5 W, respectively. Conversely, a 50% increase in load demand moves the TFE operation into the non load-following portion in Figure 5, increasing the fission power by 5% (from 1,000 W to 1,050 W), but decreasing the load electric power by ~ 3 % (from 98 W to 95 W). Results also show that in the load-following side of Figure 5, the conversion efficiency increases with increased load demand (see insert in Figure 5). However, in the non load-following portion of Figure 5, TFE efficiency decreases with increased load demand, despite the increase in fission power.

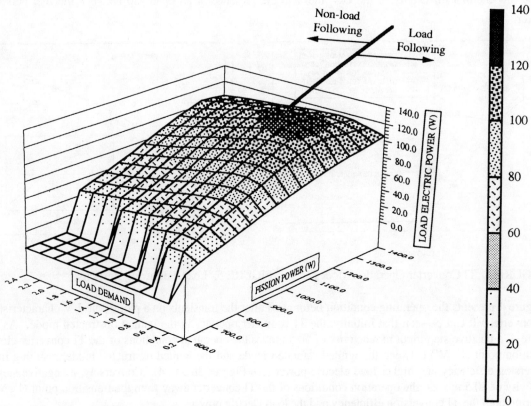

FIGURE 6. Operating Surface of the Load-Following Response of a Single-Cell TFE.

1019

Figure 6 presents an operating surface of the load following characteristics of the TFE. Because of the reactivity feedback effects, changes in load demand causes both the load electrical power and the fission power to change. Figure 6 shows that for a given load electric power there are multiple combinations of fission power and load demand. The contours in Figure 6 indicate constant load electric power regions. The dark region in Figure 6 corresponds to the maximum attainable load electric powers between 120 and 140 W for the base design TFE. The results in Figure 6 should be useful to the reactor operator to determine the combinations of load electric power and fission power needed to respond to changes in the load demand.

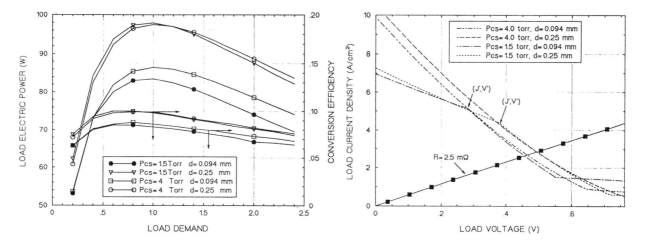

FIGURE 7. Effects of Cs Pressure and Width of Interelectrode Gap on the Load Power and Efficiency of a TFE.

FIGURE 8. Effect of Cs Pressure and Gap-Width on J-V Characteristics of a TFE.

Results presented in Figure 7 demonstrate the effect of changing the Cs vapor pressure in the interelectrode gap and/or the width of the gap on load-following characteristics and the conversion efficiency of a single-cell TFE. For the base load demand and for a gap width of 0.25 mm, increasing the Cs vapor pressure from 1.5 torr to 4.0 torr insignificantly affects the load-following response and the conversion efficiency of the TFE, since it does not change the operating condition of the TI converter (see the J-V characteristics in Figure 8). However, for a smaller gap of 0.094 mm, increasing the Cs vapor pressure to 4.0 torr shifts the operating conditions of the TI converter closer to the (J',V') transition point, thus increasing both the electric current and the output voltage, resulting in a higher load electric power and a slightly higher efficiency. It should be noted that at high Cs pressure the power losses in the electric leads increases due to the high current density, causing the load voltage to become very small

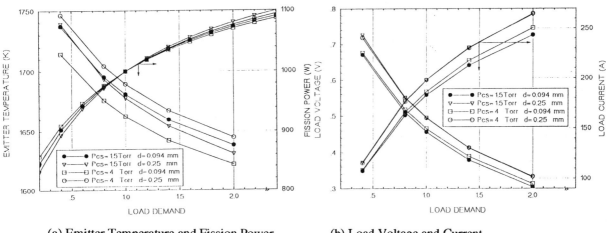

(a) Emitter Temperature and Fission Power

(b) Load Voltage and Current

FIGURE 9. Response of Changes in TFE Operation Parameters in Response to Changes in Load-Demand.

(see Figure 8). Figure 9a shows the effect of changing the Cs vapor pressure and/or the width of the interelectrode gap on the TFE fission power and the emitter temperature, while Figure 9b presents the corresponding effects on the load current and voltage.

These results suggest that in a TI nuclear power system at the begining-of-life, it is desirable to conserve Cs by lowering its vapor pressure, since increasing the Cs pressure would not affect the load electric power of the TFEs. However, should fuel swelling, after operating for an extended period of time, force the width of the interelectrode gap to decrease, the load electric power will decrease. In this case a partial compensation in the load electric power may be achieved at the expense of increasing the Cs vapor pressure in the gap or the fission power of the TFE.

SUMMARY AND CONCLUSIONS

TITAM, a dynamic model that simulates the transient and steady-state operations of a fully integrated single-cell TFE has been developed. Operation modes investigated include the transient response to a step input in reactivity, and the effect of changing the Cs vapor pressure and/or the width of the interelectrode gap on the performance of the TFE. Also investigated are the load-following characteristics of the TFE. The calculated transient parameters include the load electric power, TFE conversion efficiency, load current and voltage, and temperatures in the different regions of the TFE (fuel, emitter, collector, electric insulator, cladding and coolant). In the load-following analysis the output parameters include the changes in fission power in addition to the aforementioned parameters in response to changes in the load demand.

Results show that, although the nuclear reactor is always load following, the TFE is only partially load-following. For the base case investigated, the load electric power peaks at 100% of load demand, which corresponds to a load resistance of 2.5 mΩ. Increasing the load demand beyond this point results in a lower load electric power, (non load-following), while a reduction in load demand is accompanied by a reduction in the load electric power (load-following). Results also show that in a TI nuclear power system that employs single cell TFEs with a large interelectode gap, it is desirable to conserve Cs by lowering its vapor pressure at the begining-of-life, since increasing the Cs pressure would not affect the load electric power. However, should fuel swelling, after operating for an extended period of time, force the width of the interelectrode gap to decrease, both the conversion efficiency and the load electric power will decrease. In this case, a partial compensation in the load electric power may be achieved by increasing the fission power and/or the Cs vapor pressure in the interelectrode gap of the TFEs.

Acknowledgment
This research is sponsored by the Strategic Defense Initiative Organization and the Aero Propulsion and Power Directorate, Wright Laboratory, U. S. Air Force, Wright-Patterson AFB under Subcontract No. S-247-002-001 to the University of New Mexico's ISNPS.

References

El-Genk, M.S., C. Murray, and S. Chaudhuri (1991) "Effects of Cs Vapor Pressure on Steady-State and Transient Operation of Thermionic Converters", *Proceedings of the 26th IECEC*, vol. 3, pp. 179 - 184.

Houston, J.M. (1962) "Thermionic Energy Conversion", *Advanced Electronics*, vol. 17, pp.125-206.

Pawlowski, R. A., A. Klien, and J. B. McVey (1991) " Coupled Thermionic and Thermalhydraulic Analyses of Thermionic Fuel Elements," *Proceedings of the 26th IECEC*, vol. 3, pp. 99 -104.

Peters, R. R and Jekel, T. B. (1991) "Thermionic Diode Subsystem Model" *Proceedings of the 9th Symposium on Space Nuclear Power Systems*, M.S. El-Genk and M.D. Hoover,eds., American Institute of Physics.

Rasor, N. S. (1982) "Thermionic Energy Conversion", *Applied Atomic Collision Physics*, vol. 5, pp.170-200.

Rasor Associates Inc., (1990) *TECMDL-Ignited Mode Planar Thermionic Converter* Model, Report #E563 004-C-082988/NSR-31/90-0775, August, 1990.

Nomenclature

English

A : Coefficient in Eq (7) ($120\ A/cm^2 K^2$)

$C_i^*(t)$: Equivalent concentration of delayed neutrons precursors of the ith group

h : Heat of adsorption of cesium

J : Electric current density (A/cm^2)

J : Transition point current density in Ignited Mode (A/cm^2)

$\dot{J}$: Transition point current density in unignited mode (A/cm^2)

k_B : Boltzmann's constant (1.38×10^{-23} J/K)

P : Fission power, thermal power, or electric power (W), pressure (Torr)

R : Electric resistance

T : Temperature (K)

t : Time (s)

V : Unignited mode transition point output voltage (V)

V_i : Cs ionization energy (eV)

V_{out} : Output Voltage (V)

V'_{out} : Unignited mode transition point output voltage (V)

$\wedge$: Prompt neutron lifetime

Greek

α : Reactivity feedback coefficients

$\overline{\beta}$: Fraction of delay neutrons

δ : Width of interelectrode gap (m)

β : Delay neutrons fraction

ϕ : Work function (eV)

λ : Decay constant of the precursor of the delayed neutrons in the ith group (Eq. 1-2) / Effective electron mean free path (Eq. 10-12)

ρ : Reactivity

Subscript

c : Collector

Cs : Cesium

e : Emitter, electric, electron charge

ex : External

f : Fuel

i : Various regions e.g. fuel, cladding etc (Eq. 3-5) / Ion current (Eq. 8), ith group

in : In fuel

n : Neutral

R : Cesium reservoir

th : Thermal

Superscript

D : Doppler effect

E : Temperature effect

$^-$: Average

1022

"HPTAM" HEAT-PIPE TRANSIENT ANALYSIS MODEL:
AN ANALYSIS OF WATER HEAT PIPES

Jean-Michel Tournier and Mohamed S. El-Genk
Institute for Space Nuclear Power Studies
Department of Chemical and Nuclear Engineering
The University of New Mexico
Albuquerque, NM 87131
(505) 277 - 5442

Abstract

Space application of heat pipes in power systems, radiators and thermal management necessitates the understanding of their transient response to changes in input power or sink temperature. "HPTAM", a two-dimensional model, is developed for simulating transient performance of fully thawed liquid-metal and non liquid-metal heat pipes. The model divides the heat pipe into three radial regions: wall, liquid/wick, and vapor regions, and solves the complete form of governing equations in these regions, together with the momentum and energy jump conditions at the liquid-vapor interface. To account for the thermal expansion of the liquid during startup and/or overheating, a liquid-pooling submodel is incorporated to allow excess liquid to accumulate at the end of the condenser. Results of parametric analyses of water heat pipes are presented, illustrating the transient response to exponential rise and step decrease in input power. The effects of input power and initial liquid inventory on liquid pooling effects and both vapor and liquid pressure and temperature distributions are demonstrated.

INTRODUCTION

Heat pipes are highly reliable and efficient energy transport devices which operate passively in vacuum and zero-gravity conditions; consequently, they are considered for many space applications. In such applications, it is important to adequately simulate the operation of heat pipes during startup, shutdown and operational transients such as pulsed heat loads. Because of this wide interest in heat pipes, their steady-state operation has been modeled by numerous investigators, but much less work has been reported on transient modeling (Chang and Colwell 1985, Peery and Best 1987, Costello et al. 1986, Ransom and Chow 1987, Hall and Doster 1988, Tilton et al. 1986, Seo and El-Genk 1989, Bowman and Hitchcock 1988, Issacci et al. 1988, Jang 1988, Jang et al. 1989, Faghri and Chen 1989, and Cao and Faghri 1990). Most models have either involved a single fluid region (vapor, or liquid/wick region), or are basically one-dimensional, liquid/vapor counter-current flow models which neglect the momentum coupling at the liquid-vapor (L-V) interface. Such coupling is necessary to accurately predict the hydrodynamics of the liquid and vapor phases. Even though the time characteristic of the vapor has been found numerically to be much shorter than the characteristic time of the wick/wall region (Seo and El-Genk 1989 and Cao and Faghri 1990), both regions should be thermally and hydrodynamically coupled in the transient modeling of heat pipes. Also, during the startup, when the heat pipe is operating at low vapor density and high vapor velocity, large radial temperature gradients may develop in the wall, liquid/wick, and vapor regions (Deverall et al. 1970), hence, requiring a two-dimensional modeling approach. In previous transient models, liquid compressibility is usually neglected, an assumption which is not valid for working fluids having high thermal expansion coefficients, such as liquid metals and water. During startup, the liquid volume in the heat pipe increases, causing its excess volume to pool at the end of the condenser and reduce its effective length (Merrigan et al. 1986). Therefore, neglecting the liquid compressibility could result in erroneous predictions of the liquid and vapor pressures and of the performance of the heat pipe.

OBJECTIVES

The objectives of this work are to develop a two-dimensional transient heat pipe model, which incorporates all the important effects discussed above for simulating fully-thawed heat pipes. The model is capable of handling both cylindrical and rectangular geometries, however, the results reported herein are for a circular heat pipe with water as the working fluid. A parametric transient analysis is performed of water heat pipes, which are being considered for solar dynamics power systems and for the radiator of a nuclear power system coupled to Stirling engines for a lunar outpost. For these applications, the radiator temperature is expected to be between 300 and 500 K.

MODEL FORMULATION AND DESCRIPTION

The model divides the cylindrical heat pipe into three radial regions: wall, liquid/wick, and vapor regions, and employs the complete form of continuity, momentum and enthalpy governing equations in both fluid phases and the momentum and enthalpy jump conditions at the L-V interface. The wick is made of an annular wire-screened mesh. The governing equations and jump conditions are developed in their general and conservative form by following the procedure described by Delhaye (1974 and 1976). The model solves the following two-dimensional continuity, momentum, and enthalpy balance equations in the liquid and vapor regions:

Continuity

$$\frac{\partial \rho}{\partial t} + \text{div}\left(\rho \vec{U}\right) = 0 \tag{1}$$

Radial Momentum Balance

$$\frac{\partial (\rho U_r)}{\partial t} + \text{div}\left(\rho U_r \vec{U}\right) = \rho F_r - \frac{\partial P}{\partial r} + \left[\overrightarrow{\text{Div}}\left(2\mu \overline{\overline{D}}\right)\right]_r - \frac{2}{3}\frac{\partial}{\partial r}\left[\mu \, \text{div}\left(\vec{U}\right)\right] \tag{2}$$

where

$$\left[\overrightarrow{\text{Div}}\left(2\mu \overline{\overline{D}}\right)\right]_r = \frac{1}{r}\frac{\partial}{\partial r}\left[r \, 2\mu \frac{\partial U_r}{\partial r}\right] + \frac{\partial}{\partial z}\left[\mu \left(\frac{\partial U_r}{\partial z} + \frac{\partial U_z}{\partial r}\right)\right] - 2\mu \frac{U_r}{r^2} \tag{3}$$

Axial Momentum Balance

$$\frac{\partial (\rho U_z)}{\partial t} + \text{div}\left(\rho U_z \vec{U}\right) = \rho F_z - \frac{\partial P}{\partial z} + \left[\overrightarrow{\text{Div}}\left(2\mu \overline{\overline{D}}\right)\right]_z - \frac{2}{3}\frac{\partial}{\partial z}\left[\mu \, \text{div}\left(\vec{U}\right)\right] \tag{4}$$

where

$$\left[\overrightarrow{\text{Div}}\left(2\mu \overline{\overline{D}}\right)\right]_z = \frac{1}{r}\frac{\partial}{\partial r}\left[r \, \mu \left(\frac{\partial U_r}{\partial z} + \frac{\partial U_z}{\partial r}\right)\right] + \frac{\partial}{\partial z}\left[2\mu \frac{\partial U_z}{\partial z}\right] \tag{5}$$

Energy Balance

$$\frac{\partial (\rho h)}{\partial t} + \text{div}\left(\rho h \vec{U}\right) = \frac{\partial P}{\partial t} + U_r \frac{\partial P}{\partial r} + U_z \frac{\partial P}{\partial z} - \text{div}\left(\vec{Q}\right) + \Phi \tag{6}$$

where Φ is the viscous dissipation:

$$\Phi = 2\mu \left[\left(\frac{\partial U_r}{\partial r}\right)^2 + \left(\frac{U_r}{r}\right)^2 + \left(\frac{\partial U_z}{\partial z}\right)^2\right] + \mu \left[\frac{\partial U_r}{\partial z} + \frac{\partial U_z}{\partial r}\right]^2 - \frac{2}{3}\mu \left[\text{div}\left(\vec{U}\right)\right]^2 \tag{7}$$

and

$$\vec{Q} = \left[-k\frac{\partial T}{\partial r}, -k\frac{\partial T}{\partial z}\right] \tag{8}$$

The conservation of the normal mass fluxes at the L-V interface yields:

$$\overline{\dot{m}} = -\rho_L U_r^L = -\rho_v U_r^V \tag{9}$$

The normal momentum jump condition at the L-V interface relates the static pressure drop across the interface to the capillary pressure head and the normal viscous stress discontinuity as follows:

$$(P_L - P_V) + 2\frac{\sigma}{R_c} + \overline{m}^2\left(\frac{1}{\rho_L} - \frac{1}{\rho_V}\right)$$
$$= 2\left[\mu_L \frac{\partial U_r^L}{\partial r} - \mu_V \frac{\partial U_r^V}{\partial r}\right] - \frac{2}{3}\left[\mu_L \operatorname{div}\left(\vec{U}^L\right) - \mu_V \operatorname{div}\left(\vec{U}^V\right)\right] \qquad (10)$$

The first two terms on the left hand side represent the capillary relationship of Pascal, which has been used in almost all previous heat pipe models reported in the literature, only to check if the capillary limit has been exceeded. Pascal relationship is only valid for incompressible and inviscid phases with no mass transfer at the interface, therefore it is not necessarily applicable to heat pipe modeling. The third term on the left hand side accounts for change of momentum at the L-V interface. Other terms on the right account for radial stresses in the liquid and vapor phases.

The transverse momentum jump condition, relating the tangential velocities of the phases at the L-V interface to the axial gradient of the surface tension (Marangoni effect, $\partial\sigma/\partial z$) and the discontinuity in shear stress is:

$$\overline{m}\left(U_z^L - U_z^V\right) + \mu_L\left(\frac{\partial U_r^L}{\partial z} + \frac{\partial U_z^L}{\partial r}\right) - \mu_V\left(\frac{\partial U_r^V}{\partial z} + \frac{\partial U_z^V}{\partial r}\right) + \frac{\partial\sigma}{\partial z} = 0 \qquad (11)$$

The enthalpy jump condition, relating the enthalpy change due to evaporation/condensation across the L-V interface to the discontinuities of the conduction flux, kinetic energy and energy dissipated by viscous stress is:

$$Q_r^V - Q_r^L + \mu_L\left(\frac{\partial U_r^L}{\partial z} + \frac{\partial U_z^L}{\partial r}\right)U_z^L - \mu_V\left(\frac{\partial U_r^V}{\partial z} + \frac{\partial U_z^V}{\partial r}\right)U_z^V + \frac{\partial}{\partial z}\left(\sigma V_r^{int}\right)$$
$$+ \overline{m}\left\{ h^L - h^V + \frac{1}{2}\left[\left(U_z^L\right)^2 - \left(U_z^V\right)^2 + \overline{m}^2\left(\frac{1}{\rho_L^2} - \frac{1}{\rho_V^2}\right)\right] + 2\left[\frac{\mu_V}{\rho_V}\frac{\partial U_r^V}{\partial r} - \frac{\mu_L}{\rho_L}\frac{\partial U_r^L}{\partial r}\right]\right\} = 0 \qquad (12)$$

Equations (9) through (12) specify the discontinuities of mass, momentum and enthalpy fluxes at the L-V interface. However, two additional equations are needed at this interface to formulate a closed form solution of the governing equations and boundary conditions (see Figure 1). The first relation is obtained by assuming continuous temperature at the L-V interface (assumption valid for continuum flow, but not for transition or free molecular flows):

$$T^L = T^V \qquad (13)$$

Secondly, assuming a non-slip condition at the L-V interface, which is physically acceptable when dealing with viscous phases separated with a solid wire-screened mesh, Equation (11) is replaced by the equations:

$$U_z^L = U_z^V = 0 \qquad (14)$$

Because the meniscus is concave at the L-V interface, the wick is partially filled with vapor. For a single-screen wick, the maximum vapor volume in the wick before dryout is related to the wick porosity ε and pore radius R_p by:

$$V_p^j = \left(\frac{\varepsilon A_r^j}{\pi R_p^2}\right)\left(\frac{2}{3}\pi R_p^3\right) = \frac{2}{3}\pi R_p \cdot \varepsilon A_r^j \qquad (15)$$

Assuming hemispherical pores, the void fraction α_p in the wick (the volume of vapor in the wick over the volume of the hemispherical pores) can be expressed in terms of the cosine of contact angle $\mu_c = R_p/R_c$ as:

$$\alpha_p = \frac{1}{\mu_c^3}\left[1 - (1+\frac{\mu_c^2}{2})\sqrt{1 - \mu_c^2}\right] \tag{16}$$

For practical reasons (explained latter), both α_p and μ_c vary between 0 and 1. Equation (16) is inversed using the following approximations derived by Seo (1988) and Seo and El-Genk (1989):

$$\mu_c = \frac{8}{3}\alpha_p (1 - 1.723 \alpha_p^2) \qquad \text{for} \quad \alpha_p < 0.25 \ ,$$

$$\mu_c = \frac{8}{3}\left[0.443 - 0.0529 \alpha_p - \frac{1}{3.5} EXP(0.56 - 3.5 \alpha_p)\right] \qquad \text{for} \quad \alpha_p \geq 0.25 \ . \tag{17}$$

Since the model solves implicitly for the radius of curvature of the liquid meniscus at the L-V interface, an additional equation for the evaporation/condensation mass flux at the L-V interface, which is an extension of the kinetic theory to non-flat interfaces, is used:

$$\overline{\dot{m}} = -\rho_L U_r^L = \left(\frac{M}{4\pi RT^{int}}\right)^{1/2}[P_v^{equ} - P_v] \ , \tag{18}$$

where P_v^{equ}, the pressure of a vapor bubble in thermal equilibrium with the surrounding liquid, is given as (Defay and Prigogine 1966):

$$P_v^{equ} = P_{sat}(T^{int}) \cdot EXP\left[-\left(\frac{2\sigma}{R_c}\right)\left(\frac{M}{RT}\right)\left(\frac{1}{\rho_L}\right)\right] \ . \tag{19}$$

The governing equations and boundary conditions, together with the equations of state of liquid and vapor and thermophysical properties for the wall and both fluid phases, and the initial conditions specified by the user provide all the necessary relations to obtain a closed mathematical solution. Equations (1) through (18) are solved subject to the boundary conditions delineated on Figure 1. The jump and boundary conditions (I) are represented by Equations (9) through (14), (11) excluded, while the boundary conditions (II) and (III) are given, respectively, as:

$$Q_z = 0, \ U_r = U_z = 0, \quad \text{and} \quad Q_r^L = Q_r^W, \ T^L = T^W, \ U_r^L = U_z^L = 0 \ . \tag{20}$$

The boundary condition (IV), which can be either isoflux, isothermal, radiative or convective, is applied independently in the evaporator and condenser regions. The model calculates the wall temperatures, the temperatures, pressures, and mass fluxes of the liquid and vapor, and the radii of curvature of the meniscus at the L-V interface. Note that the model incorporates gravity effects (Equations 2 and 4), but neglects the effect of gravity on the circumferential symmetry of liquid in the annular wick because the width of the liquid annulus is very small (less than a few millimeters), hence F_r is taken to be zero.

THERMOPHYSICAL PROPERTIES AND STATE EQUATIONS

The thermophysical properties of the wall material and the liquid and vapor phases are taken to be temperature dependent (also pressure dependent when relevant). The model incorporates various working fluids such as lithium, sodium, potassium, mercury, and water, as well as various wall materials including tungsten, niobium, zirconium, stainless-steel and copper. In this paper, however, we are primarily concerned with water/copper heat pipes. Density of the liquid phase is calculated as a function of pressure and temperature to account for the liquid compressibility as:

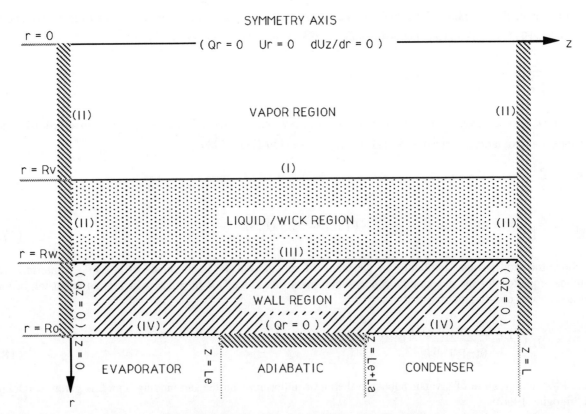

FIGURE 1. Physical Model and Boundary Conditions

$$\rho_L = \rho_{sat,L}(T) \left[1 + \beta_T \left(P_L - P_{sat}(T) \right) \right] \quad , \tag{21}$$

where β_T is the isothermal compressibility factor of the saturated liquid phase (5×10^{-10} Pa^{-1} for water). It is worth noting that this relationship underestimates the compressibility of the superheated liquid phase with respect to the van der Waals isotherms.

LIQUID - POOLING MODEL

During startup of the heat pipe, the average liquid temperature increases causing its volume to also increase due to thermal expansion. Consequently, the rising concave liquid meniscus at the L-V interface will flatten at some point along the heat pipe (the so-called wet point). At the wet point, liquid and vapor interfacial dynamic pressures become equal and the void fraction in the pores of the wick is zero. While a convex liquid meniscus in the wick is possible in principle, because of vapor shear and condensation at the L-V interface, it probably would not be very stable and the convex meniscus will be transformed into a coherent liquid plane surface (Busse and Kemme 1978). Eventually the wet point progresses to the end of the condenser, where excess liquid starts to accumulate, forming a liquid pool.

A submodel has been incorporated into HPTAM (Heat Pipe Transient Analysis Model) to handle liquid-pooling phenomena. The radial momentum jump condition at the L-V interface (Equation 10) is used to calculate the pressure in the liquid cell next to the diphasic interface. Then, using the mass balance in this cell, the mass of excess liquid due to thermal expansion is determined. This mass is then transported into the next interfacial liquid cell. Once the wet point reaches the end of the condenser, the liquid forms a pool at the end of the pipe in the vapor core region. The liquid-pooling submodel also models liquid pool recession and dewetting conditions during the cooldown of the heat pipe.

TABLE 1. Design and Operational Parameters of Heat Pipes Analyzed.

Design Parameter	Value	Operational Parameter	Value
Heat pipe length L (cm)	120	Initial pipe temperature (K)	300
Evaporator length L_e (cm)	20	Initial pore void fraction	0.02 - 0.20
Condenser length L_c (cm)	80	Startup mode	Exponential
Adiabatic length L_a (cm)	20		heating
Pipe outer diameter (mm)	25	Shutdown mode	Step function
Wall thickness (mm)	1.0	Evaporator radial heat flux	Uniform
Thickness of liquid/wick (mm)	1.0	Evaporator maximum heat flux (W/cm^2)	1 - 5
Effective pore radius (μm)	50	Evaporator exponential heating period, τ (s)	5
Wick porosity	0.7	Condenser cooling method	Convective
Wall material	Copper	Water jacket average temperature (K)	300
Working fluid	Water	Jacket heat transfer coefficient (W/m^{2}K)	1000

METHOD OF SOLUTION

The governing equations and boundary conditions are discretized using the volume integration method proposed by Patankar (1980). The numerical scheme uses the well-known Eulerian staggered-grid. The liquid and vapor velocities are determined at the cell boundaries while the other quantities, such as pressures and enthalpies, are evaluated at the cell center. The numerical solution is obtained using the predictor/corrector method (Gourlay and Morris 1968, Yanenko 1971, and Stewart and Wendroff 1984).

In the first step of the numerical algorithm, the Implicit Continuous-fluid Eulerian technique of Harlow and Amsden (1971) is used. In the conservative forms of the momentum equations, the pressure gradient is discretized implicitly, while the diffusion/convection terms are evaluated explicitly. By eliminating the advanced-time mass fluxes appearing in the mass balance equation and by considering the pressure variation with the density only, the continuity equation is reduced to the Poisson equation, which is then solved for the pressure field. During this predictor step, liquid and vapor volumes around the L-V interface are treated as functions of the pore void fractions which are linearized in terms of the cosines of the contact angles using Equation (17). The latter are implicitly related to the liquid and vapor pressures at the L-V interface through the radial momentum jump condition (Equation 10). Equation (18) is used to implicitly relate the evaporation/condensation mass fluxes and the pressures at the L-V interface. The densities, evaporation/condensation mass fluxes, and radii of curvature of the liquid meniscus for the advanced time are then obtained, with the temporary new-iteration pressures and mass fluxes.

The second step is a stabilizing step for the convective and diffusive terms. The momentum equations are solved for velocities, while the enthalpy equations are linearized and solved for temperatures. To decouple the radial and axial momentum conservation equations, the convective mass fluxes, the divergence of the velocity field and the cross-derivative viscous terms are evaluated explicitly from the previous step. These cross derivative terms are not important for the stability of the solution (Beam and Warming 1978). Also the viscous dissipation in the enthalpy conservation equation is evaluated explicitly.

After the velocities and temperatures have been corrected, the pressure field and all properties are evaluated at the advanced-time temperatures and densities. To solve the five-point linear equations resulting from the discretization, a solution routine using Gaussian elimination has been developed and tested, which includes partial pivoting and row-normalization options. The system matrix is band-stored, which considerably reduces the amount of memory storage and computational time required.

TABLE 2.   Comparison of Operational Parameters of Water Heat Pipes Investigated.

Parameter	Case 1	Case 2	Case 3	Case 4
Evaporator heat flux (W/cm^2)	1	2	2	5
Initial pore void fraction	0.02	0.10	0.12	0.20
Liquid inventory (gram)	82.805	82.332	82.295	82.148
Vapor pressure recovery at time of wet point	0.5%	36.8%	no wet point	66.0%

RESULTS AND ANALYSIS

Results of the transient operation of a water heat pipe subject to various heating conditions and having different working fluid inventories are presented. Design and operational parameters of the heat pipe for the four cases analyzed are listed in Tables 1 and 2. At the startup, the heat pipe is initially at a temperature of 300 K, and the 20-cm-long evaporator is heated uniformly with an exponential period $\tau = 5$ s, while the 80-cm-long condenser is convectively cooled by a water jacket. For the purpose of determining the temperature distributions in the various regions as well as the distributions of the vapor and liquid pressures, the numerical scheme employs 6, 2, and 2 radial cells in the vapor, liquid/wick and wall regions respectively, and 6, 6, and 18 axial cells in the evaporator, adiabatic and condenser regions, respectively. As indicated in Table 2, four cases are investigated, each with different working fluid inventory and maximum evaporator heat flux. In these cases, the maximum evaporator heat flux ranged from 1 W/cm^2 to 5 W/cm^2. At startup of the heat pipe the liquid meniscus at the L-V interface is concave everywhere so that the pressure of the vapor phase is greater than that of the liquid phase. As the water in the heat pipe heats up, the excess volume due to thermal expansion will be accommodated in the pores of the wick, and it may flatten the L-V interface at some point during the transient; such occurrence is referred to herein as the "wet point". Results on the effects of water inventory and maximum heat flux on the "wet point" and potential liquid pooling at the end of the condenser are presented in the following sections.

Liquid Pooling at End of Condenser

Figure 2a shows the interfacial dynamic pressures and void fraction distribution of Case 4 at the occurrence of the first wet point. Because the vapor pressure recovery in the condenser is relatively high (66%), the wet point is located close to the beginning of the condenser region. At the wet point, the L-V interface is flat and the pore void fraction is zero. Further thermal expansion of the liquid phase increases the liquid dynamic pressure causing the wet point to advance toward the condenser region (Figure 2b). Eventually, the wet point reaches the end of the condenser, and the excess liquid volume begins to accumulate at the end of the condenser (Figure 2c). Notice that wetting of the entire condenser occurs within a very short time, only 19 ms after the occurrence of the first wet point. Also, the maximum pore void fraction (or smallest radius of curvature of the liquid meniscus) occurs at the beginning of the evaporator, and before flooding of the wick occurs (wet point),.the void fraction recovers along the condenser.

The effects of varying the liquid inventory and the evaporator input power on the position of the wet point is delineated in Figure 3. When the vapor pressure recovery is small compared to the liquid viscous pressure drop in the condenser region, the wet point occurs at the end of the condenser region (Case 1, Figure 3a). If the pressure recovery exceeds the liquid pressure drop in the condenser, the wet point occurs at some intermediate position in the condenser (Case 2, Figure 3b). For Case 3, the mass of the working fluid was low enough that such flooding of the wick at the L-V interface does not occur even after 2 minutes into the transient, at which time steady-state operation is reached. Therefore, by decreasing the operating maximum evaporator heat flux and/or the liquid inventory in the heat pipe, pooling of liquid at the end of the condenser can be avoided.

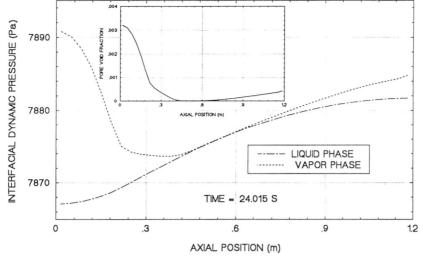

a) Occurrence of First Wet Point in the Condenser Region

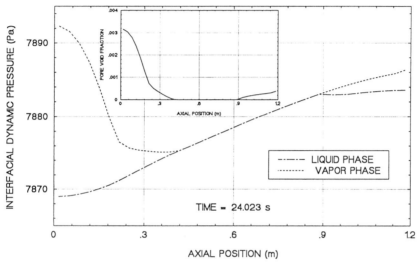

b) Propagation of the Wet Point towards the Condenser

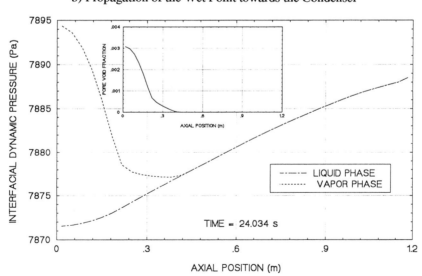

c) The Entire Condenser is Wet: a Liquid Pool Starts Forming at the End.

FIGURE 2.   Dynamic Pressure Profiles and Void Fraction Distributions at various times for the Water Heat Pipe Analyzed in Case 4.

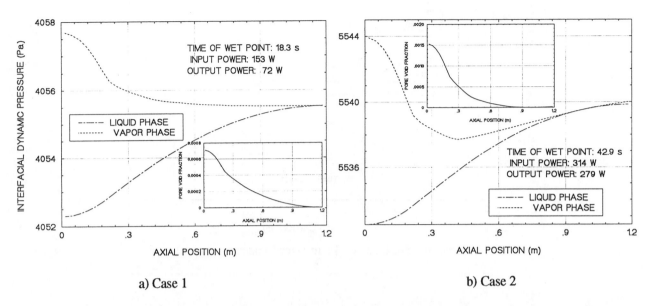

a) Case 1   b) Case 2

FIGURE 3.   Dynamic Pressure Profiles and Void Fraction Distributions for Water Heat Pipes at the Time
of First Wet Point (Cases 1 and 2).

Transient Response of Water Heat Pipe

As shown in Figure 4, in Case 4, the heat pipe reached steady-state after about 110 s. Figure 4a shows that at the beginning of the transient, the input power to the evaporator is higher than the output power to the condenser, by the amount stored in the wall, liquid and vapor regions. As the transient progresses, the output power approaches the input power and the difference between the two disappears as the heat pipe reaches steady-state, at a power throughput of 785 W. After 120 s (or 10 s after reaching steady-state), the evaporator input power is cut off, and the heat pipe begins to cool down. The heat pipe returns to its initial temperature and void fraction distribution, 100 s after the power is cut off (Figure 4). At 24 s after the initiation of the transient, a liquid pool forms at the end of the condenser (Figure 4c). This pool, however, remains relatively small, extending to a maximum of 0.58 mm at full-power (steady-state). After the input power is cut off, the pool recedes very quickly, and disappears within 11 s. Also, the pore void fraction increases and the condenser becomes fully dewetted (Figure 4d).

Heat Pipe Characteristic Period

It is interesting to analyze the transient response of the heat pipe in terms of its characteristic periods. Indeed it is found that, except during the early startup until the formation of the liquid pool (from 0 s to 24 s) and during the pool recession phase (from 120 s to 131 s), the condenser output power, evaporation/condensation mass fluxes and pore void fractions varied at a characteristic exponential period $\tau = 14.9 \pm 0.1$ s. This period applies both during the exponential heating phase and the step-function cooldown phase, and is therefore characteristic of the water heat pipe analyzed (Case 4). This period is expected to vary with the pipe geometry, temperature levels of interest (since most properties, such as heat capacity and thermal conductivity, are temperature dependent) and the type of condenser cooling (convective or radiative); further work will investigate these effects.

Transient Vapor Pressure Recovery

Issacci et al. (1988) and Faghri and Chen (1989), in their simulation of the steady-state operation of water and sodium heat pipes, have reported pressure recoveries up to 90% of the vapor pressure drop and flow reversal at the end of the condenser. This behavior intensifies as the input power, and consequently the radial Reynolds number of the vapor increases. Similar behavior is also detected by our model. Figure 5 shows the vapor pressure axial profile, normalized with respect to the maximum vapor pressure at the beginning of the evaporator, at different times

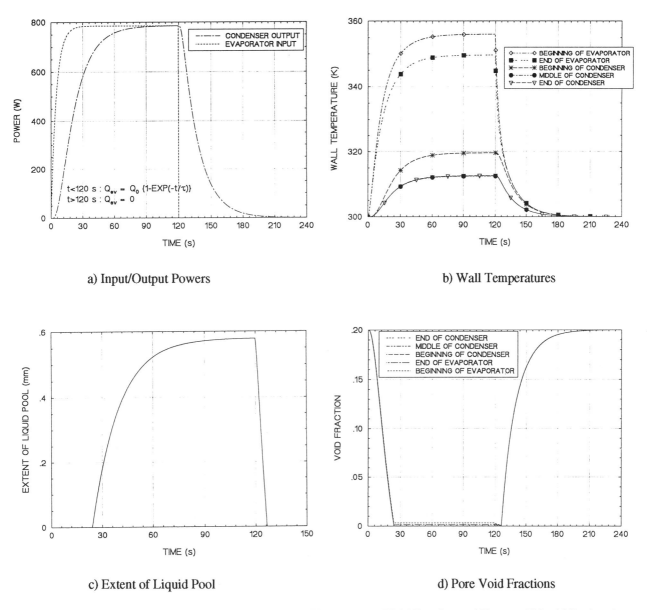

a) Input/Output Powers

b) Wall Temperatures

c) Extent of Liquid Pool

d) Pore Void Fractions

FIGURE 4.   Transient Variation of Power, Wall Temperature, Void Fraction and Extent of Liquid Pool at the End of Condenser for Case 4.

during the heatup transient. As the input power is increased with time, both the pressure level and the vapor pressure recovery in the condenser region increase dramatically. The vapor pressure recovery in the condenser reached a maximum value of 72% at steady-state, at a power throughput of 785 W (after 120 s of transient). Because the vapor pressure at this input power is 3 times that at startup, a small time step is needed to accurately predict the heat pipe transient response. The time step must be much smaller than both the exponential heating period (5 s) and heat pipe's characteristic period (14.9 s). Unfortunately, the calculations are performed with a very small time step (5 ms) in order to suppress instabilities caused by the sensitivity of the vapor and liquid pressures and densities to the temperature changes which are not accounted for implicitly during the predictor step.

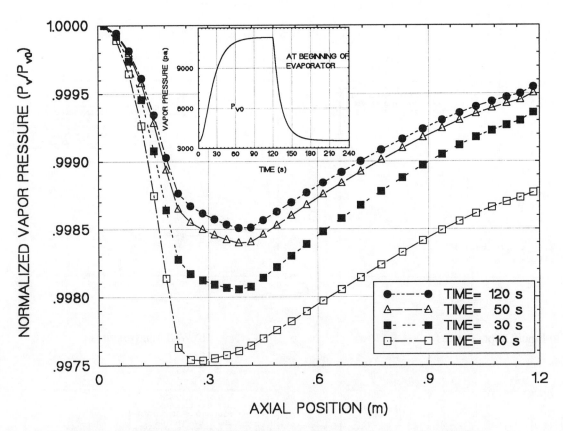

FIGURE 5.   Normalized Vapor Pressure Distribution at Various Times for Case 4.

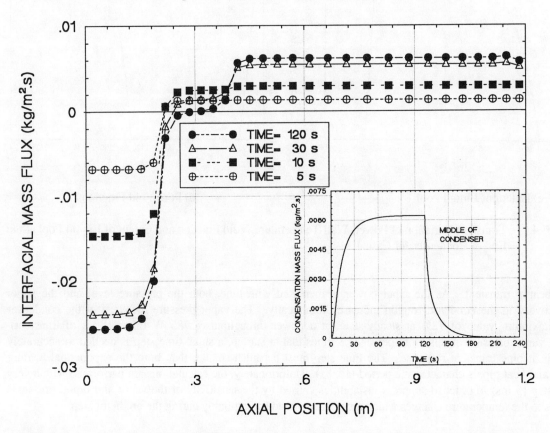

FIGURE 6.   Evaporation/Condensation Mass Flux Distribution at Various Times for Case 4.

1033

Transient Response

Figure 6 shows the variation in the radial evaporation/condensation mass flux at the L-V interface along the heat pipe at different times during the transient. After a very short transient time (less than 2 s), the radial mass flux profile becomes uniform in both the evaporator and condenser regions. However, condensation also occurs along the adiabatic section during the transient because the liquid and wall temperatures are cooler in this region. The condensation mass flux in the adiabatic region progressively disappears with time as the heat pipe reaches steady-state (at 120 s). Notice that after 24 s into the transient, a slight decrease in the condensation rate occurs at the end of the condenser because of the presence of the liquid pool there. Results in Figure 6 show that during the transient the evaporation and condensation mass fluxes in the evaporator and condenser regions, respectively, increase with time, reaching their highest values at steady-state.

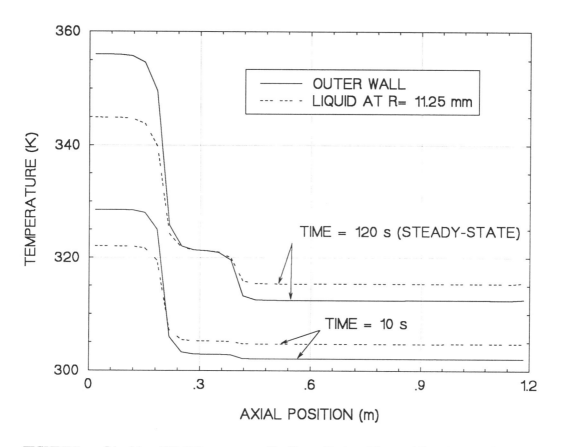

FIGURE 7.   Liquid and Wall Temperatures Profiles at Various Times of Transient for Case 4.

Figure 7 shows the calculated outer wall and liquid axial temperature profiles at different times during the heatup transient. These temperatures are axially uniform in the evaporator, adiabatic section and condenser, with sharp changes at the transitions between these regions. These results suggest that the heat is transferred mainly by radial conduction in the wall and liquid/wick regions, and that axial conduction insignificantly affects the water heat pipes transient. The radial temperature drop across the wall is also negligible (0.2 K) due to the high thermal conductivity of the container material (400 W/m.K for copper). However, the radial temperature drop in the liquid/wick region in the evaporator is significant (42 K) because of the poor thermal conductivity of water (about 0.65 W/m.K), hence justifying using a two-dimensional modeling approach of the heat pipe. With a wick porosity of 0.70, the added conductance of the copper mesh has effectively increased the conductivity of the working fluid in the wick region by only a factor of 1.85. This poor conductance of the water/wick region results in a relatively large temperature drop along the pipe (44 K at the maximum steady-state power throughput of 785 W).

Because the liquid flow in the wick is very slow, conduction is the only mode of heat transfer in this region. It is worth noting that the radial temperature difference between the outer wall surface and liquid at the L-V interface in the adiabatic section is negligible at steady-state, but it is as much as 9 K after 10 s into the transient.

SUMMARY AND CONCLUSIONS

A two-dimensional transient heat pipe model has been developed which incorporates all effects important to the accurate modeling of fully thawed heat pipes (such as liquid compressibility, viscous dissipation, energy and momentum discontinuities at the liquid-vapor interface, and the radius of curvature of the liquid meniscus at the L-V interface). The model divides the heat pipe into three radial regions: wall, liquid/wick, and vapor regions, and solves the complete form of governing equations in these regions, including the momentum and energy jump conditions at the liquid-vapor interface. To account for the thermal expansion of the liquid, a liquid-pooling submodel is incorporated to allow the excess liquid to accumulate at the end of the condenser during heatup of the heat pipe. Results are presented for water/copper heat pipes subject to exponential rise in input power with a maximum evaporator radial flux ranging from 1 W/cm^2 to 5 W/cm^2, and to a step power decrease. The effects of input power and liquid inventory on liquid-pooling phenomena and liquid and vapor pressures profiles are demonstrated, and it is shown that the characteristic period of the heat pipe can be inferred from the analysis of its transient response to these heating and cooling modes. This period is expected to vary with the pipe geometry, temperature levels of interest and the type of condenser cooling (convective or radiative); further work will investigate these effects. While it is found that radial conduction is the dominant mode of heat transfer in the liquid/wick region, it is shown that liquid flow and thermal compressibility must not be neglected to accurately predict wetting, liquid pooling and dewetting processes which control the liquid and vapor pressures profiles along the heat pipe, as well as the viscous limit and dryout conditions.

Acknowledgments

This research is funded under NASA Grant No. NAG3-941 by NASA Lewis Research Center to the University of New Mexico's Institute for Space Nuclear Power Studies (ISNPS). The authors wish to thank Mr. Albert Juhasz of NASA Lewis Research Center for his suggestions and valuable input during the progress of this work.

References

Beam, R. M. and R. F. Warming (1978) "An Implicit Factored Scheme for the Compressible Navier-Stokes Equations," *AIAA Journal*, 16(4): 393-402.

Bowman, W. J. and J. E. Hitchcock (1988) "Transient Compressible Heat Pipe Vapor Dynamics," in *Proceedings of the 1988 ASME National Heat Transfer Conference*, held in Houston, TX, 24-27 July 1988, HTD-96, vol. 1, pp. 329-337.

Busse, C. A. and J. E. Kemme (1978) "The Dry-Out Limits of Gravity-Assist Heat Pipes with Capillary Flow," in *Proceedings of the 3rd International Heat Pipe Conference*, paper AIAA-78-383, pp. 41-48.

Cao, Y. and A. Faghri (1990) "Transient Two-Dimensional Compressible Analysis for High-Temperature Heat Pipes with Pulsed Heat Input," *Numerical Heat Transfer, Part A*, 18: 483-502.

Chang, W. S. and G. T. Colwell (1985) "Mathematical Modeling of the Transient Operating Characteristics of a Low-Temperature Heat Pipe," *Numerical Heat Transfer*, 8: 169-186.

Costello, F. A., M. Merrigan, and T. R. Scollon, Jr. (1986) *Detailed Transient Model of a Liquid-Metal Heat Pipe*, Presented at the *Transient Heat Pipe Modeling Workshop*, held in Los Alamos, NM, 4-5 March 1986.

Defay, R. and I. Prigogine (1966) *Surface Tension and Adsorption*, English Edition, John Wiley and Sons, Inc., New York, NY, Chapter XV.

Delhaye, J. M. (1974) "Jump Conditions and Entropy Sources in Two-Phase Systems. Local Instant Formulation," *International Journal of Multiphase Flow*, 1: 395-409.

Delhaye, J. M. (1976) "Local Instantaneous Equations," *Two-Phase Flows and Heat-Transfer*, 1: 59-79, S. Kakac and T. N. Veziroglu, eds. *(Proceedings of NATO Advanced Study Institute on Two-Phase Flows and Heat-Transfer*, congress held in Istanbul, Turkey, 16-27 August 1976).

Deverall, J. E., J. E. Kemme, and L. W. Florschuetz (1970) *Sonic Limitations and Startup Problems of Heat Pipes*, Los Alamos Scientific Laboratory Report LA-4518 (accession number N71-18944), Los Alamos, NM, November 1970.

Fahgri, A. and M. M. Chen (1989) "A Numerical Analysis of the Effects of Conjugate Heat Transfer, Vapor Compressibility, and Viscous Dissipation in Heat Pipes," *Numerical Heat Transfer, Part A*, 16: 389-405.

Gourlay, A. R. and J. L. Morris (1968) "Finite-Difference Methods for Non-Linear Hyperbolic Systems," *Math. Comp.*, 22: 28-39.

Hall, M. L. and J. M. Doster (1988) "Transient Thermohydraulic Heat Pipe Model," in *Space Nuclear Power Systems 1987*, M. S. El-Genk and M. D. Hoover, Eds., Orbit Book Company, Malabar, FL., Vol. VI, pp. 205-218.

Harlow, F. H. and A. A. Amsden (1971) "A Numerical Fluid Dynamics Method for all Flow Speeds," *J. Computational Physics*, 8: 197-213.

Issacci, F., I. Catton, A. Heiss, and N. M. Ghoniem (1988) "Analysis of Heat Pipe Vapor Dynamics," in *Proceedings of the 1988 ASME National Heat Transfer Conference*, held in Houston, TX, 24-27 July 1988, HTD-96, vol. 1, pp. 361-365.

Jang, J. H. (1988) *An Analysis of Startup from the Frozen State and Transient Performance of Heat Pipes*, a Dissertation in partial fulfillment of the requirements for the PhD Degree in Mechanical Engineering, Georgia Institute of Technology, Atlanta, GA, February 1988.

Jang, J. H., A. Faghri, and W. S. Chang (1989) "Analysis of the Transient Compressible Vapor Flow in Heat Pipes," in *Proceedings of the 1989 ASME National Heat Transfer Conference*, HTD Vol. 110: Numerical Heat Transfer with Personal Computers and Supercomputing, pp. 113-120.

Merrigan, M. A., E. S. Keddy, and J. T. Sena (1986) "Transient Performance Investigation of a Space Power System Heat Pipe," in *Proceedings of the AIAA/ASME 4th Joint Thermophysics and Heat Transfer Conference*, held in Boston, MA, 2-4 June 1986.

Patankar, S. V. (1980) *Numerical Heat Transfer and Fluid Flow*, Hemisphere Publishing Company, Washington, DC.

Peery, J. S. and F. R. Best (1987) "Simulation of Heat Pipe Rapid Transient Performance Using a Multi-Nodal Implicit Finite Difference Scheme," in *Space Nuclear Power Systems 1986*, M. S. El-Genk and M. D. Hoover, Eds., Orbit Book Company, Malabar, FL., Vol. V, pp. 145-150.

Ransom, V. H. and H. Chow (1987) "ATHENA Heat Pipe Transient Model," in *Trans. of 4th Symposium on Space Nuclear Power Systems*, CONF-870102-Summs., held in Albuquerque, NM, 12-16 January 1987, pp. 389-392.

Seo, J. T. (1988) *Transient Analysis of Space Nuclear Power Systems*, a Dissertation submitted in partial fulfillment of the requirements for the PhD Degree in Nuclear Engineering, The University of New Mexico, Albuquerque, NM, May 1988.

Seo, J. T. and M. S. El-Genk (1989) "A Transient Model for Liquid Metal Heat Pipes," in *Space Nuclear Power Systems 1988*, M. S. El-Genk and M. D. Hoover, Eds., Orbit Book Company, Malabar, FL., Vol. IX, pp. 405-418.

Stewart, H. B. and B. Wendroff (1984) "Review Article. Two-Phase Flow: Models and Methods," *J. of Computational Physics*, 56: 363-409.

Tilton, D., J. Johnson, J. Gottschlich, and S. Iden (1986) *Transient Response of a Liquid Metal Heat Pipe*, Air Force Wright Aeronautical Laboratories Report AFWAL-TR-86-2037, Wright Patterson AFB, OH (August 1986).

Yanenko, N. N. (1971) *The Method of Fractional Steps*, Springer-Verlag, New-York.

-- **Nomenclature** --

English

A_r:	Interfacial surface area (m^2)	ρ:	Density (kg/m^3)
$\overline{\overline{D}}$:	Deformation rates tensor	σ:	Surface tension (N/m)
$\vec{F}$:	External acceleration (m/s^2	τ:	Exponential period (s)
h:	Enthalpy (J/kg)	ϕ:	Viscous dissipation
k:	Thermal conductivity (W/m.K)		
M:	Molecular weight (kg/mole)		**Superscripts**
$\overline{\dot{m}}$:	Evaporation/condensation mass flux ($kg/m^2.s$)		
P:	Pressure (Pa)	equ:	Equilibrium
$\vec{Q}$:	Conductive heat flux (W/m^2)	int:	L-V interface
R:	Radius (m), Universal gas constant	j:	Axial cell number
	(8.314 J/mole.K)	L:	Liquid phase
R_c:	Radius of curvature (m)	V:	Vapor phase
R_p:	Wick pore radius (m)		
r:	Radial coordinate		**Subscripts**
T:	Temperature (K)		
t:	Time (s)	a:	Adiabatic region
$\vec{U}$:	Velocity field	c:	Condenser region
V:	Velocity of displacement	e:	Evaporator region
Vp:	Volume of hemispherical pores at	L:	Liquid Phase
	liquid-vapor interface (m^3)	o:	Wall outer surface
z:	Axial coordinate	p:	Pore
		r:	Radial component
	Greek	sat:	Saturation
		v:	Vapor phase, liquid inner surface
α_p:	Pore void fraction at liquid-vapor interface		
β_T:	Isothermal compressibility factor (1/Pa)	v0:	Vapor phase, at beginning of evaporator
ε:	Wire-screened wick porosity		
μ:	Dynamic viscosity (kg/m.s)	w:	Wall region
μ_c:	Cosine of contact angle of liquid meniscus	z:	Axial component

MODERATED HEAT PIPE THERMIONIC REACTOR (MOHTR)

MODULE DEVELOPMENT AND TEST

Michael A. Merrigan
Vincent L. Trujillo

Los Alamos National Laboratory
P.O. Box 1663
Los Alamos, New Mexico 87545
(505)667-9834
FTS 843-2626

Abstract

The Moderated Heat Pipe Thermionic Reactor (MOHTR) thermionic space reactor design combines the low risk technology associated with the Thermionic Fuel Element (TFE) Verification Program with the high reliability heat transfer capability of liquid metal heat pipe technology. The resulting design concept, capable of implementation over the power range of 10 to 100 kWe, offers efficiency and reliability with reduced risk of single point failures. The union of TFE and heat pipe technology is achieved by imbedding TFEs and heat pipes in a beryllium matrix to which they are thermally coupled by brazing or by liquid metal (NaK or Na) bonding. The reactor employs an array of TFE modules, each comprising a TFE, a zirconium hydride (ZrH) cylinder for neutron moderation, and heat pipes for transport of heat from the collector surface of the TFE to the waste heat radiator. An advantage of the design is the low temperature drop from the collector surface to the radiating surface. This is a result of the elimination of electrical insulation from the heat transport path through electrical isolation of the modules. The module used in this study consisted of a beryllium core, an electrical cartridge heater simulating the TFE, and three heat pipes to dissipate the waste heat. The investigation was focused on the thermal performance of the assembly, including evaluation of the sodium and braze bonding options for minimizing the thermal resistance between the elements, the temperature distribution in the beryllium matrix, and the heat pipe performance. Continuing subjects of the investigation include performance of the heat pipes through start-up transients, during normal operation, and in a single heat pipe failure mode. Secondary objectives of the investigation include correlation of analytic models for the thermionic element and module including the effects of gap thermal conductances at the modules electrically insulated surfaces.

INTRODUCTION

The central design feature of the heat pipe cooled, beryllium moderated thermionic reactor is the use of a beryllium matrix to transfer heat from the thermionic fuel elements (TFEs) and stainless steel clad ZrH moderator rods to stainless steel/ potassium heat pipes. The heat pipes are routed from the modules to form the heat rejection radiator surface. Depending on the reactor output power, these beryllium modules may be either cylindrical sectors which form a right circular cylindrical reactor core when fastened together or dished, truncated triangles. In this effort the module configuration is based on the cylindrical configuration as indicated in Fig. 1. The core is segmented to simplify handling procedures during fuel assembly, particularly with regard to prevention of accidental criticality. Assembly of the core sections is done by inserting the TFEs, rods, and the heat pipes into the holes in the beryllium module and brazing or bonding them in place. The alternative method of thermal bonding employs a sodium filler between the components and the beryllium. In this method a fill stem and sodium reservoir configuration is used that allows molten sodium to be forced into the annular region between the rods and the beryllium block with subsequent pinchoff and pinchoff capping procedures being used to seal the assembly. The TFEs employed in the design are similar to those being developed for the TFE Verification Program and, hence have a niobium outer sheath. Because of the sheath thermal expansion mismatch between niobium and beryllium thermal bonding with sodium is preferred for these components. The heat deposition rate in the ZrH rods due to gamma ray and neutron heating is so low that neither brazing or sodium bonding may be needed. Cooling by thermal radiation from the surface of the ZrH rod to the beryllium module may not unduly increase the temperature of these components.

The surface of an individual beryllium module assembly is electrically insulated from the surrounding modules. Therefore, the finned heat pipe heat rejection surface for each module is at a different electrical potential. This approach has the advantage of eliminating the need for electrical insulation between the collector surface of the TFE and the radiator, resulting in very low temperature differences and consequent reduced radiator size. In the reactor configuration considered in this effort the three heat pipes used in a single module are coupled through a common radiation fin in order to improve the thermal performance for the off-design case where only two heat pipes are operational.

THERMAL ANALYSIS

Thermal analyses were conducted to determine temperature distribution occurring in a module of the reactor core for conditions of normal heat pipe operation and for the case of reduced thermal dissipation due to loss of a heat pipe or thermal bond failure. Results from the analyses were used to determine radiator area and the heat pipe configuration.

The thermal analysis considered a single, three heat pipe cooled module operating at steady-state conditions for a circular reactor design having a design power level of 30 kWe. The analysis was performed with the finite element based ABAQUS source code. The heat pipe configurations analyzed including circular heat pipes of 1.00-cm, 1.27-cm diameter, and lunette shaped cross-section heat pipes. The thermal power throughput from the thermionic fuel element was assumed to be 3 kW and the heat pipe average operating temperature was taken as 900 K. A summary of the parameters used in the analysis is given in Table 1. Module cross-section temperature gradients varied from 10 K to 21 K for normal heat pipe operation and from 69 K to 86 K for the case of a single heat pipe failure.

Geometry And Boundary Conditions

The module simulated in the analysis was from the outer tier of the circular reactor. Included in the module were the Thermionic Fuel Element (TFE) providing the heat source, and three heat pipes used to transport the thermal power to the fin radiator. Because neutronic analysis showed that the internal heat generation in the beryllium and in the zirconium hydride moderator was small it was not considered in the test module thermal analysis. The semi-circular zirconium hydride cross-sections were not included and their interface surfaces were assumed to be adiabatic. A schematic of the cross-section is shown on Fig. 1. Values and dimensions used in the analysis are listed in Table 1.

The heat conduction path begins on the surface of the thermionic fuel element and continues through the beryllium module to the three heat pipes and then to the fin radiator where the energy is radiatively dissipated to a low temperature sink. Thermal properties for commercially available beryllium were used for the core module. The heat source was modeled by a constant distributed flux on the surface of the thermionic element. Heat transport from the 2-D beryllium core cross-section to the 3-D radiator was modeled with the use of 1-D link elements. The link elements and the region of the heat pipe surface in contact with the beryllium core were given a very large thermal conductivity to simulate the isothermal characteristic of the heat pipe. The outer perimeter of the module cross-section was considered adiabatic.

TABLE 1. Parameters used for Components in Thermal Model.

COMPONENT	OUTER DIAMETER (cm)
Zirconium Hydride	2.54
Thermionic Fuel Element	1.85
Module Perimeter	
Outer boundary	38.4
Inner boundary	30.8
Included angle	18°
Thermal flux on TFE surface	28.8 W/cm^2
Heat Pipe Configuration	
Case 1	1.00-cm Diameter
Case 2	1.27-cm Diameter
Lunette Cross-section	
Inner radius	1.025-cm
Outer radius	1.660-cm
Thickness	0.635-cm

Results

Temperature distribution was determined by calculating nodal temperatures in each of the elements. In all normal operating cases the regions of highest emperature occur midway between

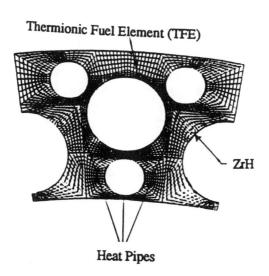

FIGURE1. Schematic of Test Module Cross-Section.

the heat pipes along the surface of the beryllium-TFE interface and the lowest temperatures occur at the heat pipe surface. In all the single heat pipe failure cases the maximum temperature occurs along the surface of the beryllium-TFE interface near the failed heat pipe. Temperature gradients in the module cross-section were lowest when lunette shaped heat pipes were used, both for the normal case and for single heat pipe failure operation. Results of the test module thermal analysis are summarized in Table 2. Although the use of lunette shaped heat pipes gave the lowest temperature difference, the increase in system temperature during off-normal operation was within allowable limits for the 1.27-cm diameter circular heat pipes and they were selected for the test module based on fabrication considerations.

TABLE 2: Summary of Temperature Gradients in the Test Module.

| Heat Pipe Geometry | ΔT (K) | |
	Normal Operation	Single Failure
1.00 cm Diameter	21	86
1.27 cm Diameter	18	80
Lunette	10	69

THERMAL BONDING STUDIES

Two methods of achieving a low thermal resistance bond between the TFEs and the beryllium module and between the heat pipes and the beryllium were investigated. The thermal bonding options included using a liquid metal to fill the void volume between the TFE or the heat pipes and the beryllium core and brazing the TFE and heat pipes directly to the beryllium core. The two methods were evaluated experimentally using tube assemblies and infra-red visualization to determine the temperature profiles.

Sodium Bonding

A preliminary sodium bonding experiment was conducted to observe the effects of using sodium to fill a small annulus gap between two stainless steel tubes. The outer surface of the inner tube was roughened by knurling to improve the surface wetting and ensure a uniform fluid film. An outer sheath tube was sized to allow a slip fit between the outer surface of the inner tube and the inner surface of the sheath. An electrical cartridge heater was inserted into the inner tube to provide the thermal input for the experiment. A thermal bond test test was performed at an average temperature of 873 K. Thermocouples were placed along the length of the outer tube and heater surface for direct temperature measurements. Surface temperature variation was monitored with a Thermatrace infrared visualization system (IR). The resolution of the IR scanner permitted direct determination of temperature variations on the surface on a scale that would be capable of indicating small void regions in the sodium bond.

Results from the IR-scanner test indicated maximum surface temperature differences of less than 5 K, within the uncertainty range of the measurements including the effect of variation in surface emissivity. The relatively small surface temperature variations indicated that the sodium filling of the void region was complete. The temperature measurements were repeated after shut-down and re-start to ensure that cycling did not affect the sodium distribution.

Brazing

Brazing the heat pipes and TFEs directly to the beryllium matrix was evaluated as an alternative method of thermal bonding. Tests were conducted to evaluate filler materials and procedures for joining of beryllium and stainless steel for the design temperature range. Because beryllium compatibility was a concern at the application temperature, silver based filler materials were considered. A review of the literature (Grant 1979) indicated that silver does not form stable beryllides and that the effect of beryllium ion diffusion-embrittlement is small at the application temperature of 900 K. Initially braze tests were run with beryllium and stainless steel flat coupons electrically heated in a vacuum. Good wetting characteristics for the braze material with both the beryllium and stainless steel surfaces were observed with a silver-copper alloy (identify). Following the coupon tests a braze test was performed with a beryllium and a stainless steel tube. The parts were fabricated to allow the beryllium tube to have a slip-fit over the stainless steel tube. After brazing the parts together the sample was sliced in both axial and radial planes. The filler material appeared to be evenly distributed in the annulus region and to have good wetting characteristics on the beryllium and stainless steel surfaces. No visible fractures due to the stress induced from thermal expansion mismatchs were observed.

BERYLLIUM-SODIUM COMPATIBITY TESTS

Literature data on compatibility between beryllium and alkali metals such as potassium and sodium is inconsistent. (Darwim 1960) The test data from the literature are based on tests conducted circa 1955 and involve materials of questionable purity. Because of advances in materials fabrication techniques and alkali metal handling methods a re-evaluation of beryllium-sodium compatibility at 900 K was initiated. The test set-up consisted of a beryllium tube sample inserted into a stainless steel capsule. Spacers were used to allow the sample to maintain a centered position in the capsule and to minimize surface contact between the stainless steel and beryllium sample. A sodium charge was transferred to the capsule by vacuum distillation and the fill tube was hermetically sealed. The capsule-sample unit was placed in a quartz tube vacuum chamber and heated by rf-induction to a temperature of 900 K. The system was maintained at 900 K for a period of 1000 hours. Radiography of the sample at the end of the 1000 hours did not show any evident material transport. Plans are to conduct destructive analysis including chemical and metallography procedures. Oxygen content and beryllium residue in the sodium medium will be determined as well as surface corrosion on the beryllium sample.

HEAT PIPE DESIGN

The Los Alamos steady-state heat pipe code (HTPIPE: Woloshun et. al. 1988) was used to simulate heat pipe behavior and determine critical parameters. Potassium was selected as the working fluid for the 900 K average operating temperature. Wicking and capillary pumping requirements were satisfied with the use of a 400 mesh screen annular wick. Stainless steel tubing was assumed for containment. Analyses were run for a range of heat pipe vapor space diameters and fluid annulus widths. The heat pipe evaporator length was assumed equal to the test module length of 18 cm and the condenser length was set at 109-cm to maintain a reasonably low length-to-diameter ratio and provide adequate radiator area. The condenser length used in the hydrodynamic analysis was based on the use of a radiation fin. The increase in cooling due to the fins was simulated in the analysis by increasing the value of the coefficients in the radiation heat transfer equation by an amount equal to the ratio of fin area to the tube area. The heat pipes were designed for thermal power throughput levels corresponding to the two operating heat pipe case. Thermal power throughput of the module was assumed to be 3 kW so the heat pipes were designed to operate with power throughput levels of 1500 W at an average temperature cf 900 K. The analysis showed that a heat pipe of 1.27-cm diameter subject to the above conditions would operate in a conservative mode with respect to the sonic, capillary, and entrainment limits. Temperature gradients along the evaporator axis were approximately 3 K with an essentially isothermal condenser. Values for the geometrical parameters used are summarized on Table 3.

TABLE 3. Summary of Parameters for Heat Pipe Fabrication.

COMPONENT	DIMENSION
Pipe Outer Diameter	1.27-cm
Pipe Inner Diameter	1.17-cm
Annulus Thickness	0.0343-cm
Distribution Screen Thickness	0.0546-cm
Evaporator Length	18.0-cm
Condenser Length	109-cm
Adiabatic Length	0.0-cm
Heat Input Rate	1500 Watts
Screen Mesh	400x400

HEAT PIPE FABRICATION AND TESTING

   Three heat pipes were fabricated. Stainless steel tubing of 1.27 cm diameter and 0.0508 cm wall thickness was used for the envelope. The annulus wick was fabricated from layers of 400 x 400 mesh stainless steel screen and was formed by sintering the layers while under compression. The effective pore size of the finished wick was determined by immersing the wick in an ethanol bath and pressurizing the internal volume with helium gas until a gas bubble formed on the wick surface. The pressure required to form the gas bubble was recorded and the corresponding pore size was then determined. A maximum pore diameter of 37 micrometers was determined for the annulus wick, a value well below the minimum design pore diameter of 63 micrometers, providing a margin in capillary pressure capacity of almost two times the design value.

   The heat pipes were charged with a predetermined volume of potassium by vacuum distillation. Heat pipe wet-in was accomplished in a vacuum furnace at a temperature of 900 K for a duration of 90 hours. Each heat pipes was operated at the design temperature of 900 K prior to assembly. These tests were conducted with the individual heat pipes in a quartz tube vacuum chamber with the thermal input provided by rf-induction heating. Heat pipe performance in these tests appeared to be in agreement with the design analysis.

MODULE DESIGN AND FABRICATION

   The beryllium test module was fabricated by a Los Alamos Laboratory beryllium machine shop. Because of the large number of curved surfaces machining of the module was accomplished by hot-wire electron-discharge-machining (EDM). Conventional machining methods were used for drilling the lead holes for insertion of the hot-wire cutter. A smooth surface finish was achieved by allowing the hot-wire cutter to make several passes along the cut surface. A tolerance of 0.005-cm was maintained along the 18-cm length of the holes in the finished part. A summary of module physical parameters are listed in Table 1.

MODULE TEST SET-UP DESIGN

   Testing of the module will be performed in vacuum chamber. An electrical cartridge heater capable of delivering 3 kW of thermal power will be used to simulate the thermal power input from a TFE unit. The beryllium matrix will be cooled by conducting heat from the TFE surface to the three heat pipes where the thermal energy will be radiatively dissipated from the condenser surface. The electrical heater and the heat pipes will be thermally coupled to the beryllium matrix by brazing. Temperature differences across the thermal bond line between the simulated TFE and the module body, and the module body and the heat pipes, as well as temperatures on the heat pipe evaporator surface, and along the condenser surface of the heat pipes will be measured with thermocouples and infrared surface scanners. In addition, the heat flow through the module and heat pipes will be measured by calorimetry using a water cooled radiantly coupled heat sink. Heat flow from the beryllium surface will be minimized with the use of several radiation shields.

Steady-state tests will be conducted with all heat pipes operating and repeated with one heat pipe not operating to simulate the case of a failed thermal bond or operationally inactive heat pipe. Heat pipe failure will be simulated by insulating one of the heat pipes from the radiant heat sink, using thermal radiation barrier materials. This will allow the inactive heat pipe to isothermalize at the module temperature with near adiabatic interface conditions. Instrumentation and recorded data for this test will be the same as for the normal operating test.

After completion of the steady-state tests a limited investigation of start-up transient conditions is planned. These tests will involve monitoring of temperatures and heat flows through the module and heat pipes during a start-up from ambient temperature conditions with selected TFE power profiles through the start. The purpose of these tests will be to ensure that a transient overtemperature condition will not occur due to start-up thermal transport limits of the heat pipes.

CONCLUSION

Based on results from thermal analyses the temperature difference from the TFE surface to the radiator surface in the beryllium matrix is expected to be less than 20 K during normal operation and about 80 K for the case of loss of operation of a single heat pipe. The system temperature increase due to an inactive heat pipe is within the allowable operating limit for the system. Results from the HTPIPE code indicate that the heat pipes will be operating in a conservative mode being well within the operational limits at the design temperature for normal and single heat pipe failure modes.

Preliminary results indicate that acceptable thermal bonds between the beryllium matrix and the heat pipes and TFEs are achievable by either sodium bonding or brazing. The sodium bond test results demonstrated that annulus regions were adequately filled and produced material interface zones with a low thermal resistance. Because of the sheath thermal expansion mismatch between beryllium and niobium the sodium bonding procedure may be preferable. Brazing the heat pipes and TFEs to the beryllium matrix is currently the most practical option for the thermal tests where stainless steel rather than niobium is used for the heater surface. Preliminary tests showed good wetting of the beryllium and stainless steel with the silver-copper filler material. No visible fractures were observed in the beryllium tube surface due to the stress induced by mismatch of the coefficient of expansion between the beryllium, filler material, and stainless steel.

Results of the analyses and tests conducted to date have indicated no adverse system effects due to single heat pipe failure, and have demonstrated the feasibility of achieving acceptable thermal coupling between the beryllium matrix and TFEs and heat pipes by either sodium bonding or brazing. The initial indications are that the MOHTR design concept is feasible and is capable of providing substantially reduced heat rejection radiator mass compared to designs requiring heat transfer across and electrical insulator in the TFE collector region.

Acknowledgments

This work was performed at the Los Alamos National Laboratory and was supported by the U.S. Air Force Phillips Laboratory. Technical diirection of the effort was provided by Dr. Michael Schuller.

References

Woloshun, K. A., M. A. Merrigan, and E. D. Best, (1988) "HTPIPE: A Steady-State Heat Pipe Analysis Program, " Los Alamos National Laboratory Manual, LA-11324-M, November 1988.

Hibbit, Karlson, and Sorenson, (1988) *ABAQUS: Finite Element Analysis,* Hibbit, Karlson, and Sorenson Inc.

Grant, L.A., (1979) "JOINING II: Brazing and Soldering," in *Beryllium Science and Technology,* Vol. 2, Edited by Dennis R. Floyd, and John N. Lowey, Plenum Publishing Corporation.

Vickers, W. (1963) " An Investigation into the Brazing of Beryllium," in *Proc. The Metallurgy of Beryllium*, Intern. Conference, Chapman and Hall, London.

Darwin, G. E. and Buddery, J. H. (1960) *BERYLLIUM,* Acedemic Press, New York.

THERMIONIC IN–CORE HEAT PIPE DESIGN AND PERFORMANCE

W. R. Determan and G. Hagelston
Rocketdyne Division
Rockwell International Corporation

6633 Canoga Avenue
Canoga Park, CA 91303
(818) 718–3375

Abstract

The heat pipe cooled thermionic reactor (HPTI) relies on in–core sodium heat pipes to provide a redundant means of cooling the 72 thermionic fuel elements (TFEs) and 36 driver fuel pins which comprise the 40 kWe core assembly. In–core heat pipe cooling was selected for the reactor design to meet the requirements for a system design with the potential to achieve a high survivability level against natural and man–made threats and one that possesses no–mission ending single point failures. A detailed study was performed to determine the potential in–core heat pipe geometries which could be developed for an HPTI concept. Requirements and performance estimates were developed for two in–core heat pipe geometries. Both nominal and faulted operating conditions were evaluated using a two–dimensional thermal model of the core to assess TFE and driver fuel pin temperature profiles. A bow tie in–core heat pipe geometry was selected as the optimum design using a HPTI honeycomb core structure.

INTRODUCTION

The U.S. Air Force Phillips Laboratory is sponsoring research studies of small, compact space nuclear power systems based upon thermionic power conversion technology for military mission applications. The requirements for these missions are numerous and quite stringent. Three of the most important requirements for such missions are the need for a power system with high survivability potential, multiple restart capability with potential start–up times as short as one hour, and a system design which possesses no–mission ending single point failures. In response to the request for proposal from the Phillips Laboratory, the General Atomics/ Rocketdyne team proposed to develop a conceptual design of a heat pipe cooled thermionic reactor concept. The HPTI concept as originally proposed is illustrated in Figure 1. The basic thermionic cell, fueled with enriched UO_2, is assembled into a six–cell series connection within a TFE. The TFE is secured within the receiving cavity of a core hexcan subassembly which has a 3.85 cm flat–to–flat dimension. The hexcan is composed of the receiving cavity for the TFE and six trapezoidal–shaped sodium core heat pipes, each of which have been subdivided into four vapor chambers. The six sodium heat pipes surrounding each

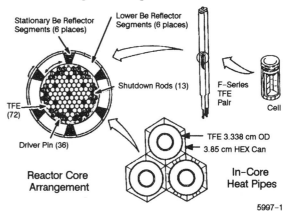

FIGURE 1. Heat Pipe Cooled TFE Concept.

TFE transfer their waste heat axially to the radiator heat pipes located above the reactor's upper axial reflector. The top cavity formed by the six heat pipes and each TFE lead provide a 2.54 cm diameter by 17.5 cm high thermal well that receives the evaporator section of a roughly four–meter–long radiator heat pipe. Thirty–six driver fuel pins with UN fuel are enclosed in their own hexcan core subassembly and located in the outer regions of the core assembly to meet criticality requirements. Figure 2 illustrates the HPTI reactor/shield assembly with its radiator interface. The hexcan subassemblies were brazed together to form the core assembly. In a two–step braze process, the TFEs would then be brazed into their respective receiver cavities during the final core assembly process. When the TFE leads are connected into a 12x6 series/parallel electrical arrangement, the HPTI design exhibits no single point failures. The in–reactor test experience with sodium heat pipes is summarized in Table 1 [Ranken 1990]. The operational test environments and heat loads are comparable to those in the proposed HPTI concept.

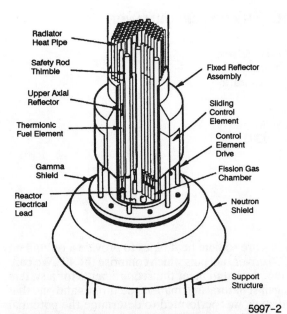

Radiator
Heat Pipe

Safety Rod
Thimble

Upper Axial
Reflector

Thermionic
Fuel Element

Gamma
Shield

Reactor
Electrical
Lead

Fixed Reflector
Assembly

Sliding
Control
Element

Control
Element
Drive

Fission Gas
Chamber

Neutron
Shield

Support
Structure

5997-2

FIGURE 2.   Heat Pipe Thermionic Reactor.

CORE HEAT PIPE GEOMETRIES AND ANALYSIS

Initial studies of the original core design indicated that at least two alternate heat pipe geometries could be used in the HPTI concept. The tricusp in–core heat pipe design, shown in Figure 3, provides a larger hydraulic diameter heat pipe design than available with the hexcan. Two tricusp heat pipes provide the same heat transport capability as the six heat pipes employed within each hexcan.

As the initial design work began in the hexcan fabrication studies, the complexity of the design became apparent. The hexcan's outer wall was found to be redundant and increased the TFE pitch and structural materials within the core. The bow tie heat pipe configuration, also shown in Figure 3, was found to be a more efficient design with a reduced number of

TABLE 1. LANL Heat Pipe Status.

- 19 Na/SS (and 8 K/SS) heat pipes tested in EBR-11 core
  a) 0.5 to 5 kWt
  b) 900–1100 K
  c) 1100 to 23,000 h (7211 h avg., 137,000 h total)
  d) 0.27 to 10 x 10^{22} n/cm^2, (E_n> 0.1 MeV)
  e) Gravity assist and gravity opposed

  No faliures

- Bends (and flexibility capability) demonstrated
  a) 2m Na/SS at 4 kWt, –1100K
  b) Bellows heat pipe

- Frozen start-up demonstrated many times

- Freeze–out shown analytically to be no problem (Experimental verification needed)

D635-0224

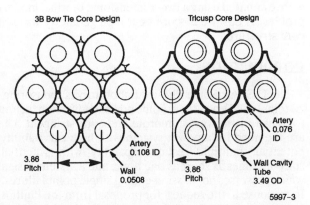

3B Bow Tie Core Design

Tricusp Core Design

3.86
Pitch

Wall
0.0508

Artery
0.108 ID

Artery
0.076
ID

Wall Cavity
Tube
3.49 OD

3.86
Pitch

5997-3

FIGURE 3.   Core Heat Transport Assembly Design Options.

parts required for core construction. The HTPIPE code [Woloshun 1988] was used to predict the performance of the bow tie and the tricusp at nominal and three upset core conditions. The analyses were based upon the pipe's hydraulic diameter and did not include two–dimensional liquid or vapor flow effects. These results are shown in Table 2 for nominal conditions and for adverse and favorable "g" environments. They indicate that both heat pipe designs can provide adequate axial heat transport for the TFEs. The driver core pin region was also analyzed in a similar fashion for both the bow tie and tricusp heat pipe geometries. There are two unique bow tie geometries and four new tricusp geometries in this driver region. Each heat pipe was evaluated at nominal operating conditions to determine its heat transport capacity versus its required cooling load. Figure 4 illustrates the design margins available in the driver core region for these heat pipes. The large margins are the result of the 3.86 cm fuel pin pitch and the 2.54 cm outside diameter of the driver pins.

CENTRAL CORE THERMAL MODEL

A two–dimensional, multinode model of the reactor core was developed using the Thermal Analyzer Program (TAP). The 2–D TAP model used a 10–pin (TFE) mock–up of the central core region. This model, shown in Figure 5, permitted investigating multiple component failures within the core and the local radiator heat pipe sinks. Core midplane and condenser midplane sections were coupled in the model by using effective heat pipe admittances (inverse resistances), which were generated from the previous heat pipe analyses. All of the TFE internal components were modeled explicitly using six circumferential nodes for the heat pipe wall, sheath tube, trilayer, and collector. A single central node was used for each TFE emitter. The purpose of the model was to

TABLE 2. Core Axial Heat Transport Capabilities (Both Designs Exceed Requirements).

Parameter	Heat Pipe Geometries	
	Bow Tie	Tricusp
Geometry type	3B	2
No per TFE	3	2
D_n(cm)	0.632	0.953
No. of arteries	1	3
Required power (kWt) at 1050 K (margin level)	2.33 (3.5)	3.5 (5.2)
Axial temperature drop (K)	5.1	3.4
Maximum power (kWt), at 1050 K	4.8	4.93
Neg 1-G power (kWt)	3.1	4.93
Pos 1-G power (kWt)	4.8	4.93
Core pitch (cm)	3.86	3.86

D635–0224

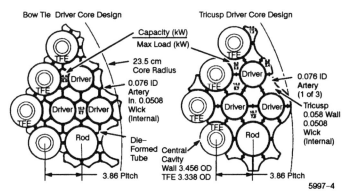

FIGURE 4. High–Power Driver Pins are Cooled by Larger Heat Transport Assemblies with Higher Design Margins.

identify which in–core heat pipe geometry provided greater lateral (across the core) heat transport (for example, lower temperature differentials) following an upset event. The three upset events evaluated were: (1) loss of a single radiator sink, (2) loss of two adjacent radiator sinks, and (3) loss of an in–core heat pipe. The nominal operating conditions evaluated with the TAP model are shown in Figure 6. Both designs exhibited a roughly 70 K temperature drop to the radiator heat pipe sink temperature. Table 3 summarizes the results from the central core thermal model. The bow tie design provided slightly lower core temperature differentials and consequential emitter temperature rises for all three upset events, but these were not considered to be significant.

DRIVER CORE THERMAL MODEL

A second TAP model was constructed for the driver core region using the tricusp heat pipe geometry. The nominal design conditions for the driver fuel pin are shown in Figure 7 along with a core midplane view. The higher power level and heat fluxes developed from the driver fuel pin resulted in a larger clad-to-radiator heat pipe temperature drop of 100 K. Increasing the condenser length by 5 cm would reduce the temperature drop to the radiator to the same 70 K value for the TFE central core section. The two upset conditions evaluated were loss of a single driver radiator heat pipe and loss of a worst–case

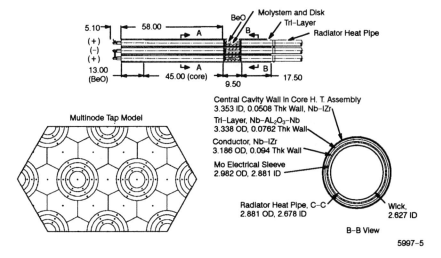

FIGURE 5. Two–Dimensional Heat Flow Within the Core Evaluated with Multinode Tap Model.

1048

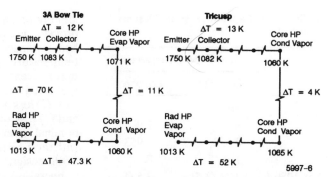

FIGURE 6.   Nominal Operation Temperature Distribution
Is Essentially the Same for Two Designs.

TABLE 3. In–Core Heat Pipe Comparative Performance.

	Bow Tie 3A)	Tricusp
Loss of single radiator heat pipe		
Core max temp differential	7 K	13 K
Emitter temperature rise	8 K	13 K
Loss of 2 adjacent radiator HPs		
Core max temp differential	12 K	16 K
Emitter temperature rise	33 K	37 K
Loss of in–core heat pipe		
Core max temp differential	2 K	11 K
Emitter temperature rise	31 K	34 K

D635–0224

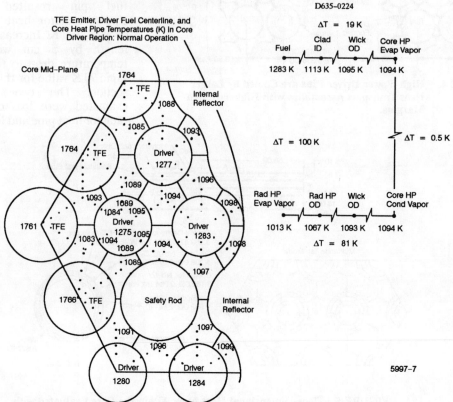

FIGURE 7.   Driver Region Tricusp Core Geometry During Normal Operation
Temperature Drop Is 100 K.

1049

driver in-core heat pipe. Table 4 summarizes the results from the driver core region model which showed that the adjacent TFE emitter temperatures would rise 31 K following the loss of a driver radiator sink. The driver pin fuel centerline temperature would rise 28 K. The loss of the worst-case driver in-core heat pipe would result in a 34 K temperature rise at the driver fuel centerline and only an 8 K emitter temperature rise. The hot-spot clad wall temperature rise would be 132 K following the loss of the in-core heat pipe due to its asymmetric cooling geometry. This event will require further stress evaluation.

SUMMARY

The results from the central core thermal model indicated that the thermal performance of the two heat pipe geometries was not a strong discriminator. Both heat pipe concepts exhibited adequate performance under normal operating conditions. Both concepts limited the TFE temperature rises following single and multiple component failures.

The bow tie heat pipe geometry provided slightly better lateral core heat transport and provides 50% more redundancy in the number of heat pipes within the core. However, the thermal model also indicated that the results for the worst-case event, loss of two adjacent radiator heat pipe sinks, were essentially the same for both geometries. The radial heat fluxes within the in-core heat pipes following the upset events were found to be well within the demonstrated sodium heat pipe technology, as indicated in Table 5.

CONCLUSIONS

The in-core heat pipes should be constructed with a 1.5 design margin at nominal operating conditions to accommodate transient heat flows following upset conditions and uncertainties in manufacturing tolerances. The

TABLE 4. Influence of Heat Pipe Failure on Temperature Differential for the Tricusp Heat Pipe Geometry.

Parameter	Loss of Single Driver Radiator Heat Pipe	Loss of Driver in Core Heat Pipe
Core maximum temperature Differential	22 K	20 K
Emitter temperature rise	31 K	8 K
Driver fuel centerline	28 K	34 K
Hot spot wall temperature rise	NA	132 K[a]

[a]NA—not applicable

D635-0224

TABLE 5. HPTI Heat Fluxes Are Well Within Sodium Heat Pipe Technology.

	Evaporator Section (Wt/cm^2)	Condenser Section (Wt/cm^2)
In-core TFE heat pipe	15	30
In-core driver heat pipe	39	100
TFE radiator heat pipe	44	11
Driver radiator heat pipe	100	11
Heat flux across arteries following failure of in-core heat pipe		
Tricusp	22	21
Bow tie	19	9
Reported test data (Dunn and Reay 1982)	200+	1,000

D635-0224

1050

design should also be capable of full load heat rejection in a negative "1–g" environment to accommodate spacecraft maneuvering loads. The preliminary evaluations of the HPTI core indicate that the core can be operated in a near isothermal condition. The two thermal models of the HPTI core indicated the importance of the core's lateral heat transport capability in mitigating core hot spot temperatures following upset events. The core design was found to be fault tolerant in that the TFE performance was degraded following multiple heat transport component failures, but the affected TFEs continued to operate at a slightly lower efficiency. The operational effects of asymmetric cooling of the TFEs and fuel driver pins will require further assessment.

Acknowledgments

This work was sponsored by the U.S. Air Force Phillips Laboratory under AFSTC Contract F29601–90–C–0059.

References

Ranken, W., 1990 Briefing to Rocketdyne, "Moderated Heat Pipe Cooled Thermionic Reactor," April 1990, Chatsworth, CA.

Woloshun, K. A., M. A. Merrigan, and E. D. Best, "HTPIPE: A Steady–State Heat Pipe Analysis Program," LA–11324 M, UC–405, Los Alamos National Laboratory, November 1988.

Dunn, P. D., and D. A. Reay (1982), Heat Pipes, third edition, Pergamon Press, New York, pp. 96–98.

TESTABILITY OF A HEAT PIPE COOLED THERMIONIC REACTOR

Richard E. Durand
Rockwell International/Rocketdyne Division
6633 Canoga Avenue
Canoga Park, CA 91303
(818) 718-3318

M. Harlam Horner
General Atomics
P.O. Box 85608
San Diego, CA 92186-9784
(619) 455-2472

Abstract

As part of the Air Force Phillips Laboratory thermionics program, Rocketdyne performed a design study for an in-core thermionic fuel element (TFE) heat pipe cooled reactor power system. This effort involved a testability evaluation that was performed starting with testing of individual components, followed by testing at various stages of fabrication, and concluding with full system acceptance and qualification testing. It was determined that the system could be thoroughly tested to ensure a high probability of successful operation in space after launch.

INTRODUCTION

One of the promising concepts for a space nuclear power system using thermionic power conversion is a reactor containing multicell thermionic fuel elements (TFE) directly cooled by heat pipes that carry thermal energy to a radiator for dissipation to space. This system has no moving parts and no single point failure modes, and by virtue of its highly integrated design, it is light in weight.

A 40 kWe version of such a thermionic nuclear power system is currently under study by the Rocketdyne Division of Rockwell International and General Atomics for the Air Force Phillips Laboratory. As an integral part of the conceptual design effort, the testability of the system is being examined in detail to ensure that the design allows for adequate testing on the ground before launch. This paper summarizes the flight qualification and acceptance test program envisioned for the system, with emphasis on those aspects presenting the greatest testability challenges.

DESCRIPTION OF SYSTEM

The basic design concept is comprised of 108 redundant fuel/thermionic power converter heat rejection modules, coupled with in-core sodium heat transfer assemblies to provide a highly survivable, redundant, passively cooled (no pumped loop, no valves) reactor power system with no single point failures. Waste heat is transferred axially from the core using in-core transport heat pipes to ex-core sodium radiator heat pipes. Figure 1 shows the reactor and shield, and Figure 2 shows the overall system configuration, including the conically shaped heat pipe radiator and the support structure that connects its power system to the payload. Figure 3 provides detail of a portion of the core assembly, showing the in-core heat pipes that remove heat from the TFEs.

QUALIFICATION AND ACCEPTANCE TEST PROGRAM

The qualification and acceptance test program for the system follows the guidelines of MIL-STD-1540B, "Test Requirements for Space Vehicles." The program, shown schematically in Figure 4, starts with qualification testing of the 19 distinct types of components that make up the system. Acceptance testing of flight components parallels the qualification testing, which is performed at the expected operating conditions rather than at the design conditions.

For system qualification, a complete nuclear flight system (a qualification unit) will be fabricated and tested for resistance to launch stresses. It will then be functionally tested in a vacuum chamber in a facility specifically designed for nuclear reactor testing. In this test, the reactor will be operated at design power levels and temperatures, and the system will produce electrical power that will be dissipated in a dummy load.

Acceptance testing of a flight system will include essentially the same tests as those run on the qualification unit, except that the reactor will be run only at "zero power" to verify criticality. Simulated DC power output of the reactor will be supplied by an external generator rather than by the reactor TFEs.

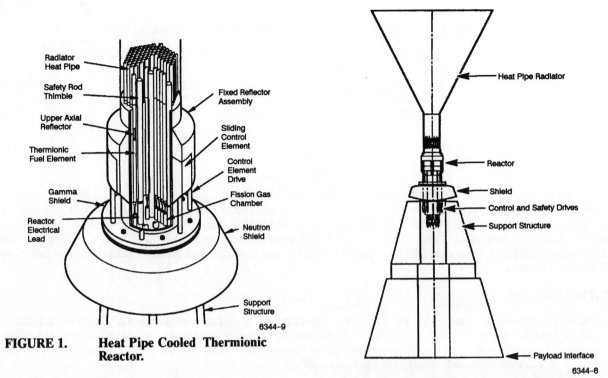

FIGURE 1.   Heat Pipe Cooled Thermionic Reactor.

FIGURE 2.   Overall System Configuration.

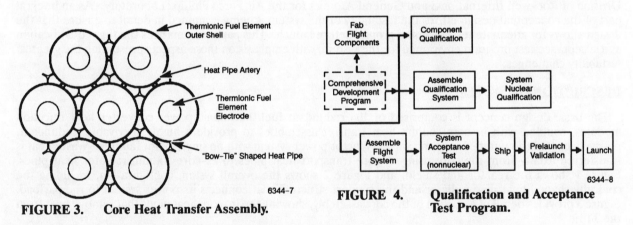

FIGURE 3.   Core Heat Transfer Assembly.

FIGURE 4.   Qualification and Acceptance Test Program.

Prelaunch validation will consist of nonnuclear mechanical and electrical functional tests.

COMPONENT QUALIFICATION

The heat pipe thermionic (HPTI) hardware tree, shown in Figure 5, has been marked with shadowed boxes to indicate which of the hardware items can and should be considered as components to establish the scope of component-level qualification and acceptance-level testing.

Table 1 shows the test matrix that resulted after applying the MIL-STD-1540B requirements to component qualification testing. Note that a burn-in acceptance test is conducted as part of the initial functional test for all electrical and electronic components.

The closely integrated assembly of TFEs, the in-core heat pipes (core assembly), and the radiator heat pipes introduce some special testability considerations. The assembly and testing sequence of these elements is shown in Figure 6 and discussed further in the following sections.

1053

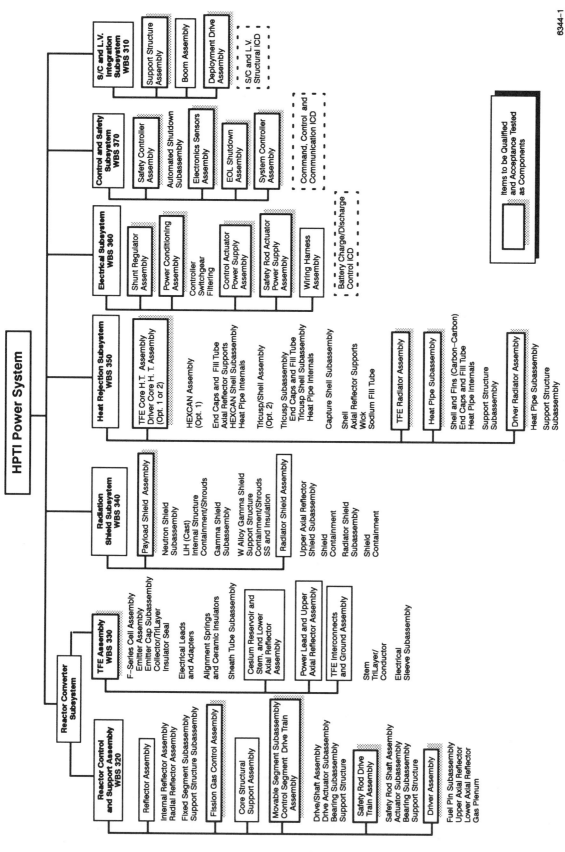

FIGURE 5. Heat Pipe Cooled Thermionic Reactor System Hardware Tree.

6344-1

1054

TABLE 1. Component Qualification Test Matrix.

Component or Assembly	Functional[a]	Leak	Pyroshock	Functional[b]	Random Vibration	Functional[b]	Leak	Radiation	Functional[a]	Thermal Cycling	Functional[b]	Thermal Vacuum	Functional[b]	Pressure	Leak	EMC	Life	Functional[a]
Control segment/drivetrain	x		x	x	x	x		x	x	x	x	x	x			x	x	x
Fission gas control assembly					x	x	x							x	x			
Safety rod drivetrain	x		x	x	x	x		x	x	x	x	x	x			x		x
Driver assembly		x			x		x											
Thermionic fuel element	x	x	x	x	x		x											x
Payload shield	x	x			x	x	x					x	x		x			
Core assembly					x							x	x		x			
TFE radiator heat pipe	x				x							x	x					
Driver radiator heat pipe	x				x							x	x					
Electronics radiator	x				x							x	x					
Shunt regulator/power conditioner	x		x	x	x	x				x	x	x	x			x		x
Parasitic load radiator	x		x	x	x	x				x	x	x	x			x	x	x
Control actuator power supply	x		x	x	x	x				x	x	x	x			x	x	x
Safety rod actuator power supply	x		x	x	x	x				x	x	x	x			x		x
Electronic sensors	x		x	x	x	x		x	x	x	x	x	x			x		x
EOL shutdown assembly	x		x	x	x	x				x	x	x	x			x		x
System/safety controller	x		x	x	x	x				x	x	x	x			x		x
Deployment drive	x		x	x	x	x				x	x	x	x			x		x
Boom	x				x	x						x						x

D635–0230

[a]Complete performance test
[b]Abbreviated functional test

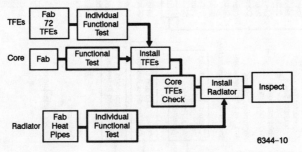

FIGURE 6.   Reactor Assembly and Test Sequence.

TFE TESTING

Just as for the current generation of TFE Verification Program TFEs, individual functional testing of the HPTI TFEs will take place when the graphite integral cesium reservoir is loaded with cesium. At that time, the devices are finally degassed but are still attached to the evacuation and cesium loading equipment. Cesium discharge testing is performed between 0.1 and 2.0 torr cesium pressure to characterize the cesium vapor voltage breakdown behavior of the device at 60 Hz, with varying cesium pressures using an attached liquid cesium reservoir. The circuitry used for the testing is illustrated in Figure 7.

Once the graphite reservoir is loaded with cesium, the liquid reservoir is isolated from the TFE and graphite reservoir with a high temperature valve, and discharge behavior is again examined using only the graphite cesium reservoir to vary cesium pressure. The voltage breakdown data with respect to liquid cesium and cesium–loaded graphite temperatures are used to define the cesium pressure–temperature relationship for the graphite reservoir, and the characteristics of the discharge are used to determine the quality of the performance of the device. Figure 8 shows a typical I–V trace taken during cesium breakdown testing of a three–cell TFE at about 0.6 torr cesium pressure.

For the HPTI reactor, cesium discharge testing of individual TFEs will be performed in a completely analogous way. Performance characteristics determined from testing will be based upon experience gained from tests of single cell, electrically heated thermionic devices built to replicate the identical electrode materials and geometric conditions inside a TFE. Deviations from ideal conditions will be interpreted as shown in Table 2.

Acceptable TFEs will be assembled into the reactor core and cross connected in pairs at the radiator heat pipe end. Circuitry similar to that shown in Figure 7 will be attached to each TFE pair and cesium breakdown testing performed with the reactor core assembly held at a temperature (800 to 1000 K) to provide a cesium pressure of 0.1 to 2.0 torr from the cesium reservoir. Test voltages will range from about 30 to 40 volts for this test series. A small thermal gradient (on the order of 25 K) will be imposed axially on the reactor core to cause a variation

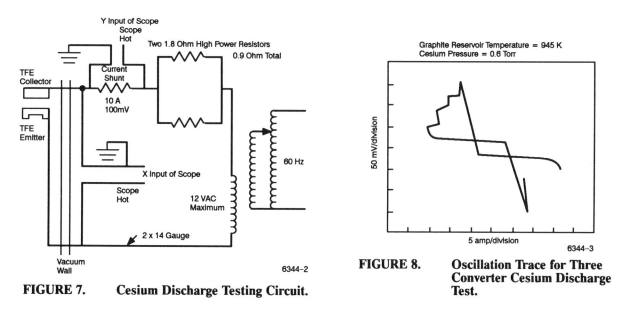

FIGURE 7.   Cesium Discharge Testing Circuit.

FIGURE 8.   Oscillation Trace for Three Converter Cesium Discharge Test.

TABLE 2.   Interpretation of Cesium Breakdown Test Results.

Observation	Problem
Flat I–V curve High ignition voltage	Leak
Low forward voltage Ignitions missing	Shorted converters
No current	Open current
Low saturation current	Contaminated electrodes
High ignition voltage	Defective Cs reservoir
Low resistance to ground	Defective sheath or stem insulator

D635–0230

in discharge voltage from converter to converter. Six pairs of discharge inflections from the 12 converters in each TFE pair will be resolvable and will be used to verify that the TFE characteristics have not been changed by the installation process. A similar test is planned once the six parallel strings of 12–series connected TFEs are connected by welding leads and cross connections at the shield end of the reactor. Further work is needed to define the exact testing conditions required. At this time, the radiator heat pipes and supports will also be installed so that no further assembly operations that would jeopardize TFE quality will be performed.

CORE ASSEMBLY TESTING

The core heat transfer assembly, shown in Figure 3, is assembled by a hot isostatic processing (HIP) method, described by Horner et al. (1991), to form the cylindrical structure of the reactor core assembly. A series of inspections and thermal performance tests are conducted throughout the assembly process to qualify the assembly for installation of the 72 TFEs and 36 driver elements (108 total). Figure 9 illustrates the bonded reactor core structure, which is an isometric section through the region containing TFEs near the center of the core. Inspections of the bonds between the TFE tubes and septums that make up the bow tie shaped heat pipes will be performed by ultrasonic testing once the core structure undergoes the HIP process. At that time, bonds between the tube sheets at each end of the reactor core, and the heat pipe septums and tube walls will be similarly tested to qualify the entire structure for final machining before heat pipe processing.

A thermal functional test, illustrated in the sequence chart of Figure 7, will be conducted after heat pipe processing to verify performance of the assembly. The test will be performed with electric heaters installed into the core TFE and driver positions (illustrated for TFEs in Figure 9) to dump heat equivalent to the waste heat load along the length of the receiving portion of the core heat pipes. Calorimeters will be installed at the radiator heat pipe end of the core below the region being heated to extract and measure the heat flowing into the core heat pipes from the electrical heaters. Both the calorimeter end of the assembly and the heated region will be instrumented with thermocouples to map the core temperature distribution at the correct heat load. Figure 10 schematically illustrates the test and instrumentation arrangement.

RADIATOR HEAT PIPE TESTING

Each radiator heat pipe is subjected to random vibration testing at ambient temperature and pressure. The heat pipe is then placed in a vacuum chamber with cooled walls. Heat is supplied to the evaporator section of the heat pipe electrically, possibly by induction heating. The heat pipe radiates to the inner, cooled walls of the chamber. Performance is verified by measuring temperature along the length of the heat pipe. Shutdown and startup cycles are simulated by reducing electrical power input to the heat pipe and allowing the heat pipe to gain equilibrium at the lower temperature; the power is then restored and the response of the heat pipe observed.

SYSTEM QUALIFICATION AND ACCEPTANCE TESTING

Two flight systems will be fabricated: one for qualification, including a nuclear ground test, and the other for the actual mission. The units are identical except for additional data–gathering instrumentation on the qualification unit.

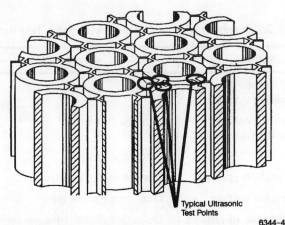

Typical Ultrasonic
Test Points

6344–4

FIGURE 9.   Typical Locations for Ultrasonic Testing.

1057

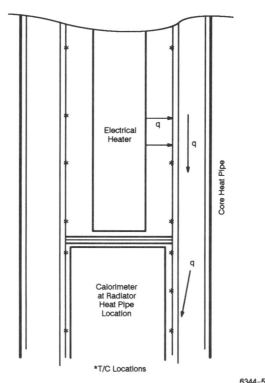

Figure 10. **Electric Heater/Calorimeter Schematic for HPTI Core Assembly Thermal Testing.**

As shown in Figure 11, the qualification unit goes through the uppermost test sequence, culminating in a nuclear performance test, and the flight unit goes through the lower test sequence, culminating in installation in the launch vehicle. The first test in both cases is a functional test in a vacuum chamber, with the reactor, shield, and control/safety rod drives heated electrically to their expected operating temperatures. Mechanical and electrical checks are performed, including a cesium discharge test, as previously described, on the complete assembly of the in-core TFEs. The system is positioned with the radiator heat pipes pointed downward to inhibit the heat pipes from removing heat as fast as it is put in. Qualification testing then proceeds as shown through electromagnetic compatibility (EMC); shock; acoustic; a repeat of the initial functional test; and, finally, a nuclear performance test at full power. In this latter test, the radiator is oriented with the heat pipes pointed upward to allow them to operate essentially as designed.

The flight unit test sequence is similar, except that EMC testing is bypassed, and nuclear testing is not performed except for a criticality check at the fabrication site. Nuclear operation takes place after the system has reached its planned stable orbit.

CONCLUSIONS

The heat pipe cooled thermionic nuclear power system can be thoroughly tested at the component and system levels to ensure high probability of successful operation in space after launch. Although the multicell TFEs cannot be operated before launch, the combination of a complete nuclear ground test and an electrical cesium discharge test on all TFEs provides very high confidence in their operability. The in-core heat pipe assembly is testable using electric heaters in place of the TFEs.

References

General Atomics (1990), *TFE Verification Program Semiannual Report for the Period Ending September 30, 1989*, GA-A-19876, San Diego, CA.

General Atomics (1991), *TFE Verification Program Semiannual Report for the Period Ending September 30, 1990*, GA-A-20335, San Diego, CA.

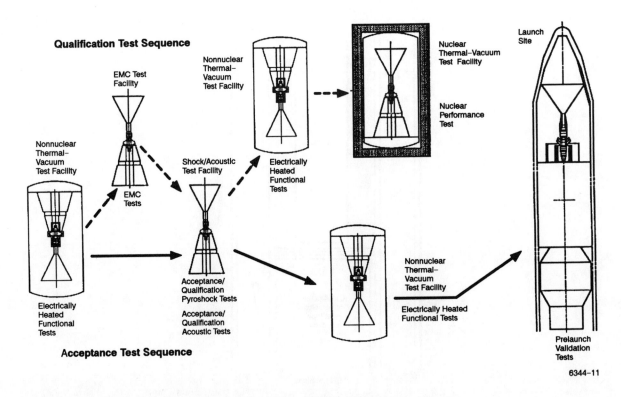

Qualification Test Sequence

Acceptance Test Sequence

6344–11

FIGURE 11.   System Qualification and Acceptance Test Sequence.

Horner, M. H. (1991) "Heat Pipe Cooled Thermionic Reactor Core Fabrication," in *Proceedings of Ninth Symposium on Space Nuclear Power Systems*, CONF-920104, M. S. El-Genk and M. D. Hoover, eds., American Institute of Physics, New York.

MIL–STD–1540B (1982) *Test Requirements for Space Vehicles*.

Acknowledgments

This work was performed under contract to the U.S. Department of Defense, Headquarters Air Force Phillips Laboratory, PL/STPP, Contract NO. F29601–90–C–0059.

D635–0230/sjh

1059

TEST DEVELOPMENT FOR THE
THERMIONIC SYSTEM EVALUATION TEST (TSET) PROJECT

D. Brent Morris and Vaughn H. Standley
Phillips Laboratory
New Mexico Engineering Research Institute
901 University SE, Rm. 135
Albuquerque, NM 87106
(505) 272-7266/-7243

Michael J. Schuller
Phillips Laboratory
Space and Missiles Technology Directorate
Space Nuclear Power Branch
Kirtland AFB, NM 87117-6008
(505) 846-2878

Abstract

The arrival of a Soviet TOPAZ-II space nuclear reactor affords the US space nuclear power (SNP) community the opportunity to study an assembled thermionic conversion power system. The TOPAZ-II will be studied via the Thermionic System Evaluation Test (TSET) Project. This paper is devoted to the discussion of TSET test development as related to the objectives contained in the TSET Project Plan (Standley et al. 1991). The objectives contained in the Project Plan are the foundation for scheduled TSET tests on TOPAZ-II and are derived from the needs of the Air Force Thermionic SNP program. Our ability to meet the objectives is bounded by unique constraints, such as procurement requirements, operational limitations, and necessary interaction between US and Soviet scientists and engineers. The fulfillment of the test objectives involves a thorough methodology of test scheduling and data management. The overall goals for the TSET program are gaining technical understanding of a thermionic SNP system and demonstrating the capabilities and limitations of such a system while assisting in the training of US scientists and engineers in preparation for US SNP system testing. Tests presently scheduled as part of TSET include setup, demonstration, and verification tests; normal and off-normal operating tests, and system and component performance tests.

INTRODUCTION

The compact nuclear reactor has fostered interest in space power development. Issues associated with space nuclear power systems generally deal with balance of plant and not the core. For example, the choice of an energy conversion subsystem depends on many considerations including the power requirements of the satellite. For the range of 5 to 40 kWe, a thermionic conversion subsystem is attractive (Kiernan 1991). The USSR has used thermionic technology in at least two operational missions and has tested thermionic components and subsystems for many years (Bennett 1989). The US has no comparable experience in thermionics technology.

If the unfueled Soviet TOPAZ-II thermionic space power system is approved for purchase it will be delivered to the US and tested in the modified New Mexico Engineering Research Institute (NMERI) facility as part of the TSET project. The primary objective of the TSET project is to assess the performance of a thermionic conversion system and to apply the knowledge gained toward the US SNP program. The endeavor is intended to "jump-start" the Air Force thermionic SNP program (Henderson 1991). Testing will be planned and conducted through the combined efforts of five organizations: the Phillips Laboratory, Los Alamos National Laboratory, Sandia National Laboratories, the University of New Mexico (collectively known as the New Mexico Strategic Alliance), and the Strategic Defense Initiative Organization (SDIO) which sponsors the program (Standley et al. 1991).

BACKGROUND

Recent programs have spurred renewed interest in space nuclear power and particularly in thermionic SNP systems. A unique opportunity exists to acquire and study an available state-of-the-art system and direct the results and conclusions into the current Air Force Thermionic SNP program.

Historically, spacecraft power requirements have been less than 5 kWe, with design lifetimes of less than 5 years. These requirements depend on the available conventional subsystem power, launch vehicle, and mission payload technologies. Launch vehicle constraints, packaging volume and mass are concerns above 5 kWe. The constraint on power restricts the mission payload, which limits operational mission capabilities.

The Phillips Laboratory has formulated a technology program designed to advance the level of maturity of SNP far enough to provide a real option for meeting the power requirements of the user community. The cornerstones of this program are to pursue a systems-driven approach rather than a technology approach, identify and demonstrate key materials and components of the most promising options, apply sufficient resources to a concept in order to address technological concerns, and maintain an adequate technology base program to meet future needs.

The Phillips Laboratory, the Department of Energy (DOE) and SDIO have formulated a Thermionic SNP Systems Technology Program designed to advance the level of maturity of thermionic SNP far enough to provide a real option for meeting the near-term power requirements of the user community. To this end, a Memorandum of Understanding (MOU) exists between the Air Force, DOE, and SDIO. Figure 1 illustrates the managerial structure agreed upon in the MOU. The target date for completing this effort is the end of fiscal year 1995. This approach will assist the government in making an informed decision to pursue a flight demonstration program.

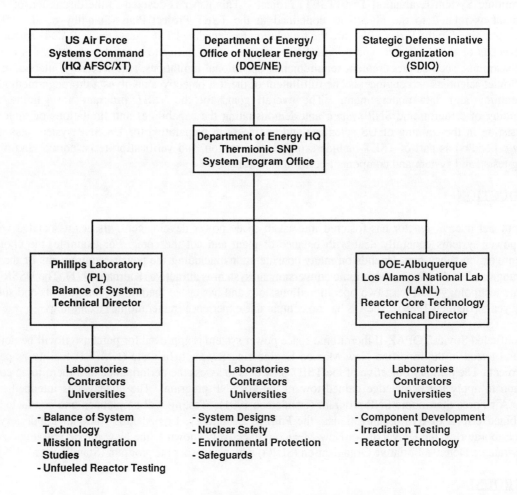

FIGURE 1.   Management Structure for the Thermionic Space Nuclear Power Systems Technology Program.

The TSET program originated with a Soviet offer to sell space nuclear power technology to the US. With direction from the US Congress to study thermionic technology more closely, the US is pursuing the offer as part of an overall thermionics development program.

The role of TSET in the overall thermionics program is to evaluate and learn from Soviet technology and experience, benchmark US codes for space power system modelling, and train US engineers and technicians in non-nuclear component and subsystem ground testing procedures. Moreover, TSET provides both a developmental methodology for Air Force thermionic subsystem design and information to support a SNP flight decision.

The TSET program involves testing and evaluating an unfueled Soviet TOPAZ-II reactor and associated components. The reactor uses single-cell thermionic fuel elements (TFEs). For TSET, the UO_2 fuel is replaced with special tungsten heaters. Components provided by the USSR include individual thermionic fuel elements (TFEs), an accompanying test stand, a cesium reservoir apparatus, an electromagnetic pump, and a radiator segment. In addition, electrical hardware and control room components are included in the purchase.

Equipment delivery will begin shortly after approval for purchase from SDIO. Acceptance testing and hands-on training from the Soviets should take three months. After acceptance, six months of tests are planned, with further testing to be scheduled depending on funds, proposals, and the results of the initial tests.

TSET PROJECT OBJECTIVES & PHILOSOPHY

The overall objective of the TSET project is to expedite the thermionic SNP system design and technology development program by improving US government and industry understanding of the design and operation of a space-based thermionic reactor. Emphasis will be on short duration tests that probe for system limits regarding reliability, transient response, temperature, etc. rather than routine steady-state life testing. The goal is to conduct tests that define technology limits in the most effective manner in terms of both cost and time. In regard to this, a set of project objectives has been developed in the TSET Project Plan (Standley et al. 1991). The objectives help in selecting tests. The tests will be weighed against the project objectives; that is, those tests meeting the most objectives will be given priority. However, the process of distinguishing which objectives are met by a given test involves special consideration and analysis (Schuller et al. 1991).

TSET will also be used to train US personnel in ground test operations. This will include government and industry teams as well as students in university engineering programs. In addition, information from all tests, including those sponsored by laboratories, industry, or universities, will be available to all participants and documented in a timely fashion.

TOPAZ-II TEST ARTICLE DESCRIPTION

The TOPAZ series of reactors exploit in-core thermionic conversion of nuclear heat directly into electrical power. There are 37 special tungsten electrical heaters used to simulate the heat produced from the fissioning fuel. Thirty-four of the 37 electrical heater-TFE units are connected in series. The other three units are connected in parallel and are dedicated to powering the DC electromagnetic conduction coolant pump.

The TOPAZ-II single-cell TFEs use mono-crystal molybdenum alloy substrate with a chemical vapor deposition (CVD) coating of tungsten as the emitter material. Niobium (Nb) is used as the collector tube material which is coaxial with the fueled emitter. The collector-emitter gap is filled with cesium vapor which optimizes thermionic performance, and sweeps out fission gases and oxygen which leak from the fuel.

An electric insulator isolates the TFE from the coolant and a sheath supports the TFE and presents a heat transfer surface to the NaK coolant. In TOPAZ-II the collector-insulator-sheath structure is formed from a stainless steel sheath tube coaxial with, but larger than, the niobium collector tube. The annular gap is filled with helium which insulates the collector from the coolant and calandria electrically while still providing a good thermal bond.

The TOPAZ-II reactor requires a helium gas supply to provide helium to the annular space between the sheath and collector of the TFE. The helium gap serves the dual function of electrical insulation and thermal bond. The gas is supplied to an axial plenum on the lower (shield) end of the reactor. The TOPAZ-II also uses carbon-dioxide (CO_2) which is supplied to the interior of the moderator calandria. This cover gas is apparently used to help seal the coating on the zirconium hydride moderator. In addition, argon (Ar), helium (He), and CO_2 are supplied to the lithium-hydride (LiH) shield (Gunther 1990).

CONSTRAINTS

The TSET project will be executed under some unique constraints that must be considered in the selection of tests and test procedure development. There are several constraints which revolve around the total system performance that pertain to the electric heaters and the TFEs. For example, during operation, total electric power is reduced by 15% due to resistive heating by the molybdenum lead wires. Therefore, 85% of the electrical power is provided along the active length of a heater. Maximum steady-state electrical power to the heaters is 115 kWe. Power levels greater than this will cause the molybdenum leads to the electrical heaters to overheat. Also, the temperature ramp rate during cycling cannot exceed 10 A/min for one heater and 100 K/hr for the system (Gunther 1990).

The primary constraint relating heater operation to TFE operation occurs when a heater is inoperative. If an electrical heater is turned off or fails, its diode stops producing power and that diode appears as a high resistance in the series string of 34 diodes. Soviet experience indicates that with "one heater off" there is no damage and there is still current through the series string. With two or more heaters off there will no longer be output from the TFE string, there will be an over-voltage condition on the string that poses the danger of collector insulator breakdown, and high temperature drops that are tolerable but not desirable. Because of this, the power to two or more heaters cannot be turned off at the same time (Gunther 1990).

FACILITY DESCRIPTION

The TOPAZ-II will be tested at the NMERI complex in Albuquerque, New Mexico. The complex consists of two large high bay areas designated Workspaces 1 and 2, respectfully, and outdoor courtyard and equipment rooms. Workspace 1 houses the TOPAZ-II, its vacuum chamber (5.8 m high, 2.8 m wide) and test scaffold. In addition, Workspace 1 will house a TFE test stand, a tungsten heater outgassing stand, and miscellaneous support equipment. The test control room and machine shop are located in Workspace 2. The outdoor facilities contain an auxiliary building to house the cooling water system and vacuum pumps; an uninterruptible power source (UPS) building for the Soviet transformers, an UPS system and backup batteries, breaker panels, and both a 300 kVA motor generator set (208 V, 60 Hz input; 380 V, 50 Hz output) and engine generator (380 V, 50 Hz) (Thome 1991).

TSET TESTING SUPPORT

The TSET project team at NMERI is the group responsible for administering all scheduled and proposed tests. The team is and will be available for technical consultations on various test considerations. The TSET project is designed to be a cooperative effort between the TSET project team and the space nuclear power community. In consideration of the TSET/NMERI facility, the TSET project team can provide a well equipped testing facility for both full system and individual component testing. The facility is equipped with an overhead bridge type crane rated at 9.1 metric tons with a crane hook height of 5.8 m. The control room is equipped with a complete data acquisition system. Because both American and European power are available, test equipment of either designation can be used during any particular test. In addition, the UPS allows for reliable long duration testing. The machine shop in Workspace 2 contains tools and equipment for manufacturing special equipment necessary for some tests.

SUMMARY

The TSET Project will boost both the technological understanding of thermionic space nuclear power systems and the development, organization, and administering of tests on such systems. The well equipped TSET facility housing the TOPAZ-II and its support equipment provides an exceptional testing environment and resources. Above all, the space nuclear power community will benefit by contributing their technical ideas, assistance, and overall thermionics program desires.

Acknowledgments

The TSET Project is sponsored by SDIO. Additional technical assistance was received from the TSET personnel at the Phillips Laboratory and associated personnel from Space Power Incorporated. Special thanks to General Atomics, Rocketdyne, Martin Marietta, Space Power Inc., SDIO, DOE, and the Phillips Laboratory for their feedback to the TSET Project Plan preliminary draft which contributed to the development of this paper.

References

Air Force-DOE-SDIO Memorandum of Understanding (MOU), Internal Memo, 1991.[*]

Bennett, G. L. (1989) "A Look at the Soviet Space Nuclear Power Program," in *Trans. 24th Intersociety Energy Conversion Engineering Conference*, held in Washington, DC, 6-11 August 1989.

Fairchild, J.F., J.P. Koonmen, F.V. Thome (1992), "Thermionic System Evaluation Test (TSET) Facility Description," in *Trans. 9th Symposia on Space Nuclear Power Systems*, held in Albuquerque, NM, 12-16 January 1992.

Gunther, N. G. (1990) *Characteristics of the Soviet TOPAZ-II Space Power System*, F29601-90-C-0048, Space Power Inc., San Jose, CA.[*]

Henderson, B. W. (1991) "US Buying Soviet TOPAZ-II to Boost Space Nuclear Power Program," Aviation Week and Space Technology, 14 January 1991, pp. 54-55.

Kiernan, V. (1991) "US Plan Would Test Nuclear Reactor in Space," Space News, 16-22 September 1991, p. 10.

Lawler, A. (1991) "Defense Teams Leading Way to Soviet Store," Space News, 23-29 September 1991, pp. 3, 45.

Schuller, M. J., V. H. Standley, D. B. Morris (1991), "Thermionic System Evaluation Test (TSET) Planning," Unpublished Paper, prepared for The 2nd Specialist Conference "Nuclear Power Engineering in Space. Physics of Thermionic Energy Converters," held in Sukhumi, USSR, 28-31 October 1991.

Standley, V. H., D. B. Morris, M. J. Schuller (1991), "Preliminary Draft of the Thermionic System Evaluation Test Project Plan," Phillips Laboratory, Kirtland AFB, NM, 14 August 1991, p. 2.[*]

Thome, F. V. (1991) "Specifications for the Creation of a Special Test Facility in the New Mexico Engineering Research Institute (NMERI) Building in Preparation for the TOPAZ (TSET) Test," Internal Document, Phillips Laboratory/Space Power Branch, Kirtland AFB, NM.[*]

REACTOR TEST FACILITIES FOR IRRADIATION NPPP FUEL COMPOSITIONS, MATERIALS AND COMPONENTS

E. O. Adamov, V. P. Smetannikov, V. I. Perekhozhev, V. I. Tokarev
Research and Development Institute of Power Engineering
Moscow, 101000

Sh. T. Tukhvatulin, O.S. Pivovarov
Joint Expedition of SIA "Luch"
Semipalatinsk, 490021

INTRODUCTION

When creating space nuclear power propulsion plants (NPPP) one of the key issues is to develop the reactor fuel and fuel elements for the flow parameters (temperature, pressure, coolant velocity) and life required. To solve this issue designers should have at their disposal the instruments - special irradiative devices and research reactors where the conditions characteristic for future standard items can be simulated to a certain approximation extent. Their availability will effect a saving of financial contributions and reduce terms to solve the global task. The basis of the NPPP development methodology adopted in the USSR includes a sequent running other tests of the fuel compositions, fuel elements and fuel assemblies (FA). Following this the assembled reactor is developed for the life and "compatibility" of its separately tested components. Further the NPPP is tested as a whole.

To realize this methodology on NPPPs development the USSR possesses of sufficiently equipped potential incorporating typical materials test IVV-2M - type reactors and special research reactors such as IGR, RA and EWG1 allowing to perform sequently the reactor component development under conditions corresponding to the operating conditions in standard items.

RESEARCH TRENDS

The major research trends in the reactors mentioned are as follows:

1. Investigations of fission products release from fuel elements, processes of their transportation and filtration;
2. Investigations of thermal strength, radiation and chemical resistance of fuel elements and fuel assemblies as a whole;
3. Investigations of hydrogen penetration of structural materials under irradiation conditions;

While performing reactor investigations there are carried out pre-reactor analysis of specimens under test for geometric dimensions, composition, structure, mass, electrical conductivity, strength, etc., irradiation tests for the life at working medium pressure (hydrogen, nitrogen, helium, argon or other gases) up to 20 MPa and temperatures up to 3000 K and then, following the irradiative devices separation (ampoules or fuel channels), post-reactor analysis.

Hydrogen penetration is studied in specially developed test facilities permitting to measure coefficients of hydrogen diffusion, penetration and solubility in structural materials within the range of pressure 0.1-20.0 MPa and specimens temperature 300-1100 K and above under irradiation and laboratory conditions.

IVV-2M REACTOR

In research IVV-2M reactor fuel compositions development to involve them into future space NPPPs is performed in the specially developed gas-vacuum test facility RISK. The test facility RISK intends to investigate gaseous fission products (GFP) released from fuel under constantly controllable irradiation parameters (temperature, thermal neutron flux density, gas flow rate) under both stationary irradiation conditions and thermocycling. The GFPs concentration is determined by the gamma-spectrometric method.

Irradiation is performed in ampoules providing the required temperature conditions and located in the channel which allows to carry out thermocycling not influencing the reactor and adjacent irradiative devices. Ampoules cooling is performed by the reactor primary circuit water. A schematic diagram of the device is shown in Figure 1. The characteristics of the research cells in the IVV-2M reactor core are shown in the Table 1. The gas-vacuum test facility provides transportation of the GFPs releasing from fuel with the specified gas-carrier flow rate up to the measuring volume, and further sample localization in exposure gas holders. Fuel temperature regulation is provided by both maintenance of the specified gas mixture in thermal regulative gap and power density change due to fuel movement along the core height. Helium, neon, argon, nitrogen and their mixtures are employed as gas-carrier. Gas pressure is 0.02-0.3 MPa. The transportation time of the sample from fuel to the measuring volume depends on gas flow rate and it is varied from 2 to 30 minutes.

TABLE 1. Cell Characteristics

Cell	Dimension, mm	Flux density $n/cm^2.s.MW$	
		thermal	fast
Central trap	120	3×10^{13}	2×10^{12}
Side trap	60	2×10^{13}	8×10^{12}
Periphery trap	60	1×10^{13}	2×10^{12}

Core length is 500 mm.

Semiconducting gamma-spectrometer and chosen measurement methodology allow to determine GFPs release from fuel within range $R_i = 10^3$ at/s at generation rate $B \sim 10^{10}$ at/s. The methodology provides measurement of xenon isotopes ^{138}Xe, ^{137}Xe, ^{135}Xe, ^{135m}Xe, ^{133}Xe and crypton ^{89}Kr, ^{88}Kr, ^{87}Kr, ^{85m}Kr. Measurement of xenon isotopes ^{135}Xe and ^{133}Xe requires special irradiation modes.

In the test facility the fuel thermocycling systems by both its "slow" movement along the core height by electrodriver with rate 0.05 mm/s and "fast" movement with rate 20 mm/s and power density change by a factor of 5-7 are envisaged. In the test facility monitoring of temperature, thermal neutron flux density, gamma-irradiation dose power is ensured. Impurities (H_2O, O_2 and oth.) in gas-carrier are determined periodically by special instrumentation including the laser spectrography system. The operating ampoule sections are being developed for concrete irradiation conditions and fuel geometry.

The channel for irradiation represents aluminum casing where three or six ampoules are arranged, electromovement and pneuma-movement systems are installed. Thermocycling is provided by absorptive shield. Each ampoule is connected to test facility by gas tubes, and it allows to keep individually necessary temperature mode and perform measurement of GFPs release, thermal neutron flux density is measured by direct charging sensors or thermoneutron sensors. Dimensions and geometry of fuel specimens determining the ampoule size. The ampoule structures developed allow to irradiate fuel in the form of pins, pebbles, pellets, within temperature range of 700-2400 K and power density of 100-700 W/ampoule. Thermocycling is provided with different duration cycles from hundreds of seconds up to several hours at temperature variation rate from levels of 100-1000 K up to 2000-2200 K.

IGR REACTOR

IGR reactor (Figure 2) is impulse uranium-graphite homogeneous thermal self-shutdown nuclear one of heat capacity type with graphite moderator and reflector. The rector core represents the stack with uranium-containing graphite blocks assembled as columns. The reactor reflector is made of similar blocks without uranium. The stack has the reactor control and protection system rods and experimental channels to arrange test objects. The stack is embraced by leak-proof vessel filled up with helium and arranged in the rector water cooling tank. The reactor structure does not envisage a possibility to cool uranium-graphite stack compulsorily. Heat released during the reactor operation is accumulated by the stack and then it is gradually transferred into the cooldown circuit water through the reactor vessel walls.

The reactor operates under two major modes;

1) "Flash" - mode of unregulable neutron impulse with amplitude and duration to be determined by the magnitude of initial reactivity imparted to the reactor. In this case power suppression is implemented due to appearance of negative reactivity conditioned by availability of negative temperature effect;

2)   "Impulse" - regulable operation mode. Under this mode realization the reactor power is changed according to the pre-specified law, and compensation of negative temperature effect is provided by control withdrawal from the core.

RA REACTOR

RA reactor (Figure 3) is research high-temperature gas-cooled heterogeneous thermal one with zirconium hydride moderator and neutron beryllium reflector. The reactor incorporated 37 leak-tight ampoules with fuel which are arranged in three-circular-raw cells and central cell. Ampoules are filled up with gas environments at pressures up to 20 MPa. Fuel temperature inside of ampoules is up to 2400 K. Ampoules cooling is provided by outer air flow. Twelve control drums are arranged in beryllium reflector. One-two experimental FAs can be installed into the reactor instead of operating ampoules. These FAs can be both unflowing and cooled by gas coolant flowing directly through FAs.

EWG1 REACTOR

EWG1 reactor is research heterogeneous thermal one with light-water moderator and coolant and beryllium neutron reflector. The reactor core includes 30 water-cooled fuel channels arranged in three-circular-raw cells. Beryllium reflector contains 10 control drums. In the reactor central part the loop channel surrounded by beryllium moderator is located. In this channel there can be arranged an experimental system containing one or several investigated FAs cooled by hydrogen, nitrogen or other gas. Twelve rods of the reactor reactivity compensation system are mounted in beryllium moderator. The major reactor characteristics are shown in Table 2.

As one can see from Table 2, the characteristics spectrum of the reactors considered is such it is feasible to provide tests of fuel elements and FAs of space NPPPs under different modes of their operation. In particular, in the IGR reactor it is reasonable to research fuel elements behavior under the modes simulating NPPP startup or NPPP transition of power (small power) mode into propulsion mode. In the RA reactor there is a possibility for the life-time tests of fuel elements and FAs under the modes corresponding to the NPPP operation power mode, and with regard to the EWG1 reactor this will be performed at the power corresponding to the NPPP operation propulsion mode.

CONCLUSION

Thus, the available in the USSR potential of specialized research reactors and materials test IVV-2M-type reactors and feasibility to develop fuel elements, FAs and other core components of the gas-cooled reactor represent a basis for development of activities on space power propulsion systems.

TABLE 2. Main Characteristics of IGR, RA and EWG1 reactors.

Parameter	IGR	RA	EWG1
Core height, mm	1400	600	800
Cross-section core dimensions, mm	square, 1400x1400 mm	square, D-339 mm	square, D-548 mm
Uranium-235 loading, kg	9.1	8.3	4.6
Maximum reactor power, MW	1000-control mode; 10000-mode of uncontrollable neutr. impulse with max. semi width 0.1s	0.5	35
Operation duration at maximum power	limited by temperature stack and is 5 s and 0.5 s corresponding	limited by fuel burn up effects and is 4000 hours	limited by coolant margins and is 2 hours
Loop channel dia., mm	228	41	164
Maximum thermal flux in loop channel, n/sm^2s	$5 \cdot 10^{17}$	$5 \cdot 10^{12}$	$5 \cdot 10^{14}$
Temperature reactivity coefficient	negative	negative	negative

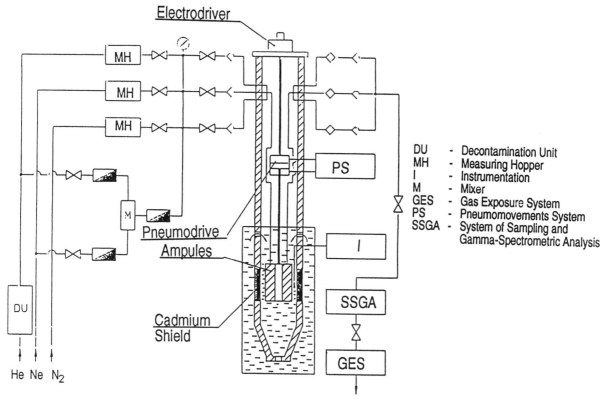

FIGURE 1. Gas - Vacuum Test Facility RISK   To Special Vent

DU   - Decontamination Unit
MH   - Measuring Hopper
I   - Instrumentation
M   - Mixer
GES - Gas Exposure System
PS   - Pneumomovements System
SSGA - System of Sampling and
       Gamma-Spectrometric Analysis

1. Reactor Vessel
2. Reactor Core
3. Reflector
4. Movable Core Part
5. Central Experimantal Channel
6. CPS Rod
7. Pipings to Supply Helium into Vessel
8. Ionization Chamber
9. Side Experimental Channel
10. Reactor Water Cooldown Tank

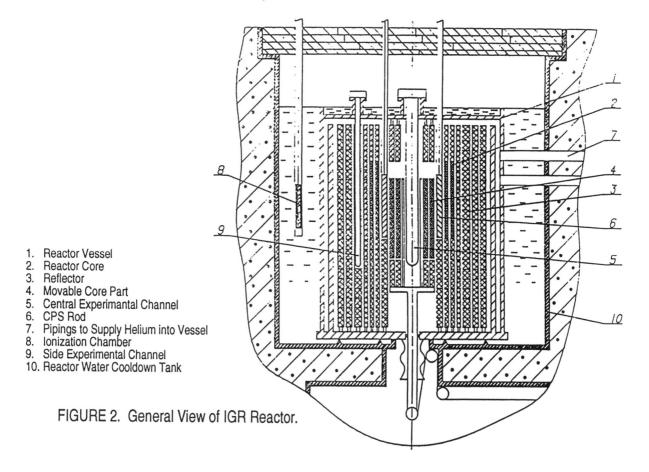

FIGURE 2. General View of IGR Reactor.

1065

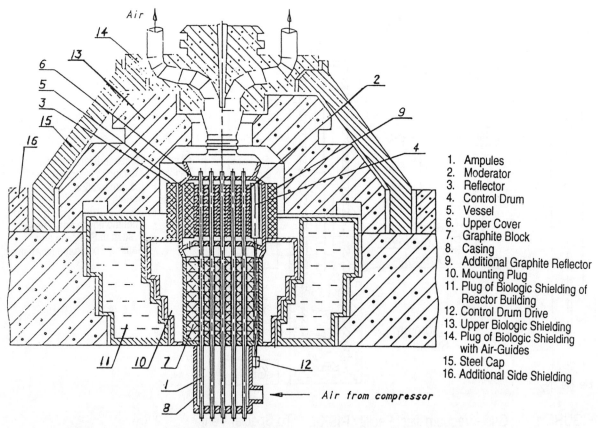

1. Ampules
2. Moderator
3. Reflector
4. Control Drum
5. Vessel
6. Upper Cover
7. Graphite Block
8. Casing
9. Additional Graphite Reflector
10. Mounting Plug
11. Plug of Biologic Shielding of Reactor Building
12. Control Drum Drive
13. Upper Biologic Shielding
14. Plug of Biologic Shielding with Air-Guides
15. Steel Cap
16. Additional Side Shielding

FIGURE 3. General View of RA Reactor.

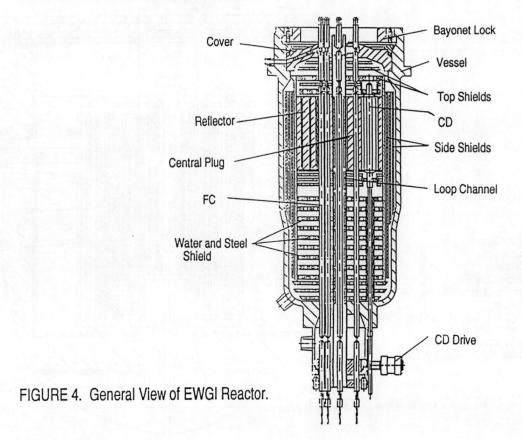

FIGURE 4. General View of EWGI Reactor.

1066

NUCLEAR FACILITY LICENSING, DOCUMENTATION AND REVIEWS, AND THE SP-100 TEST SITE EXPERIENCE

Bruce C. Cornwell, Ted L. Deobald, and Ernest J. Bitten
Westinghouse Hanford Company
P. O. Box 1970
Richland, Washington 99352
(509) 376-3837

Abstract

The required approvals and permits to test a nuclear facility are extensive. Numerous regulatory requirements result in the preparation of documentation to support the approval process. The principal regulations for the SP-100 Ground Engineering System (GES) include the National Environmental Policy Act, Clean Air Act, and Atomic Energy Act. The documentation prepared for the SP-100 Nuclear Assembly Test (NAT) included an Environmental Assessment, state permit applications, and Safety Analysis Reports. This paper discusses the regulation documentation requirements and the SP-100 NAT Test Site experience.

INTRODUCTION

Ground test preparation for a space reactor is a lengthy process. In addition to system and component design, component development and testing, and procedure preparation, approvals must be obtained from regulating organizations to actually perform reactor testing. Obtaining the approvals may be the most challenging step. In the approval process one must demonstrate that the environment can be protected and that there are no undue risks to the health and safety of the public. Knowledge of the SP-100 GES experience with the approval process is beneficial for other major projects or new facilities.

In parallel with design and construction, numerous other activities are required. Some of these activities support obtaining approvals required by policy and law. There are many regulatory requirements to be met. Among them are the National Environmental Policy Act (NEPA), Clean Air Act, Clean Water Act, and Atomic Energy Act.

The objective of this paper is to present an overview of the process and a brief discussion of SP-100 Test Site experience. This paper will discuss the NEPA, safety analysis reports and state permitting activities required to support an authorization to construct and test the SP-100 GES at Hanford, WA.

Preparations for SP-100 GES testing have included many major decisions. In support of these decisions many documents were prepared. The completed documentation is an indicator of the status of the overall process. Documentation timing is critical, as each documentation builds on prior decisions and supporting documentation. The general flow of a program is shown in Figure 1, with relative timing of individual activities.

Obtaining the Department of Energy (DOE) approvals is very similar to a Nuclear Regulatory Commission (NRC) licensing process. Both include sitting studies, preconstruction, construction licensing, and operation licensing. The major difference is that the DOE grants approval, where as the NRC issues a license.

DISCUSSION OF REQUIRED DOCUMENTATION AND EXPERIENCE

National Environmental Policy Act

The National Environmental Policy Act requires federal agencies to assess the potential impacts of their actions on the quality of the human environment. The NEPA process is intended to help public officials make good decisions, based on an understanding of environmental consequences.

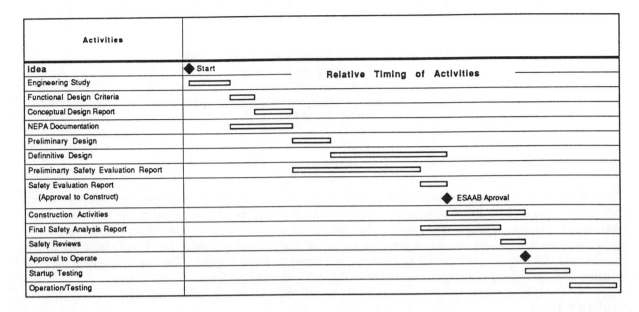

Activities	Relative Timing of Activities
Idea	◆ Start
Engineering Study	▭
Functional Design Criteria	▭
Conceptual Design Report	▭
NEPA Documentation	▭
Preliminary Design	▭
Definnitive Design	▭
Preliminarty Safety Evaluation Report	▭
Safety Evaluation Report (Approval to Construct)	▭ ◆ ESAAB Aproval
Construction Activities	▭
Final Safety Analysis Report	▭
Safety Reviews	▭
Approval to Operate	◆
Startup Testing	▭
Operation/Testing	▭

FIGURE 1. General Project Flow and Principal Activities

The SP-100 GES NEPA documentation discusses consideration given to environmental impacts during concept and test site selection, and examines the environmental effect of the proposed ground test of the nuclear subsystem. It also describes alternatives to the proposed action and examines the potential radiological risks of use in space.

The DOE used the information and analysis to determine whether the proposed action would have significant impacts on the environment. An Environmental Assessment was published, notice was given in the Federal Register, and public comment solicited. Based on the results, DOE published in the Federal Register a finding of no significant impact (FONSI). If the impact had been determined to be significant, an Environmental Impact Statement (EIS) would have been required. Table 1 contains a brief chronology of the SP-100 GES experience with the NEPA process.

The NEPA procedures and required documents are defined in regulations issued by the Council on Environmental Quality in Title 40 Code of Federal Regulations (CFR), Parts 1500 through 1508 (CEQ 1978). The DOE guidelines for compliance with NEPA are found in 10 CFR 1021 (DOE 1980). Figure 2 contains a flow chart of NEPA documentation and decision process currently used. The chart illustrates the required steps and decision points. A brief description of the steps and document content follows.

Environmental Decision Process and Documents

Categorical Exclusion. Categorical exclusions are listed in section D of the DOE NEPA guidelines. (DOE is actively pursuing changes in the NEPA guideline regulations.) The excluded activities do not violate laws or regulations, require expansion of waste disposal facilities, nor adversely affect environmentally sensitive areas such as threatened wildlife habitat; historic property; floodplain and wetlands; special water sources; and federal and state parks, wilderness areas or wildlife refuges. A categorical exclusion may take one to nine months to obtain.

Information Bulletin. An information bulletin is issued by DOE. It contains the basis for decision regarding the categorical exclusion and a brief description of the appropriateness of a Section D categorization.

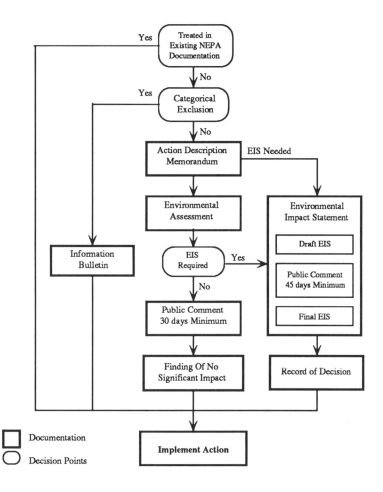

FIGURE 2. Flow Diagram of Environmental Decision Process

Action Description Memorandum (ADM). An ADM serves as the basis for determining the required level of NEPA documentation. It is a brief document that usually contains no analysis, but includes a description of the purposed action, location and affected environment, and potential environmental issues. An action description memorandum may take from one month to a year and occasionally longer.

Environmental Assessment (EA). An EA assesses whether a proposed action is a "major Federal action, significantly affecting the quality of the human environment." It serves as the basis for determining if an environmental impact statement is required. The EA provides an interdisciplinary review of proposed actions, identifies and provides a description of alternatives, assesses environmental consequences, and states required mitigation measures needed if an EIS is not prepared. An EA generally takes 18 months for preparation and approval. The Test Site experience was that it took much longer. Refer to Table 1 which contains a Test Site chronology of events.

Public Comment. The EIS is prepared in draft form for comment. In the case of an EA, a proposed FONSI is published in the Federal Register for comment. Comments are invited from federal agencies with jurisdiction and appropriate state and local agencies (including Indian tribes). Comments are also requested from the public, particularly interested or affected people and organizations.

Finding of No Significant Impact (FONSI). A FONSI records a decision that the environmental impacts are not significant and that an EIS is not required for a proposed action. The FONSI is published in the Federal Register.

Environmental Impact Statement (EIS). An EIS is a public document, prepared independently. It assess the environmental impact of a proposed action. The EIS includes an activity purpose and need description, alternatives considered, proposed action, the affected environment, and environmental consequences. An EIS is prepared in accordance with the requirements of 40 CFR Section 102(2)(C) of the National Environmental Policy Act. Preparation, review, and approval may take from two to three years and cost two to four million dollars.

Record of Decision (ROD). An ROD is a concise public record of a decision on a proposed action for which an EIS was prepared. It includes the alternatives considered, the environmentally preferable alternative, factors balanced in the decision, and mitigation measures and monitoring to minimize harm. The record of discission is published in the Federal Register.

Table 1. SP-100 Sequence of Events for NEPA Process .

Date		Event
Nov.	85	Hanford selected as SP-100 GES Test Site
Jan.	86	**Start Work on Environmental Documentation**
June	86	Action Description Memorandum issued, direction to prepare an EA
Nov.	86	GE Concurrence on Draft EA
Sept.	86	Draft EA to DOE-RL
Jan.	87	Draft EA forwarded to DOE-NE Office of NEPA Project Assistance (ONPA) for review.
July	87	First Release of Environmental Assessment
Aug.	87	EA forwarded to ONPA for approval
Sept.	87	Comments from ONPA
Jan.	88	DOE-RL decision that EIS is not needed
June	88	Expand Scope of EA per DOE-NE and DOE-EH
Dec.	88	Revised Environmental Assessment Published
Dec.	88	**Publish proposed FONSI in Federal Register**
Jan.	89	Extended Public Comment Period - Publish in Federal Register
Sept.	89	**Finding of No Significant Impact Published in Federal Register**
June	90	Supplemental Analysis Draft to DOE-RL. Design evolution required review of EA (Cell inerting, and deletion of tritium removal.)
Oct	90	DOE-HQ Supplemental Analysis review requested by DOE-RL

Safety Analysis Reports

Safety analysis is a documented process to systematically identify the hazards of an operation, to describe and analyze the adequacy of the measures taken to eliminate, control, or mitigate identified hazards, and to analyze and evaluate potential accidents and their associated risks.

The safety analysis report (SAR) is a key document in the approval (or licensing) process. The SAR plays a similar role for both DOE and NRC. The DOE requirements for safety analysis are located in DOE Order 5481.1B (DOE 1986a) and overall nuclear reactor safety requirements are listed in DOE Order 5480.6 (DOE 1986b). Specific guidance for reactor siting, design and licensing requirements are found in the federal regulations, licensing requirements in 10 CFR Part 50 (NRC 1988a) and siting requirements in 10 CFR Part 100 (NRC 1988b), NRC Regulatory Guide 1.70 (NRC 1978), and the Standard Review Plan (NRC 1981).

The role of the SAR is to document major decisions and commitments made, and provide sufficient design detail of all systems to allow for review and approval. The SAR must accurately reflect the facility, the design, and performance characteristics. Safety analysis report preparation begins with the start of project activities. The development of requirements and implementing them into a design is a lengthy process, which the SAR parallels and documents. The SAR is first issued in a preliminary form to support the authorization to begin construction. Following completion of design and construction, it is issued in a final form. The authorization to begin operation is based on the Final Safety Analysis Report (FSAR).

Specific NRC nuclear facility license requirements are found in 10 CFR 50, Parts 30 through 39. There are two basic licenses: a license to construct and a license to operate. The DOE follows a similar pattern, requires the same basic types of documentation, and addresses the same basic types of issues. DOE requires the use of the NRC SAR format and content. For a license, the NRC seeks assurance that adequate financing is available to construct, operate and decommission, and emergency response plans are in place, and that the facility design and siting are technically adequate. There is also concern for physical security and safeguard contingencies. The application technical content differs for the construction and operating license.

Preliminary Safety Analysis Report (PSAR) The PSAR contains the following types of information: description and safety assessment of the site, with attention to details affecting facility design; analysis and evaluation of the major structures, systems, and components that bear significantly on the acceptability of the site; and a summary description of the facility, the principal design criteria developed for application, and the materials to be used for construction. The PSAR also contains design and performance assessments of risks to public safety and health, and information necessary to determine safety margins and accident prevention and mitigation and identification of probable technical specifications. It identifies needed research and development to confirm the adequacy or resolve safety question relative to the facility structure systems and components.

Final Safety Analysis Report (FSAR) The FSAR expands on the PSAR, and includes discussions and descriptions of the facility; design basis; limits of operation; all major systems; and a safety analysis of the facility as a whole. It includes current results of environmental and meteorological monitoring, which support site evaluation factors. A description and analysis that show safety functions will be accomplished. The radioactive materials produced and their control are assessed. The assessed risks to public safety and health are discussed to demonstrate adequate safety margins for accident prevention and mitigation. Safety questions asked at the construction stage must be resolved.

The FSAR, as it relates to operation, includes a statement of the organization, the structure, and resources. Also discussed are controls to assure safe operation, and plans for preoperational testing, conduct of operations, operator qualification and training, and coping with emergencies; and identification of proposed technical specifications.

Test Site Experience with Safety Analysis.

The sequence of events of safety documentation at the test site to date is as depicted in Table 2. Due to the number of organizations involved, the review and approval process has been extensive. Development of the requirements has been greatly aided by the close proximity of other reactor facilities. During the reviews by the various groups and organization the following have been identified as significant.

- Selection of appropriate principal design criteria,
- Core disruption events (Beyond Design Basis Events),
- Safety analysis keeping pace with evolving design,
- Protection of the NAT refractory materials,
- The presence of water in or near the vacuum vessel,
- Tritium generation in liquid metal coolants, and
- Application of ASME Sec. III to the NAT materials and design.

TABLE 2. SP-100 GES Safety Analysis Documentation, Chronology and Reviews .

Date		Event
Jan.	87	Start Preparation: A joint Westinghouse Hanford Company (WHC) and General Electric Aerospace (GE) effort to outline safety analysis report, and identify responsibilities.
Jan.	88	PSAR Synopsis: Reviewed by and approved by WHC Test Site, and WHC Safety.
May	88	PSAR Rev A: Reviewed and approved by WHC GES Test Site, WHC Safety, and GE.
Nov.	88	PSAR Rev B: Reviewed and approved by WHC, WHC Test Site, WHC Safety, WHC Safety and Environment Advisory Council (SEAC), Reactor Subcouncil, and GE; reviewed by Los Alamos National Laboratories (LANL). Incorporates comments on Rev A and design refinements.
May	89	PSAR Rev C: Reviewed and approved by WHC, WHC Test Site, WHC Safety, WHC SEAC Reactor Subcouncil, and GE; reviewed by LANL and DOE-RL. Incorporates comments on Rev B and design refinements.
July	89	PSAR Rev 0: Reviewed and approved by WHC, WHC Test Site, WHC Safety, WHC SEAC Reactor Subcouncil, and GE; reviewed by LANL, NUS, and DOE-RL. Incorporates WHC, GE, and LANL Comments on Rev C, and design refinements.
April	90	PSAR Rev 1: Reviewed and approved by WHC, WHC GES Test Site, WHC Safety, WHC SEAC Reactor Subcouncil, GE, DOE-RL and DOE-HQ. Incorporates comments from earlier reviews and design refinements.
Sept.	90	DOE-HQ issued the Safety Evaluation Report (SER).
Oct.	90	Topics identified for potential PSAR revision.

Clean Air Act Construction Permits

Before construction may begin, at the Hanford Site, permits are required from the Washington State Department of Ecology and Department of Health. These permits result from the Clean Air Act. Specific construction applications are required from the Environmental Protection Agency (EPA) for release of hazardous air pollutants, per 40 CFR 61.07, Washington State Department of Health (W-DOH) application for emission to the air, per WAC 402-80-70, and Washington State Department of Ecology W-DOE) for prevention of significant deterioration for a new source of airborne radioactive emissions, per WAC 173-403-080.

Construction permit applications were submitted to the EPA (DOE-RL 1990a), W-DOH (DOE-RL 1990b), and W-DOE (DOE-RL 1990c). These applications document the releases anticipated to the air resulting from operation. During operation the SP-100 GES may release radioactive materials, thus, documenting and getting approvals for the new source of airborne radionuclide emissions is required. The permits are required even though the releases will be very small, about 0.5% of the regulatory limit. For SP-100 the primary release is anticipated to be tritium, from the reactor shielding and lithium coolant.

The state agencies have been thorough in their permit applications reviews and assessment for use of best radionuclide control technology. Their concerns have not been limited to the scope of the particular regulation, but for all aspects of potential operation. For future planning it is important to realize that the regulatory climate is constantly changing. For example, recent changes to the Clean Air Act make obtaining permits even more difficult.

CONCLUSIONS

Obtaining the needed approvals (licenses) and permits to construct and operate a nuclear test facility is a lengthy process. The process details are described in various state and federal regulations. Even if the risks and consequences are low, the full rigor of the regulations and review and approval cycles are required. In fact, preparation for a future nuclear test may be even more difficult, due to increasing regulations and public concerns. The approval and permitting activities must be included in planning activities, in order to avoid construction and testing program delays. The approval process is complex when many organizations are involved, as was the case with SP-100. Environmental activities may take up to three years or longer. Safety analysis and documentation, which parallel design and construction, may take from three to five years to complete.

Acknowledgments

This work was supported by the US Department of Energy under contract DE-AC06-87RL10930.

References

CEQ (1978) National Energy Policy Act, Title 40 Code of Federal Regulations, Parts 1500 through 1508, Council on Environmental Quality, Washington, DC.

DOE (1980), National Environmental Policy Act Implementation Procedures, Title 40 Code of Federal Regulations Part 1021, US Department of Energy, Washington DC.

DOE (1986a) Safety Analysis and Review System, DOE Order 5481.1B, US Department of Energy, Washington DC, September 1986.

DOE (1986b) Safety of Department of Energy-Owned Nuclear Reactors, DOE Order 5480.6, US Department of Energy, Washington DC, September 1986.

DOE-RL (1988) Environmental Assessment of SP-100 Ground Engineering System Test Site, DOE/EA-0318 US Department of Energy, Richland, WA.

DOE-RL (1990a) National Emissions Standards for Hazardous Air Pollutants, Application for Approval of Construction SP-100 Ground Engineering System Test Site, DOE/RL-90-15, US Department of Energy, Richland, WA.

DOE-RL (1990b) Prevention of Significant Deterioration Application for Approval to Construct SP-100 Ground Engineering System Test Site, DOE/RL-90-14, US Department of Energy, Richland WA.

DOE-RL (1990c) Department of Health Application for Approval of Construction SP-100 Ground Engineering System Test Site, DOE/RL-90-16, US Department of Energy, Richland WA.

NRC (1978) Standard Content and Format of Safety Analysis Reports for Nuclear Power Plans, US Nuclear Regulatory Commission Regulatory Guide 1.70, US Nuclear Regulatory Commission, Washington, DC, November 1978.

NRC 1981) Standard Review Plan for the Review of Safety Analysis Reports for Nuclear Power Plants, US Nuclear Regulatory Commission, NUREG 0800, US Nuclear Regulatory Commission, Washington, DC, July 1981.

NRC (1988a) Domestic Licensing of Production and Utilization Facilities, Title 10 Code of federal Regulations, Part 50, US Nuclear Regulatory Commission, Washington DC.

NRC (1988b) Reactor Site Criteria, Title 10 Code of federal Regulations, Part 100, US Nuclear Regulatory Commission, Washington DC.

SP-100 THERMOELECTRIC ELECTROMAGNETIC PUMP DEVELOPMENT -- ELECTROMAGNETIC INTEGRATION TEST PLAN

Regina S. Narkiewicz, Jerry C. Atwell, John M. Collett, Upendra N. Sinha

General Electric Company
6835 Via Del Oro P.O. Box 530954
San Jose, CA 95153-5354
(408) 365-6442

Abstract

SP-100 is a space nuclear power system that meets the needs of a wide spectrum of potential applications requiring electrical power in the range of 10's to 1000's of kilowatts. The thermoelectric electromagnetic (TEM) pump is a key component of the system that efficiently and reliably pumps lithium coolant through both the heat transport and heat rejection subsystem of SP-100. The adequacy of key analytical models used in the performance prediction of the TEM pump have been experimentally demonstrated in previous tests. These tests were based on simplified characteristics of the pump. The Electromagnetic Integration Test (EMIT), described in this paper, has been planned as the next step in achieving full prototypicality in TEM pump testing to reconfirm previous test results, and verify other critical aspects of the pump performance. Continued thermoelectric cell development coupled with data gathered from EMIT, and the development of advanced materials, will ultimately lead to the Converter/Pump Assembly Test (CPAT) -- a fully prototypic test of the TEM pump.

INTRODUCTION

SP-100 is a space nuclear power system that meets the needs of a wide spectrum of potential applications requiring electrical power in the range of 10's to 1000's of kilowatts. A key component of this system is the thermoelectric electromagnetic (TEM) pump. The TEM pump can efficiently and reliably pump molten lithium coolant through both the heat transport and heat rejection subsystem of the SP-100 power system. A particular TEM pump assembly is shown in Figure 1, which provides a view of the TEM pump inlet and outlet piping in relation to the SP-100 space power system.

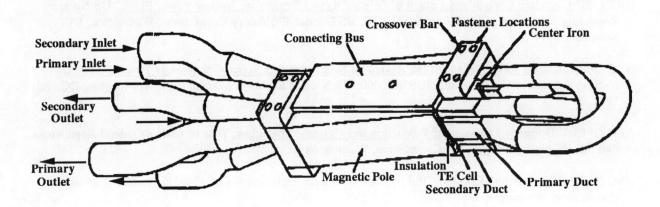

Figure 1. SP-100 Space Power System TEM Pump Assembly.

The primary inlet and outlet ducts refer to the primary heat transport system; and, the secondary inlet and outlet ducts refer to the heat rejection subsystem.

The development of the TEM pump includes analyses and predictions of performance, along with development testing of the subcomponents and specific phenomena. From these sources of information and experience, simulation models are being developed to determine and produce an effective TEM pump for any system arrangement. The Electromagnetic Integration Test (EMIT) is a current component level testing program for the TEM pump development. Its objective is to quantify the effects of liquid metal motion on the non-uniform magnetic flux and provide experimental verification of the TEM pump performance model.

Other chief factors that establish the TEM pump development program include the reference pump design and Magnetic Bench Test. This test provided the verification of the magnetic profile of a specified TEM pump geometry. Following the EMIT program, thermoelectric (TE) cell and pump fabrication development will continue, along with advanced materials development and pump assembly. The final testing for the pump will include all elements and components of a prototypic pump in the Converter/Pump Assembly Test (CPAT) program. This will ultimately produce the fundamental TEM pump geometry, material selection, and simulation model.

TEM Pump Design

The TEM pump utilizes thermoelectric cells that consist of semiconductor materials and are subjected to a temperature gradient between primary and secondary ducts. These thermoelectric cells generate electric current which induces a magnetic flux in the Z-shaped magnetic structure. The electric current and magnetic flux pass through the liquid lithium perpendicularly to each other to develop the pumping force. Figure 2 presents simplified axial and cross sectional views of the TEM pump.

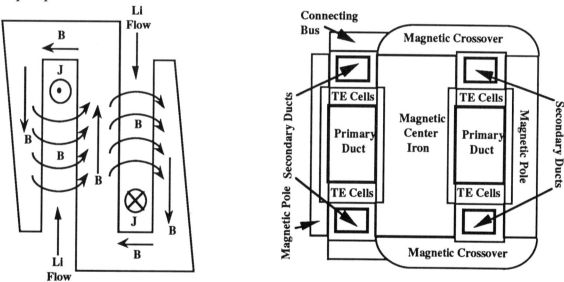

Figure 2. Simplified Axial and Cross Sectional View of TEM Pump.

This design provides a minimum mass, self-regulated pump, with no moving parts, and self-powered by an internal temperature gradient.

TEM Pump Development

The milestones of the TEM pump development program consist of the reference pump design, the Magnetic Bench Test, the Electromagnetic Integration Test (EMIT), the thermoelectric cell development, pump fabrication development, and the TEM pump performance test under prototypic conditions in the Converter/Pump Assembly Test (CPAT).

The Magnetic Bench Test confirmed the validity and accuracy of the magnetic performance prediction method devised for the TEM pump. It also aided in the mass minimization efforts for the magnetic structure of the pump. The evaluation of the test results showed good correlation with the analytical predictions, including the effects of fringing and axial current. The magnet used in the test was designed to be prototypic so that it could also be used in the EMIT pump test article.

The objectives of the EMIT encompass providing experimental verification of analytical performance models for the SP-100 TEM pump. EMIT is a continuation of the magnetic bench test, which characterized only the electromagnetic performance. This component level test will confirm analytical prediction prior to testing with a TEM pump that includes thermoelectrics. EMIT will be an externally powered test of flowing sodium-potassium (NaK) in an electromagnetic configuration at room temperature and non-prototypic conditions, i.e. stainless steel instead of PWC-11 ducts. The non-prototypic test configuration will reduce the risk of testing with high temperature lithium; and, it will also reduce costs and scheduling while maintaining desired reliability of results.

The Converter/Pump Assembly Test (CPAT) program will examine and evaluate the thermoelectric converter assembly and the TEM pump under somewhat prototypic conditions. CPAT will also demonstrate each components performance under steady state and selected transient conditions, and, interface compatibility. One of the principal objectives of CPAT is to validate the prototypic TEM pump, and its simulation modeling.

Electromagnetic Integration Test Plan

The test article for EMIT, shown in Figure 3, displays that the thermoelectric cells and secondary ducts have been removed from the prototypic configuration. This will allow reasonably accurate testing of the flowing liquid metal in the magnetic flux. The electrically conducting connecting buses provide the path for the external direct current to flow through the moving test fluid and primary ducts. The EMIT test pump will be scaled to equal the performance of the SP-100 TEM pump,and will be powered by an external electrical supply. NaK at room temperature will be the test fluid. A series of tests will be performed to characterize pump performance and provide data for determining electrical current distribution and magnetic field profile from dead head (no flow) to 120% of design current. Figure 3 presents a simplified schematic of the proposed EMIT.

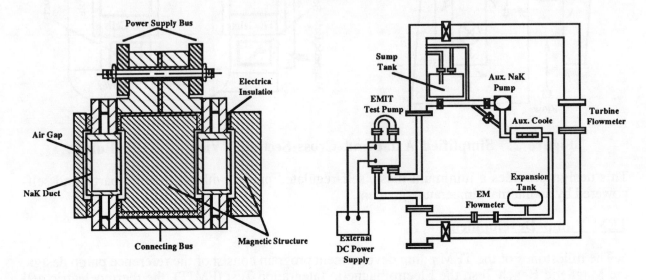

Figure 3. Test Article Cross Section and Simplified Schematic of EMIT.

The specific test objectives for EMIT include: quantify the electromagnetic induction effects of moving liquid metal in a nonuniform magnetic flux density profile; measure current fringing in the pump end regions; measure and quantify the distortion of the magnetic field in the duct due to the applied electric current and induced axial electric current; measure the magnetic flux density along both sides of the duct in the insulation gap; determine the opposing electromagnetic pumping effect in the end region; characterize the pump performance as a function of central and sidewall current; determine start-up characteristics and stability of pump.

A prediction simulation model has been formulated to predict the magnetic and electrical responses of the test article, along with predictions of the developed and delivered pressures, with respect to electrical, magnetic, and hydraulic losses. Upon the completion of EMIT, this simulation model will be revamped to include the liquid metal flow interaction in the magnetic structure for a prototypic TEM pump.

CONCLUSIONS

The thermoelectric electromagnetic pump development can be viewed as a procedure, starting with the Magnetic Bench Test which defined and validated the magnetic profile of a specified TEM pump geometry. The acquired data and information was utilized for a preliminary prediction simulation model by recreating magnetic profiles for different imposed currents in a specific TEM pump geometry. EMIT is a component level test to confirm analytical predictions prior to testing the pump which includes thermoelectric cells for reduced risk to the prototypic pump performance test in the high temperature lithium. The data acquired from EMIT will be used to simulate the flow and interaction of liquid metal fluids in a magnetic field, and the prediction simulation model will be updated. Thermoelectric cell development tests will provide the TE cell performance characteristics data for the TEM pump performance model. CPAT will be the final test which will yield information of how the thermoelectrics perform and interact in a system of a flowing liquid metal in a magnetic field under high temperatures, and how they affect the performance of the TEM pump. Pump fabrication, advanced materials, and pump assembly will also be ascertained. The analytical performance simulation model will be reviewed and remodeled with this current information and data.

Acknowledgments

This work was supported by the U. S. Department of Energy under DOE contract: DE-AC03-865F16006. The authors would also like to thank Les Dahl for his tremendous amount of assistance in formulating the simulation model of the TEM pump, along with the model for the EMIT test article. Thanks also goes out to Samir Salamah for his constructive, beneficial, and practical counsel toward composing this document.

References

Atwell, J., Salamah, S., Sinha, U., Zarghami, F., *Thermoelectric Electromagnetic Pump Design for the SP-100 Reference Flight System,* Proceedings of the 6th Symposium on Space Nuclear Power Systems, January 1989, Albuquerque, New Mexico.

Collett, J., Kugler, W., Sinha, U., Surjadi, T., *Thermoelectric Electromagnetic Pump Design for SP-100,* Proceedings of the 23rd Intersociety Energy conversion Engineering Conference, July 31 - August 5, 1988, Denver, Colorado.

Gamble, R., Choe, H., Kirpich, A., Greenwood, F., Narkiewicz, R., Collett, J., Sinha, U., *SP-100 Heat Transport System Performance During a Hypothetical Failure of One Radiator,* Proceedings of the 8th Symposium on Space Nuclear Power Systems, January 1991, Albuquerque, New Mexico.

A PRELIMINARY COMPARISON OF GAS CORE FISSION AND INERTIAL FUSION FOR THE SPACE EXPLORATION INITIATIVE

Terry Kammash and David L. Galbraith
Department of Nuclear Engineering
The University of Michigan
Ann Arbor, MI 48109
(313) 764-0205

Abstract

Potential utilization of fission and fusion-based propulsion systems for solar system exploration is examined using a Mars mission as basis. One system employs the open cycle gas core fission reactor (GCR) as the energy source, while the other uses the fusion energy produced in an inertial confinement scheme known as the Magnetically Insulated Inertial Confinement Fusion (MICF) concept, to convert thermal energy into thrust. It is shown that while travel time for each approach may be comparable, the GCR must overcome serious problems associated with turbulent mixing, fueling and startup, among others, while the fusion approach must find ways to reduce the driver energy required for ignition.

INTRODUCTION

The Space Exploration Initiative calls for, among other things, a manned mission to the planet Mars sometime in the early part of the next century. Since space travel is hazardous and man is unable to endure long journeys without experiencing physical and mental degradation, it is imperative that such missions be completed in the shortest possible time. This in turn means that one or more "advanced" rocket propulsion schemes must be developed to meet these objectives. Two promising approaches in this regard are the open cycle gas core (Ragsdale 1990) fission reactor (GCR) and an inertial confinement fusion scheme known as the Magnetically Insulated Inertial Fusion (MICF) concept (Hasegawa et al. 1986). The principle of operation in GCR involves a critical uranium core in the form of a gaseous plasma that heats, through radiation, a hydrogen propellant which exits through a nozzle, thereby converting thermal energy into thrust as demonstrated in Figure 1. The MICF is a fusion scheme that combines the favorable aspects of inertial and magnetic fusions into one where physical confinement of the plasma is provided by a metal wall, while its thermal energy is insulated from that wall by a strong self-generated magnetic field, as illustrated in Figure 2.

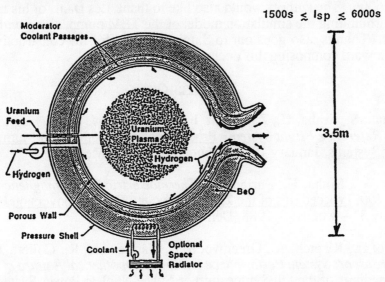

$1500s \leq Isp \leq 6000s$

~3.5m

FIGURE 1. High Specific Impulse, Porous Wall Gas Core Engine (Courtesy of NASA, Lewis Research Center).

In contrast to solid core reactors where temperature limitations, imposed by material melting, place severe constraints on rocket performance, the gas core concept circumvents these limitations because the nuclear fuel is allowed to exist in a high temperature (10,000-100,000 K) partially ionized state commonly referred to as the uranium plasma. Nuclear heat released as thermal radiation from the surface is absorbed by a surrounding envelope of seeded hydrogen propellant which is then expanded through a nozzle to generate thrust. With this scheme specific impulses of several thousand seconds appear to be feasible (Ragsdale 1990).

By comparison, the deuterium-tritium (DT) fusion plasma in MICF is generated inside a metal shell through ablation by an incident laser beam that enters the target through a hole as depicted in Figure 2. In addition to creating the plasma, the laser beam gives rise to an instantaneous, strong magnetic field which serves as a thermal insulator of the metal shell from the hot plasma during the fusion burn. The burn time in MICF is about one tenth of a microsecond (Kammash and Galbraith 1989) at the end of which a hot plasma consisting of the DT and metal shell particles is exhausted through a magnetic nozzle to produce the propulsive thrust. This approach to propulsion is also capable of producing specific impulses of several thousand seconds as will be noted shortly.

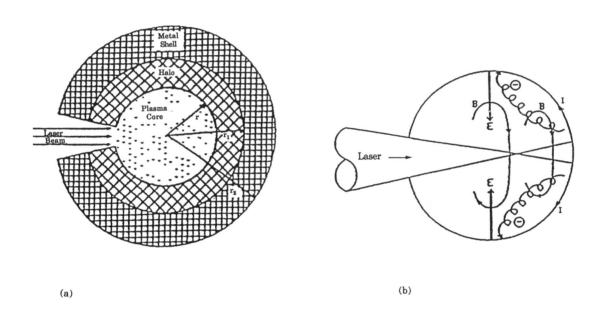

(a)   (b)

FIGURE 2.   Schematic Diagram of (a) Plasma Formation and (b) Magnetic Field Formation in MICF.

MAJOR PHYSICS AND ENGINEERING ISSUES

While possessing unique propulsion capabilities, each of the above approaches must overcome some serious technological obstacles to make them suitable for space exploration. To highlight some of these issues we choose in each case a preliminary design for which relevant parameters are available. In the case of the open cycle gas core reactor we select a 7500 MW design (Borowski 1987) in which the radius of the uranium core is 1 meter; the pressure in the system is 1000 atm; and the hydrogen temperature is about 17,500 K which suggests that the fuel temperature is about 35,000 K (Kascak and Easely 1968). Our preliminary analysis of this system (Poston and Kammash 1992) shows that the mean velocity of the propellant which is commensurate with a cited mass flow rate of 4.5 kg/s is about 5m/s. The mean velocity of the uranium in the core is generally taken to be 10-15 times smaller than that of the propellant, and as a result it could be considered stationary. It is a known fact that when a fluid of density ρ_2 moves with velocity v_2 past another fluid of density ρ_1, which is stationary, in the presence of a gravitational force, the (sharp) boundary between them will, upon perturbation, undergo oscillations which under certain conditions can become unstable. Known as the Kelvin-Helmholtz instability (Chandrasekhar 1961) the sufficient condition for the occurrence of this instability is given by:

$$v_2^2 > \frac{g\rho_1}{k\rho_2} \qquad (1)$$

where g is the gravitational acceleration (associated with flow curvature) and k is the wave number of the oscillation taken along the boundary that separates the two regions. In writing the above equation we made use of the fact that the uranium density ρ_1 for the conditions at hand is much larger than that of the propellant, ρ_2. The growth rate γ of the instability may be expressed as

$$\gamma = \frac{g}{v_2}\sqrt{\frac{\rho_1}{\rho_2}} \qquad (2)$$

which, along with Equation (1) allows us to calculate the diffusion coefficient and the particle flux across the spherical surface of the uranium core. At a pressure of 1000 atm, the hydrogen and uranium temperatures cited above, and a core radius of 1 meter, the Kelvin-Helmholtz instability results in a loss of about 10 kg/s of uranium which is unacceptably high and much larger than the 1% of hydrogen mass flow rate (45 g/s) often cited as the loss due to turbulent mixing in gas core reactors. To see how such a loss impacts the travel time, we apply these results to a Mars mission for which we will consider a continuous burn acceleration/deceleration trajectory profile that assumes constant specific impulse (I_{sp}) and thrust (F). Using a dry vehicle mass of 123 Mg as specified in the design cited earlier, we calculate the round trip times for various ratios of uranium mass flowrate ($\dot{m}_u$) to hydrogen mass flowrate ($\dot{m}_H$ = 4.5 kg/s). The results are given in Table 1.

TABLE 1.   Effect of Turbulent Mixing on Travel Time.

$\dot{m}_u/\dot{m}_H$	F(KN)	I_{sp}(s)	τ_{RT}(days)
0	87.6	1987	197
0.01	87.7	1970	198
0.1	88.5	1820	213
0.5	92.2	1390	280
1.0	96.8	1098	344
2.0	106.02	940	398

These results are to be contrasted with those produced by the MICF fusion propulsion system (Kammash and Galbraith 1991) summarized in Table 2.

1080

TABLE 2.   MICF Propulsion Capabilities.

Specific Impulse, I_{sp}	4.51×10^3 s
Thrust (at 100 rep rate) F	41.2 KN
Laser Input Energy	2.59 MJ
Energy Gain Factor	724
Vehicle Dry Mass	664 mT
Round trip time to Mars (τ_{RT})	138 days

The loss of uranium due to turbulent mixing and burn up requires fueling the core at a rapid rate in order to maintain criticality. We propose fueling by means of pellet injection. Noting that the ionization potential "ε" of uranium is 6.18 eV, we can estimate the pellet ablation time t_A from

$$t_A = \frac{r_p n_s \varepsilon}{q_e} \qquad (3)$$

where r_p is the radius of the injected pellet, n_s is the solid state density, and q_e is the heat flux, which in the case of uranium plasma at 35,000 K, has a value of 6.2×10^{28} eV/cm^2 - s. For a pellet of 0.5 cm radius whose mass is about 10 g, the ablation time is about 1.5 microseconds, and the injection velocity needed to reach the center of the core would be about 670 km/s. This is a very high speed indeed and is perhaps out of reach for current or near term technology. But it should not be taken too seriously since a "bare" pellet does not remain bare once it enters the hot uranium core. Depending on what processes are postulated to take place in the "halo" region, a survival enhancement of the pellet of several hundred is possible and that may reduce the injection speed to several kilometers per second which puts it perhaps well within reach of technology in the not too distant future. Since the uranium burnup rate for the 7500 MW reactor under consideration is quite small (~ 0.1 g/s) it is clear that an injection rate of about 1000 per second of such pellets will be required to make up for the loss of uranium due to the turbulent mixing phenomenon described earlier. Once again such an injection rate may be truly formidable, and the force needed to accelerate such particles to the desired velocity over a reasonable distance may be equally as challenging!

Although two or three major problems for GCR have been briefly highlighted in this paper, there are several others that are equally as critical but will not be addressed due to space limitations. These include the question of uranium confinement hydrodynamically or by magnetic fields, thermal and plasma instabilities, and questions associated with startup, shutdown and control. In the case of MICF, the major problem centers around target designs and fuel composition that would allow ignition at modest input laser energies.

CONCLUSION

Preliminary designs of a gas core fission reactor and an inertial fusion system have been examined in terms of their propulsive capability for a round trip mission to Mars. It is shown, on the basis of present day understanding of these approaches, that while trip times may be comparable, the fission system has many serious physics and engineering issues to overcome, while the fusion system, though better understood, must find means to reduce the input laser energy needed for large energy multiplication, and correspondingly a significant reduction in the dry weight of the vehicle.

Acknowledgments

This work was carried out under the auspices of the Nuclear Engineering Department of the University of Michigan.

References

Borowski, S. K. (1987) "Nuclear Propulsion—A Vital Technology for the Exploration of Mars and the Planets Beyond," *NASA Technical Memorandum*, 101 354, NASA Lewis Research Center, Cleveland, OH, 18 July 1987.

Chandrasekhar, S. (1961) *Hydrodynamic and Hydromagnetic Stability*, Dover Publications, N.Y.

Hasegawa, A., et al. (1986) "Magnetically Insulated Inertial Fusion: A New Approach to Controlled Thermonuclear Fusion," *Phys. Rev. Letters*, 56:139.

Kammash, T. and D. L. Galbraith (1989) "A High Gain Fusion Reactor Based on the Magnetically Insulated Inertial Confinement Fusion Concept," *Nuclear Fusion* 29:1079.

Kammash, T. and D. L. Galbraith (1991) "An Inertial Fusion Propulsion Scheme for Solar System Exploration," *Proc. Eighth Symposium on Space Nuclear Power Systems*, CONF-910116, Albuquerque, NM, 6-9 January 1991.

Kascak, A. F. and Easely, A. J. (1968) "Effect of Turbulent Mixing on Average Fuel Temperatures in a Gas-Core Nuclear Rocket Engine," *NASA TN D-4882*, NASA Lewis Research Center, Cleveland, OH, 6 August 1968.

Poston, D. I. and T. Kammash (1992) "Heat Transfer Model for an Open-Cycle Gas Core Nuclear Rocket," Ninth Symposium on Space Nuclear Power Systems, CONF-920104, held in Albuquerque, NM, 12-15 January 1992.

Ragsdale, R. G. (1990) Nuclear Thermal Propulsion, A Joint NASA/DOE/DOD Workshop, Cleveland, OH, 10-12 July 1990.

HEAT TRANSFER MODEL FOR AN OPEN-CYCLE GAS CORE NUCLEAR ROCKET

David I. Poston and Terry Kammash
Dept. of Nuclear Engineering
University of Michigan
Ann Arbor, MI 48109
(313) 747-0900

Abstract

A heat transfer model is developed to assess the propulsion capability of the open-cycle gas core nuclear rocket. The model is used to determine the maximum specific impulse achievable without violating the wall material temperature and heat flux limits. For a 3000 MW reactor with a wall heat flux limit of 100 MW/m^2, it is shown that a specific impulse of 3160 s and a thrust of 125 kN can be obtained.

INTRODUCTION

The Gas Core Nuclear Rocket (GCR) is a propulsion scheme that could allow manned missions to Mars in relatively short times. In the GCR a fissioning uranium plasma heats (primarily by radiation) a hydrogen propellant, which is exhausted through a choked nozzle. In this paper we carry out a heat transfer analysis to determine the important propulsion parameters, namely specific impulse and thrust, and establish their dependence on various parameters and limitations of the system.

The propulsion capability of the rocket is dependent on the state of the propellant in the reservoir proceeding the nozzle throat. A computer code was written to calculate the propellant temperature in a high power density nuclear rocket as a function of power, dimensions, mass flow rate, and initial conditions. The propellant (hydrogen) flows with a constant mass flow rate through a cylindrical annulus surrounding the reactor core. The code offers the option of including the fuel region (uranium) in the heat transfer model (in which the power is a function of the fuel density), or simply specifying a heat flux at the fuel/propellant interface. In both cases it is assumed that the fuel and propellant do not mix. The outer cavity wall is assumed to remain at a constant temperature due to transpiration cooling.

The thermal radiation absorption coefficient of the propellant is sufficiently high so that a diffusion heat transfer analysis can be used (the absorption cross section of hydrogen is relatively low at temperatures less than 10,000 K, thus the hydrogen is seeded with solid particles to aid in the absorption of radiation at low temperatures). The diffusion method models radiative heat transfer in the form of conduction, with a conduction coefficient depending on T^3.

The code uses an iterative finite difference scheme to calculate propellent temperatures as a function of radial and axial position. An iterative solution is required because the radiative 'conductivity' coefficient is temperature dependent, making the set of heat transfer equations non-linear. The results of these calculations are used to evaluate rocket performance for various reactor designs.

ANALYSIS

Temperature Solution

The fundamental conservation equations of mass, momentum, and energy are solved numerically.

$$\frac{\partial \rho}{\partial t} + \nabla \cdot \rho \underline{v} = 0 \ , \tag{1}$$

$$\rho \frac{D\underline{v}}{Dt} = -\nabla P - \nabla \cdot \underline{\underline{\tau}} + \rho \underline{g} \ , \ and \tag{2}$$

$$\rho C_p \frac{DT}{Dt} = -(\nabla \cdot q) + \frac{DP}{Dt} - (\tau : \nabla v) + Q_v \quad . \tag{3}$$

The continuity equation is used to determine the flow velocity. The momentum equation is nullified by the following assumptions: steady inviscid flow, constant pressure, and no external forces. To simplify the energy equation, it is also assumed that: fluid properties do not change significantly over one mesh interval, there is no axial conduction, and that azimuthal symmetry applies. The steady state energy equation becomes:

$$\rho C_p \left(v_z \frac{\partial T}{\partial z} + v_r \frac{\partial T}{\partial r} \right) = -\frac{1}{r} \frac{\partial}{\partial r}(r q_r) + Q_v \quad . \tag{4}$$

The volumetric power generation Q_v is dependent on the fuel density, and the radial velocity v_r is assumed to be zero except when transpiration cooling is required. Using the diffusion method of heat transfer:

$$q_r = -k \frac{\partial T}{\partial r} \qquad \left(k = k_{cond} + k_{rad} = k_c + \frac{16 \sigma T^3}{3 a_r} \right) \quad , \tag{5}$$

where a_r is the Rosseland mean absorption coefficient. The value of a_r includes the absorptivity of solid seed particles inserted in the flow to aid radiation absorption at low temperatures.

The diffusion approximation is valid in all regions except near the wall, where the optical thickness is low. An exponential absorption integral is evaluated to more accurately predict wall heat flux:

$$q_{wall} = \frac{1}{r_w} \int_{r_{int}}^{r_w} r a_p \sigma T^4 e^{-\int_r^{r_w} a_p dr'} dr + \frac{r_{int}}{r_w} q_{int} e^{-\int_r^{r_w} a_p dr'} \quad . \tag{6}$$

The conductivity coefficient at the wall is iterated until the diffusion solution produces a wall heat flux consistent with the absorption integral.

The above equations are solved by the method of finite differencing. The boundary conditions are determined by the wall temperature, and either the fuel/propellant interface heat flux (if the fuel region is not modelled, or symmetry at the centerline (if the fuel region is modelled). An iterative solution is required because the fluid properties and the radiative conduction coefficient are temperature dependent.

If a maximum wall heat flux is specified, then the boundary condition at the outer wall changes. In addition to specifying the wall temperature T_w, the temperature at the node adjacent to the wall is limited to a maximum value of:

$$T_{cool} = T_w + \frac{\Delta r q_{max}}{2 k_w} \quad . \tag{7}$$

The transpiration cooling flow rate required to maintain this temperature is found via the energy equation. It is assumed that the cooling fluid (also hydrogen) is instantaneously heated from T_w to T_{cool}, and that the cooling fluid does not interact with the principle flow (the flow entering through the core inlet) until it reaches the reservoir.

In order to evaluate rocket performance, the reservoir temperature must be determined. The reservoir is considered to be the region between the core exit and the nozzle throat. Here it is assumed that the propellant (including the cooling fluid) comes to a uniform temperature, which is determined by the average enthalpy of the fluid.

Rocket Performance

To calculate rocket performance the core exit region is modelled as a reservoir (the kinetic energy of the fluid is negligible) from which the propellant flows through a chocked nozzle to a vacuum. The nozzle expansion ratio is defined by user input. The thrust and specific impulse are both a function of the equivalent velocity.

$$F = \dot{m}v_{eq} \qquad and \qquad I_{sp} = \frac{v_{eq}}{g} \tag{8}$$

$$where: \qquad v_{eq} = v_e + \frac{P_e A_e}{\dot{m}} \ . \tag{9}$$

Therefore the exit velocity, pressure, and area must be calculated.

Modelling the propellant flow through the nozzle is difficult since the fluid may dissociate, ionize and/or recombine. As a simple approach, the standard isentropic compressible flow equations are used, which yield T,P, and ρ downstream in terms of M. The exit velocity and area are (note that $C_p = \gamma R/[\gamma-1]$):

$$v_e = M_e \sqrt{\gamma R_e T_e} = M_e \sqrt{(\gamma-1)C_{p_e}T_e} \ , \ and \tag{10}$$

$$A_e = \frac{\dot{m}}{\rho_e v_e} \ . \tag{11}$$

It was found that this approach yields values of I_{sp} 10% to 20% lower than calculated by Patch (1971), in which the flow of high temperature hydrogen through a choked nozzle is modelled much more thoroughly. Therefore, it was decided to use values of thrust and specific impulse based on data from Patch (1971), rather than using the above equations. However the above approach does provide a means of estimating the effect of uranium flow on rocket performance. The specific impulse of multi-species flow is:

$$I_{sp} = \frac{1}{g} \frac{\sum\limits_i \dot{m}_i v_{eq_i}}{\sum\limits_i \dot{m}_i} \ . \tag{12}$$

Fluid Properties

Fluid properties were obtained from various reports. For hydrogen the specific heat and density were taken from Patch (1971), while the thermal conductivity (which is important only at low temperatures) was found in Incropera and DeWitt (1981), and the Rosseland and Planck mean absorptivities were taken from Patch (1969). All of the required data for uranium was given in Parks et al. (1968). The hydrogen is assumed to contain .7 weight percent seed material with an absorption cross section of 5000 m^2/kg. This value was experimentally achieved by Williams et al. (1969).

RESULTS AND DISCUSSION

Temperature Solution

The input parameters of a large GCR, similar to one referenced by Borowski (1987), are listed below along with the corresponding results.

Input Parameters		Results	
Reactor Power	3000 MW	Fuel reservoir temp.	66320 K
Reactor Pressure	1000 atm	Prop. reservoir temp.	17880 K
Max. Wall Heat Flux	100 MW/m^2	Wall cooling flow	0.4 kg/s
Inlet Prop. Flow Rate	3.7 kg/s		
Inlet Temperature	2200 K	Specific Impulse	3160 s
Wall Temperature	2200 K	Thrust	125 kN
Fuel/Prop. Flow Ratio	10/1		
Core Length	2.0 m		
Fuel Region Radius	0.8 m		
Outer Wall Radius	1.0 m		

Figures 1 and 2 contain the axial and radial temperature distributions for the case listed above.

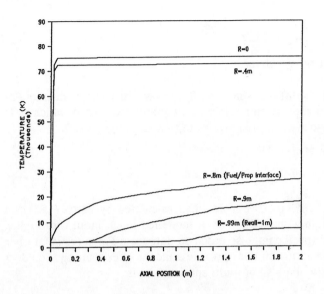

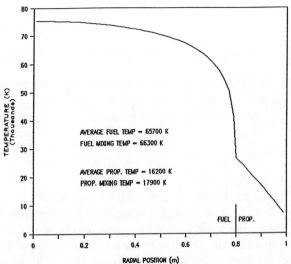

FIGURE 1. Axial Temperature Profile.   FIGURE 2. Radial Temp. Profile at Core Exit.

In Figure 1 it can be seen that the fuel reaches its equilibrium temperature very quickly. This is due to the low velocity and the relatively low specific heat of the uranium. Once the fuel reaches its equilibrium temperature profile, all of the energy generated in the fuel is transferred to the propellant. Therefore the fuel region temperature and interface heat flux are essentially constant, except at locations very near the core inlet. Consequently there is almost no difference between modelling the fuel region, or simply specifying the fuel/propellant interface heat flux. So unless fuel temperatures are desired, it is more convenient to specify the interface heat flux and ignore the fuel region.

Since the fuel quickly reaches its equilibrium temperature, the overall thermal conductivity between the fuel and propellant does not significantly impact the propellant temperature. If the conductivity is reduced, then the fuel temperature will simply increase to a temperature which will produce the same interface heat flux. Therefore if fuel/propellant mixing were to be modelled, the increase of the heat transfer coefficient at the interface would not significantly change the results. This does not diminish the overall significance of fuel/propellant mixing, which is of primary importance when calculating containment and criticality conditions.

Rocket Performance

The effects of various reactor parameters on rocket performance are discussed below.

Inlet Mass Flow Rate - Figure 3 shows how specific impulse varies with the inlet mass flow rate. If there is no maximum wall heat flux specified, the specific impulse increases as the flow rate decreases until a maximum of 4000 s is reached. This maximum indicates the propellant has reached an equilibrium temperature, meaning that any additional energy generated in the fuel will be conducted to the wall. If a maximum wall heat flux is specified, then the specific impulse reaches a peak at some optimum flow rate. If the inlet flow is reduced below this value then transpiration cooling is required, thus lowering the specific impulse.

Max. Wall Heat Flux - The specific impulse and thrust of the reactor described above are plotted versus maximum wall heat flux on Figure 4. As q_{max} is increased the maximum obtainable I_{sp} increases until it reaches a value of 4000 s, this indicates that the propellant has reached its equilibrium temperature. At low values of q_{max} the I_{sp} levels off around 2200 s. This value of I_{sp} can be obtained without any heat reaching the outer wall.

1086

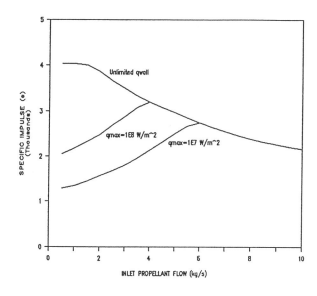

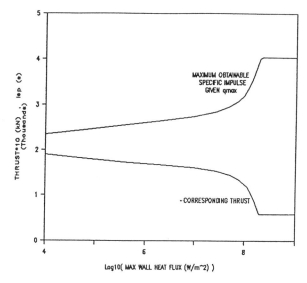

FIGURE 3. I_{sp} Dependence on Inlet Flow Rate.   FIGURE 4. I_{sp} Dependence on Max. Wall Heat Flux.

Dimensions - The dimensions of the system have little effect on the propellant temperature, and thus specific impulse. This is also due to the fact that nearly all of the energy generated flows to the propellant. However, increasing the dimensions can help reduce the wall heat flux by increasing the wall surface area. Yet this benefit may be nullified by the resulting mass penalty.

Power - Increased reactor power improves overall rocket performance, but not necessarily specific impulse. If reactor power is increased, the flow rate must be increased to avoid exceeding the critical wall heat flux, resulting in the same specific impulse but higher thrust. Likewise if power is reduced, the flow rate can be reduced, yielding the same specific impulse but lower thrust.

Uranium/Seed Particle Flow - Using the standard isentropic compressible flow equations, the effect of uranium flow on rocket performance was estimated. For a 10/1 propellant to fuel mass flow ratio, the specific impulse decreased 6% and thrust increased 2%. This decrease in specific impulse is relatively small because the uranium is at such a high temperature (~66,000 K). The seed particle amounts to only .7% of the propellant mass, thus should not significantly effect rocket performance.

Inlet/Wall Temperature - The temperature of the inlet fluid and wall has a minor effect on rocket performance. An increase in the wall and inlet temperature from 1100 K to 2200 K causes less than a 1% increase in I_{sp} and thrust. Possibly more important than the increase in performance, is the increase in the regenerative capacity of the propellant before entering the core. Thus eliminating the need for, or reducing the size of a possible radiator.

CONCLUSIONS

A heat transfer model is developed to assess the propulsion capability of the open-cycle gas core nuclear rocket. Using this model the open-cycle GCR is found to yield a very attractive combination of thrust and specific impulse. The major limiting factor in achieving high specific impulse is the maximum wall heat flux. The results of this and other studies indicate that although there are several major technical issues that must be addressed, the open-cycle gas core nuclear rocket merits future consideration as a high performance propulsion scheme..

Acknowledgments

All work was done at the University of Michigan. The work was sponsored by NASA Lewis Research Center as part of the NASA Graduate Researchers Program. The authors would like to acknowledge John Clark and Stan Borowski of NASA LeRC for their assistance.

References

Borowski, S. K. (1987) Nuclear Propulsion - A Vital Technology for the Exploration of Mars and the Planets Beyond, NASA TM-101354, NASA/Lewis Research Center, Cleveland, OH.

Incropera F. R. and D. P. DeWitt (1981) Fundamentals of Heat Transfer, John Wiley and Sons, New York, pp. 777-778.

Parks, D. E. et al. (1968) Optical Constants of Uranium Plasma, NASA CR-72348, Gulf General Atomic, San Diego, CA.

Patch, R. W. (1969) Interim Absorption Coefficients and Opacities for Hydrogen at High Pressure, NASA TM X-1902, NASA/Lewis Research Center, Cleveland, OH.

Patch, R. W. (1971) Thermodynamic Properties and Theoretical Performance of Hydrogen to 100000 K and $1.01325 \times 10^8 \ N/m^2$, NASA SP-3069, NASA/Lewis Research Center, Cleveland, OH.

Williams, J. R., J. D. Clement, A. S. Shenoy, W. L. Partain (1969) The Attenuation of Radiant Energy in Hot Seeded Hydrogen, Quarterly Status Report 2, Georgia Institute of Technology, Atlanta, GA.

Nomenclature

English

A:	Flow area (m)
a_p:	Planck mean absorptivity (1/m)
a_r:	Rosseland mean absorptivity (1/m)
C_p:	Specific Heat (J/kg·K)
F:	Thrust (N)
g:	Gravitational constant (9.81 m/s)
h:	Enthalpy (J/kg)
I_{sp}:	Specific impulse (s)
k:	Thermal conductivity (W/m·K)
M:	Mach number
$\dot{m}$:	Mass flow rate (kg/s)
P:	Pressure (Pa)
q:	Heat flux (W/m^2)
Q_v:	Volumetric heat generation (W/m^3)
r:	Radial position (m)
T:	Temperature (K)
v:	Flow velocity (m/s)
z:	Axial position (m)

Greek

γ:	Ratio of specific heats
ρ:	Density (kg/m^3)
σ:	Stefan-Boltzmann constant (5.67 x 10^{-8} W/m^2·K^4)
τ:	Stress tensor

Subscripts

e:	Nozzle exit
in:	Reactor inlet
int:	Fuel/prop interface
w:	Reactor wall

NUCLEAR THERMAL ROCKET CLUSTERING I:
A SUMMARY OF PREVIOUS WORK AND RELEVANT ISSUES

John J. Buksa, Michael G. Houts and Richard J. Bohl
Nuclear Technology and Engineering Division
Reactor Design and Analysis Group
Los Alamos National Laboratory
Los Alamos, NM 87544
(505) 665-0534
FTS 855-0534

Abstract

Clustering of rocket engines refers to the parallel connection of two or more small engines so that the performance of the propulsion system is superior to that of a single large engine. A general review of the technical merits of nuclear thermal rocket clustering is presented. A summary of previous analyses performed during the Rover program is presented and is used to assess clustering in the context of projected Space Exploration Initiative missions. A number of technical issues are discussed including cluster reliability, engine-out operation, neutronic coupling, shutdown core power generation, shutdown reactivity requirements, reactor kinetics, and radiation shielding.

INTRODUCTION

Clustering of rocket engines refers to the parallel connection of two or more small engines so that the performance of the propulsion system is superior to that of a single large engine. Chemical propulsion stages have employed clustering with great success for many years. Clustering of Nuclear Thermal Rocket (NTR) engines is of particular interest for those Space Exploration Initiative (SEI) missions that require high thrust (such as a piloted Mars mission). Clustering is additionally advantageous as it may lead to (1) reduced single-engine ground test facility size and cost because of the smaller size of individual engines, (2) the ability to meet a wider range of mission profiles through varying the number of engines in the cluster, (3) increased propulsion system reliability through redundancy, (4) lower engine development costs resulting from the reduced size of each engine, and (5) reduced flight safety concerns through an engine-out operating capability. Figure 1 presents a sketch of a conceptual cluster configuration containing three engines. This figure identifies several important clustering issues: propellant heating, reactivity coupling, and payload dose resulting from scattering among the engines and off the nozzles. Control (startup and shutdown), engine-out performance, power tilting, and stability are several other issues not depicted in this figure. Note that this concept utilizes engine/tank modules, as opposed to a single-tank configuration in which all engines share a common large tank. The modular approach was investigated in much detail during the Rover program under Westinghouse Astronuclear Laboratory's (WANL) Project NE 1840 (Kim 1966). Key topics addressed by WANL included preliminary internal shield optimization, neutronics, kinetics, and control. A single-tank configuration may be preferable because it minimizes tankage mass fraction and eliminates any streaming pathways from the engine cluster to the spacecraft payload.

In addition to the work done by WANL, other studies performed by Aerojet General as part of Rover (Houghton 1965) and Douglas Aircraft Company through an internal R&D program (Woyski and Langley 1968) analyzed clustered NTRs. Nuclear interactions between coupled reactors were also investigated in detail at Los Alamos National Laboratory (LANL) and a benchmark subcritical experiment was performed between two KIWI class reactors (KIWI-TNT and PARKA) in September and October of 1964 (Chezum et al. 1967). The reference section contains a concise listing of relevant reports that came out of these efforts.

The departure point for this paper is an abbreviated argument for the use of a clustered NTR propulsion system. The probability of mission success rests heavily on the reliability of the spacecraft's propulsion system. A highly reliable clustered engine system is more easily attainable than a highly reliable single large engine because of the redundancy offered by the engine-out capability of a clustered propulsion system. This relationship is characterized by:

$$R=\sum_{k=0}^{n} \frac{N!}{(N-k)!k!} R_E^{N-k} \left[R_s \left(1-R_E \right) \right]^k,$$ (1)

where R is the overall propulsion system reliability, R_s is the engine-out safety (or shutdown cooling) subsystem reliability, R_E is the single-engine reliability, N is the number of engines in the cluster, and n is the number of failed engines acceptable with mission success (Woyski and Langley 1968). Figure 2 shows the overall propulsion system reliability as a function of single-engine reliability and the reliability of the engine-out safety subsystem (for a three-engine

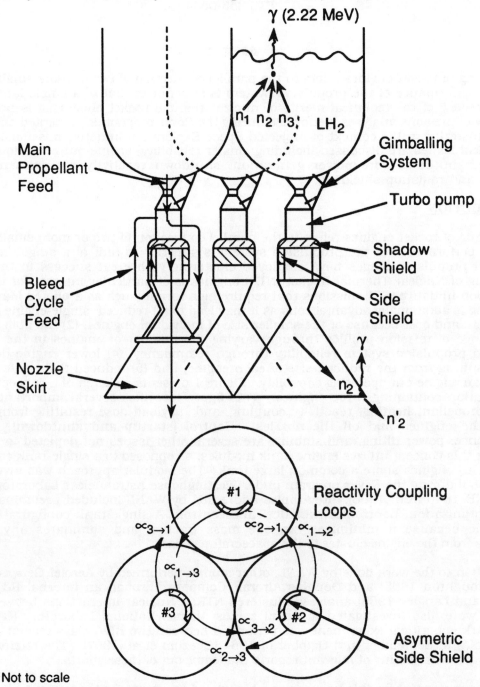

FIGURE 1.   Historical Cluster Configuration employing Separate Engine/Tank Modules and Asymmetric Side Shields for Propellant Heating Minimization.

cluster with a one engine-out capability). The safety system is needed in order to assure that the balance of the engines are not affected by any one failed engine. The advantage of a cluster system over that of a single-engine system results from the ability of the cluster system to perform sufficiently when one of the engines fails. To emphasize this fact, Figure 2 shows the case where only a single large engine is employed (system reliability equals engine reliability). For an arbitrary system reliability goal of 0.991 and a 100% reliable safety subsystem, the reliability requirement for a cluster of three engines is 0.943 for any single engine in the cluster, whereas a single large engine must be 0.991 reliable to attain the same 0.991 goal. Over 500 successful demonstrations of engine performance with one failure would be required to assure the 0.991 goal while less than 100 similar demonstrations would be needed to show 0.943 reliability (Woyski and Langley 1968). Consequently, the potential savings of a smaller testing program make clustering an attractive option worth pursuing. The disadvantage of clustering is a lower thrust/weight compared with a single large engine. Aerojet General estimated an overall mass penalty for clustering of around 30%, but went on to identify a number of significant development challenges for the large NTR (Houghton 1965). The objective of this paper is to address the technical feasibility of clustering by grouping the majority of relevant issues into three areas: operational issues, reactor nuclear issues, and spacecraft shielding issues. The remaining sections address each of these areas by summarizing (where possible) past analyses and comparing them with present-day technical, safety, and performance requirements.

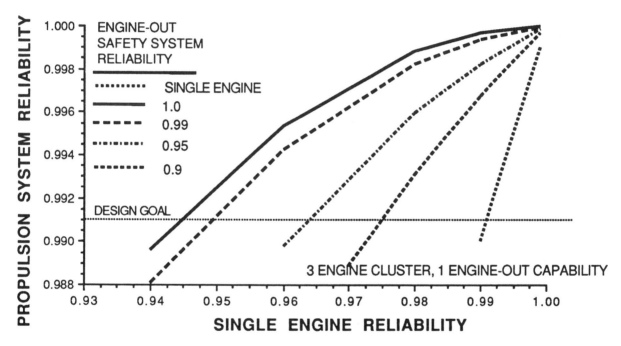

FIGURE 2.   Propulsion System Reliability as a function of Single-Engine and Engine-out Safety System Reliabilities for a Three-Engine Cluster.

OPERATIONS

Perhaps the most important near-term issue deals with the operation of the cluster and the philosophy by which an engine-out accident (EOA) would be handled. The EOA refers to any situation where an engine fails to rise to full power upon demand (regardless of the cause of failure). In order to assist in this discussion, Table 1 describes six possible operational schemes for a cluster system. The schemes use either 1) a pulsed cooling engine-out safety subsystem or 2) an engine-out jettison safety subsystem. A pulse cooling subsystem uses pulses of LH_2 propellant to cool a shutdown engine, removing any decay heat or coupling fission power generated in the core. The shutdown coupling power is present only if the adjacent reactors are still operating and is a consequence of the neutronic coupling between the cores. More will be said about steady state and shutdown neutronic coupling later. The jettison safety subsystem consists of some mechanism to rapidly disconnect the failed engine and to eject it from the vicinity of the spacecraft. The need to either cool or jettison a failed engine is required in order to protect the remainder of the spacecraft from the possible consequences of reactor core melting. Furthermore, because trans-Mars injection may incorporate a triple perigee burn, jettison of a used engine before the third burn would entail jettison into Earth orbit. Even though the jettison system is essentially 100% reliable, top-level safety requirements may preclude the incorporation of a jettison system. Consequently, the three operational schemes in Table 1 that use a jettison system are no longer considered.

The concept of shutdown engine cooling by pulse cooling has been studied in fair detail (Retallick 1971) and was employed in many of the engine tests during Rover. Immediately after scram, full propellant flow is maintained until the decay power level is low enough to allow pulse cooling. Short bursts of propellant are then pumped through the reactor when core temperatures are sufficiently high. It was shown that propellant usage can be minimized if each pulse is long enough to sub-cool the core slightly. Three of the schemes in Table 1 use this technique to cool a shutdown engine. Of these, scheme 1 is undesirable because of the range of possible thrust/burn time combinations that result from not knowing when an engine will fail during the mission. Schemes 2 and 3 are attractive because they offer near constant propulsion system thrust/weight ratios in the event of an EOA. The throttle scheme (#3) offers a slightly higher reliability because the nominally operating engines are at a reduced thrust level (lower power, propellant flow rate, axial pressure drop, and axial stress). Note, however, that if the nominal thrust level is 75 klb (1500 MW) then the EOA (or design) thrust level is 112.5 klb (2250 MW) for a three-engine propulsion system nominally rated at 225 klb thrust. A larger cluster of smaller engines will reduce the difference between the nominal and EOA power levels. Unlike the throttle scheme, the spare scheme (#2) would require that only one engine size be developed but that a cooled spare engine be carried along and possibly never used. Consequently, both the throttle and spare schemes should be investigated further.

TABLE 1.   Descriptions of Possible Engine Clustering Schemes including Engine-out Operation.

ENGINE-OUT SAFETY SYSTEM	ENGINE CLUSTERING SCHEMES		
	1	2	3
PULSE SUB-COOLING	A D E	B D F (spare scheme)	C D F I (throttle scheme)
JETTISON	A E G H	B F G H	C F G H

A- Nominal: all engines full thrust. Engine-out: remaining engines continue for longer burn.
B- Nominal: all engines full thrust, spare not run but cooled. Engine-out: spare engine stepped to full thrust.
C- Nominal: all engines at throttled thrust. Engine-out: remaining engines stepped to full thrust.
D- Propellant reroute if separate engine/tank modules are employed.
E - Optimal trajectory must envelope all possible combinations of nominal and engine-out thrust and burn time.
F- Near constant thrust/weight ratio maintained.
G - Safety concern related to jettison of used engine in Earth orbit.
H- Higher single-engine reliability required to account for jettison system reliability.
I - Lower single-engine reliability required due to reduced power during nominal operation.

NEUTRONICS/KINETICS

The primary difference between the behavior of an autonomous nuclear rocket engine and the behavior of an engine that is a member of a propulsion system cluster is the neutronic coupling that occurs between engines. This interaction can be described by the coupled reactor point kinetics equations:

$$\frac{dn_1(t)}{dt} = \frac{\left(\rho_1(t) - \bar{\beta}\right)}{L_1} n_1(t) + \sum_g \lambda_{g1} C_{g1}(t) + \sum_{i=2}^{m} a_{i \to 1}\left[\frac{n_i(t - \tau_{i \to 1})}{L_i}\right], \qquad (2)$$

$$\text{and} \quad \frac{dC_{g1}(t)}{dt} = \beta_{g1}\frac{n_1(t)}{L_1} - \lambda_{g1} C_{g1}(t), \qquad (3)$$

where n is reactor power, L is average neutron lifetime, ρ is reactivity, β is the total delayed neutron fraction, λ_g is the gth group precursor decay constant, C_g is the gth group precursor concentration, $\alpha_{i \to 1}$ is the coupling reactivity at reactor 1 due to reactor i, and $\tau_{i \to 1}$ is the transfer time delay of neutrons leaving reactor i and reaching reactor 1. If all reactors in the cluster are identical at the same steady state power level, and $\tau_{i \to 1}$ is neglected, the last term of Equation (2) becomes a total reactivity coupling coefficient of reactor 1 due to all other reactors in the cluster (Woyski and Langley 1968, Chezem et al. 1967, and Mowery and Romesburg 1965). In general, the coupling coefficient is the amount of reactivity that must be added to reactor 1 in order to restore its criticality when the ith reactor is removed from the cluster. Mathematically, the coupling reactivity between engine 1 and the ith engine is the product of three factors: a leakage factor for the ith engine, a geometric attenuation factor between the two reactors, and an effectiveness or reactivity factor for engine 1. Individual coupling reactivities can be superimposed to get the last term of Equation (2) for clusters of greater than two engines.

Several methods have been developed to estimate the three coupling factors for a cluster of NTRs. WANL employed multigroup discrete ordinate calculations to estimate the driver core leakage and driven core effectiveness factors. Previously, Douglas developed a simple method for estimating the geometric attenuation factor but, unfortunately, it did not consider the angular importance of incident neutrons at the driven core boundary. Both WANL and Douglas later made adjustments to account for this deficiency. WANL predictions of the coupling reactivity between two identical cores at criticality are presented in Figure 3 along with the experimental results of the KIWI-PARKA experiment at Los Alamos. As expected, the coupling reactivity between cores is small and is inversely proportional to separation distance and to core size. It is important to realize that coupling is relative; it makes sense only when talking about the coupling at one reactor due to another. This idea is graphically depicted in Figure 1, which shows the six coupling coefficients involved in a three-engine cluster. In the nominal operating case, all of the coefficients are equal, and coupling is straight forward. Only in the

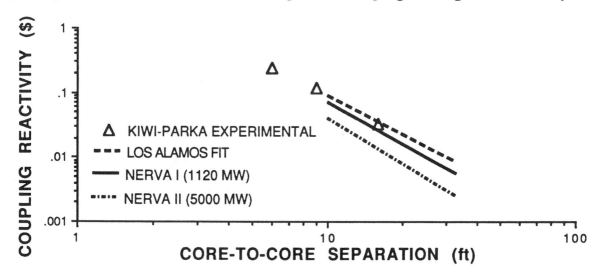

FIGURE 3.   Experimental and Past Analytical Predictions of Coupling Reactivity as a function of Core Separation(from Kim 1966).

EOA case are there more than one coefficient. For example, in the case where engine #3 is out: $\alpha_{1\to2} = \alpha_{2\to1}$ = several cents, $\alpha_{3\to1} = \alpha_{3\to2}$ = zero , and $\alpha_{1\to3} = \alpha_{2\to3}$ = engine #3 shutdown margin, which is typically minus several dollars. The fact that coupling is not reversible between the out engine and any operating engine is of little consequence except for the shutdown engine's power tilting, which is discussed next.

One unwanted consequence of neutronic coupling is the subcritical fission power generated in an out engine during an EOA. Recall that the coupling coefficient is the product of leakage, geometric attenuation, and incident neutron fission effectiveness factors. Between a shutdown engine core and an operational core, the leakage and geometry factors are constant from nominal to EOA conditions. Relative to the shutdown engine, however, the effectiveness factor scales linearly with the subcritical multiplication factor, which in turn scales linearly with the shutdown margin of the shutdown core. Even though the shutdown reactor is far from critical, the nearby operating engines act as a relatively significant source of neutrons, which can undergo multiplication and can produce power. Furthermore, the driver core leakage term scales linearly with power level and the geometric term between cores varies little for constant core separation distances (because of small changes in the view factors between reactors). WANL calculated the shutdown power level as a function of coupling coefficient for both the NERVA I (1120 MW) and the NERVA II (5000 MW) engines. Table 2 presents several shutdown power levels calculated by WANL along with other scaled estimates for several shutdown margins and for a 1500 MW nominal power level. Also shown in this Table are the 4% decay power levels (following scram) for comparison. Note that even for minimal shutdown margins, pulse cooling systems designed for decay power levels should be more than adequate to remove EOA coupling powers (assuming the original failure mechanism does not propagate to the shutdown safety subsystem).

TABLE 2.   Cluster Shutdown Engine Heat Generation Rates resulting from Decay Power and Neutronic Coupling to Adjacent Operating Engines.

Nominal Engine Power Level (MW)	Maximum Decay Power Level (MW)	Power Level in Single Shutdown Coupled Core[a]		
		Shutdown[b]= $2	$4	$25
1100	44	31[c]	15[c]	2.5
1500	60	41	21	3.3
5000	200	110[c]	60[c]	8.8

[a]Multiply by number of active engines in cluster to get total heating in shutdown engine
[b]$\beta = 0.007$
[c]From Mowery and Romesburg 1965.

The subcritical fission power generated in the shutdown reactor is not evenly distributed. During nominal operation, power tilting is very small. Both LANL and WANL calculated nominal core power tilting levels of less than 1% and the KIWI-PARKA experiment demonstrated the same (Kim 1966, Chezum et al. 1967). During an EOA, however, power tilting is proportional to the shutdown reactivity. For large shutdown margins, a peak-to-average heating ratio on the order of 4 or 5 at the interface with an adjacent operating reactor may be experienced. Although not presented here, WANL clearly showed this effect as a function of radial position and shutdown margin (Mowery 1965). LANL experimentally showed these behaviors, and more recently we have used a Monte Carlo neutron/photon transport code (MCNP) to show this effect. Results are presented in a companion paper (Houts et al. 1991). Note that as the shutdown margin increases in magnitude, the total amount of fission power generated in the shutdown core decreases, but preferentially peaks at the peripheral interface of the core. Table 2 indicates that the peak heating rate resulting from neutronic coupling power is much less than the decay power level immediately after shutdown. In fact, the 4% decay power level shown corresponds to the heating rate immediately after scram, and drops off with a stable period of 80 s soon afterwards. This decay power level is determined by the prompt drop approximation:

$$\frac{P}{P_0} = \frac{\beta}{\rho_s + \beta} \tag{4}$$

where P is the power level, β is the total delayed neutron fraction and ρ_s is the shutdown reactivity. Consequently, during an EOA, ρ_s influences the integrated energy generated in the scrammed reactor, the amount of coupling between reactors in a cluster, and the power tilting in the shutdown core. The $25 margin indicated in Table 2 is used to estimate the maximum decay power level and is an estimate based on preliminary water immersion criticality calculations using a detailed MCNP model. This large shutdown margin will substantially reduce shutdown core coupling power but will strongly perturbate the power generation towards the adjacent reactor interfaces. Once all engines in the core are shut down, the coupling power level goes to zero and the shutdown cooling system need only remove decay heat, which drops to about 0.1% of the nominal power level in about 1 hour.

Several other issues are relevant to any technical discussion of clustered NTRs but, unfortunately, are beyond the scope of this paper. In particular, stability, control, and detector positioning have been identified as points requiring further analysis. Both WANL and Douglas derived the transfer function for a cluster of nuclear rockets and found that no inherent instabilities exist. In fact, open loop stability was found to be higher for the cluster than for a single engine because of the neutronic coupling loops. LANL experimentally determined the zero-power transfer function of the KIWI-PARKA system using an oscillating rod technique, and similar conclusions were drawn. Both WANL and Douglas predicted that sequential startup of the cluster poses no obvious problems. Douglas did, however, show that simultaneous startup, or for that matter any common mode transient, would lead to power density spikes resulting from inherent feedback reactivity exacerbated by neutronic coupling. Staggered startup tends to smooth the rise of all engines and should be considered (Kim 1966, Chezum et al. 1967, and Woyski and Langley 1968). In order to facilitate a rapid and smooth simultaneous startup, a reactivity based controller should be considered in order to close the control loop. In this case, detector location may play a role in the effectiveness of the control system, especially if in-line or asymmetric engine configurations are employed.

SHIELDING

The technical challenge of shielding a cluster of NTRs is only marginally more difficult than that of shielding a single large engine. Of particular concern is the shielding mass penalty associated with using a cluster instead of a single large engine. The discussion here is general in nature and a more detailed discussion is presented in a companion paper (Houts et al. 1991). Shielding may be needed to shield one reactor from another or the spacecraft from the cluster as a whole. For example, the shielding requirement for spacecraft components (such as turbopumps, valves, propellant tanks, or electronic equipment) is significantly different from that for shielding a crew from reactor generated radiation. In both cases it is important to account for both primary (originating from the reactor) and secondary (from primary collisions) sources. Figure 1 depicted two of the most important secondary sources: the (N, γ) reaction in liquid H_2, and nozzle scattering. Neutron scattering rates from the nozzle will be small because the nozzle contains very little mass. Gammas scattered back towards the payload will have low energies and can be attenuated by residual propellant or by spot shielding of the crew. Primary neutrons and gammas can best be shielded at the reactor where the shield cross-sectional area is the smallest. A single-tank configuration will offer additional shielding from the reactors and nozzles because its diameter would shadow-shield the entire propulsion system. From a crew dose standpoint, an internal neutron/gamma shield may not be required because of the good attenuation of hydrogen propellant and the presence of a reconfigurable crew storm shelter. An internal shield may, however, be required to reduce tank propellant heating, the radiation dose to components located near the engine inlet, and to reduce the time that the crew has to spend in the storm shelter following the final burn if the engines are not jettisoned (not in Earth Orbit). A preliminary internal shield design was attempted by WANL in order to arrive at an minimum shield mass (Kim 1966). Throughout these analyses, the top-level requirement was to minimize propellant heating through the use of internal shields and included rough estimates of the propellant heating caused by the energetic (N, γ) reaction, which has a large thermal neutron cross section. WANL looked at using a top shadow shield, top shield extensions, and side shields (see Figure 1). Analyses indicated that sufficient attenuation (90% of unshielded value) is possible through the use of a top shadow shield with small vertical extensions. Although asymmetric side shields can lead to nearly 100% attenuation levels, they are only marginally mass effective and complicate the thermal-hydraulic design of the core periphery. As indicated earlier, the reactivity coupling between operating engines is very small and no neutron shielding between reactors will be needed for the sake of neutronic isolation.

CONCLUSIONS

As a consequence of their increased reliability, economy, and flexibility, clustered NTR propulsion systems should be considered for SEI missions requiring high thrust levels. As a first step, an engine-out accident operational scheme should be identified as soon as possible and should take into consideration top-level safety requirements and engineering practicalities. The neutronic interaction between reactors in a cluster does not impact the performance of the cluster during nominal operation. During an engine-out accident, however, the coupling is manifested as a power source in the shutdown engine (in addition to decay heat) and is preferentially located at the interface with the adjacent operating engine. Clustered NTRs are inherently stable, can be sequentially started, and (with small internal shields only) do not require large amounts of additional shielding in order to limit dose rates to engine components, propellant, or the crew. Several areas have been identified where further research and development may be required, including simultaneous startup, reactivity based control, detector positioning, shutdown heat removal system design, and the design of a (possibly in-core) large reactivity shutdown system. Contemporary computational techniques need to be developed to verify cluster neutronic, kinetic, and thermal hydraulic behaviors, particularly for the engine-out accident case which involves reactors at unequal power levels.

Acknowledgments

This research was performed at Los Alamos National Laboratory under internal program development funding support. Special thanks go out to everyone who worked in the Rover program for giving us a great starting point for the exploration of the solar system. Thanks also go to B. Milder and G. Mirabal for their artistic and editorial contributions.

References

Chezum, C.G., G.H. Hansen, H.H. Helmick, and R.L. Seale (1967) "The Los Alamos Coupled Reactor Experience," Los Alamos report no. LA-3494, Los Alamos National Laboratory, Los Alamos, New Mexico, March 1966.

Houghton, W.J. (1965) "Clustering of Nuclear Engines," Aerojet General report no. RN-PA-0001, Rev A, Aerojet General Corporation, October 1965.

Houts, M.G., J.J. Buksa and R.J. Bohl (1991) "Nuclear Rocket Clustering II: Monte Carlo Analyses of Neutronic, Thermal, and Shielding Effects," in Proc. Ninth Symposium on Space Nuclear Power Systems, CONF-920104, to be held 12-15 January 1992.

Kim, C-K (1966) "Clustered NERVA II Engine System Studies," Westinghouse Astronuclear Laboratory report no. WANL-TME-1361, Westinghouse Electric Corporation, Pittsburgh, Pennsylvania.

Mowery, A.L. and C. Romesburg (1965) "Reactivity Coupling of Clustered NERVA Reactors," Westinghouse Astronuclear Laboratory report No. WANL-TME-1152, Westinghouse Electric Corporation, Pittsburgh, Pennsylvania.

Retallick, F.D. (1971) "Cooldown and Thrust Nulling Requirements," Westinghouse Astronuclear Laboratory report No. WANL-TME-2749, Westinghouse Electric Corporation, Pittsburgh, Pennsylvania.

Woyski, J.S. and R.W. Langley (1968) "Clustered Nuclear Engines," Douglas report No. DAC-608790, Douglas Missile and Space Systems Division, Santa Monica, California.

NUCLEAR THERMAL ROCKET CLUSTERING II: MONTE CARLO ANALYSES OF NEUTRONIC, THERMAL, AND SHIELDING EFFECTS

Michael G. Houts, John J. Buksa, and Richard J. Bohl
Los Alamos National Laboratory
Reactor Design and Analysis Group
Los Alamos, NM 87545
(505) 665-4336

Abstract

Monte Carlo analyses of a cluster of nuclear thermal rockets have been performed using the code MCNP (Briesmeister 1986). Three effects of clustering have been analyzed: neutronic coupling, nuclear heating in a shutdown engine, and radiation scattering. Preliminary analyses indicate that clustering phenomena will not have significant effects on engine design or operation. More detailed analyses should be performed as engine and vehicle designs progress, although the basic conclusions of this study are not expected to change.

INTRODUCTION

Clustering nuclear thermal rocket (NTR) engines may yield safety, reliability, cost, and schedule benefits. However, clustering NTRs raises several issues, including neutronic coupling among the cluster of engines, nuclear heating in a shutdown engine, and radiation scattering effects on shielding design. This paper presents the preliminary results of Monte Carlo analyses aimed at investigating these issues.

A cluster of three 330,000 N (75,000 lb) thrust engines appears suitable for manned Mars missions. This configuration is being baselined by the NASA/DOE/DoD Nuclear Propulsion Mission Analysis Panel (Wickenheiser 1991). An engine sized to provide 330,000 N of thrust is the 1500 MW Phoebus 1B, tested in 1967 during the space nuclear rocket program (Rover). Although improvements may be incorporated into future nuclear rockets, the Phoebus 1B design is representative of a 330,000 N thrust nuclear engine, and is used in this study.

An MCNP (Briesmeister 1986) model of a cluster of three Phoebus 1B engines has been developed. The engines are arranged in an equilateral triangle with a center-to-center spacing of 5 m. The engines are divided radially into four regions: the core, the graphite reflector cylinder, the beryllium reflector, and the pressure vessel. Axially, the engines are divided into five regions: the nozzle, the hot-end support block, the core, the cold-end support plate, and the internal shield. Engine length (excluding nozzle) is 1.63 m, and engine diameter is 1.29 m. A schematic of the Phoebus 1B engine model is shown in Figure 1, and a schematic of the cluster/vehicle assembly is shown in Figure 2. Although not a feature of the Phoebus 1B, the radial surface of each engine's pressure vessel is assumed to be coated with 0.5 mm of boral to absorb thermal neutrons exiting or entering the engine. Future analyses will assess the importance of this coating. Nozzles are modeled as carbon cones with an opening angle of 45 degrees and a wall thickness of 5.0 mm.

NEUTRONIC COUPLING

The effect of neutronic coupling on reactor control is one potential concern with clustering nuclear thermal rocket engines. The MCNP engine cluster model was used to determine the fraction of the core flux in each engine caused by neutrons originating in other engines. For the configuration analyzed (a cluster of 3 engines operating at identical powers), less than 0.01% of the core flux in an engine is caused by neutrons that originated in one of the other two engines.

This result leads to several observations. First, the minimal coupling among engines indicates that neutronic effects will be negligible if all engines are operating at roughly the same power (within one or two orders of magnitude). Second, a minimum power level will exist for a shutdown engine in a cluster (a few orders of magnitude lower than the maximum power engine in the cluster), as discussed later in this

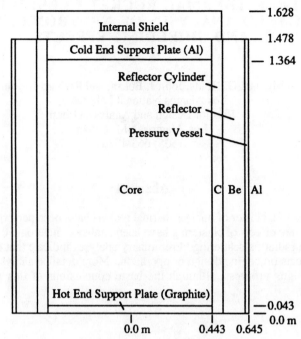

- 1.628
- 1.478
- 1.364

Internal Shield

Cold End Support Plate (Al)

Reflector Cylinder

Reflector

Pressure Vessel

Core C Be Al

Hot End Support Plate (Graphite)

- 0.043
- 0.0 m

0.0 m   0.443 0.645

FIGURE 1. Axial Cross Section of an MCNP Model
of Phoebus 1B (Nozzle not Shown).

——— Dose Plane
at 100 m

——— Dose Plane
at 10 m

NTR
Cluster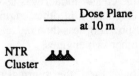

FIGURE 2. Schematic of the MCNP Cluster/Vehicle Assembly Model (to Scale).

paper and in a companion paper (Buksa et al. 1991). Finally, if an engine is operating near its minimum power, neutronic coupling among the cluster should be taken into account during multiple engine transients.

An example of when neutronic coupling among the cluster should be taken into account is as follows. Suppose two engines in the cluster are operating at high power while the third engine is being brought from subcritical to critical. If the two operating engines are rapidly shut down, the flux in the third engine will decrease because of the decrease in source strength. An unsupervised reactivity-based controller could then underestimate reactivity and request a reactivity insertion. Normal limits on maximum reactivity and rate of change of reactivity should preclude any safety implications. It may also be desirable, however, to have a supervisory controller take into account the possibility of erroneous reactivity measurements caused by neutronic coupling. Although neutronic coupling should be taken into account, it does not appear that coupling will have a significant effect on engine design or operation.

NUCLEAR HEATING IN A SHUTDOWN ENGINE

As noted above, neutronic coupling among a cluster of nuclear thermal rockets results in a minimum power level at which an engine in the cluster can operate. The following method was used to quantify nuclear heating in a shutdown engine. First, the Phoebus 1B engine model was modified to yield an engine with a k_{eff} of 0.95, simulating a shutdown engine. This greater than $7 shutdown was simulated by adding a small amount of boron to the radial reflector. The shutdown engine was then placed in a three engine cluster (with a center-to-center spacing of 5 m), with the other two engines each having a k_{eff} of 1.00. Finally, MCNP was used to calculate nuclear heating in each engine's core.

The MCNP results indicate that for a three engine cluster, power generation in a shutdown engine will be about 0.1% of the total power generated by the two operating engines. This result compares well to the shutdown power predicted during the Nuclear Engine for Rocket Vehicle Application (NERVA) program (Kim 1966). As expected, the power in the shutdown engine is peaked towards the two operating engines, and towards the edge of the core.

Figure 3 shows peak-to-average heating in the shutdown NTR. The results are accurate to within $\pm$ 20% based on MCNP statistics. The maximum peak-to-average power ratio of 2.7 occurs between a radius of 0.396 and 0.443 m, on the side of the shutdown core facing the two operating cores. Peaking in an actual core may be less severe because of in-core shutdown rods and power flattening. A radially power-flattened NTR would have a lower-than-average fuel loading near the radial reflector, reducing peaking both during operation and during shutdown.

The primary observation that can be made based on the MCNP results is that in the evaluated configuration, nuclear heating in a shutdown engine will be an order of magnitude less than the decay heat in an engine shortly after shutdown from full power. Systems designed to remove decay heat after prolonged full power operation should thus also be capable of removing heat generated in a shutdown engine if the engine has not been run. If an operating engine is shut down during a burn and the remaining engines in the cluster continue to operate, slightly more heat removal capability would be required for the clustered engine than for a nonclustered engine. If the capability to pump propellant still exists, the engine may be cooled by cyclically allowing the engine to heat up and then subcooling it with hydrogen. Additional heating caused by clustering could be removed by reoptimizing the schedule and duration of hydrogen cooling bursts. If the engine is designed for dual-mode operation, it may also be possible to remove significant power via closed-loop cooling. Power peaking in a shutdown engine may affect the optimal design of the shutdown cooling system.

SHIELDING EFFECTS OF CLUSTERING

Radiation shielding of the payload may be complicated by the use of a cluster of nuclear thermal rockets as opposed to a single large engine configuration. Radiation scattering among engines will allow some radiation to bypass engine shadow shields, limiting their effectiveness.

The MCNP model of the three engine cluster was modified to allow an analysis of the effects of clustering on radiation shielding. A 0.15-m thick, 0.5-m radius internal shield was placed above the cold-end support plate of each engine. The shield is composed of borated zirconium hydride, and has a theoretical density of 0.80 (to allow for cooling channels). Flux tally cells were added at distances of 10 m and 100 m above each engine, and cylindrical flagging cells were added around each engine.

MCNP was then used to calculate energy dependent neutron and gamma fluxes at distances of 10 m and 100 m above the cluster of engines. The flux was also divided into four components: the total flux, the flux that interacted with a nozzle, the flux that passed through the flagging cells that separate the engines, and the flux that either passed through the flagging cells that separate the engines or interacted with the nozzles. Shielding from the propellant tanks, propellant, structure, crew module, and other components was not taken into account in this calculation. The MCNP results were then used to calculate the relative radiation dose caused by radiation coming directly from the engine, radiation scattered off of the nozzles, and radiation scattered among the cluster.

Gamma radiation appears to be of primary concern, since even 2 m of liquid hydrogen (propellant) will attenuate virtually all neutrons coming from the engines. Even on the final engine burn, enough propellant should remain in the tank at shutdown (remaining propellant would be used to remove decay heat from engine) to shield the crew from neutrons. Also, a 10-m diameter tank will shadow the entire crew module (in the proposed configuration) from all engine components (including the nozzle), thus scattered neutrons should also be attenuated. Liquid hydrogen provides considerably less attenuation of gamma rays, thus gamma radiation may be a concern to the crew or to components located near the engine inlet.

The MCNP calculations show that at a 100-m separation distance (in the stated configuration), gamma rays scattered from nozzles or among engines in the cluster account for less than 3% of the dose to the crew. Moreover, nearly all of the scattered gamma rays that reach the dose plane have an energy of less than 0.5 MeV, and should thus be relatively easy to attenuate. Most of the radiation dose from scattered gamma radiation comes from gamma rays that interact with the nozzles.

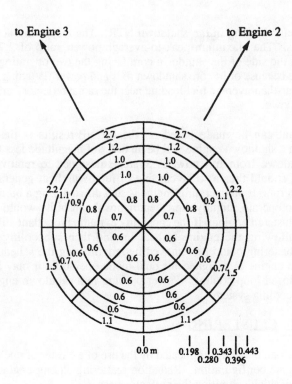

FIGURE 3. Peak-to-Average Heating in the Shutdown Nuclear Thermal Rocket with Keff = 0.95. Two Driver Engines located 5 m from Shutdown Engine.

FUTURE RESEARCH

A more detailed MCNP NTR model that includes variations in fuel loading (for power flattening) and design features typical of currently proposed NTRs should be developed. After the model is developed, the following research should be performed.

Crew Shielding

Clustered nuclear thermal rocket shielding should be optimized using Monte Carlo methods. The primary astronaut shielding concern will be shielding the engine cluster shortly before and for a few hours after (if the engines are not jettisoned) the final burn. Prior to the final burn, the 30-m (or more) column of hydrogen between the engines and crew should provide adequate gamma attenuation. Although neutron shielding will not be a concern after shutdown, gamma radiation will continue to be emitted from the core.

The Synthesis Group (Synthesis Group 1991) has recently stated that a crew "storm cellar" shielded by 16 g/cm^2 of water should provide adequate shielding against anomalously large solar flare events. The Synthesis Group also noted that the water used for the shielding would enhance safety by providing a backup for the water loop closure systems. The storm cellar may also provide adequate shielding during and after the final NTR burn, especially if it can be reconfigured to place more of the water between the engines and crew. In general, it may enhance safety to have the crew inside the reconfigured solar flare storm cellar during and shortly after all NTR burns.

Engine Internal Shield

The NTR internal shield reduces nuclear heating of components and propellant located near the inlet to the engine. The shield also reduces radiation dose near the engine inlet and at the crew dose plane. However, shrinking the NTR internal shield could significantly reduce NTR mass. Future research should examine the potential of turbopump cavitation caused by nuclear heating near the engine inlet, the radiation dose tolerance of components located near the engine inlet (such as turbopumps and actuators), and the effect of internal shield size on crew dose following engine shutdown. Internal shield design can then be optimized.

Quantification of Neutronic Coupling Effects

The effects of neutronic coupling should be quantified by developing a point kinetics model for each engine, and (using the MCNP-generated coupling data) examining the interaction among engines during various transients.

Quantification of Temperature Profiles in a Shutdown Engine

The power profile within a non-operated, shutdown engine in an NTR cluster has been calculated using MCNP. Also, decay heat generation rates have been calculated as a function of engine run time and time after shutdown. Decay power and power generation from clustering effects should be used to calculate the transient temperature profile within a shutdown engine for various scenarios. Potential scenarios include shutdown a few minutes prior to scheduled shutdown (in which case decay power will be significant) and failure of an engine to start up (in which case the engine would be at ambient temperature, and much of the induced heating could be stored via thermal inertia).

CONCLUSIONS

Monte Carlo analyses of a cluster of nuclear thermal rockets (NTRs) have been performed using the code MCNP (Briesmeister 1986). Three effects of clustering have been analyzed: neutronic coupling, nuclear heating in a shutdown engine, and radiation scattering.

Preliminary results indicate that while clustering effects should be taken into account, they will not have a significant effect on engine design or operation. Neutronic coupling may need to be taken into account if two engines operating at significantly different powers simultaneously undergo transients. Significant heat will be generated in a shutdown engine; however, decay heat removal systems should be capable of removing the heat generated by clustering effects. If 10-m diameter propellant tanks are used (as currently envisioned) radiation scattering from nozzles and among the cluster of engines should not be of additional concern since the entire payload will be shadowed from the engine cluster. Potential problems will be further mitigated if a reconfigurable solar flare shelter is available to the crew. More detailed analyses should be performed as engine and vehicle design progresses, although the basic conclusions of this study are not expected to change.

Acknowledgments

This study was supported by internal funding. The authors are grateful to the individuals who made the Rover Program a success and laid the foundation for future nuclear rocket development.

References

Briesmeister, J.F. (1986) *MCNP - A General Monte Carlo Code for Neutron and Photon Transport*, LA-7396-M, Rev. 2, Los Alamos National Laboratory, Los Alamos, NM.

Buksa, J.J, M.G. Houts, and R.J. Bohl (1991) "Nuclear Thermal Rocket Clustering I: a Summary of Previous Work and Relevant Issues," in Proc. Ninth Symposium on Space Nuclear Power Systems, CONF-920104, to be held 12-16 January 1992.

Synthesis Group (1991) *America at the Threshold, Report of the Synthesis Group on America's Space Exploration Initiative*, Superintendent of Documents, US Government Printing Office, Washington, DC.

Kim, C.K. (1986) "Clustered NERVA II Engine System Studies, "WANL-TME-1361, Westinghouse Electric Corporation, Pittsburgh, PA.

Wickenheiser, T. (1991) "Nuclear Propulsion Mission Analysis Panel Steering Committee Report," presented 23 April 1991 at NASA Lewis Research Center, Cleveland, OH.

A FISSION FRAGMENT REACTOR CONCEPT FOR NUCLEAR THERMAL PROPULSION

Ahti J. Suo-Anttila, Edward J. Parma, Paul S. Pickard, Steven A. Wright, and Milton E. Vernon
Sandia National Laboratories
P.O. Box 5800
Albuquerque, NM 87185-5800
(505) 845-3046

Abstract

The Space Exploration Initiative requires the development of nuclear thermal and nuclear electric technologies for space propulsion for future Lunar and Mars missions. Sandia National Laboratories has proposed a new nuclear thermal propulsion concept that uses fission fragments to directly heat the propellant up to 1000 K or higher above the material temperatures. The concept offers significant advantages over traditional solid-core nuclear rocket concepts because of higher propellant exit temperatures, while at the same time providing for more reliable operation due to lower structure temperatures and lower power densities. The reactor can be operated in either a steady-state or pulsed mode. The steady-state mode provides a high thrust and relatively high specific impulse, as compared to other nuclear thermal concepts. The pulsed mode requires an auxillary radiator for cooling, but has the possibility of achieving very high specific impulses and thrust scaleable to the radiator size. The propellant temperatures are limited only by thermal radiation and transient heat conduction back to the substrate walls.

INTRODUCTION

The basic concept of the fission-fragment reactor or foil reactor has been outlined in previous work (Wright 1990 and Suo-Anttila, et al. 1991). The concept utilizes thin coatings of UO_2, 1 to 4 μm thick, placed on a substrate with suitable neutronic and mechanical properties. The substrates are arranged into tubular modules, each with its own pressure boundary and rocket nozzle. The propellant is heated to temperatures 1000 K or higher above the substrate and fuel-coating temperature by direct energy deposition of the fission fragments which escape the fuel coating. This allows for the possibility of achieving a very high specific impulse (Isp>1000) as compared to traditional solid-core nuclear rocket concepts. It also allows for other significant advantages such as more reliable operation due to lower structural temperatures, lower power densities, loss of coolant survivability, and low fuel inventory.

There are a number of parameters which may be varied to optimize the design to a given performance level or mission requirement. These parameters include engine size, fuel type, coating thickness, substrate material, module length, and mode of operation. Two concepts, a steady-state design and a pulsed-mode design, which are presented in this paper, have been optimized neutronically to produce a small efficient system for a Mars-type mission.

GENERAL DESIGN CONCEPT AND TRADEOFFS

Figure 1 shows the calculated particle and energy escape fractions for average fission fragments (Z=46, A=118, E=81 MeV) emitted from a UO_2 coating. The curves are equally valid for PuO_2. For each fission event two fragments are emitted isotropically in opposite directions. As the coating thickness decreases to zero, both the particle and energy escape fractions approach 1/2. Hence the maximum energy that could be deposited in a gas for a very thin coating is 50% of the total energy produced in the coating. As the thickness goes to infinity, both the particle and energy escape fractions go to zero. The range of average fission fragments in UO_2 is 7.4 mg/cm² (~ 7.4 μm); for coatings thicknesses greater than the range, the fragments that are born beyond the range cannot escape into the gas. The optimum coating thickness is a tradeoff between the largest possible escape fraction (very thin coatings) and the reactor size required for criticality. For coating thicknesses of 1, 2, and 3 mg/cm^2, the energy escape fractions are 0.33, 0.24, and 0.19, respectively.

Once the fragments are in the propellant gas, there must be enough gas to allow the energy of the fragment to be deposited. The fission-fragment range is inversely proportional to the gas density. For H$_2$ gas, the range of a 81 MeV fragment is 10 cm at 101 kPa (14.7 psia) and 300 K. At 5068 kPa (735 psia) and 3000 K, the range is 2 cm.

Two modes of operation are possible for the foil reactor concept, a steady-state mode and a pulsed mode. The steady-state mode requires the substrate material to be porous tubes or plates, constructed of BeO or other ceramic with suitable neutronic and thermal properties, coated on one side

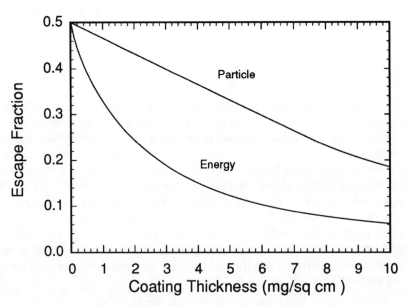

FIGURE 1.   The Particle and Energy Escape Fractions for Average Fission Fragments in a UO$_2$ Coating.

with a thin coating of UO$_2$. Other fuel coating materials such as PuO$_2$, carbides, nitrides, or metals are also possible depending on their thermal and adhesive properties. The coating would be 1 to 4 µm thick with 33% to 15% of the total energy deposited directly into the gas. As seen in Figure 2, the hydrogen propellant is pumped through the porous ceramic, absorbing the fission energy deposited in the coating, and then flows into the exhaust stream where it is heated further by the fission fragments emitted from the coating.

Shown in Figures 3, 4, and 5, the porous tubes (or plates) can be assembled into a self-contained rocket module complete with its own nozzle if desired. The rocket modules are operated at high pressures (typically 6895 kPa (1000 psia)) and include a pressure boundary (carbon-carbon or similar material). The rocket modules are clustered, as shown in Figure 6, to form a critical reactor. D$_2$ or D$_2$O is required as a moderating material (along with the BeO substrates which act both as a moderator and neutron multiplier), and a reflector of H$_2$ or H$_2$O. Cooling of the moderator and reflector can be accomplished by flowing the hydrogen propellant through the reflector prior to injecting it into the rocket modules. The constraints of criticality, heat transfer, flow through porous media and gaseous absorption of fission fragments requires that designs of this type be optimized to minimize engine size.

An alternative concept to the steady-state design is the pulsed-mode concept. Here the reactor would operate in a pulse/cooldown mode utilizing the thermal inertia of the substrates to control the maximum structural temperature. The substrate material would not be porous; instead the propellant would enter the channel from the top of the module. The substrate would be cooled between pulses by a cooling system which would utilize an external radiator. The pulsed-mode design offers the advantage of much higher propellant temperatures, since the heat capacity of the substrate is used to allow for more energy deposition into the gas.

For either design there exists a variety of design parameters that can be optimized depending on the engine and mission requirements. The fuel thickness has two tradeoffs. A thick fuel coating reduces the engine size because more fuel is present per unit volume so that criticality requirements can be more easily achieved. However, a thick coating is less energy efficient since more fission energy is absorbed in the coating rather than directly in the propellant.

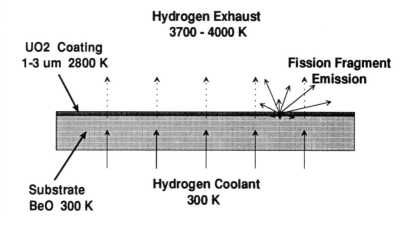

Other design parameters include the substrate spacing and the operating pressure. High pressures allow high thrust but increase the gas density which reduces the fission fragment absorption range in the propellant. Too short of a range causes excessive heat transfer back to the coating, whereas too long of a range allows the fission fragments to pass through to the opposite wall without having sufficient energy absorption by the intervening propellant gas.

FIGURE 2.   Schematic Diagram of Hydrogen Propellant Flowing through a Fissile Coated Substrate.

An attempt has been made to develop designs for both the steady-state and pulsed-mode concepts for a Mars-type mission. The first step was to design a module with a high efficiency by trading off the parameters described above. The next step was to design the reactor such that the minimum number of modules was required. A representative design for both the steady-state and pulsed-mode concept are presented below along with some of the major characterisitics.

STEADY-STATE MODE

The representative design for the rocket module consists of 6 pairs of annular rings which make up the substrate, fuel, inlet channel, and exhaust channel. Shown earlier in Figures 4 and 5, the overall module size has a diameter of 35.6 cm and length 150 cm. Between each pair of annular rings is a thin inlet channel approximately 2 mm in width. Separating the pairs of rings is an exhaust channel of approximately 2 cm in width. The inner six rings are made of a porous BeO ceramic, 3 mm in thickness and coated on one side with UO_2 to 2 μm. The outermost ring is a pressure boundary constructed of carbon-carbon or similar high strength/light weight material 3 mm thick. The hydrogen flow rate through a module is ~1 kg/s. The hydrogen propellant heats up to 2800 K when flowing through the micron-size pores in the ceramic. The fission fragments further heat the gas to 3600 K at the top end of the rocket module. Heat transfer back to the porous walls causes the gas to cool to 3410 K as it exits the module and enters the rocket nozzle. The effect of thermal radiation and H_2 dissociation was not included in these calculations. It is expected that at these temperatures, the conduction loss from the gas is the dominant loss mechanism. At higher temperatures, such as presented in the pulsed-mode design, radiation and dissociation will have a much larger effect. The rocket nozzles are constructed of porous BeO, coated with UO_2, and are transpiration cooled by additional propellant.

With this module design, parameters associated with optimizing the reactor for criticality versus weight were evaluated. These parameters include the number of modules, the module spacing, the moderator and its density, the reflector, and the fuel type (U-233, U-235, or Pu-239). The fuel coating thickness can be varied with the restriction that the module efficiency changes.

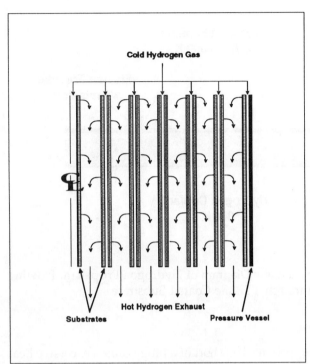

FIGURE 3. Schematic Layout of the H_2 Flow and Substrates.

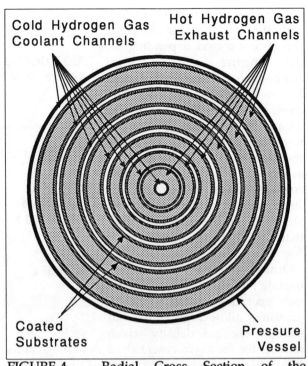

FIGURE 4. Radial Cross Section of the Substrates and Channels in Module.

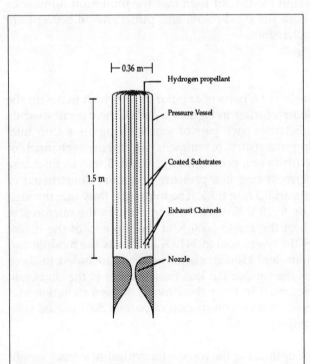

FIGURE 5. Axial Cross Section of a Rocket Module.

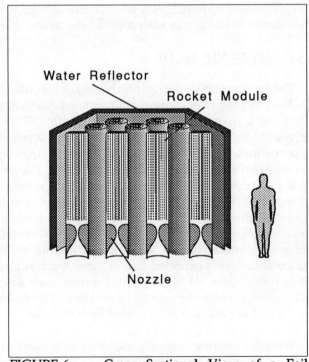

FIGURE 6. Cross Sectional View of a Foil Reactor Rocket Concept.

The neutronic calculations were performed using the Monte-Carlo code KENO and a 27-group, 300 K, cross-section set. The initial calculations were performed to determine the moderator type, density, module spacing, and reflector type, with the goal to minimize the total reactor mass. It was determined that for the given module specifications, the optimum design parameters are a D_2O moderator at nominal density, a H_2O reflector 10 cm in thickness, and a module configuration on a square pitch with the modules touching. The reflector covered all sides of the reactor except the bottom; a 15-cm BeO plate was used to mockup the rocket nozzles.

Figure 7 presents the effect on k_{eff} of varying the fuel type (oxides of fully enriched U-235, U-233, and Pu-239) and coating thickness. Shown in the figure is the number of modules required for criticality ($k_{eff}=1$) versus the coating thickness in mg/cm^2 (note that since the density for UO_2 and PuO_2 is ~11 g/cm^3, then 1 mg/cm^2 is approximately 1 µm thick). Although the parameter η (the number of neutrons emitted per absorption in the fuel) is the largest for U-233 in this configuration, the leakage is the smallest for Pu-239. The leakage is a much more dominant effect for a dilute reactor such as this and has a more significant effect on k_{eff}. The PuO_2 fuel allows the most compact system; a 1-µm coating can achieve criticality using 72 modules. For U-235 and U-233, a 1-µm coating

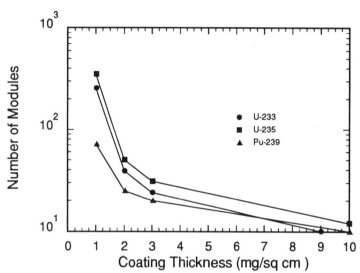

FIGURE 7. Number of Rocket Modules Required for Criticality as a Function of Coating Thickness for U-235, U-233, and Pu-239.

would require a reactor with greater than 300 modules. For a 2 or 3-µm coating, a reactor of reasonable size (20 to 50 modules) can be achieved for any of the three fuels, with the order choice being Pu-239, U-233, and U-235. Although a 10-µm coating would not be efficient for directly heating the gas, calculations were performed to determine the size effect. For a 10-µm coating, a 10 module reactor is achievable using U-233, U-235, or Pu-239.

The mass of fuel required for a given reactor can be calculated by multiplying the coating thickness in mg/cm^2 (or µm) by the number of modules and dividing by 10 to give the result in kg of oxide. Hence for a 39 module reactor and a coating thickness of 2 mg/cm^2, the total amount of fuel required for the reactor is 7.8 kg of oxide.

Figure 8 presents the reactor mass in metric tonnes as a function of the number of modules. The mass includes the substrate, fuel, moderator, and reflector. The mass does not include the propellant gas, nozzles, shielding, turbo-machinery, support structures, payload or control mechanisms. For a 39 module reactor, the reactor mass is ~7 tonnes; for a 25 module reactor, the mass is ~4.6 tonnes.

A representative steady-state engine design could have a 2-µm coating of U-233, corresponding to criticality with 39 modules. The overall size of the 39 module engine has a length of 160 cm (not including shielding or nozzles), and width of 270 cm. The thrust of such a design is 365,000 N (82,000 lbf) with an Isp of 1070 seconds using a nozzle expansion ratio of 106. The reactor power is 1800 MW. The thrust to reactor weight of such a design is 52 to 1. The total weight of the reactor portion of the engine is 7 tonnes.

For comparison, a 1-µm thick PuO_2 fueled engine would have 72 modules and a width 305 cm. The thrust is 663,000 N (149,000 lbf) with an Isp of 1150 seconds using a nozzle expansion ratio of 113. The reactor power is 3555 MW. The thrust to reactor weight of would be 53 to 1. The total weight of the reactor portion of the engine is 12.5 tonnes.

There exist some natural safety features of a fission fragment reactor design. The most important feature is the low structure temperature compared to the propellant gas temperature. Figure 9 is a plot of the predicted temperature profile in the porous substrate at representative conditions. As shown in the figure, 90% of the structure mass is at a temperature less than 2000 K with the bulk of it less than 1000 K. Another safety feature is the modularity of the system. A power to flow mismatch in one rocket module could cause some of the fuel to melt or vaporize, depending on the severity of the

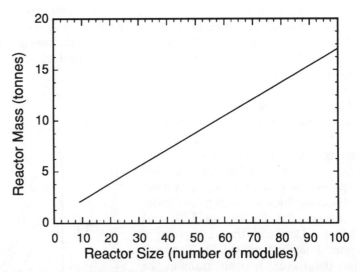

FIGURE 8.   Total Mass of the Reactor, Excluding Propellant, Nozzles, and Shielding.

mismatch. Because of the modularity the overall effect on the engine due to the loss of one module would be relatively small. A loss of propellant flow type of accident would not result in any damage to the reactor if the reactor shut down, because there is adequate heat capacity in the substrate to accommodate such excursions. Although for the representative design presented here, complete loss of propellant in the reactor has been calculated to have a positive reactivity effect on the order of $1.00, a more thorough study could result in a design with a negative worth. Other safety features are low fissile inventory (less than 8 kg of UO_2), and low power densities (~500 W/cm^3).

PULSED-MODE CONCEPT

As discussed earlier, an alternative concept for operating a fission-fragment rocket is to operate in a pulse/cooldown mode utilizing the thermal inertia of the substrate material to control the maximum structural temperature. The advantage of the pulse mode is that the gas can be heated to much higher temperatures than the steady-state concept. The disadvantage is that an external heat radiator must be used to cool the structure.

The overall design and configuration of the reactor is shown in Figure 10, and would be very similar to the steady-state design. Some design constraints can be relaxed, but other design features

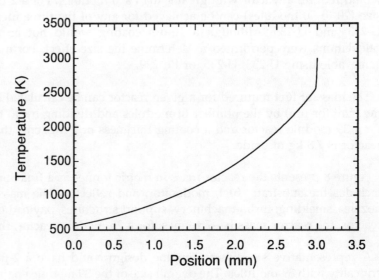

FIGURE 9.   The Temperature Distribution in a 3-mm BeO Substrate with a UO_2 Coating. The Power Generation rate is 500 W/cm^2.

must be added. The requirement for porous substrate rings can be eliminated and solid metallic or ceramic materials can be used instead. The hydrogen would now flow down from above rather than through the

1108

tube wall. The primary disadvantage of the pulsed mode is that a substrate cooling system is required with an external heat exchanger/radiator.

A representative pulse-mode reactor has the same configuration as for the representative steady-state design, 39 modules with a 2-μm coating of U-233 oxide. The substrates consist of non-porous BeO with imbedded cooling tubes through which H_2 or He coolant flows. The thickness of the substrates is approximately 1 cm in order to keep the temperature rise down to approximately 300 K per pulse. Hydrogen propellant enters the rocket modules from the top (rather than through porous walls as in the steady-state engine). Although it is not essential, improved efficiency would be obtained by installing a check valve in the throat of the rocket nozzle to restrict gas flow until heated to the maximum temperature during the reactor pulse. Surrounding the reactor would be a 6-m diameter by 3-m high heat radiator. The gas which cools the substrates would circulate through the heat radiator at a temperature near that of the substrates. The heat loss from such a radiator depends upon its temperature which in turn is ultimately determined by the choice of materials that are used in the construction of the radiator, substrates and other components of the coolant flow loop.

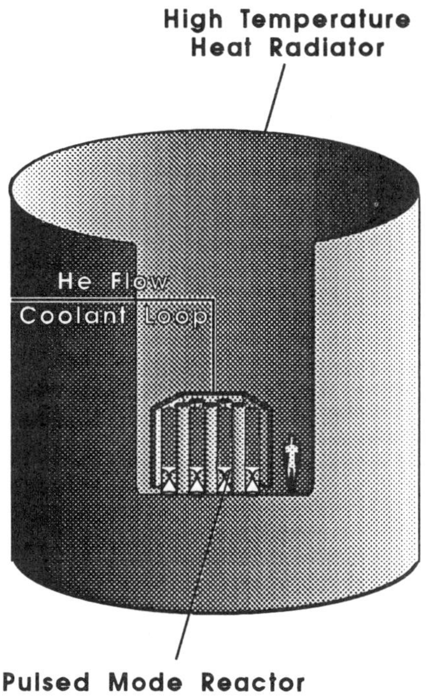

High Temperature Heat Radiator

Pulsed Mode Reactor

FIGURE 10. Foil Reactor Concept for Pulse Mode Operations Showing Relative Sizes of Components.

Figure 11 is a plot of the total heat lost from a heat radiator of the size described above. In order to maximize the average thrust of a pulse-mode engine the radiator heat loss must be as large as possible in order to cool the substrates as fast as possible so that the next reactor pulse can be fired in the minimum amount of time. Clearly tradeoffs exist in the choice of desired thrust, choice of materials, mission objectives, and cost.

Figure 12 is a plot of the predicted pressure in a pulse-mode engine as a function of time using a relief valve in the nozzle as described above. The start of the calculation corresponds to the point where the rocket module has just been filled with hydrogen propellant at 1500 K and 3448 kPa (500 psia). The gas is then heated in a 100 ms reactor pulse at a power level of 12,000 MW to a temperature of approximately 10,000 K. The nozzle valve is then opened to allow the high-pressure gas to escape over the next few hundred milliseconds. During the pulse, the average temperature rise in the substrates is 315 K.

In the calculations it has been assumed that the gas can be heated to 10,000 K without radiative losses to the surrounding media. The radiative emissivity of hydrogen propellant under the conditions of fission-fragment bombardment is not known at the present time; it is known that at very high temperatures, it does become a good radiator/absorber. These unknowns could be measured by performing reactor experiments on representative channels of UO_2 coated substrates.

100% dissociation of the H_2 gas into H atoms has been assumed to occur during the heat-up to 10,000 K. (The amount of dissociation which would occur under these conditions is not clearly known. It is expected that these measurements could also be made by performing subscale reactor experiments.) The rather large amount of energy associated with the dissociation process has been included in the calculations. Although the dissociation process does increase the specific impulse by about 40% the energy cost of doing so is quite high. It is possible that some of the energy can be recovered in the expansion nozzle by recombination. However, no credit for this phenomenon was included in the calculations. The Isp of monatomic hydrogen gas at 10,000 K is approximately 2150 seconds, assuming perfect gas properties.

Figure 13 is a plot of the time averaged thrust of a pulse-mode engine as a function of radiator temperature. As shown in the figure, the thrust of the representative engine is on the order of a hundreds of lbf. This puts it in a category of approximately two orders of magnitude less than the steady-state design but still orders of magnitude higher than conceptual nuclear electric designs (ion-accelerator engines).

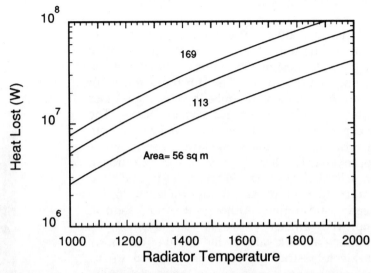

FIGURE 11.   Heat Radiated to Space for Various Size Radiators.

Figure 14 presents the period or cooldown time required after each pulse to cool the substrate down to the starting temperature, assuming a radiator with diameter 6 m and length 3 m (56 m^2). A typical pulse-mode engine may operate with a radiator temperature of 1500 K, which requires a cooldown time of 120 s. The time averaged thrust of such an engine would be 530 N (120 lbf). The run time required for a Mars type mission would be on the order of days as opposed to minutes for the steady-state design.

COMPARISON TO OTHER PROPULSION CONCEPTS

Specific impulse (Isp) as a function of thrust and rocket-engine design is depicted in Figure 15. As shown in the figure, the pulse-mode engine concept fills a gap between nuclear-electric and nuclear-thermal rocket engine designs. A specific impulse greater than 2000 s is possible at a thrust of ~900 N (200 lbf), which is more than adequate for a Mars type mission. The steady-state concepts fits within the Isp and thrust range of traditional nuclear concepts.

CONCLUSIONS

The fission-fragment reactor concept provides an alternative to traditional nuclear propulsion concepts which rely strictly on the thermal properties of materials to achieve high temperatures of the propellant. This design may also bridge the gap between nuclear electric and conventional nuclear thermal engines, allowing for the possibility of achieving a very high specific impulse without a significant loss in thrust or a large increase in weight. Although there remains a number of unknown phenomenological questions associated with heating a propellant to such very high temperatures, a significant amount of answers could be derived by performing subscale tests in research reactors.

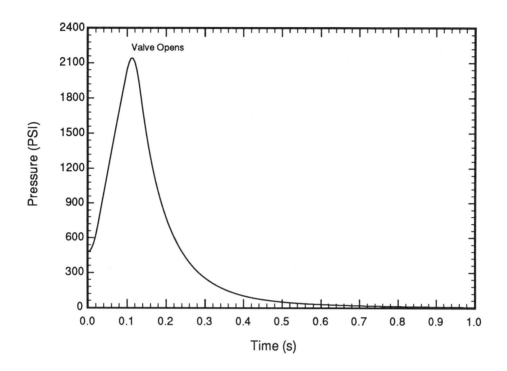

FIGURE 12. Typical Rocket Module Gas Pressure when Operating in a Pulse Mode.

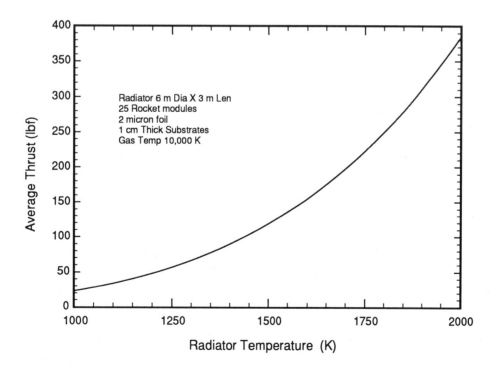

FIGURE 13. The Time Average Thrust of a Pulse Mode Engine.

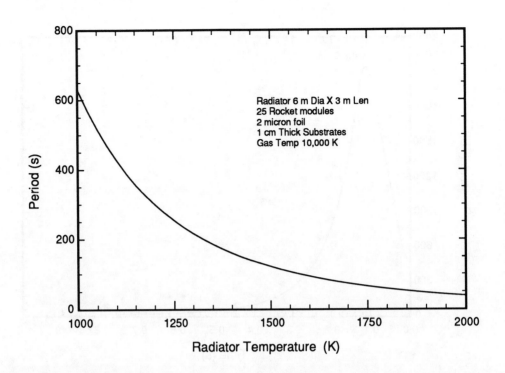

FIGURE 14.   The Pulse Period for Pulsed Mode Operations as a function of Radiator Temperature.

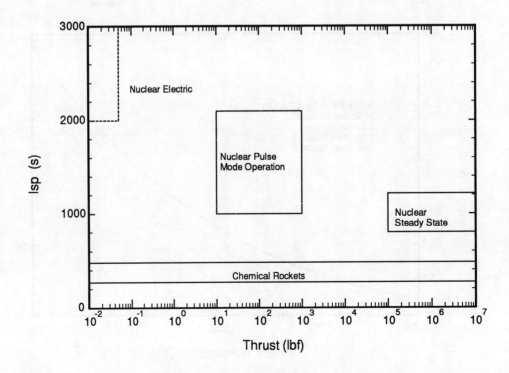

FIGURE 15.   Isp as a function of Thrust for Various Engine Types.

1112

Acknowledgments

This work was performed at Sandia National Laboratories, which is operated for the U.S. Department of Energy under Contract DE-AC04-76DP000789.

References

Siegel, R. and J. R. Howell (1981) Thermal Radiation Heat Transfer, Second Edition, McGraw Hill Book Co., New York, p. 763.

Suo-Anttila, A. J., E. J. Parma, S. A. Wright, M. E. Vernon, and P. S. Pickard (1991) "A Fission Fragment Reactor Concept For Nuclear Space Propulsion," to be published in Fusion Technology.

Wright, S. A. (1990) viewgraph presentation at the "NASA Conference on Advanced Nuclear Propulsion Concepts", Cleveland, OH, July 1990.

THERMIONIC DIODE SUBSYSTEM MODEL

Ralph R. Peters
Division 6465
Sandia National Laboratories
P.O. Box 5800
Albuquerque, NM 87185
(505) 845-8712 FTS 845-8712

Todd B. Jekel
Mechanical Engineering Department
University of Wisconsin
Madison, WI 53705
(608) 263-3346

Abstract

A model has been developed that predicts the electrical currents in thermionic diodes and heat transfer rates across the thermionic diode gaps for large networks of thermionic diodes. A computer code that implements this model has been written; eventually, this code will be incorporated into overall thermionic reactor system codes. The thermionic diode subsystem (TDS) model accounts for spatially-dependent radial and axial temperatures in the thermionic diodes. Standard techniques for iteratively solving this system of equations have not been successful so a procedure based on optimization techniques has been developed to solve this problem. Results of TDS calculations for TOPAZ II compare favorably with the available data. The TDS model will be validated against future TOPAZ II test results.

INTRODUCTION

A variety of reactor power systems have been proposed to provide electrical power for civilian and military space-based missions in the tens to hundreds of kWe range. One system from this group, which is currently receiving much attention, uses the thermionic emission of electrons to directly convert reactor heat into electricity. This type of system has a number of advantages including relatively high conversion efficiencies (5 - 15% depending on the design), relative simplicity of design, small radiator area because the system operates with a high radiator temperature, and the direct conversion of heat into electricity. The disadvantages include high temperatures for a few system components that require the use of expensive materials, and the relative sensitivity of system operation and efficiency to changes in design and operating conditions. For example, a drop of 80 K in the temperature of the thermionic diode subsystem component that emits electrons may decrease the thermal-to-electric conversion efficiency by one-half. Thus, accurate modeling of the thermionic diode subsystem is an important part of designing a thermionic reactor system.

Modeling the thermionic diode subsystem is difficult. For this paper, the thermionic diode subsystem (TDS) is defined to consist of the emitter, the gap between the emitter and the collector, the collector, the electrical load, and the wiring that connects these components together. (The Physical Model section will discuss the TDS in greater detail.) Models of thermionic emission processes (for example, TECMDL by Rasor Associates, Inc.) are available that can be used to predict the performance of a thermionic diode at a point on its surface. For example, the TECMDL code will calculate the current density if the component temperatures and gap voltage are known. However, the component temperatures vary spatially and the gap voltage varies along the thermionic diode because of resistance losses. Thus, it is difficult to obtain an accurate estimate of thermionic diode subsystem performance using a "gap" model like TECMDL even if the component temperatures are known. Modeling the TDS requires an electrical circuit model to calculate the gap voltages (and voltages throughout the TDS) that are used by the gap model to calculate the gap current. The gap model and the electrical circuit model must interact (for example, via iteration) because TDS voltages vary with the distribution of current flowing across the gap.

This paper will briefly discuss thermionic physical processes. It will then discuss the model that predicts the output of the subsystem, given the system design and operating conditions (usually defined by temperatures). Finally, the results of calculations done for a subsystem having characteristics similar to the TOPAZ II are presented.

Physical Model

Figure 1 schematically shows a single thermionic diode converter and its connection to a load (modeled as a resistor). The emitter is heated to a very high temperature by fission energy (for example, 1900 K) causing electrons to boil off the emitter surface. These electrons flow across the gap to the collector and return to the emitter via an external circuit providing electrical energy to the load. Many types of thermionic converters fill the gap with low-pressure cesium vapor to increase the performance of the converter by reducing the work function of the collector and neutralizing the space-charge effect of the electrons flowing across the gap. Heat is transferred to the collector by the flow of electrons, by thermal radiation and conduction through the cesium vapor in the gap; this heat may be removed from the collectors by circulating liquid metal coolant or by heat pipes.

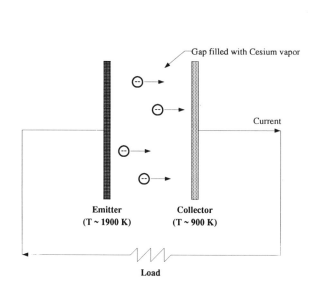

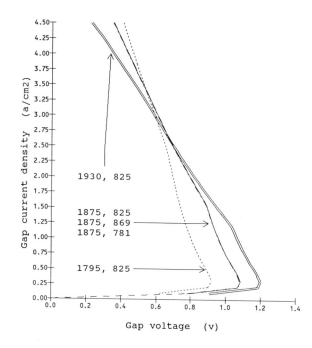

FIGURE 1. Schematic of a Thermionic
Diode Subsystem.

FIGURE 2. Current Density Versus Gap
Voltage for Different Emitter and
Collector Temperatures Based on TECMDL
Calculations. Curve Labels Are Emitter
and Collector Temperatures.

Performance of the thermionic diode is highly variable. To illustrate this point, calculations predicting the gap performance of TOPAZ II were made with TECMDL using properties that are thought to be representative of TOPAZ II and conditions that may be seen during normal operation. TOPAZ II provides power to the payload using 34 thermionic fuel elements (TFEs) in series. A TFE is a cylindrical unit with nuclear fuel in the center surrounded by the emitter, then the gap, and finally the collector. Each TFE extends the full length of the core and so substantial variations in component temperatures occur along the TFE. (More information about TOPAZ II may be obtained from a variety of sources including Nickitin et al. (1991), Glushkov et al. (1991), and Ya et al. (1991). Figure 2 shows the current density versus gap voltage for a number of emitter temperatures and collector temperatures; the individual values chosen for the emitter and collector temperatures are their peak value, average value and low value. Figure 2 shows that the current is a double valued function with the upper branch representing the "ignited" mode and the lower branch representing the "unignited" mode of operation. The transition occurs in the range of 0.8 to 1.0 volts with the electric current changing rapidly in that range. Figure 2 shows the current density, in the ignited mode, is strongly dependent on the emitter temperature, but is unaffected by the collector temperature for the values plotted. Figure 2 indicates the electric current flowing through the external circuit decreases as the gap voltage is increased if the TFE is operating in the ignited mode. (Gap voltage is defined as the emitter potential with respect to the collector. Therefore, as the gap voltage increases, the voltage of the collector is being biased negative with respect to emitter, and the flow of electrons from the emitter to the collector is being retarded.) Figure 3 shows the electric power transferred across the gap versus gap voltage. Figure 3 indicates the power delivered to load will first increase and then decrease to zero as the the gap voltage is increased. The important thing to note from these figures is that the current density and electric power produced along the TFE may be highly variable because the voltage and temperature vary considerably along the TFE. Additional detailed information concerning thermionic systems may be found in many reports and texts including Angelo and Buden (1985) and Hatsopoulos and Gyftopoulos (1973).

Calculational Model

The TDS may be represented as a simple electrical ladder circuit with resistances and variable current (or voltage) sources. The electrical circuit model for an example TFE system with a single cell" electrical connection is shown in Figure 4. The TFEs in Figure 4 have been divided into 3 equal regions for the purposes of calculations. (In a later section, we will show that results may vary as the number of TFE calculational regions increase.) Both the emitter and the collector are modeled as a number of resistors in series (4 resistors for each TFE emitter and collector in Figure 4). The resistance of each resistor is a function of the material

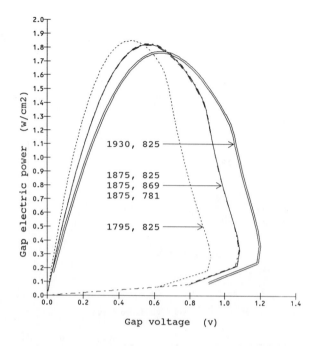

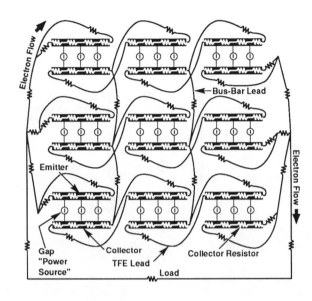

FIGURE 3. Electric Power Density Versus Gap Voltage for Different Emitter and Collector Temperatures Based on TECMDL Calculations. Curve Labels Are Emitter and Collector Temperatures.

FIGURE 4. Sample TDS Circuit Drawing.

properties, temperatures, and physical dimensions. The thermionic processes that occur across the gap are modeled as variable electrical power sources (shown as current sources in this figure) whose values are determined by TECMDL. Two pairs of leads (called "Single cell" leads) connect the TFEs (3 TFEs per string in this figure) together in series to provide higher voltage at the load. Bus-bar leads interconnect the series-strings to increase system reliability by allowing current to bypass a TFE that is not passing current (or "open"). Leads at both ends of the TFE strings connect them together to supply power to the leads that are attached to the "load" resistance.

The TDS calculational model has two distinct parts. The first part models the physical processes occurring in the gap; the code used for this modeling is TECMDL. The second part models the movement of electricity throughout the rest of the TDS; this code will be referred to as the "Circuit Solver" or CS.

These two parts may be linked together in a variety of ways to model the entire system. Typically, the linkage between these two would be through iteration. For example, an initial estimate for each TFE in the system of one gap voltage and gap currents for the remaining variable power sources is input to CS to calculate the voltages and currents throughout the remainder of the system. (One gap voltage per TFE must be input to CS to obtain unique voltages in the solution.) For the case shown in Figure 4, the input is 9 sets (1 set per TFE) of initial estimates of one gap voltage and two gap currents. The CS solution includes one gap current and gap voltages for the remaining variable power sources for each TFE in the system. TECMDL uses the CS information to calculate one gap voltage and gap currents for the remaining variable power sources for each TFE. The TECMDL solution is input to CS to begin another iteration cycle. A solution for the entire TDS is obtained when the parameter values (for example, one gap voltage and the remaining gap currents for each TFE) are nearly identical from one iteration cycle to the next. However, this type of iteration scheme does not work for cases that model TOPAZ II. An iteration scheme that has tight controls on the manner and rate that variables (for example, gap voltages) are allowed to change from one iteration cycle to the next and an initial estimate that is very close to the final solution will diverge rapidly in a few cycles and "blow-up."

The second way that was tried to obtain a solution uses an optimization approach. An initial estimate for each TFE in the system of one gap voltage and gap currents for the remaining variable power sources is input to both CS and TECMDL to calculate one gap voltage and gap currents for the remaining variable power sources for each TFE. The normalized difference between the CS and TECMDL output (the "error") is calculated. An optimization routine attempts to reduce the error using some algorithm. MINA, the routine used in our calculations, is a "stepping" routine. First, MINA sequentially tries increasing and decreasing the input values to find which direction causes the error to decrease. It then takes steps in that direction as long as the error continues to decrease. If the error starts to increase after a few steps, it goes back to the location of the minimum error and again tries to find the direction that will minimize the error and takes steps in that direction. Step size may have to be decreased at some point so that MINA is not taking such a large step that it can't find any direction that causes the error to decrease. Eventually, MINA will stop because it has either found an error less than the input tolerance, or it will reach the limit on the number of times it can decrease the step size.

Optimization techniques, like iterative techniques, require reasonably good starting points to begin their calculations. One method of obtaining a good starting point uses a bootstrap method. First, one uses the TDS code to obtain a solution for a system that has only one node per TFE. In general, MINA can find a solution for this case because there are only a few variables that strongly affect the error. Next, this solution is used as a starting point for calculations for the same system where there are now two nodes per TFE. The solution for this two-node-per-TFE system is used as a starting point for a system with four nodes per TFE. The bootstrap sequence continues increasing the number of nodes per TFE until it is obvious that the solution is not changing as number of nodes per TFE is increased.

Sample Problem

The TDS code was tested using a using a thermionic diode subsystem that is based on the portion of TOPAZ II that provides power to the payload. The general characteristics of the system, as we have modeled it, are listed in Table 1. The values in this table are based the draft report by Gunther (1990) that summarizes the information available from a dozen Soviet papers.

TABLE 1. TOPAZ II Payload Power TDE Description.

---

34 Thermionic Fuel Elements (TFE) are connected in series in a single string using "single-cell" connections

---

	TFE properties

Active length 35 cm

Emitter	
outside diameter	1.96 cm
thickness	0.115 cm
material	Molybdenum with Tungsten overcoat
work function	4.95 eV
resistivity function	-22.7 + 0.043*Tem (micro-ohm-cm)
	(54.7 micro-ohm-cm @ 1800 K)

Gap	
size	0.05 cm
Cesium reservoir temperature	500 K

Collector	
inside diameter	2.06 cm
thickness	0.140 cm
material	Molybdenum
work function	1.7 eV
resistivity function	-1.83 + 0.025*Tcol (micro-ohm-cm)
	(20.7 micro-ohm-cm @ 900 K)

Individual lead resistance	500 micro-ohm
Load lead resistance	100 micro-ohm
Load resistance	.126 ohm

The TFE temperature profile used in these calculations is plotted in Figure 5. It is based on information from Gunther (1990).

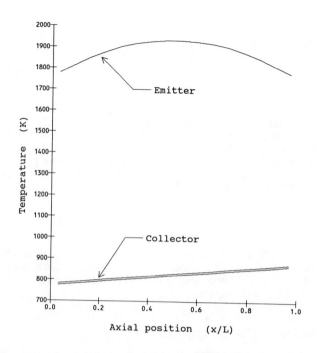

FIGURE 5. Emitter and Collector Temperature Along the TFE. The Axial Position of 0 is at the Core Inlet.

The TDS code was used to calculate voltages and currents throughout the TDS including those at the load. These results are shown in Table 2. These values lie within the ranges listed by Gunther (1990) for the TOPAZ II system indicating the model is valid. The error (defined as the average of the normalized differences between the TECMDL and CS variable power supply values) was slightly less than 0.001.

TABLE 2. TDS Model Solution for TOPAZ II.

Voltage across the Load	27.8 V
Current passing the Load	220 A
Electric power dissipated in the Load	6.13 kW
Total thermal power crossing the Gap	130 kW

TDS code solutions for TOPAZ II (and probably many other systems) are a function of discretization of the TFE. The solution shown in Table 2 was for a calculation that divided each TFE into 16 pieces. The effect of discretization on the load voltage, load electric power, and gap total thermal power is shown in Figures 6-8. In these figures, the term "Number of Nodes" refers to the number of discrete pieces that make up a single TFE. The "Number of Nodes" is also equal to the number of variable power sources that are used to model each TFE (see Figure 4). Figures 6-8 indicate that each TFE must have at least 4 nodes (or must be divided into 4 pieces) in order to accurately model TOPAZ II performance.

Calculations for thermionic diode subsystems occasionally use an average property and average temperature approach. This is exactly the same as a 1 node solution with the TDS code. Examination of Figure 7 indicates that 1 node solutions predict electric power outputs that are 20% less than that predicted using 16 nodes. For this case, it appears that using average property and average temperature approach is very inaccurate.

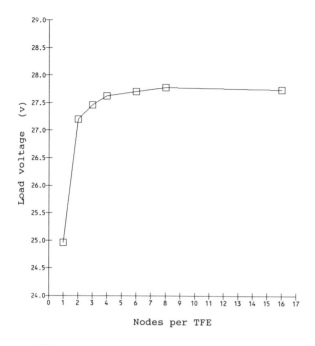

FIGURE 6. Load Voltage Versus Number of Nodes.

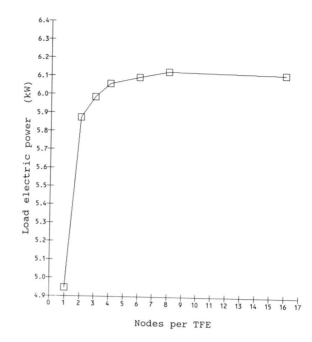

FIGURE 7. Load Electric Power Versus Number of Nodes.

The reason that 4 nodes are required to get an accurate solution while a 1 node solution will not get an accurate solution can be investigated by examining axial profiles of TFE internal parameters. Figures 9 and 10 show axial profiles of the gap voltage, electric current density, electric power density, and thermal power density (divided by 20) for the 16 node and 4 node solutions for a TFE in the middle of the string of 34 TFES; TFEs near either end of the string may have slightly different profiles because the resistance in the external circuit is usually different from that in the internal wiring. The first thing to notice is that the gap voltage is fairly constant with a peak value of about 1 V. The electric current density and electric power profiles are strongly peaked because the emitter temperature at the ends are low (see Figure 5), and so produce little electric power at gap voltages of about 0.95 V (see Figures 2 and 3). The thermal power is also peaked at the center of the TFE, but because it is the sum of electron cooling (proportional to the electric current density), radiation, and conduction it is not as peaked as the current or electric power density. Figures 9 and 10 indicate that as the number of nodes is decreased the TDS code models the conditions inside TFE less accurately. If only 1 or 2 nodes are used, the parabolic shape of the TFE internal parameters is completely lost and the solution is much less accurate than if 4 or more nodes are used.

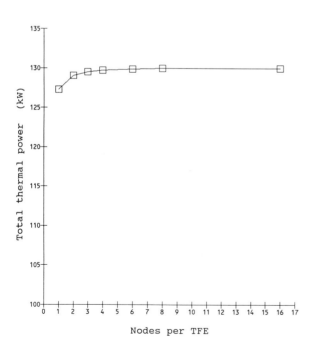

FIGURE 8. Total Thermal Power Versus Number of Nodes.

The gain in solution accuracy is obtained at a cost. The CPU time required to obtain a solution on a SPARC 1+ workstation versus the number of nodes per TFE is plotted in Figure 11. All the calculations were started with an initial estimate of a constant voltage and current density across the gap that was determined by the 1 node solution. This figure indicates that

1119

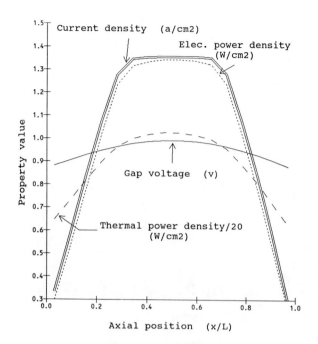

FIGURE 9. Parameter Profiles for a 16 Node Solution.

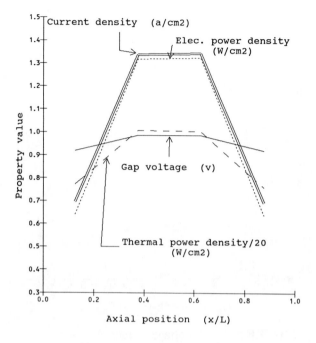

FIGURE 10. Parameter Profiles for a 4 Node Solution.

increasing the number of nodes from 4 to 8 per TFE for each of the 34 TFEs in series increases the CPU time by a factor of 50. If only slight changes are made in TDS conditions (for example, as a result of a time step in a transient code) the solution time is much shorter. For example, if the emitter temperature is uniformly increased by 0.2 K, then the time to reach a solution for a TOPAZ II TDS model with 4 nodes is about 10 seconds.

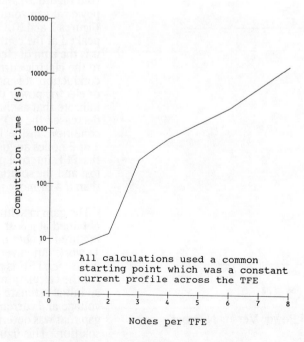

FIGURE 11. Sparc 1+ Computer Time

These calculations indicate that about 40% of the TFE is delivering very little power. According to Figure 3, reducing the gap voltage will allow these areas, which have a relatively cool emitter surface, to provide power. The results of 8 node calculations that increased the load conductance are shown in Figures 12 and 13. Figure 12 indicates that increasing the load conductance by a factor of 3 (decreasing the resistance listed in Table 1 by a factor of 3) will increase the electric power delivered to the load by about 30% while increasing the thermal power requirement by about 12%. According to Figure 13, the load voltage would drop from 28 V to 19 V. (Operation of TOPAZ II under this condition might require the addition of a power conditioner to step up the voltage to a satellite payload.) Figure 14 shows that the gap voltage has decreased from about 0.95 volts to about 0.7 resulting in a considerable flattening of the axial profiles.

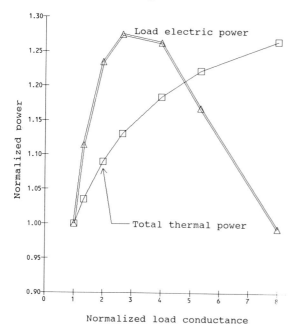

FIGURE 12. Load Electric Power and Total Thermal Power Versus Load Conductance.

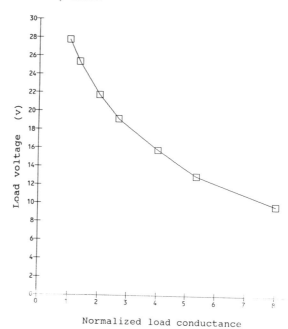

FIGURE 13. Load Voltage Versus Load Conductance.

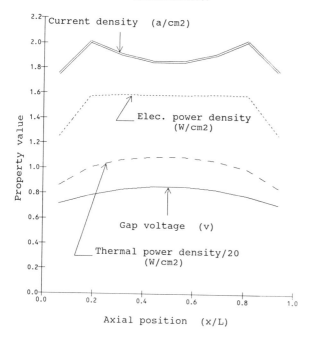

FIGURE 14. Parameter Profiles for a Normalized Load Conductance of 2.66.

The calculations, with variable load conductance, are somewhat inaccurate because they assume that the emitter and collector temperature profiles remain the same; in reality, there may be considerable change in temperatures because heat transfer rates have changed and there may be reactor power changes because of temperature-induced reactivity changes. However, the trend of increasing power as the load voltage drops should occur. A system model of TOPAZ II that contains the TDS code will be required to investigate these and other interesting questions concerning this system.

CONCLUSIONS

A computer code based on fundamental physical principles has been developed that can accurately model the TFEs of the TOPAZ II system and other types of thermionic systems if the system design, material properties, and temperatures are known. If a TDS contains thermionic diodes (for example, TFEs) that have substantial variations in temperature or gap voltage (because of resistance losses) in the element, then modeling this element using average properties (exactly the same as making a "1 node" calculation) may be inaccurate. A system model that includes the TDS code will be required to model situations where interactions between components occur that may cause temperature changes.

Acknowledgments

This work was performed at Sandia National Laboratories and was funded by Phillips Laboratory, Kirtland AFB, New Mexico. Sandia National Laboratories is operated for the U.S. Department of Energy under contract number DE-AC04-76DP00789.

References

Angelo, J. A. and D. Buden (1985) Space Nuclear Power, Orbit Book Company, Malabar, FL

Glushkov, E. S. et al. (1991) "The Neutron-Physical Aspects of the Converter Reactor with the Single-Element TFE," in Proc. Eighth Symposium on Space Nuclear Power Systems, CONF-910116, M. S. El-Genk and M. D. Hoover, eds., American Institute of Physics, New York.

Gunther, N. G. (1990) Characteristics of the Soviet TOPAZ II Space Power System -- DRAFT, SPI-52-1, Space Power, Inc., San Jose, CA.

Hatsopoulos, G. N. and E. P. Gyftopoulos (1973) Thermionic Energy Conversion, The MIT Press, Cambridge, MA.

Nickitin, V. P. et al. (1991) "TOPAZ-2, Thermionic Space Nuclear Power System and Perspectives of Its Development," in Proc. Eighth Symposium on Space Nuclear Power Systems, CONF-910116, M. S. El-Genk and M. D. Hoover, eds., American Institute of Physics, New York.

Ya, A. G. et al. (1991) "The Temperature Effect and Safety of the Topaz Reactor," in Proc. Eighth Symposium on Space Nuclear Power Systems, CONF-910116, M. S. El-Genk and M. D. Hoover, eds., American Institute of Physics, New York.

MODELING THE ENERGY TRANSPORT THROUGH A THERMIONIC FUEL ELEMENT

Ronald A. Pawlowski[†] and Andrew C. Klein
Department of Nuclear Engineering
Oregon State University
Corvallis, OR 97331-5902
(503) 737-2341

[†]Currently with Battelle, Pacific Northwest Laboratories, Richland, WA.

Abstract

A thermionic fuel element is complex both in terms of its construction and in its operating principles. An energy transport analysis of such a device must consider the heat transfer properties of the different materials as well as the modes of heat transfer from the various interfaces. A computer code (TFEHX) to model the thermal and electrical performance of a TFE has been developed. This code uses finite element methods to compute the temperature profile within the TFE components, as well as the electrical output from the TFE. The thermionic emission properties are computed using a phenomenological model for the cesium diode converter. The code allows the specification of various TFE design parameters, including the spatial dependence of the thermal power and the TFE geometric and material properties.

INTRODUCTION

A Thermionic Fuel Element (TFE) is a nuclear fuel pin with built-in provisions for the conversion of heat into electrical energy. A cross sectional view of a TFE is shown in Figure 1. The fueled region contains uranium dioxide (UO_2) and is covered by a refractory metal sheath (the emitter). Fission energy generated within the fuel is conducted outward and heats the surface of the emitter; the resulting high temperatures allow electrons to escape from the emitter surface via the thermionic emission process. These electrons traverse a narrow gap to another refractory metal cylinder (the collector). Cesium vapor within the gap assists in the emission of the electrons and in their transport across the gap. The collector is held at a lower temperature to allow the emitted electrons to condense on the collector surface, resulting in a net electrical current J flowing from the emitter to the collector. By holding the collector at a voltage V relative to the emitter, a net electrical power output

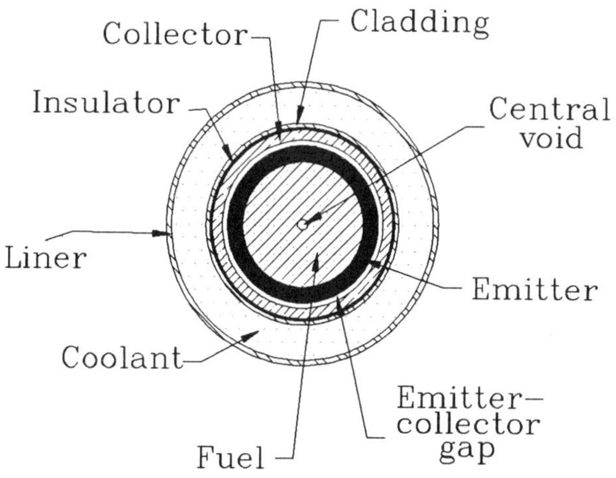

FIGURE 1.   Cross-sectional View of a Thermionic Fuel Element.

JV is produced. The outside surface of the collector is coated with an electrical insulator, which is in turn covered by a cladding of refractory metal. Liquid metal coolant flows along the outside of the cladding. If the TFE is placed within a block of neutron moderating material, then a liner of refractory metal protects the moderator block from the liquid metal coolant. The central void within the fuel prevents swelling of the UO_2 by allow gaseous fission products to be removed from the fuel to an external fission gas trap.

An analysis of the thermal and electrical behavior of a TFE must consider several factors. The generation of heat within the fuel is not uniform along the length of the TFE due to the axial variation of the neutron flux within the reactor core. The conduction of heat from the fuel to the emitter and from the collector to the coolant occurs through a variety of materials which may have vastly different thermal conductivities. Heat is transferred across the emitter-collector gap via three processes: (1) emitter cooling and collector heating due to the transport of the emission electrons, (2) radiative heat transfer between the emitter and collector surfaces, and (3) conduction through the cesium vapor in the gap. Convection of heat from the TFE cladding to the coolant must also be modeled; the convective heat transfer coefficient at the cladding-coolant interface varies with temperature and with the composition of the coolant. Furthermore, the electrical resistance of the electrodes will lead to voltage drops and ohmic heat generation within the emitter and collector. All of these processes are closely coupled, in that they both affect and are affected by the temperature distribution within the TFE. Therefore, a heat transfer analysis of a TFE must necessarily employ an iterative procedure, in which the temperature distribution is used to recompute physical properties (such as thermal conductivities and heat transfer coefficients) which are in turn used to update the temperature distribution.

This paper discusses a heat transfer analysis of a single-celled TFE; that is, a TFE that is composed of a single emitter and collector which cover the entire length of the UO_2 fuel (approximately 25 cm). The electrical conversion performance of such a TFE is investigated for a variety of design and operating conditions. The dependence of maximum fuel temperature on the current output and coolant inlet temperature is also discussed.

METHOD OF ANALYSIS

In this analysis, the geometry of the TFE is assumed to be symmetric in the θ- (azimuthal) direction. The solid regions of the pin are divided into a number of finite volumes, and an energy balance is performed on each volume. Depending on its location within the TFE, the boundaries of a particular finite volume may be adiabatic, or heat may be transferred across them due to conduction, convection, radiation or thermionic electron transport.

TABLE 1. Generic Design Parameters for a Single-Celled TFE.

Fuel	Uranium dioxide (UO_2). ID: 0.05 cm OD: 0.55 cm
Emitter	Tungsten. ID: 0.55 cm OD: 0.75 cm
Interelectrode gap	ID: 0.75 cm OD: 0.80 cm
Collector	Niobium. ID: 0.80 cm OD: 1.00 cm
Insulator	Alumina (Al_2O_3). ID: 1.00 cm OD: 1.005 cm
Cladding	Niobium. ID: 1.005 cm OD: 1.05 cm
Coolant	Eutectic NaK (78% K by weight) Flow channel: ID: 1.005 cm OD: 1.05 cm Coolant flow rate: 0.12 kg/s

Heat Transfer Models

The conduction of heat through the solid regions of the TFE is modeled using the conductive heat transfer equation:

$$\nabla \cdot [k(r) \nabla T(r)] + q(r) = 0 \tag{1}$$

where q is the volumetric heat generation rate within the finite volume. This equation accounts for the spatial dependencies of both q and the thermal conductivity k. An empirical relation for the amount of heat conducted through the cesium vapor in the interelectrode gap was taken from Kitrilakis and Meeker (1963).

The electrical output from the thermionic converter is computed using the phenomenological model of a cesium vapor diode as described by Rasor (1982) with improved treatments of both the ignited and unignited plasma modes of operation. This model assumes that the electron temperature and density are invariant across the emitter-collector gap. The formulation of the model is based on a planar converter geometry; however, it is assumed to be valid for the cylindrical converter, since the thickness of the emitter-collector gap is typically so much smaller than the radius of the emitter surface.

The axial temperature variations along the converter electrodes are considerable. This results in significant variations in the thermal and electrical conductivities of the electrodes along the length of the TFE. To account for these effects, the converter is divided lengthwise into smaller converter segments consisting of finite elements of the emitter and collector surfaces. Each converter segment is electrically connected to any adjacent segments; the resistances of these connections are temperature dependent and are determined through empirical correlations. Thermionic electron transport is assumed to occur only between the finite surface elements in a given segment; therefore only radial transport of the thermionic electrons is modeled.

The energy transfer across the interelectrode gap due to thermionic emission is computed using an iterative procedure.

1124

The temperatures of the emitter and collector surfaces are provided as input to the thermionic module. An initial axial voltage profile is assumed, and the interelectrode current and thermionic heat transfer across each segment are then evaluated for both the ignited and unignited modes. The appropriate mode for a segment is determined as the one with the greatest interelectrode current at the given temperatures and voltage. The voltage drops along the electrodes are determined so that they are consistent with the currents across the converter segments, and with the resistances between the segments. These results are used to repeat the procedure until the voltage profile converges, after which the net output currents and voltages are determined.

The transfer of radiant heat is modeled between finite elements on the emitter and collector surfaces. The amount of energy transferred is computed using the Stefan-Boltzman law. It is assumed that an element on the emitter surface exchanges radiant heat only with the closest element on the collector surface. This assumption is valid due to the great length of these surfaces compared to the spacing between them. Therefore only radial transport of radiant heat across the emitter and collector surfaces is modeled. The effective emissivity ϵ between these two surfaces is required as input to the code. For this study, a typical value of $\epsilon = 0.2$ between the refractory metal electrodes of a cesium vapor diode was used.

Correlations for liquid metal flow in concentric annuli are used to compute the convective heat transfer coefficient from the cladding surface to the coolant. Eutectic sodium-potassium alloy (NaK with 78 weight percent potassium) is used as the coolant. Empirical correlations are used to obtain the density and thermal conductivity of the NaK coolant.

Finite Element Model of a Thermionic Fuel Element

The spatial temperature distribution within the TFE is computed through finite element analysis. The materials within the TFE are subdivided into a series of small control volumes upon which energy balances are performed. This results in a set of first order differential equations which have variable coefficients, and which may be inhomogeneous and nonlinear, depending on the location of the finite element within the TFE. Variable coefficients occur due to the temperature dependence of the thermal conductivities of the TFE materials. Inhomogeneous terms are associated with heat generation within the finite element. Nonlinear terms arise due to the heat transfer processes across the interelectrode gap and due to convection to the coolant. The nonlinear and inhomogeneous terms are lumped together into a source term, and difference approximations are used to cast the set of differential equations into matrix form. The matrix is used to

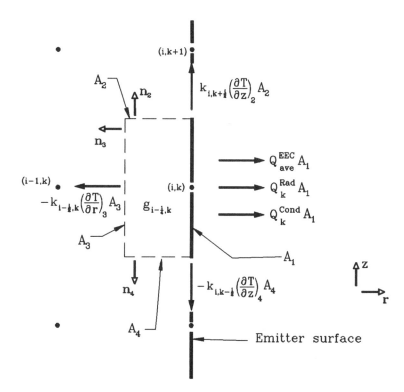

FIGURE 2.   Energy Balance for the Mesh Points (i,k) located on the Surface of the Emitter.

compute the temperature profile, from which the variable coefficients and source terms are updated. This procedure is repeated until the temperature profile converges.

The thermionic fuel element is modeled at several discrete mesh points in the r- and z- (radial and axial) directions. The control volume about a mesh point located on the emitter surface is shown in Figure 2. Heat is transferred across surfaces A_2, A_3, and A_4 via conduction. Heat is transferred from surface A_1 (which corresponds to the surface of the emitter) through conduction into the cesium vapor (Q^{Cond}), by thermal radiation (Q^{rad}), and by the electrons emitted from the emitter surface (Q^{EEC}). Energy is also produced within the control volume at a rate of $g_{i-\frac{1}{2},k}$ watts per cm^3, due to ohmic heating and due to neutron and gamma ray collisions in the control volume.

RESULTS

The TFEHX code was used to perform parametric studies on a long, single-celled thermionic fuel element which is typical of the Advanced Thermionic Initiative (ATI) concept. The design values for the TFE are given in Table 1.

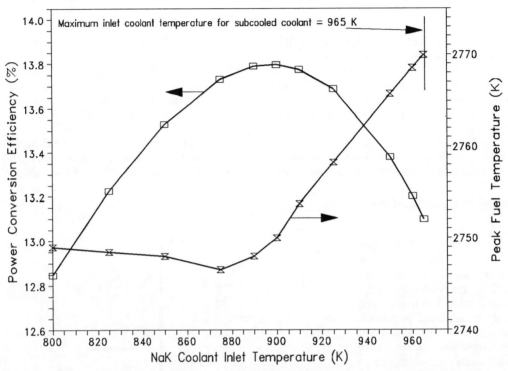

FIGURE 3.   Power Conversion Efficiency and Peak Fuel Temperature as a Function of Inlet Coolant Temperature.

The temperature of the coolant at the flow inlet has a significant impact on the efficiency of the TFE. Figure 3 shows the results of calculations which were performed for a TFE with a total heat generation rate of 3 kWt and an output current of 490 A. There is a peak in the conversion efficiency at an inlet coolant temperature of 895 K. Note also that the peak fuel temperatures are relatively unaffected by the inlet coolant temperature until the latter increases beyond 895 K. Presumably the increase in conversion efficiency is accompanied by an increase in the electron cooling of the emitter surface. The resulting increase in heat transfer from the fuel offsets any change in the peak fuel temperature due to the rising coolant temperature. However, the intensity of the electron cooling begins to decrease for coolant inlet temperatures above 895 K, and in this operating region the fuel temperature increases with the coolant temperature.

Figure 4 shows how variations in the output current can affect the TFE conversion efficiency and the peak fuel temperature. The figure indicates that the optimum efficiency for this particular TFE is achieved at an output current of 240 A. Note also that as the output current increases, the peak fuel temperature is greatly decreased. This is due to the increased electron cooling of the emitter surface at the higher emission currents.

The thickness of the electrode materials is also important to the efficiency of the TFE. Figure 5 shows that a thicker emitter results in a higher electrical power output for a given thermal power input. The reason for this effect is that a thicker conductor has a lower electrical resistance than one of smaller cross section; therefore there is less ohmic loss in the thicker electrode and more of the electrical power can be delivered to the load.

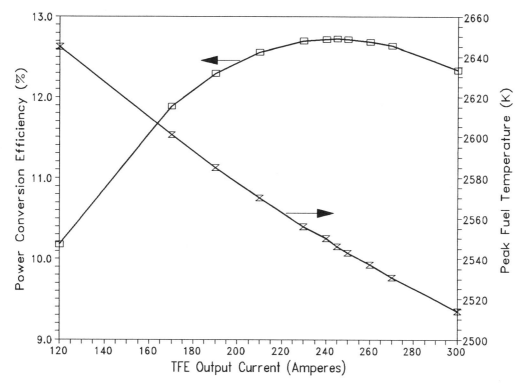

FIGURE 4.   Effect of the Output Current on the TFE Conversion Efficiency and Peak Fuel
Temperature.

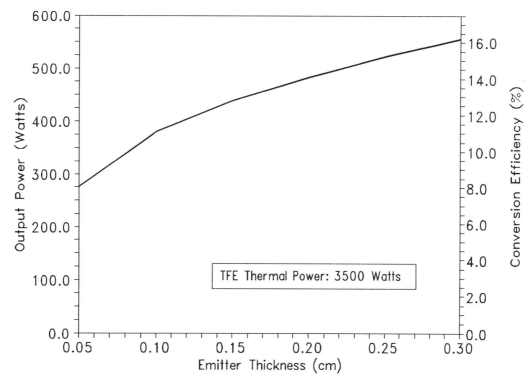

FIGURE 5.   The Efficiency of a Thermionic Fuel Element increases as the Thickness of its
Emitter Material is increased.

CONCLUSIONS

The energy transfer through a thermionic fuel element has been successfully modeled using a finite element approach with detailed treatments of the various heat transfer modes and materials properties. A computer code (TFEHX) which employs this model is capable of analyzing a wide variety of TFE materials and operating parameters. Results from this code include electrical performance data for the TFE as well as the temperature distribution within the device.

An optimum value for the coolant inlet temperature exists at which the efficiency of the TFE is maximized while the peak fuel temperature is at a minimum. The efficiency of a TFE can also be optimized through a judicious choice of the output current; however, the increase in heat transfer from the fuel as the output current is increased may make it beneficial to operate the TFE at higher than optimum output currents. The efficiency of a TFE also increases as the thickness of its emitter is increased, since the ohmic losses are less for current flowing through a thicker emitter. (However, if natural tungsten is used as the emitter material, a thicker emitter will increase the amount of neutron poison in the reactor to the point where it may become difficult to achieve criticality. This effect must be weighed against any possible gain in efficiency by using an emitter of larger cross section.)

Acknowledgments

This work was supported by Universal Energy Systems, Inc., Dayton, Ohio, and the Advanced Thermionic Initiative of the Wright Research and Development Center, Wright-Patterson Air Force Base, Ohio; T. Lamp, Program Director.

REFERENCES

Kitrilakis, S.S. and M. Meeker (1963) "Experimental Determination of the Heat Conduction of Cesium Gas" in Advanced Energy Conversion, 3:59-68.

Rasor, N.S. (1982) "Thermionic Energy Conversion" in Applied Atomic Collision Processes, vol. 5, pp. 169-200,.

Rasor Associates (1990) "Planar Converter Standard Model Documentation - Supplementary Description of TECMDL Converter Physics", E-563-002-B-063087.

Rasor Associates (1990) "TECMDL - Ignited Mode Planar Converter Model", E-563-004-C-082988.

McVey, J.B. (1990) "START_1HX Program Description and User's Manual", Rasor Associates, E-561-058-A-061190.

Rasor Associates (1990) "Preliminary Technical Report - CYLCON Semi-2D Cylindrical Converter Model".

G.N. Hastopoulos and E.P. Gyftopoulos (1973) Thermionic Energy Conversion, p. 200, MIT Press.

el-Wakil, M.M. (1981) Nuclear Heat Transport, third ed., p. 269, American Nuclear Society.

(1972) Sodium-NaK Engineering Handbook, O. Faust, ed., vol. 1, pp. 52-53.

Lewis, B.R., R.A. Pawlowski, K.J. Greek and A.C. Klein (1991) "Advanced Thermionic Reactor Systems Design Code", in *Proceedings of the Eighth Symposium on Space Nuclear Power Systems*, CONF-910116, M.S. el-Genk and M.D. Hoover, eds., American Institute of Physics, New York.

CENTAR MODELLING OF THE TOPAZ-II:
LOSS OF VACUUM CHAMBER COOLING DURING FULL POWER GROUND TEST

Vaughn H. Standley and D. Brent Morris
Phillips Laboratory
New Mexico Engineering Research Institute
901 University SE, Rm. 135
Albuquerque, NM 87106
(505) 272-7240/-7266

Michael J. Schuller
Phillips Laboratory
Space and Missiles Technology Directorate
Space Nuclear Power Branch
Kirtland AFB, NM 87117-6008
(505) 846-2878

Abstract

The Code for Extended Non-linear Transient Analysis of Extraterrestrial Reactors (CENTAR) was used to model an electrically heated TOPAZ-II thermionic space reactor operating at full power following a loss of coolant to its enclosing vacuum chamber. The purpose of the work was to quantify the response time available to Thermionic System Evaluation Test (TSET) operators following an interruption of vacuum chamber cooling and to test the utility of the CENTAR code for modelling a true-to-life application. A parametric study was done to test key assumptions and to refine the TOPAZ-II input deck being used. The vacuum chamber temperature history was then solved for under the assumption that full power would be maintained (at 115 kW_{th}) during the loss of vacuum chamber cooling. The vacuum chamber temperatures were substituted into the CENTAR input deck for the space temperature variable. Each space temperature was associated with a point in time to simulate transient conditions in the electric heaters, thermionic elements, liquid metal coolant, and radiator. It was verified that the TOPAZ-II equilibrated fast enough such that CENTAR could run in steady-state mode to generate a quasi-transient solution. Results indicated that TSET operators would have several minutes to regain total or partial cooling and that drastic action (emergency shutdown of the TOPAZ-II electric heater power, for example) would not be required.

INTRODUCTION

Compact nuclear reactors are appropriately the heart of many space power systems, planned or demonstrated. Issues associated with space nuclear power systems often concern the balance of plant and not the core. For example, the choice of an energy conversion subsystem depends on many considerations including the amount of electrical power needed by the user satellite. For electricity needs of between 5 to 40 kW_e, a thermionic conversion subsystem is attractive (Kiernan 1991). The USSR has tested thermionic technology since 1960 and has used thermionic conversion in at least two operational missions (Bennett 1989). The US has no comparable experience.

An unfueled Soviet TOPAZ-II thermionic space reactor power system is to be purchased, delivered to the US, and tested in the modified New Mexico Engineering Research Institute (NMERI) facility as a part of the Thermionic System Evaluation Test (TSET) project. The endeavor is intended to help "jump-start" the US thermionic space reactor program (Henderson 1991). Testing will be planned and conducted through the combined efforts of four organizations (also known as The New Mexico Strategic Alliance): the Phillips Laboratory, Los Alamos National Laboratory, Sandia National Laboratories, and the University of New Mexico (Standley et al. 1991). The TSET project is sponsored by the Strategic Defense Initiative Organization.

BACKGROUND

Testing of the TOPAZ-II is conducted using tungsten electrical heaters in place of nuclear fuel. Also, testing is conducted in a vacuum chamber to approximate the radiative heat transfer conditions in space. Normally, the radiated heat from the TOPAZ-II is removed from the chamber walls by water (kept under 333 K) which is pumped to a cooling tower. The amount of heat (P_{th}) needing to be removed is normally 85 kW_{th} but may be as high as 115 kW_{th} for the electrically heated system.

The loss of vacuum chamber cooling water while testing the TOPAZ-II is a credible accident. Though power to the electric heaters can be terminated at any time, their complete and immediate (emergency) shutdown would result in a destructive cooling rate. Soviet experience suggests that cooling rates resulting from emergency shutdown (dT/dt > 100 °C/hr) will induce mechanical failure of critical welds. Also, no more than twelve normal startups and shutdowns (dT/dt > 15 °C/hr) of the system should be experienced (Schuller 1991). The preferred response to a vacuum chamber loss of coolant accident (LOCA) is to regain chamber cooling.

The Code for Extended Non-linear Transient Analysis of Extraterrestrial Reactors (CENTAR) was developed specifically for the analysis of space nuclear reactor power systems. The code solves the transient nonlinear partial differential equations for liquid metal fluid flow, solid conduction with internal heat generation, neutron kinetics, decay heat, and allows for convective and radiative boundary conditions for each computational cell. In addition to the fundamental conservation equations, CENTAR contains models for accumulator, heat pipe, and thermionic fuel elements (Nassershariff 1991). The thermionic fuel element (TFE) model uses a look-up table for calculating TFE performance and is based on an input emitter-collector gap voltage drop (ΔV_{gap}).

OBJECTIVES

The vacuum chamber LOCA was modelled in order to quantify the response time available to TSET operators following an interruption of vacuum chamber cooling. Within the Phillips Laboratory, this work is the first application of the CENTAR code toward addressing a problem of high consequence (namely, possible damage to the TOPAZ-II). Also of interest was how straightforward the code was to use and how amenable it was to perturbations in the input deck.

APPROACH

Using an existing TOPAZ-II integral system input deck, a preliminary study was done to determine which parameters in CENTAR had the greatest effect on output variables of interest (including core and radiator temperatures and electric power production). Key approximations that could be made precipitated from this study. Also, results of the parametric study were used to refine the input deck being used. Vacuum chamber temperatures were determined for each point in time and substituted for the space temperature variable (T_{space}) in the CENTAR input deck as the boundary condition. CENTAR was then run to obtain temperature-, power-, and pressure-time curves for each space temperature.

The CENTAR input deck modified for this study was developed by Science Applications International Corporation (SAIC) and is summarized in Figure 1. The model collapses components of a given type into one. The collapsing scheme preserves flow area, volume, and wall area whereas hydraulic diameter is kept to that of an individual pipe (Giguere 1991). Except for the radiator, all outer surfaces were assumed adiabatic in the original SAIC input deck.

The effective radiator area εA, in $\dot{Q} = \sigma \varepsilon A(T^4_{radiator} - T^4_{space})$, was varied to determine its importance on core temperature, where $\dot{Q}$ is the electric heater power. Effects of temperature on electric power production were tested by varying ΔV_{gap} from .88 V in the SAIC model. In general, effects of temperature on ΔV_{gap} are complicated but are conservatively bounded by $\Delta V = k\Delta T$, or about .05 V. Also, the gap emissivity, ε_{gap}, was adjusted to determine its importance on electrical production in the CENTAR model.

The SAIC TOPAZ-II input deck assumes a constant pressure drop across the electromagnetic (EM) pump. More correctly, head generated by the pump depends on the potential of three TFE elements connected in parallel (dedicated to the pump). The EM pump voltage was varied (indirectly, by varying ΔP) to estimate what effect the coolant flow rate would have on TFE temperatures and, in turn, how the TFE supplied pump voltage would change. This voltage/temperature feedback process needed to be characterized in order to neglect it in the accident problem. The present version of CENTAR does not model the feedback process.

The electric power calculated for the TOPAZ-II (operating at 135 kW$_{th}$) by SAIC is 4 kW. The gap voltage, effective radiator area, and emitter emissivity were adjusted to bring the electrical power up to the Soviet reported value of 6 kW. Pipes not a part of the radiator were given an emissivity of $\varepsilon = 0.3$ to approximate parasitic radiative heat losses. Once the CENTAR input deck satisfactorily duplicated the known performance of the TOPAZ-II according to Soviet literature, it was used in connection with the arguments below to model the vacuum chamber LOCA problem.

During a vacuum chamber LOCA, the output of the electric heaters in the TOPAZ-II does not change. The heat emitted from the TOPAZ-II radiator is absorbed by the vacuum chamber inner wall, conducted to the outside wall, and convected away by the surrounding air. The vacuum chamber absorbs heat and rises in temperature with time proportional to the chamber heat capacity (mc_p). Heat convected away is proportional to the heat transfer coefficient (h), the chamber outside area (A), and the drop in temperature from the chamber surface to the surrounding ambient air ($T - T_a$). Conduction effects were neglected. The temperature of the vacuum chamber is described by Equation 1, the solution of which is given by Equation 2.

$$\dot{Q} = mc_p \frac{dT}{dt} + hA(T - T_a) \tag{1}$$

$$T(t) = \frac{\dot{Q} + hAT_a}{hA} \left(1 - e^{-\frac{hA}{mc_p}t} \right) + T_0 e^{-\frac{hA}{mc_p}t} \tag{2}$$

The heat transfer coefficient for the vertical surface of the vacuum chamber ($h_{Vertical}$) is computed using the Nusselt number ($Nu_L = hL/k$) given by Equation 3 (Churchill and Chu 1975):

$$Nu_{L,Vertical} = \left[0.825 + \frac{0.387 Ra_L^{1/6}}{(1+(0.492/Pr)^{9/16})^{8/27}} \right]^2 \tag{3}$$

where

$$Ra_L = \frac{g\beta(T - T_a)L^3}{\alpha\nu} \tag{4}$$

For the case of air evaluated between the hot chamber wall temperature and that of the surrounding ambient air (about 400 K), $\beta = 1/400$ K, $\alpha = 38.3\times10^{-6}$ m^2/s, $\nu = 26.4\times10^{-6}$ m^2/s, Pr = 0.690, and k = 33.8$\times10^{-3}$ W/mK. For the horizontal circular top of the chamber the Nusselt number is given Equation 5 (McAdams 1954):

$$Nu_{L,Horizontal} = 0.15 Ra_L^{1/3} \tag{5}$$

where the length L, used to compute Ra_L for the horizontal (top) portion of the vacuum chamber, is an effective length equal to the top's exposed area divided by its perimeter (Goldstein et al. 1973). The vacuum chamber/TOPAZ-II arrangement is described in Figure 1.

The current version of CENTAR (4.0) does not directly provide for entering a continuously changing temperature boundary condition. Instead, each temperature (computed using the above scheme) was substituted for the space temperature variable in the CENTAR input deck. The code was then run in steady-state mode until a solution using the new space temperature was computed. To determine if the TOPAZ-II could be modelled in this way, the most abrupt vacuum chamber temperature rise resulting from the postulated accident was entered as the space temperature in CENTAR and the time needed to reach equilibrium was determined.

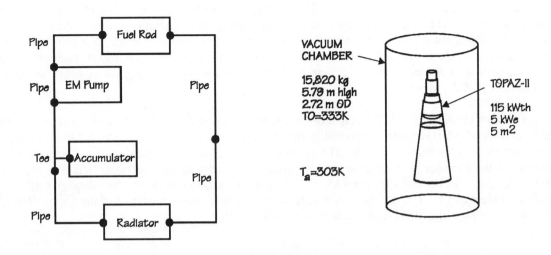

FIGURE 1. CENTAR Input Deck for TOPAZ-II.   FIGURE 2. Accident Configuration.

RESULTS

Table 1 is a summary of the parametric study conducted. Electrical power (P_e), radiator (T_{rad}), core inlet (T_{in}), and core outlet (T_{out}) temperatures are all sensative to the effective radiator area. Electrical power production and temperatures changes only slightly for small perturbations in DV_{gap} (< .05V). Emissivity in the TFE gap (ε_{gap}) has a dramatic effect on the system. Pump head (also, pump voltage) only slightly influences the emitter temperature. Case 27 is most representative of to-life TOPAZ-II performance.

Vacuum chamber temperatures for the LOCA problem were computed on a Microsoft Excel spread-sheet (using circular references). Hourly temperatures for the vacuum chamber wall interior are reported in Figure 3. Temperatures that would be observed in the TOPAZ-II (as computed using CENTAR) are shown in Figure 4. Output parameters for the TOPAZ-II were shown to be fully equilibrated for each chamber wall temperature in less than five simulated minutes (that is, quasi-equilibrium was satisfied).

CONCLUSIONS

The inability to couple TFE to thermal-hydraulic performance in this computer simulation leaves some doubt about the accuracy achieved but has little impact on accident case modelled. The emitter-collector gap voltage is closely tied to emitter temperature. System performance is also not effected significantly by the inability to couple pump head and TFE voltage. By adjusting the parameters available in the present version of CENTAR, a reasonable approximation of a TOPAZ-II system was developed.

For the conditions asserted in this study, CENTAR was used to bound the transient response of the TOPAZ-II. The code was very amenable to changes to the input deck. Consequences of the LOCA are limited in time by the boundary condition (the vacuum chamber inner wall temperature). Operators managing this ground-test accident have minutes, not seconds, to reverse the conditions, depending on the initial vacuum chamber temperature.

CENTAR is a robust system for modelling the thermal-hydraulic performance of space reactor systems. A good follow-on application of CENTAR will be to model a rapid transient in the TOPAZ-II that is caused by a water leak into the vacuum chamber from the chamber's cooling jacket. CENTAR modelling of this accident will answer critical safety questions.

TABLE 1. Parametric Study Summary.

Case #	P_{th} (kW)	T_{space} (K)	εA (m^2)	TFE ε_{gap}	Pump ΔP (kPa)	TFE ΔV_{gap} (V)	ε_{pipes}	P_e (kW)	Core T_{in} (K)	Core T_{out} (K)	T_{rad} (K)
0	135	298	5.0	0.3250	65.0	0.880	0.0	3.73	802	881	826
1	135	298	5.0	0.3250	65.0	0.880	0.0	3.82	797	876	821
2	135	298	5.0	0.3250	65.0	0.880	0.3	4.15	778	858	800
3	135	555	5.0	0.3250	65.0	0.880	0.3	3.46	816	895	838
4	135	255	4.5	0.3250	65.0	0.880	0.3	3.83	797	856	818
5	135	333	4.0	0.3250	65.0	0.880	0.3	3.94	818	896	839
6	135	300	5.5	0.3250	65.0	0.880	0.3	4.10	762	842	783
7	135	300	6.0	0.3250	65.0	0.880	0.3	4.87	747	828	768
8	135	255	5.0	0.3250	65.0	0.800	0.3	4.15	780	860	801
9	135	255	5.0	0.3250	65.0	0.700	0.3	4.16	778	857	800
10	115	255	5.0	0.3250	65.0	0.880	0.3	3.30	751	820	770
11	115	300	5.0	0.3250	65.0	0.880	0.3	3.29	753	822	772
12	135	255	5.0	0.2000	65.0	0.880	0.3	8.00	766	840	786
13	135	255	5.0	0.3000	65.0	0.880	0.3	4.49	772	856	799
14	135	255	5.0	0.2500	65.0	0.880	0.3	6.28	773	850	793
15	135	255	5.0	0.2750	65.0	0.880	0.3	5.07	776	854	693
16	115	300	5.0	0.2500	65.0	0.880	0.3	4.60	747	815	765
17	135	255	4.5	0.2750	65.0	0.880	0.3	5.00	794	872	815
18	135	255	4.1	0.2500	65.0	0.825	0.3	5.39	809	885	830
19	135	255	4.3	0.2250	65.0	0.825	0.3	6.69	797	872	817
20	135	255	4.3	0.2500	65.0	0.825	0.3	5.49	800	877	821
21	135	255	4.3	0.2375	65.0	0.825	0.3	6.09	798	874	819
22	115	255	4.3	0.2375	65.0	0.825	0.3	4.39	773	839	790
23	85	255	4.3	0.2375	65.0	0.825	0.3	2.87	725	776	738
24	115	255	4.3	0.2375	71.5	0.825	0.3	4.21	764	874	792
25	115	255	4.3	0.2375	58.5	0.825	0.3	4.16	764	879	794
26	135	255	4.3	0.2375	65.0	0.825	0.3	6.00	805	881	825
27	115	300	4.3	0.2375	65.0	0.825	0.3	4.39	774	841	792

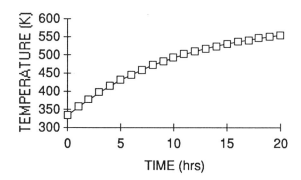

FIGURE 3. Vacuum Chamber Temperature History.

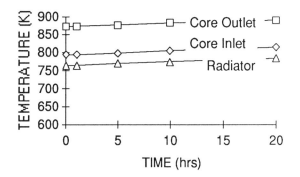

FIGURE 4. TOPAZ-II Temperature History.

Acknowledgments

This work was sponsored by the United States Air Force. Special thanks go to B. Nassersharif of Texas A&M University for his assisstance in using CENTAR.

References

Bennet, G (1989) "A Look at the Soviet Space Nuclear Power Program," *in Trans. 24th Intersociety Energy Conversion Engineering Conference*, **2**, Institute of Electrical and Electronics Engineers, United Engineering Center, 345 East 47th Street, New York, New York, 1989, p. 1187.

Churchill, S. W., and H. H. S. Chu (1975) "Correlating Equations for Laminar and Turbulent Free Convection from a Vertical Plate," *Int. J. Heat Mass Transfer*, **18**, 1323, 1975.

Giguere, P. T. (1991) *CENTAR Model for TOPAZ-II Integral System*, SAIC Report No. 91-6513, Science Applications International Corporation, 2109 Air Park Road, SE, Albuquerque, NM, 4 June 1991[1].

Goldstein, R. J., E.M. Sparrow, and D. C. Jones (1973) "Natural Convection Mass Transfer Adjacent to Horizontal Plates," *Int. J. Heat Mass Transfer*, **16**, 1025, 1973.

Henderson, B. W. (1991) "US Buying Soviet TOPAZ 2 to Boost Space Nuclear Power Program," *Aviation Week and Space Technology*, 14 January 1991, pp. 54-55.

Kiernan, V. (1991) "U.S. Plan Would Test Nuclear Reactor in Space," *Space News*, 16-22 September 1991, p. 10.

McAdams, W. H. (1954) *Heat Transmission*, 3rd ed., McGraw-Hill, New York, 1954, Chapter 7.

Nassersharif, B., M. J. Gaeta, and M. R. Gaeta (1991) *CENTAR User's Guide and Input Description*, TAMU-TEES-NSCL-90-30, Texas A&M University, College Station, Texas, March 1991, p. 3.

Schuller, M. J. (1991) USSR/US "Technical Interchange Meeting Notes," Internal Memo, March 1991[2].

Standley, V. H., D. B. Morris, M. J. Schuller (1991) "Preliminary Draft of the Thermionic System Evaluation Test Project Plan," PL/STPP, Phillips Laboratory, Albuquerque, NM, 14 Aug 1991[3].

---

[1]Distribution authorized to U. S. Government agencies only; Proprietary Information, June 1991. Other requests for this document must be referred to PL/STPP, Kirtland AFB, NM 87117-6008.

[2]Distribution authorized to U. S. Government agencies only; Proprietary Information, March 1991. Other requests for this document must be referred to PL/STPP, Kirtland AFB, NM 87117-6008.

[3]Distribution authorized to U. S. Government agencies only; Proprietary Information, August 1991. Other requests for this document must be referred to PL/STPP, Kirtland AFB, NM 87117-6008.

Research sponsored by SDIO with Phillips Laboratory acting as agent.

MODELING SPACE NUCLEAR POWER SYSTEMS WITH CENTAR

Mark J. Dibben, Taewon Kim, and Ronald F. Tuttle
Air Force Institute of Technology
AFIT/ENP
Wright-Patterson AFB, OH 45433
(513)255-8989

Abstract

The Code for Extended Nonlinear Transient Analysis of Extraterrestrial Reactor Systems, CENTAR, is a simulation code for space nuclear reactors. This paper describes some of the individual components of this code and then combines them into a model of the TOPAZ system. Problems with the individual components and the system that is modeled are discussed. Unsatisfactory results are related to the lack of TOPAZ data and the beta version of the code utilized for this project.

INTRODUCTION

The Code for Extended Nonlinear Transient Analysis of Extraterrestrial Reactor Systems (CENTAR) is a simulation code for space nuclear reactors. It is under development at the Numeric/Symbolic Computations Laboratory at Texas A & M University (Nassersharif et al. 1991). CENTAR solves for different parameters associated with both liquid metal cooled and solid core reactors. Version 2.1 beta was used for this work.

The objectives of this project were to install the CENTAR code, become familiar with its operation and attempt to model a working space nuclear reactor called TOPAZ.

CENTAR DESCRIPTION AND EXECUTION

A basic description of the working environment of CENTAR is as follows. Iterative schemes are used to solve nonlinear equations. A Newton-Raphson method is used for solving the mass and energy systems. Partial differential equations are represented by finite differences, which are then solved by Alternating Direction Implicit method. For solving the matrix solution, either sparse Gaussian, row Gaussian, or column Gaussian elimination can be used (Nassersharif et al. 1991). The conservation equations solved by different iterative schemes are based on several assumptions. First, the equation of states are based solely on temperature. Second, the gravity term is neglected. Finally, heat is allowed to transfer only radially through the walls of components to the fluid. These conditions and other minor assumptions for models, like axial conduction, are briefly described in the manual (Nassersharif et al. 1991).

Execution of CENTAR requires creation of an input deck that contains information about the system. To use CENTAR, the input deck, CENTIN, is executed by the CENTAR4.X file to create the output files, CENTOUT and CENTECHO. Next, mc_convert4 processes the output of CENTAR4.X for CIGS4.X, which produces user defined graphical output from the results.

The input deck, CENTIN, is created by combining different types of components. CENTAR provides a reactor kinetics model, accumulator, fuel rod, heat exchanger, pipe, plenum, pressure boundary conditions, electromagnetic pump, and tee as the components to model a space nuclear system (Nassersharif et al. 1991).

The input deck begins with header information providing a general description of the system and is required for any system that is modeled. If kinetics information is used, it must be entered right after the header information. All other components may be entered in any order.

In the process of creating the input deck, the following steps help locate errors. The first step is to check the CENTECHO file which is created by CENTAR. By checking this file, the location of the errors can be found. It will also inform the user of input format errors. The CENTOUT file which contains the output information can also help in locating errors. Some other errors like a floating point error can indicate conditions that are outside of the material's property range.

The initial buildup of the input deck can be difficult if some of the unusual requirements of the code are overlooked. Some of the idiosyncracies of the code are as follows:

1. Some parameters that appear to be related do not match, for example, the flow area and the hydraulic diameter are not related in the code's operation;

2. Position of the data is important when building an input deck. For example, the pressure table from the pressure boundary condition component must be entered in a specific order;

3. The rectangular components can be attached directly to the cylindrical components, and;

4. Restart files can not be made directly into CENTIN files. The inlet and outlet conditions must first match all the components.

TESTING OF CENTAR COMPONENTS

One component tested was the electromagnetic (EM) pump. The pump was attached to a pipe system with NaK as the working fluid. Currently, CENTAR only supports a rectangular pipe as a carrier for the EM pump. The model of the EM pump was divided into five different sections to show pressure change through out the pipe. The pump was arbitrarily attached to the middle section. The axial cell pressures are in Table 1.

The pump increased the pressure by 10000 Pa. The velocity in each axial interface in relation to pressure changes, from inlet to outlet, were 4.22, 4.14, 4.07, 4.02, 3.85, and 3.72 m/s, respectively. Based on these numbers, velocity and pressure were independent of each other.

TABLE 1.   Axial Cell Position versus Pressure and Temperature of
            the EM Pump Model.

Axial Cell	Pressure (Pa)	Temperature (K)
1	6016	1351
2	5097	1307
3	4152	1270
4	14152	1158
5	13473	1056

It was also noted that the temperature profile was identical for a pipe with and without an EM pump, which indicates that the EM pump does not affect the heat balance. The temperature in each axial cell is also shown in Table 1.

Another component tested was the fuel rod. The code had the capability to model a liquid metal cooled or solid core reactor. Briefly, the liquid metal cooled and the solid core reactor models performed as expected. However, the addition of the thermionic option to these models only worked for the solid core reactor. The liquid metal cooled reactor with the thermionic option was unable to converge to a steady state solution. More detailed information will be provided in the model for TOPAZ.

TOPAZ MODEL

The second goal of our project was to combine these components into a working reactor system. The TOPAZ was the system selected. The TOPAZ is a thermionic space nuclear power system capable of producing 5 to 7 kWe (Nickitin et al. 1991).

A complete description of the TOPAZ with all of information required by CENTAR was not available. By combining several different sources, a partial system was designed and made into an input deck. The unknown information was filled in based on other systems that were available. Table 2 lists the known, calculated and assumed values.

Because of limitations of the code, TOPAZ was modeled as shown in Figure 1. The modeled reactor contained 37 thermionic fuel elements (TFE) based on available literature (Glushkov et al. 1991). Each TFE was composed of a uranium dioxide fuel rod surrounded by a tungsten emitter and a niobium collector. The radiator was the only component to radiate heat. All other components were thermally insulated.

Using CENTAR, the closed loop model did not produce results. Floating point errors occurred with every modification of the closed loop model attempted. Some results were obtained using an open loop system, such as the temperature profiles in the reactor. Figure 2 shows the temperature of the cooling fluid as a function of time and location in the reactor.

1137

TABLE 2.   TOPAZ Dimensions and Characteristics

Parameter	Value		Source
fuel rod length	375	mm	AFPL 1991
fuel diameter	15	mm	calculated
emitter O.D.	20	mm	AFPL 1991
emitter wall thickness	1.5	mm	AFPL 1991
gap between fuel and emitter	1	mm	assumption
electrode gap	0.4064	mm	AFPL 1991
collector wall thickness	1.5	mm	AFPL 1991
pitch	37	mm	Glushkov et al. 1991
pressure drop across reactor	9000	Pa	assumption
gap conductance	7800		assumption
voltage at reactor	29.5	V	Nickitin et al. 1991
maximum working fluid temperature	873	K	Nickitin et al. 1991
heat power	135	kWt	Nickitin et al. 1991
electrical power	6	kWe	Nickitin et al. 1991

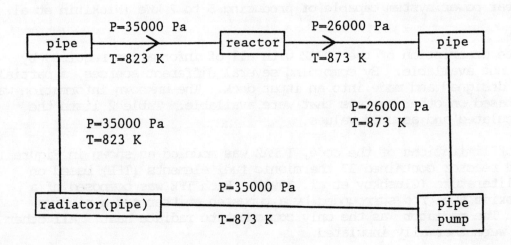

FIGURE 1.   Schematic of the Major Components of the TOPAZ System.

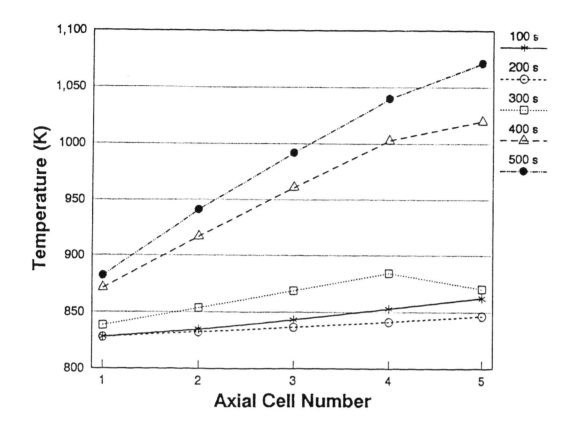

FIGURE 2. Axial Temperature of Reactor with Time.

As can be seen from this graph, the temperature begins to converge and then increases dramatically. The maximum temperature of the working fluid is supposed to be only 873 K. Obviously, one of the parameters is incorrect. Either the pressure drop is larger, which would allow the flow rate to increase, or the volume of the coolant should be larger, which would keep the temperature within specifications. More information about the system was not available to determine the correct simulation model.

Other problems occurred with the TOPAZ simulation. Neither the closed loop system nor the open loop system with pressure boundary conditions was able to reach a steady state solution. The closed loop system was unable to start after reading the input deck. A possible explanation for this is the input file conditions were not close enough to steady state for the initial time step to converge. Since the open system was able to run for 500 seconds without a fatal error, an attempt was made to close the system by modifying the boundary conditions until they matched each other. The boundary conditions now reflected the conditions of the components to which they were attached. The elimination of boundary conditions should have resulted in a working closed system. However, this method did not work. The method of solution was also varied to get a running system. All combinations of explicit, implicit and sparse Gaussian, row Gaussian elimination and column Gaussian elimination were attempted without success.

CONCLUSIONS

Within the time and resources made available, it was determined that some of the individual components of CENTAR, such as the pipe, gave reasonable results while others, such as the heat exchanger, did not function properly. Some of the difficulties in utilizing the CENTAR code are due to lack of experience, an inadequate component description for building the input deck, and the undocumented errors in the Version 2.1 beta of CENTAR. Understandably, the code is not ready for production work and needs further development and testing. Regardless, this evaluation indicates that some of the mathematical and engineering models are not functioning at all while others violate the conservation laws.

RECOMMENDATIONS

CENTAR needs further development and testing. The manual can be improved by including better diagrams of the components with better component descriptions. Working examples of all the components should be added to the manual's appendix. The newer versions of the code must be checked to see if the heat exchanger is working as implied by the authors. Comprehensive documentation and a user-friendly graphical environment would significantly enhance this code and make it possible for simulating space nuclear power concepts and aid their evaluation and comparison. After the implementation of the recommendations, another evaluation is required to determine the usefulness of the code in modeling the space nuclear power systems.

Acknowledgments

This work was done at the Air Force Institute of Technology in partial fulfillment of the requirements of the Master of Science degree in Nuclear Engineering and received CENTAR code support from the Air Force Phillips Laboratory.

References

Air Force Phillips Laboratory (1991) Space Power System Core Length Thermionic Converter System Design Study, Air Force Phillips Laboratory Quarterly Briefing, Kirtland AFB, NM, April 1991.

Glushkov, E. S., G. V. Kompanietz, V. G. Kosovskii, N. E. Kukkharkin, N. N. Ponomarev-Stepnoi, O. N. Smirov, and V. A. Usov (1991) "The Neutron-Physical Aspects of the Converter Reactor with Single-Element TFE," in Proc. 8th Symposium on Space Nuclear Power Systems, held in Albuquerque, NM, 15-18 January 1991.

Nassersharif, B., M. J. Gaeta, and M. R. Gaeta (1991) CENTAR User's Guide and Input Description version 2.2, Texas A&M University, College Station, TX.

Nickitin, V. P., B. G. Ogloblin, and A. N. Luppov (1991) "Topaz-2 Thermionic Space Nuclear Power System and the Perspectives of its Development," in Proc. 8th Symposium on Space Nuclear Power Systems, held in Albuquerque, NM, 15-18 January 1991.

SYSTEM MODELLING WITH
SALT GPS

Daniel J. Robbins and Ronald F. Tuttle
Air Force Institute of Technology
Department of Engineering Physics
AFIT/ENP
Wright Patterson Air Force Base, OH 45433
(513) 255-4498

Abstract

Plans for thermionic space nuclear power systems have been in development for many years but their applicability and analysis are incomplete on a system level. Argonne National Laboratory has developed an in-house code, the Systems Analysis Language Translator (SALT), that can be used to model a wide variety of systems on today's workstations. This work explored the intricacies of SALT with specific applications aimed at modelling thermionic space nuclear power systems, including the Soviet TOPAZ. Using results from previous studies, reports, designs and other information, models were constructed and evaluated. Results indicated that SALT is a powerful modelling tool with the potential to critically compare different types of thermionic nuclear power systems with substantial accuracy after based component models are further developed for the SALT library.

INTRODUCTION

SALT/GPS, a preprocessor/translator developed by Argonne National Laboratory (ANL) for use with the Systems Analysis Language Translator (SALT) system analysis code, also written by ANL, was used to analyze the power demand of the primary coolant pump on the Soviet built TOPAZ space nuclear power system. SALT uses a quasi-Newtonian update/steepest descent technique, not further described in the SALT/GPS system description (Geyer and Berry 1985) or in the SALT Programmer's Guide (Geyer and Berry 1985), for solving systems of linear or nonlinear algebraic equations. It uses either Gear's backward differences method or the Adams-Bashford-Moulton (sic) method for solving ordinary differential equations. The algorithms used are presumably similar to those methods described in typical numerical methods texts. GPS allows the user to write SALT task models, simple programs linking component models together to represent a particular subsystem or system and the input values and constraints for the component models. With GPS the user can manipulate the components of a system and do parametric studies on a system without writing more complex programs in SALT itself. The architecture of GPS also permits the user to write additional component models with GPS rather than edit existing SALT component models or write new programs from scratch.

SCOPE

The purpose of this work was to gain familiarization with SALT/GPS and assess the practicability of using it to analyze specific space nuclear power systems. The theory, methods, and accuracy of the numerical methods used by SALT were not evaluated. It was assumed that the numerical methods used were sound and produce precision and accuracy to a degree acceptable to the developers of the program. Herein, no attempt is made to explain how to use or write SALT/GPS programs.

GPS FAMILIARIZATION

Familiarization with GPS was gained in stages. First, some of the examples included with the SALT/GPS user guide were executed to learn basic control commands. Next, simple problems were solved making use of the algebraic and differential equation solvers in SALT. Third, the variation of efficiency of a hypothetical nuclear Brayton cycle system with compressor ratio was modeled and plotted so that additional features of GPS/SALT could be exercised across the network. Lastly, an existing GPS/SALT nuclear thermionic space power system model was used to develop task models, which were run, varying input parameters and studying the results.

Intrinsic Examples

All of the SALT component models, divided into functional sets according to the system model to which they applied, had SALT/GPS task examples of a system to be modelled. There were also some examples of direct command programs which solved algebraic or differential equations and, in one case, plotted the results. These examples and variations of them served to demonstrate the mathematical and functional capabilities of SALT and were reported by Robbins (1991). Running the examples proved easy, even during the first session in which several of the examples were modified and successfully executed.

Direct Problem Solution

Two problems were solved. The first was to find the pressure drop of water with a known flow rate through a pipe of known material, length, and diameter. Using the Darcy equation

$$\Delta p = -\frac{\rho v^2 f l}{2 g_c D_h} \tag{1}$$

where
p is the pressure drop through the pipe
ρ is the density of the fluid
v is the fluid velocity, averaged across the flow area of the pipe
f is the friction factor from a Moody diagram (Rust 1979)
l is the pipe length
g_c is the gravitational acceleration constant and
D_h is the hydraulic diameter of the pipe.

In the SALT/GPS environment, the task model was easily written and the solution was found very quickly. The second problem solved a system of two first order ordinary differential equations.

$$\frac{dz}{dt} = -2z - 4(w+2) \tag{2}$$

$$\frac{dw}{dt} = -z \tag{3}$$

with the initial conditions z(0) = w(0) = 0.

This initial value problem was easily coded and solved by SALT/GPS. In addition, the variation of the dependent variables as a function of time were easily plotted with the aid of GPS model class denoted as **plot**. Although the graphics currently available within GPS are limited, GPS graphics are sufficient to generate two-dimensional plots and system diagrams of component layouts.

Brayton Cycle Efficiency Variation

Initially the authors desired to modify a steady-state gas turbine model already resident in SALT to model a gas turbine system using a nuclear heat source. It was soon recognized that this steady-state model actually represented a once through, open-ended system, a rocket motor. Modification of the steady-state model was considered but quickly abandoned because of uncertainties and a more straight forward alternative. As an alternative a task model was written to calculate and plot the thermodynamic efficiency for varying compressor ratio of a specific hypothetical Brayton cycle system. The system and the equations can be found in <u>Nuclear Power Plant Engineering</u> (Rust 1979). Modelling of this system took considerably longer to write successfully. The main effort involved the correct representation the equations as a GPS class model. Constructing the correct GPS statements to access the plot routines was relatively easy. The numerical and graphical results, together with the system model and the GPS model, were reported by Robbins (1991).

SALT/GPS THERMIONIC MODEL

Thermionic power system model classes written within the SALT/GPS structure were component models and included a reactor, a thermionic converter, radiator, boost converter, flow splitter, resistor (load), bus, and mass. The mass model performed a global sum over all the other component models to provide the total system mass. The program modelled the General Atomics STAR-C power system using data and equations provided by the Air Force. Most of the modelling was done with simple correlations resulting in a first-order approximation of component and total system masses. The reactor was treated as a black box heat source. No variations on reactor fuel, or for that matter, thermionic material, or radiator shape and material was possible. The user can vary the power required, the thermionic cell efficiency, and several other input values. Task models were easily constructed and modified within the limits described. The original program was modified to vary the thermionic cell efficiency over a specified range and system mass was plotted against efficiency as Figure 1. The program was also modified to provided Figure 2, a diagram of the system modelled. The SALT/GPS routines created PostScript files. The modifications consisted of adding calls to these programs to the task model. Note that the system diagram represents flow of data through the SALT/GPS component models. In this case, it also accurately depicts the power flow through the system because it is the power flow that the SALT/GPS model class uses to calculate component and system masses.

1143

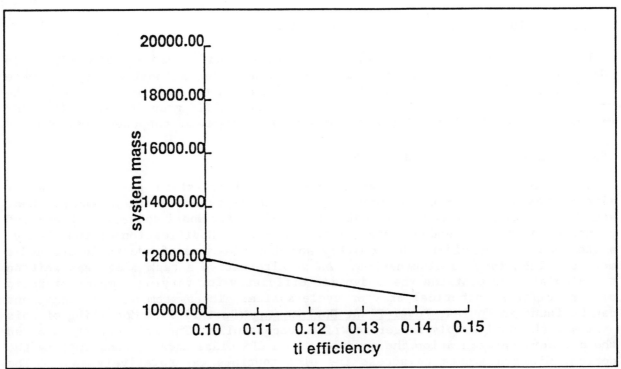

FIGURE 1.   Plot of STAR-C System Mass Variation with Thermionic Cell Efficiency
Produced by SALT/GPS.

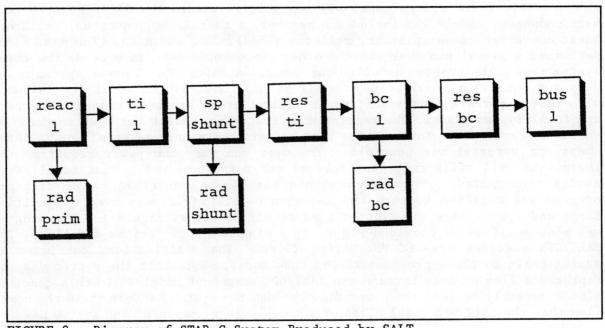

FIGURE 2.   Diagram of STAR-C System Produced by SALT.

MODELLING THE SOVIET TOPAZ REACTOR SYSTEM

The original goal for this project was to write a program containing all the necessary components to model the Soviet TOPAZ nuclear power system. In particular, it was desired to model the coolant flow and power flow. This proved impossible for two reasons. First, complete data on the TOPAZ system was not available to the degree necessary to construct the component models and the SALT/GPS coding required to do a first-order approximation would not have given results significantly different from the built-in SALT/GPS thermionic model. Rather than being able to modify the existing model with GPS commands to represent the new system, it would have been necessary to write a completely new system model in SALT using C++ language. It was decided, because of the lack of substantive TOPAZ subsystem and system information, not to extend the project beyond the time available for this project. As an alternative, it was decided modify the existing code using GPS commands to calculate the power necessary to drive the primary coolant pump for the Soviet TOPAZ system and subtract this power from what was initially available at the main bus. Some input parameters to the existing SALT thermionic model, reactor thermal power, required power at the bus, thermionic cell efficiency, coolant temperature at reactor outlet, and bus terminal voltage, appropriate to the TOPAZ system were known and entered in the SALT model. Component models that were not needed, a shunt power splitter, voltage boost converter, resistance for the boost converter, and radiators for waste heat from both components, were deleted from the flow description in GPS. Two approaches were used to subtract out the pump power. Both methods defined two new variables, a mass flow rate of the coolant, NaK, and a pump power. The first method was to create a second bus model which had as input the power at the first bus minus the pump power calculated above. The second method was to create a pseudo resistance using the resistance model and calculate a resistance sufficient to consume power equivalent to the pump requirement. It can easily be shown that this resistance is:

$$R = \frac{V^2}{P_{input}^2} P_p \qquad\qquad (4)$$

where
R is the resistance in ohms
V is line voltage
P_p is the pump power in watts and
P_{input} is the power from the thermionic in watts.

The power flow was sent from the thermionic model through this resistance model then on to the bus model.

RESULTS

The system model with a second bus would not converge. Only when the bus models were separated and the bus with pump power included (before being subtracted), as the basis for comparison, would the model converge and give an answer. By setting bus two to bus one minus the pump power and manually iterating, it was possible to get a solution giving power at the bus and pump power. This manual iteration was done by running the modified GPS task requiring the bus power to be the desired level (6 kilowatts) and subtracting the pump power. The later was subtracted from the former and the difference added to the desired power. This difference turned out to be 135 watts so the code was changed to require convergence on bus power equal to 6.135 kilowatts. The other method failed to converge and no modifications were found which produced convergence. It must be pointed out here that to get satisfactory answers the value for thermionic efficiency was calculated by dividing the output power at the bus by reactor thermal power and using this as an input parameter. That is, the system efficiency was calculated off-line and used as the thermionic efficiency rather than creating a model that calculates system efficiency (Robbins 1991).

CONCLUSIONS

SALT/GPS is powerful tool with a lot of potential but requires a working knowledge of C++ before an operator can use all its capabilities. SALT/GPS was described as requiring only slight acquaintance with any computer language prior to use and modification of the models. In fact, using existing models can be learned very quickly, however, available models are system specific and writing new models requires thorough knowledge of C or C++ on the part of the programmer/user. New models can be written for GPS, sometimes with considerable effort, and software engineering design principles should be applied before component models are constructed using C++ to represent the final system with the SALT/GPS environment.

The modified SALT programs probably failed to converge because the modifications were not included in the convergence algorithm but were appended outside the model flow. A thorough model of the TOPAZ reactor, accurately portraying fluid and power flows, is better written in SALT. It should also be possible to write a generic nuclear-thermionic based system model which can be accessed through GPS and appropriate components and input values specified using the GPS task model.

Acknowledgments

This research was conducted under the auspices of the Air Force Institute of Technology in partial fulfillment of the requirements of the Master of Science degree in Nuclear Engineering and received SALT/GPS code support from the Air Force Phillips Laboratory.

References

Geyer, H. K. and G. F. Berry (1985) The Systems Analysis Language Translator (SALT): Programmer's Guide, Argonne National Laboratory, Chicago, IL.

Geyer, H. K. and G. F. Berry (1985) The SALT: User's Guide, Argonne National Laboratory, Chicago, IL.

Robbins, D. J. (1991) "System Modelling with SALT/GPS," Report AFIT/GNE/92M-06/07/91, Air Force Institute of Technology, Wright-Patterson AFB, OH 45433.

Rust, J. H. (1979) Nuclear Power Plant Engineering, S. W. Holland Company, Atlanta.

AN EXPERIMENTAL COMPARISON OF WICKING ABILITIES
OF FABRIC MATERIALS FOR HEAT PIPE APPLICATIONS

Timothy S. Marks and Andrew C. Klein
Department of Nuclear Engineering
Radiation Center, C116
Oregon State University
Corvallis, OR 97331-5902
(503) 737-2341

Abstract

Replacement of components of a space reactor heat pipe by advanced ceramic fabrics will decrease system mass considerably. Replacement of the metal wick by a fibrous materials makes calculation of the wicking ability difficult. An experimental approach has been designed and implemented to allow for determination of the wicking ability of ceramic fabrics. This experiment's uniqueness lies in its attempt to simulate operating conditions of the heat pipe. Variables such as material composition, surface preparation, weave type and density, and pressure/temperature variations have been cursorily examined. More extensive testing is being performed to allow for optimization of wick structure.

INTRODUCTION

The SP-100 space nuclear reactor program is well underway, and should be able to develop a usable reactor system by the turn of the next century. However, since the figure of merit used in defining any space power system is the specific power (in electrical power output per unit power system mass), the decrease in the mass of any reactor system component will yield a tremendous benefit to the overall system performance. The heat rejection system of any power system can make up a large portion of the total system mass, thus any reduction in the radiator mass is of significant interest.

A continuing research project at Oregon State University focuses on the use of lightweight fabric composite material (such as SiO_2, SiC, or alumina-boria-silica fabrics) in heat pipes. Fabric materials can be used to replace the metal wall and wick found in the condenser section of a normal heat pipe, as seen in Figure 1, providing a significant reduction in weight. Here, the weave of the fabric and the high strength capability of the individual fibers can provide the supporting structure, while a highly conductive and chemically stable metallic liner bonded to this fabric allows for fluid containment. Layers of fabric can also replace the traditional wire mesh wick. This should result in a significant reduction of overall system weight. This project is currently focusing on the use of fabrics as the wick structure.

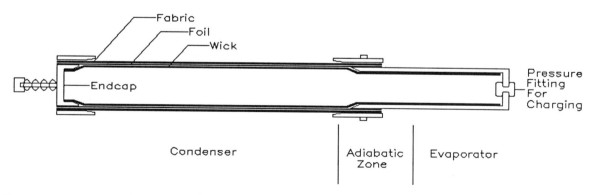

FIGURE 1. Fabric Composite Heat Pipe.

DISCUSSION

It is very difficult to develop a model that will allow calculation of wicking capability of a fibrous medium because of the consideration of all the possible fluid flow channels within a fibrous wick (Clark and Miller 1978). The only assured way of deciding between the available materials is through experimentation. It is found that two fairly independent parameters need to be measured to provide comparison of the material's ability to wick. These parameters are the wettability (wicking height) and the capillary capability (wicking rate) of the material. Since the wicking is a strong function of the liquid properties of the working fluid (specifically the surface tension), it is important to examine the effect of various working temperatures and pressures as well as the effect of different materials, weave structures, and weave densities on these parameters. Two experiments have accordingly been designed to measure these. These experiments were presented previously (Marks and Klein 1991). Slight modifications have been made, and experimentation is being performed examining various fabrics. This paper presents some of the findings from the wettability experiment.

EXPERIMENTAL METHOD

The wettability experiment measures the ability of a dry fabric to take up liquid. It determines the velocity of the advancing liquid front in a unwetted sample and the maximum capillary height. This test is performed horizontally to eliminate the effect of gravity, allowing the results to be extrapolated to space applications (Symons 1974). As seen in Figure 2, a long rectangular strip of the fabric is stretched uniformly across a horizontal table with springs. This table is inserted into a pressure vessel, which is charged to the desired

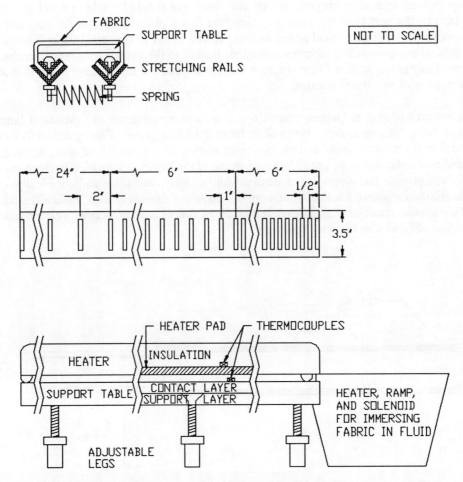

FIGURE 2.   Wettability Experiment Support Table Detail.

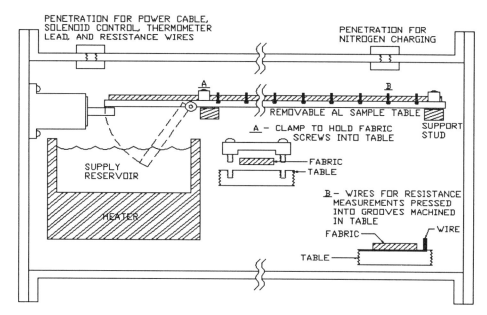

FIGURE 3. Wettability Experiment Pressure Vessel Detail.

operating pressure with nitrogen, similar to figure 3. A supply reservoir containing the working fluid is heated to a temperature which is slightly subcooled (to prevent vapor transport of fluid). The temperature on the table is maintained close to the reservoir temperature through use of the heater sitting on top of the fabric. The heated fluid is then pumped from the reservoir into a bath in which one end of the fabric sample is immersed. The position of the liquid front is then measured by use of a computer data acquisition system. This system takes conductivity measurements through evenly spaced contacts etched onto the table surface, allowing the time at which the liquid front reaches discrete distances to be measured. The instantaneous velocity can be determined by determining the time interval required for the liquid to travel between any two contacts. The liquid front will only advance a certain distance through the fabric before the pressure differential across its length becomes too great to take up any more liquid.

RESULTS

The wettability experiment has been built and experimentation is underway, using water as the working fluid. It is intended that comparisons be made with various material types while holding constant the fabric weight and weave density, with various fabric weaves while holding constant the material type and fabric weave, and at elevated pressures and temperatures for the same fabric. Table 1 displays the test matrix used for these tests.

Cases have been run to date at room temperature and pressure for Silicon Dioxide, S-glass, Silicon Carbide, and Graphite. All these materials are plain weave, yet with different weave sizes and fiber bundle size. Preliminary analysis has indicated the importance of the weave density. The Silicon Dioxide used has a weave density of 42x32 picks per inch. The weave is tight enough to inhibit liquid transport vertically through the wick. Wicking in one layer of this fabric is much slower then that observed in the other fabrics, however use of two layers provides a faster wicking rate then what has been seen for the other variations. Figure 4 graphs time required to wick a given distance against discrete liquid front positions. From this figure, it can be seen that the initial liquid front velocity is relatively large, and decreases in a logarithmic fashion as it penetrates deeper into the unwetted fabric. This decrease is do to the increased pressure drop the liquid supplying the front encounters as it is forced to travel a greater distance.

The Silicon Carbide examined has larger yarn sizes, and a weave density of 18 picks per inch. Wicking in this substance is quite slow for plain weave, with apparent wicking heights of 10 cm observed. Two layers performs somewhat better, with wicking heights around 30 cm. Comparison tests with a Crowfoot-Satin weave show a

TABLE 1. Test Matrix for Wettability Experiment. Variation column describes "independant" parameter to be examined by a particular test grouping. (Note: Kevlar is a registered trademark of DuPont).

Variation	Material Type	Weight $(g/cm^{2)}$	Weave	Pressure (kPa)	Temp (C)
Material	Kevlar	2.7	Plain, 17x17	101	20
	S-glass	2.7	Plain, 17x17	101	20
	SiC	5.2	Plain, 17x17	101	20
Weave Structure	S-glass	2.7	Plain, 17x17	101	20
	S-glass	2.7	4H Satin, 32x32	101	20
	Kevlar	2.7	Plain, 17x17	101	20
	Kevlar	5.2	4H Satin, 17x17	101	20
Pressure	SiO_2	2.7	Plain, 42x32	101	20
	SiO_2	2.7	Plain, 42x32	1000	20
	SiO_2	2.7	Plain, 42x32	1500	20
	Kevlar	2.7	Plain, 17x17	101	20
	Kevlar	2.7	Plain, 17x17	1000	20
	Kevlar	2.7	Plain, 17x17	1500	20
Temperature	SiO_2	2.7	Plain, 42x32	101	20
	SiO_2	2.7	Plain, 42x32	101	95
	Kevlar	2.7	Plain, 17x17	101	20
	Kevlar	2.7	Plain, 17x17	101	95
Pressure and Temperature	SiO_2	2.7	Plain, 42x32	101	20
	SiO_2	2.7	Plain, 42x32	1500	130
	Kevlar	2.7	Plain, 17x17	101	20
	Kevlar	2.7	Plain, 17x17	1500	130

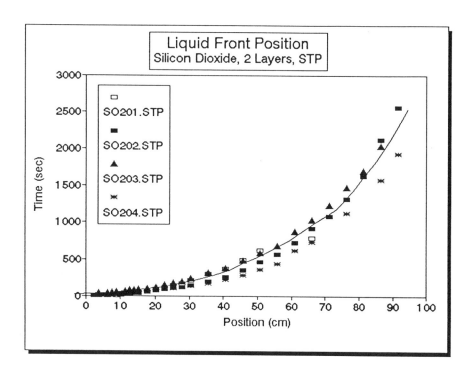

FIGURE 4. Graph of Time Required to Reach Discrete Wicking Heights.

considerable difference. The latter weave results in initial uptake rates about five times those of the former weave. The principle route that this wicking takes has not yet been determined.

The S-glass has a much courser weave, with a density of 18 picks per inch. This fabric has been treated with a sizing which resists wetting, which effectively stops any wicking from occurring. It is thus necessary to heat clean this fabric. The treatment of the fabric is thus an important parameter to consider when determining wicking ability.

Elevated pressure tests have been performed on Silicon Dioxide. Variation of the operating pressure has revealed consistently that the wicking rate is affected strongly by this parameter. Repeated tests have been performed on the Silicon Dioxide, with pressures ranging from atmospheric pressure to 1500 kPa. The time to reach one meter of wicking height at 1500 kPa has been seen to be approximately 25% of that seen at atmospheric pressure (see figure 5).

CONCLUSIONS

The wettability experiment has initially shown that operating pressure is an important parameter to consider when examining wicking rate. It is important to expand this test to determine if some particular parameter of a given material makes it more susceptible to variation of temperature and pressure when examining wicking rate. Preliminary tests have indicated the wide variation of wicking abilities between different materials and weave types. It has been demonstrated that this testing is very important to perform before even beginning to build a heat pipe. It is expected that a small database will be built with the results of these experiments allowing individual comparison of the effects of variation of material types, weave types, and weave densities. This experiment is performed to initially narrow the selection of available materials. Some variation is expected between the wettability determined here and that which might be determined by zero gravity tests. It is thus important to compare those fabrics which have been deemed desirable by these tests under low gravity conditions.

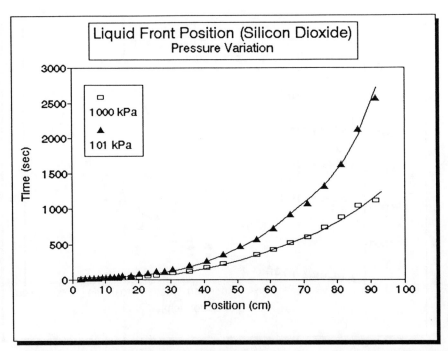

FIGURE 5. Comparison of Elevated and Atmospheric Pressure Wicking Rates.

Acknowledgments

This work was supported by the United States Department of Energy, grant number DE-FG07-89ER12901. Such support does not constitute an endorsement by DOE of the views expressed herein.

References

Clark, D.B. and B. Miller (1978) "Liquid Transport Through Fabrics; Wetting and Steady-State Flow. Part II: Fabric Wetting," Textile Research Journal, 48:256-260.

Dunn, P.D. and D.A. Reay (1982) Heat Pipes, Third Edition, Pergamon Press, Oxford, U.K.

Harnett, P.R. and P.N. Mehta (1984) "A Survey and Comparison of Laboratory Test Methods for Measuring Wicking," Textile Research Journal, 54:471-578.

Hollies, R.S. et al. (1956) "Water Transport Mechanisms in Textile Materials. Part I: The Role of Yarn Roughness in Capillary-Type Penetration," Textile Research Journal, 27:8-13.

Hollies, R.S. et al. (1957) "Water Transport Mechanisms in Textile Materials. Part II: Capillary-Type Penetration in Yarns and Fabrics," Textile Research Journal, 27:8-13.

Jahn, S. (1990) Private Communications, Account Representative, 3M Ceramic Materials Department, Seattle, WA, 22 May 1990.

Marks, T.S. and A.C. Klein (1991) "An experimental Approach to Compare Wicking Abilities of Fabric Materials for Heat Pipe Applications,", Proceedings of the Eighth Symposium on Space Nuclear Power Systems, CONF-910116, M.S. El-Genk and M.D. Hoover, eds., American Institute of Physics, New York.

Symons, E.P. (1974) "Wicking of Liquids in Screens," NASA-TN-7657, May 1974.

SODIUM HEAT PIPE WITH SINTERED WICK AND ARTERY: EFFECTS OF NONCONDENSIBLE GAS ON PERFORMANCE

Robert M. Shaubach and Nelson J. Gernert
Thermacore, Inc.
780 Eden Road
Lancaster, PA 17601
(717) 569-6551

Abstract

High performance heat pipes with arteries are being designed into space nuclear power systems. Artery depriming and subsequent heat pipe failure can occur if noncondensible gas comes out of solution in sufficient quantities to terminate artery liquid return to the evaporator. This paper describes results of analyses and tests to quantify the conditions for which dissolved gas will come out of solution and shutdown the artery. Results show that a stainless steel/sodium heat pipe will operate reliably when gas concentrations in the liquid are kept at a partial pressure of 0.01 bar or less. For example, the heat pipe survived over 25 cycles of start-up from the frozen condition being heated to over 800°C with an evaporator heat flux of 45 W/cm^2 with the evaporator elevated 40-cm above the condenser.

INTRODUCTION

Advanced space nuclear power systems may require high performance heat pipes for efficient transport of thermal energy to fulfill their increased power generation and thermal management requirements. For example, one design for the Idaho National Engineering Laboratory, Small Ex-core Heat Pipe Thermionic Reactor requires radiator heat pipes with an axial heat flux of 5400 watts/cm^2 using sodium as a working fluid (Thermacore 1991). These high performance heat pipes require sintered wick structures with arteries to meet these challenges. The arteries serve as conduits providing low flow resistance to liquid returning from the condenser to the evaporator.

Thermacore has been developing high performance sintered powder metal heat pipe wicks with arteries since 1977. Technology has been developed for in situ sintering of powder metals and base metals ranging from the more easily sintered copper and stainless to refractories, silicon, aluminum, and titanium. These sintered wicks provide a much higher radial and axial heat flux capability than that made possible with screen or groove wicks. Several examples of the performance achieved by Thermacore with sintered powder wicks with arteries are mentioned in the following paragraph.

At room temperatures, heat fluxes of 18 W/cm^2 have been demonstrated in a 3.8-cm diameter copper/water heat pipe with sintered arteries with the evaporator elevated 5-cm above the condenser (Gernert 1990). At temperatures of 1000 K, the sodium heat pipe discussed below operates at 50 W/cm^2 with the evaporator elevated 40-cm above the condenser.

Each of the above mentioned heat pipes operate with significant boiling occurring within the evaporator wick structure next to the arteries. Special design features are required to prevent this vapor from entering the arteries causing them to deprime. Such features include the use of a patented artery liner provided with a pore size smaller than the surrounding coarse sintered powder wick. The surface tension forces provided by the smaller pores force the vapor generated by boiling to flow through the coarse powder around the artery into the heat pipe vapor space, as shown in Figure 1 (Eastman et al. 1989). Artery depriming from vapor penetration into the artery will not occur in a properly designed heat pipe, as shown by tests of heat pipes such as those mentioned above.

Artery depriming can occur, however, if noncondensible gas comes out of solution in the arteries. This occurrence can be avoided most of the time by degassing the wall and working fluid (Dunn 1984), proper cleaning, and selection of a compatible materials system. However, in practice during operation, some gas will most likely show up in the heat pipe vapor space to the degree that the above precautions have been neglected or from such phenomenon as H_2 permeation through the heat pipe wall.

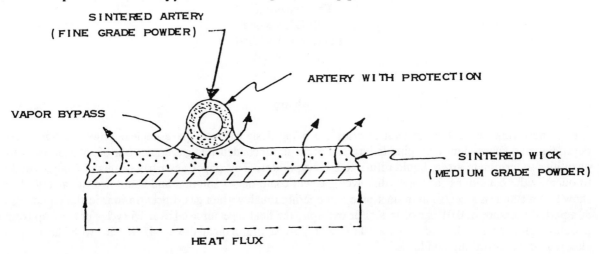

FIGURE 1. Boiling Tolerant Artery.

This paper describes work conducted to determine the effects of noncondensible gas on artery performance. Analyses are provided to establish the conditions for which dissolved gas will come out of solution and shutdown the artery. A series of tests are conducted to demonstrate heat pipe performance with and without gas in the heat pipe vapor space. Conclusions are provided at the end of this paper.

APPROACH

Experimental and analytical approaches were taken to determine the effects of gas on maintaining artery priming. These approaches are outlined, below:

- A representative single artery, stainless steel/sodium heat pipe was designed and fabricated for use as a test article. Geometric parameters are discussed under APPARATUS, below.

- The wick was designed to carry over twice the power with the artery primed as unprimed. The onset of artery depriming is visibly evident from the occurrence of hot spots in the induction heated evaporator when operating at power levels greater than 50 percent. Heat pipe performance with and without the artery primed was predicted by analyses and verified by test. Results are covered under TEST RESULTS, below, and are used to establish the heat pipe operating conditions for the artery depriming tests with and without gas in the condenser.

- Predictions of the dissolved gas concentration that will cause artery depriming are discussed under THEORY, below. Results are used to establish the gas charge and operating conditions for the artery depriming tests.

- The heat pipe was operated with the evaporator elevated 40-cm above the condenser for a number of start-up, steady state and shutdown cycles that included melting and freezing of the working fluid. Power was set at a level for which hot spots would occur in the evaporator at the onset of artery depriming. Data for operation with and without gas are covered under TEST RESULTS, below.

APPARATUS

The test apparatus is shown schematically in Figure 2. The major components are described briefly below:

- The heat pipe was made from 3.81-cm (1½") diameter, schedule 40, 304L stainless steel pipe, 51-cm long. Figure 2 shows wick details including the -150, +200 mesh sintered nickel powder with one 0.318 cm diameter artery surrounded by 600 mesh 304 stainless steel screen. The artery pressure after sintering was 14-cm water column using methanol as a test fluid. About 120 grams of sodium serve as a fluid charge.

- The heat pipe was mounted on a tilt table for purposes of measuring performance as a function of the elevation of the evaporator above the condenser. Liquid and vapor purge vessels were attached to the vapor space and artery penetrations as shown in Figure 2. These vessels were used to add and remove working fluid and gas from the vapor space and arteries.

- A Lepel 20 kW radio frequency generator was used as an induction heat source. A 9 turn, 5.46-cm I.D., 6.35-cm long coil made from 0.476-cm diameter copper tubing was used to couple the RF energy to the heat pipe evaporator. Heat flux levels as high as 60 W/cm² were achieved.

- Instrumentation included a MIKRON, M90-0 pyrometer for measuring temperatures; RF plate voltage and current were used to determine plate power into the coil. Average coil efficiency was measured at 25 percent when accounting for radiant losses directly under the coil.

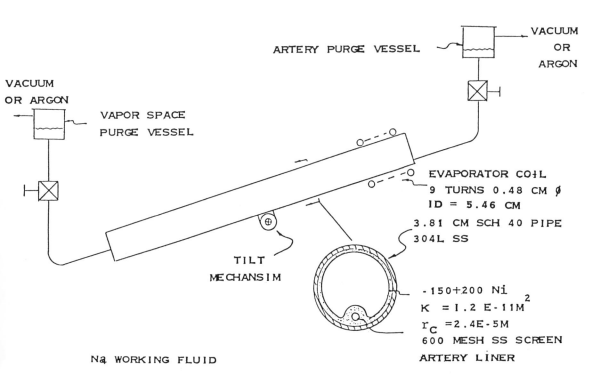

FIGURE 2. 304 SS Heat Pipe Test Apparatus.

THEORY

Heat pipe arteries serve as low pressure drop conduits for transferring liquid condensate from the condenser to the evaporator. Their use allows a significant increase in heat pipe power for a given vapor space diameter.

The artery liquid pressure of an operating heat pipe is generally lower than the heat pipe vapor space pressure due to pressure drops in the condenser wick and artery, as well as from any gravity head. This is shown in Figure 3. This reduced artery liquid pressure plus local liquid heating in the evaporator results in a superheated liquid ready to degas and vaporize at the larger nucleation sites located along the artery wall. Any nucleation site will generate a gas/vapor bubble if its pore radius (r_p) is greater than (Dwyer 1976):

$$r_p > \frac{2\sigma(1-\zeta)}{P_v - P_L}, \quad \text{where}$$

$$
\begin{aligned}
\sigma &= \text{surface tension of the working fluid,} \\
P_v &= \text{saturation vapor pressure of liquid at the liquid temperature,} \\
P_L &= \text{artery liquid pressure,} \\
\zeta &= r_p \, P_g / 2\sigma, \text{ and} \\
P_g &= \text{partial pressure of gas in artery liquid.}
\end{aligned}
$$

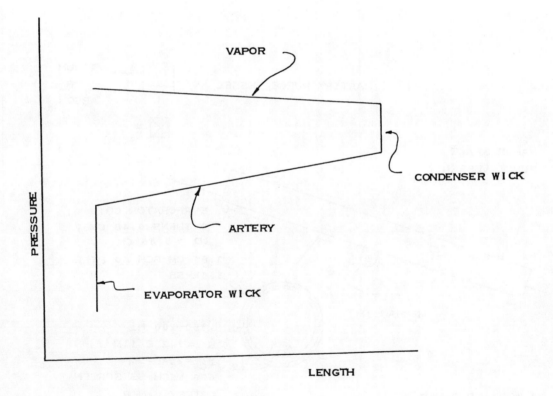

FIGURE 3. Heat Pipe Liquid and Vapor Pressures.

After forming, the vapor/gas bubble will grow quickly until the artery is deprimed terminating the transfer of artery liquid to the evaporator. The term ζ is a dimensionless parameter, which can vary from 0 to over 1 and is considered to be a measure of the dissolved gas content of the liquid. At ζ equal to 1, the partial pressure of dissolved gas, would equal $2\sigma/r_p$. An artery heat pipe will not operate with $\zeta = 1$ because at the slightest superheat the gas will come out of solution and deprime the artery. At $\zeta = 0$, the gas partial pressure would be zero, therefore the artery liquid superheat $(P_v - P_L)$ can reach $2\sigma/r_c$ before vapor depriming occurs.

Only a small fraction of gas is required to dissolve in sodium to achieve a partial pressure equal to $2\sigma/r_p$ or $\zeta = 1$. This is shown for argon in Figure 4 which represents argon pressure versus argon mole fraction for several operating temperatures. Notice that the solubility of argon increases with increasing temperature; more argon will be dissolved in sodium as temperature is increased. Conversely, dissolved argon will tend to come out of solution at nucleation sites in the arteries causing depriming when the heat pipe temperature is reduced.

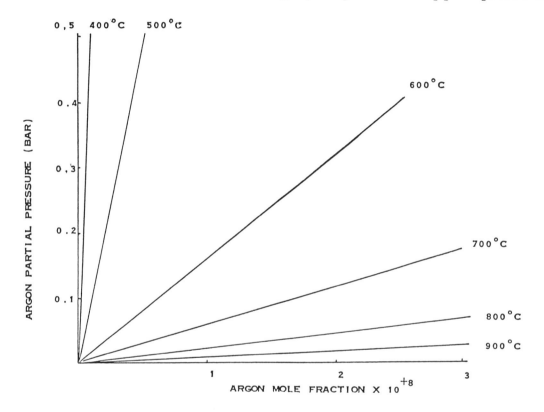

FIGURE 4. Solubility of Argon in Liquid Sodium (Dwyer 1976).

Two competing events occur that set the concentration of inert gas in a heat pipe artery as follows:

- Gas dissolved in the working fluid comes out of solution as the liquid enters the evaporator and is vaporized. The gas is swept by the vapor to the condenser where it collects in the coldest region of the heat pipe at a pressure equal to the heat pipe vapor pressure.

- Some of the gas collecting at the condenser is redissolved in the liquid condensate at the vapor/gas interface. The amount of gas that contacts liquid and is dissolved into the condensate is a function of the amount of gas/liquid contact area in the condenser. This contact area is set by the amount of gas diffusion and turbulence in this region of the heat pipe. Gas that redissolves in the condensate will return to the evaporator via the artery to come out of solution either in the artery causing depriming or during evaporation in the evaporator, as discussed above.

In the limit, it appears that the maximum partial pressure of dissolved gas that could occur in the working fluid would equal the pressure of the gas swept to the end of the condenser. This gas pressure is nearly equal to the saturation pressure of the heat pipe vapor space.

Figure 5 shows a plot of ζ versus temperature, assuming the partial pressure of dissolved gas equals the sodium heat pipe saturation pressure. ζ would equal 1 at a heat pipe operating temperature near 640°C. The heat pipe artery would not function properly at this condition because any artery liquid superheat would result in the formation of a gas/vapor bubble causing artery depriming.

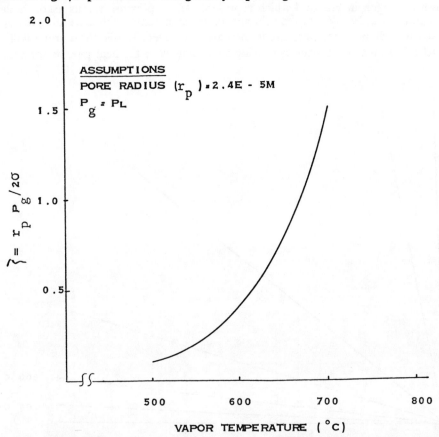

FIGURE 5. Argon Partial Pressure Fraction versus Sodium Vapor Temperature.

The amount of gas/liquid contact area that occurs in the condenser region has not been computed as a part of this effort. Instead, tests have been run with and without gas located in the heat pipe condenser. The purpose of these tests is to experimentally determine if significant quantities of gas are redissolved in the heat pipe working fluid. The effects of any dissolved gas on artery performance will be established by these tests.

TEST RESULTS

Tests were conducted on the stainless steel heat pipe, as follows:

- Peak power carrying capability was measured as a function of evaporator elevation above the condenser with and without the arteries primed.

- Heat pipe performance was measured with gas saturation pressures below 0.01 bar and between 0.01 bar and 1.2 bar.

Results for each of the tests are described separately, below:

Peak Power versus Elevation

Calculations and measurements were made for the 304L stainless steel heat pipe peak power carrying capability with and without the artery primed. Gas concentration levels in the vapor space were maintained below 0.01 bar. These results, shown in Figure 6, provide a baseline for comparison to peak power levels measured during the test program to establish when artery depriming occurs. Notice that the heat pipe transfers almost twice the power with the artery primed as unprimed.

The results, in Figure 6, show that when the artery is primed, the heat pipe evaporator can dissipate over 50 W/cm² when elevated 40-cm above the condenser. Heat flux levels of this magnitude cause nucleate boiling to occur within the evaporator sintered powder metal wick. This boiling condition requires that maximum power be predicted using an analytical tool modeling two phase flow effects. The Thermacore evaporator wick boiling model (Shaubach et al. 1990) was used for this effort. Good agreement exists between measured and predicted results.

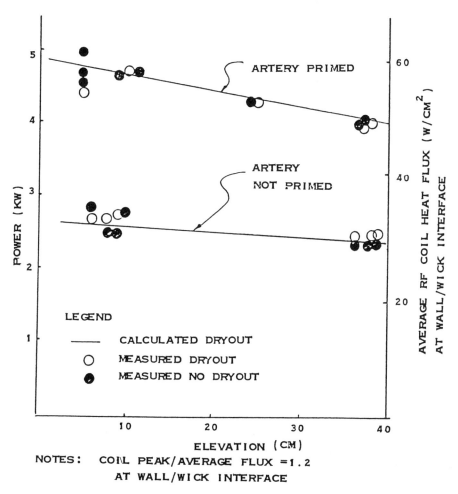

FIGURE 6. Maximum Power with and without Artery Primed.

Effects of Gas Concentration on Power

Over seventy start-up and shutdown cycles were conducted on the stainless steel heat pipe. Test conditions were varied to cover a wide range in parameters, as follows:

Parameter	Range
Power (kW)	0 to 5
Temperature (°C)	20 to 900
Elevation of Evaporator above Condenser (cm)	0 to 40
Argon Pressure in Condenser (bar)	<0.01 to 1.25

The argon pressure of less than 0.01 bar was achieved by operating the heat pipe at 500°C and venting any gas collecting in the condenser vapor space. Room temperature was nominally 20°C. Venting was repeated until steady state conditions were reached.

Argon pressures greater than 0.01 bar were achieved by adding an argon charge to the heat pipe. During operation, this argon was swept by the vapor into the condenser vapor space; a gas column length of 4-cm was visible as a cold end when operating at a vapor temperature of 700°C. The pressure of the argon nearly equaled the saturation pressure of sodium corresponding to its vapor temperature.

The heat pipe was subjected to many different start-up transients; the most severe included twenty-five cycles of the following:

- Heat pipe warmup at a heat flux of 12 W/cm^2 from room temperature to 150°C with the heat pipe evaporator elevated 5 cm above the condenser.

- The evaporator was then elevated 40-cm above the condenser and power was stepwise increased to 45 W/cm^2 (3700 watts).

- After reaching steady state for about 10 minutes, power was terminated and the heat pipe was allowed to cool to room temperature with the evaporator elevated 5-cm above the condenser.

The above transient was selected because it represents an actual artery heat pipe application for a solar receiver that supplies thermal power to a Stirling engine.

Test results are summarized, below:

1. The heat pipe, charged with <0.01 bar argon, operated as designed. The artery did not deprime when the heat pipe was operated within the design limits of Figure 6. It survived all the cold start transients described above, as well as warm restarts from 300 to 400°C. The artery required repriming when the limits of Figure 6 were exceeded.

2. The heat pipe, charged with argon, operated satisfactorily in the steady state but the artery would frequently deprime during start-up. In the steady state, after repriming the artery, the pipe reliably transported power as defined in Figure 6. However, during warm restarts from 300°C, the artery would frequently deprime regardless of the elevation.

CONCLUSIONS

Several conclusions were reached as a result of this test program, as follows:

1. The stainless steel/sodium heat pipe with a sintered wick and artery operated reliably when gas concentrations in the liquid were kept at low levels (ζ <0.1). For example, the heat pipe survived over twenty-five cycles of start-up from the frozen condition being heated to over 800°C with an evaporator heat flux of 45 W/cm² with the evaporator elevated 40-cm above the condenser.

2. The heat pipe, when charged with argon, operated reliably in the steady state with the artery primed. Apparently, the degassing of the working fluid that takes place in the evaporator prevents significant gas buildup from gas/liquid contact in the condenser.

3. The artery of the argon charged heat pipe usually deprimed during warm restart attempts. Apparently, significant mixing of the working fluid and gas occurs during shutdown when sufficient vapor flow is not available to keep the gas trapped in the condenser. The artery liquid could become supersaturated with gas such that bubbles are formed during cool down. These bubbles would remain during heat up resulting in artery depriming.

4. An analytical expression has been identified for estimating when gas will come out of solution in an artery. This expression will be useful in future work to better quantify the amount of gas that can be tolerated in artery heat pipes. Meanwhile, proper attention to cleaning, degassing, and material selection is mandatory.

Acknowledgments

Thermacore, Inc. funded the work described in this paper.

References

Dunn, P. D., and D. A. Reay, (1982) Heat Pipes, Pergamon Press, Elmsford, NY.

Dwyer, O. E., (1976) Boiling Liquid Metal Heat Transfer, American Nuclear Society, LaGrange Park, IL, pg. 9.

Eastman, G. Y., D. M. Ernst, R. M. Shaubach, and J. E. Toth (1989) Advanced Heat Pipe Technology for Space Heat Transport and Rejection Technologies, Space Power Conference, Cleveland, OH, June 1989.

Gernert, N.J. (1990) High Performance Copper/Water Heat Pipe, by Thermacore, Lancaster, PA, Report prepared for NASA Lewis Research Center, April 1990.

Shaubach, R. M., P. M. Dussinger, and J. E. Bogart, (1990) Boiling in Heat Pipe Evaporator Wick Structures, *Proceedings Seventh International Heat Pipe Conference* held in Minsk, USSR, May 1990.

Thermacore (1991) Thermionic Heat Pipe Module, C91-103269-DAJ-17-91, Thermacore, Lancaster, PA, April 1991.

HIGH TEMPERATURE CAPILLARY PUMPED LOOPS
FOR THERMIONIC POWER SYSTEMS

William G. Anderson
Thermacore, Inc.
780 Eden Rd.
Lancaster, PA 17601
(717) 569-6551

Jerry E. Beam
Wright Laboratory
Wright-Patterson AFB, OH 45433
(513) 255-6241

Abstract

A high temperature capillary pumped loop (CPL) is being developed for thermal control applications in space power systems. Space power systems where capillary pumped loops can be used include the SP-100 system, the SDIO Multi-Megawatt system, and the Air Force Thermionic Nuclear Power System. The advantages of a CPL include the ability to subcool liquid before it enters the evaporator, reducing the chance of boiling in the evaporator artery. A second major benefit is that only the evaporator and condenser need a wick, simplifying fabrication of the adiabatic section. Finally, it is easy to connect multiple evaporators and condensers with a single vapor and liquid line. Several high temperature capillary pumped loops were fabricated with sodium as the working fluid. With a 1.75 m long vapor line, a CPL carried 5300 W near 1000 K, with an evaporator heat flux of 60 W/cm^2. The evaporator was elevated 15 cm above the condenser, which was the maximum possible due to mechanical limitations of the experimental setup.

INTRODUCTION

Thermacore and Wright Laboratory are currently working to develop a high temperature capillary pumped loop (CPL) for thermal control applications. Space power systems where capillary pumped loops can be used include the SP-100 system, the SDIO Multi-Megawatt system, and the Air Force Thermionic Nuclear Power System. This paper will examine the application of capillary pumped loops in thermionic nuclear power systems.

Capillary Pumped Loops

A capillary pumped loop is a device for efficiently transferring heat with a low temperature drop, by evaporating and condensing a fluid in a closed loop. A typical capillary pumped loop is shown in Figure 1. Heat supplied to the evaporator vaporizes the liquid working fluid located in the evaporator wick. The vapor travels through the vapor line to the condenser. Due to heat losses, a portion of the vapor condenses in the vapor line, while the remainder condenses in the condenser. Shear stresses applied by the cocurrent vapor flow drag the liquid film on the walls toward the end of the condenser.

The liquid collecting in the condenser is returned to the condenser through a liquid return line, which feeds the working fluid into the evaporator wick. The liquid flow is driven by capillary forces at the wick/liquid/vapor interface, which pump the liquid back into the wick, overcoming the pressure drops in the CPL.

As can be seen in the brief description above, a capillary pumped loop is similar to a heat pipe in many respects. However, there are several advantages to a loop design for transferring heat over long distances. One advantage is the ability to subcool liquid before it enters the evaporator, thus reducing the chance of boiling in the evaporator artery.

A second major benefit is the simplification of the wick design, with the associated benefit of easier fabrication. In a typical long heat pipe, the vapor space between the evaporator and condenser is surrounded by some type of wick for the countercurrent flow of liquid from the evaporator to the condenser. Typical wicks are grooves, or sintered porous metal wicks with an artery. Since the liquid flow is opposite to the vapor flow, the wick must be

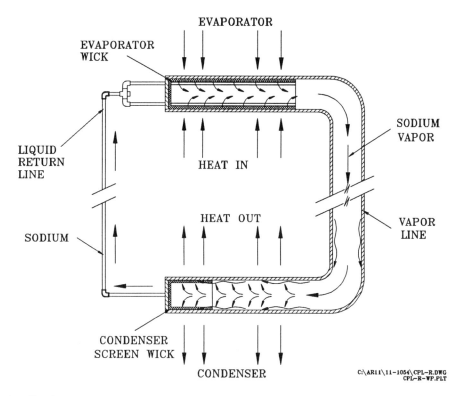

FIGURE 1. Capillary Pumped Loop.

designed to prevent the possibility of entrainment. On the other hand, a CPL requires a wick only in the evaporator and condenser. A wick is not required in the vapor line, since the vapor drags any condensed liquid toward the condenser. In contrast to the relatively complicated wick in a heat pipe, the vapor line and liquid line in a CPL are only "tubes", and can be easily fabricated with bends in the line. Since a wick is not required in the vapor line, the mass of the CPL is also reduced when compared to a typical CPL with a screen or sintered powder metal wick.

THERMIONIC SYSTEM COOLING

There are two basic types of thermionic systems used with nuclear reactors; (1) an in-core conversion configuration and (2) an out-of-core configuration. In one out-of-core design, the core is radiatively cooled through flat-plate thermionic devices that are located around the core. This out-of-core system is not considered further here, since no heat transfer loop is required for operation with this configuration.

The in-core thermionic system employs cylindrical fuel/converter elements that are stacked on end in a manner similar to flashlight cells. The converter stack is encased in a metallic cladding to form a thermionic fuel element (TFE). During power operation, heat from the fuel boils electrons off the emitter surface (1800 K). These electrons are captured on the collector (1000 K) to provide thermal-to-electric power conversion at about 8.5 percent efficiency. The waste heat must then be carried away from the collector and radiated to space.

There are several different concepts for transporting the waste heat, including electromagnetic (EM)-pumped loops, heat pipes, and capillary pumped loops. In an EM-pumped loop, the waste heat is carried away by flowing liquid NaK coolant. The advantages of replacing the EM-pumped loop with a CPL (or heat pipes) include the reduction of mass by eliminating the EM pump, the expansion compensator, and most of the liquid metal inventory. The loop temperature drop is also reduced, increasing the efficiency and reducing the size of the radiator. The CPL requires no electrical power for operation.

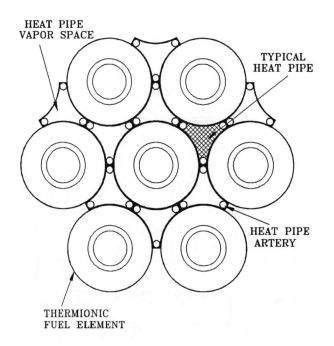

FIGURE 2. Rockwell Heat Pipe Cooled TFE Concept (From Rockwell 1991).

An example of a heat-pipe cooled system is the Rockwell/General Atomic heat-pipe-cooled TFE design (Rockwell 1991). A cross-section is shown in Figure 2. In this design, each TFE is surrounded by six heat pipes to form a hexagonal shaped heat transfer module. Outside of the core, these heat pipes are connected to a secondary heat pipe which transfers heat to the radiator.

One advantage of a capillary pumped loop is that it is easy to connect multiple evaporators to a single vapor line. Each heat pipe in the Rockwell concept could act as a CPL evaporator. Multiple evaporators, on the order of six, would be connected in parallel to a single vapor line that would deliver the heat to the radiator. A single liquid return line would branch into multiple arteries just before reaching the evaporators. An advantage of this concept is that it eliminates the temperature drop between the primary and secondary heat pipes.

To insure redundancy, the six evaporators connected to the vapor line would cool six different thermionic fuel elements. The failure of a CPL would only reduce the cooling for each TFE by 1/6.

HIGH TEMPERATURE CPL DESIGN

To date, most capillary pumped loops have been built for operation near room temperature. Ku et. al. (1987) and Gottschlich (1989) provide a brief overviews of capillary pumped loop technology. A typical ambient temperature CPL contains a reservoir attached to the liquid return line. The reservoir is used for priming the evaporators. By heating or cooling the reservoir, the CPL can be operated as a variable conductance device, controlling the liquid inventory and CPL temperature. Increasing the reservoir temperature moves fluid from the reservoir to the condenser, partially blocking the condensers and reducing the heat removal rate. More recently, several ambient temperature capillary pumped loops have been designed that do not require a reservoir(Maidanik 1990 and Gottschlich and Richter 1991). While the system is no longer a variable conductance device, the advantages of removing the reservoir include decreased system complexity, and decreased mass.

Thermacore and Wright Laboratory are developing high temperature capillary pumped loops for eventual application in space power systems. In contrast to previous ambient temperature capillary pumped loops, these CPLs use an alkali metal as the working fluid. Typical operating temperatures range from 975 to 1075 K, which

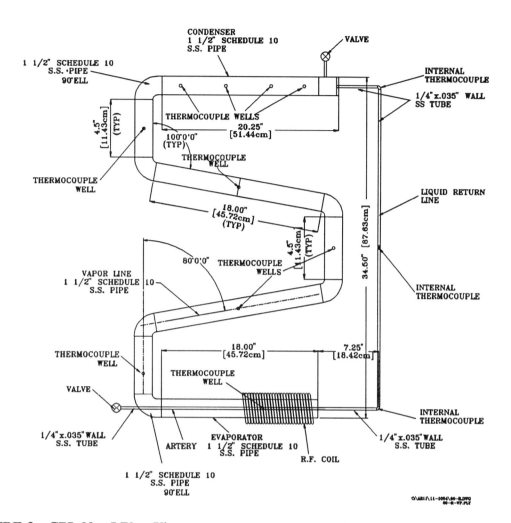

FIGURE 3. CPL No. 5 Plan View.

is in the range required for cooling thermionic power systems. A number of high-temperature, bench-scale capillary pumped loops have been fabricated and tested. A plan view of one CPL is shown in Figure 3. Heat is supplied with a radio frequency (RF) coil to the CPL evaporator. Sodium is evaporated, and travels through the vapor line to the condenser, where heat is removed by radiation and natural convection. Liquid returns from the condenser to the evaporator through a 4.6 mm ID tube. These proof-of-concept loops use stainless steel tube, for ease in fabrication, and low cost. No attempt was made to minimize mass, which would be required for space applications.

The CPL evaporator, shown in Figure 4, uses a boiling tolerant artery design initially developed by Thermacore for ambient temperature heat pipes(Thermacore 1989). In the CPL program, this artery design was successfully applied to heat pipes and CPL evaporators. In a boiling tolerant artery, the artery is surrounded with a sintered powder, screen, or slotted tube with pore radii that are substantially smaller than the pores in the bulk distribution wick. At high heat fluxes, vapor generated by boiling in the bulk wick is diverted by capillary forces around the artery, preventing artery blockage, see Figure 5. The current evaporator design uses a fine mesh screen artery to provide boiling tolerance, with a single powder size used for the remainder of the wick. The wick is sintered from -150+200 nickel powder, and the artery is formed from **a 600-mesh 316-stainless-steel screen.**

The CPL was fabricated with the screen wick condenser design shown in Figure 6. In the condenser, the walls near the liquid return line are covered with 5 layers of 100 x 100 mesh, 316 stainless steel screen. Liquid is wicked into this screen, and returned to the liquid line, while non-condensible gas is trapped in the condenser. The walls

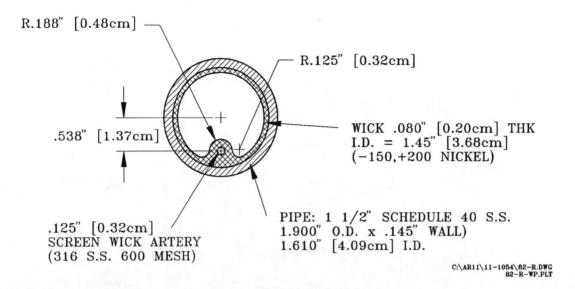

FIGURE 4. Boiling-Tolerant CPL Evaporator Cross-Section.

in the remainder of the condenser are bare. In a CPL designed for space, these walls would contain grooves to wick the fluid toward the liquid return line. The bench-scale CPL does not contain these grooves, to reduce fabrication costs.

Heat to the evaporator is supplied by a radio frequency (RF) induction heating coil (450 kHz), placed around the evaporator. Supplemental heating is provided by beaded resistance heaters wrapped around the vapor line and condenser, and cable heaters wrapped around the liquid return line. The supplemental heaters are used during start-up, and are shut-off during operation.

The vapor line (adiabatic section) is roughly 1.75 m long, and is covered with a 2.54-cm layer of fiberfrax insulation to minimize heat loss. Most of the heat supplied to the CPL is rejected in the condenser, by convection and radiation. The vapor line is shaped like a "W" to reduce the size of the overall layout. It also demonstrates that the CPL can be bent, if desired. The elbows used in the center of the vapor line are 80° elbows. This tilts the horizontal runs in the vapor line slightly, to reduce liquid holdup if the system is operated as a gravity-aided heat pipe during start-up.

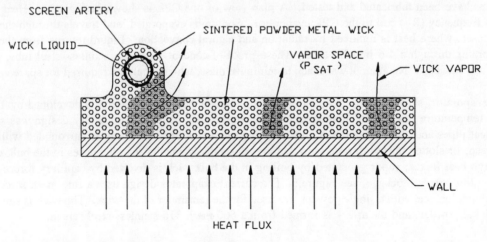

FIGURE 5. Boiling Tolerant Artery Design for CPL Evaporators.

1166

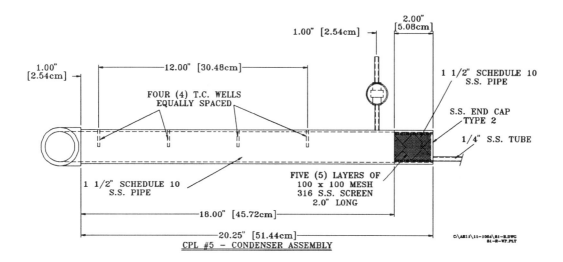

FIGURE 6. CPL No. 5 Condenser, with Screen Wick.

The liquid return line is completely separate from the vapor line. In a space application, the liquid line would probably travel along the vapor line, providing trace heating for the line. In this demonstration CPL, the liquid line is separated to more easily determine when sodium is flowing in the liquid line.

HIGH TEMPERATURE CPL OPERATION

Figure 7 shows temperature profiles, power, and evaporator elevation for the CPL during high heat flux tests. The evaporator was initially elevated 2.54 cm above the condenser. The power supplied to the evaporator was increased to 5300 W, then the evaporator was elevated above the condenser in 1 inch increments, as shown in Figure 7. The elevation was increased to 15 cm (6 in), which is the maximum possible due to mechanical limitations of the experimental setup. With the evaporator elevated 6 inches above the condenser, the CPL carried 5330 W with a heat flux into the evaporator of 60 W/cm². The length of the CPL adiabatic section is 1.75 m, giving a heat transport factor of 9300 W-m.

Temperature in the liquid return line, and RF heat flux, are plotted in Figure 8. The temperature in the liquid line near the condenser is slightly less than the evaporator (and condenser) temperature, dropping to 910 K near the first elbow. As the liquid travels through the line, it gradually cools, until it enters the evaporator with a temperature near 660 K. Note that there was no electric resistance heat on the liquid return line or the vapor line during these tests.

Using supplemental electric heaters, the CPL has been successfully started from room temperature, with the evaporator elevated above the condenser. This demonstrates that gravity is not required for operation of the CPL. The CPL has also been operated with the evaporator level with the condenser. In this orientation, liquid condensing in the vapor line is driven to the condenser by shear forces from the cocurrent vapor flow.

FUTURE WORK

Now that several bench-scale loops have been built and operated successfully, construction has started on a 20-kW CPL using sodium as the working fluid. The design heat flux in the evaporator is 50 W/cm², at an operating temperature of about 1000 K. The evaporator and condenser are each 1 meter long, and the vapor line and liquid return line are 6 meters long. The evaporator has 4 boiling tolerant arteries, which are identical to the arteries used

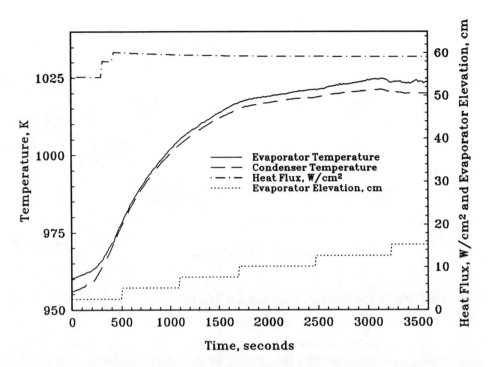

FIGURE 7. CPL Temperatures, Evaporator Heat Flux and Elevation for the Bench-Scale CPL with a Screen Wick Condenser. Evaporator Power: 5300 W; Evaporator Heat Flux: 60 W/cm².

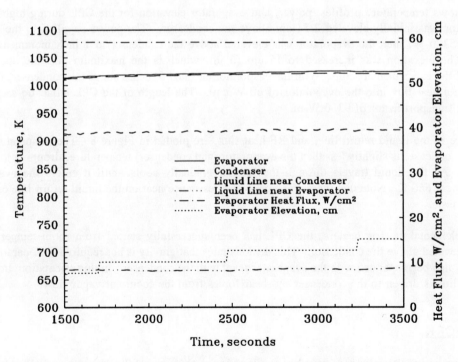

FIGURE 8. Evaporator, Condenser, and Liquid Return Line Temperatures for the Bench-Scale CPL with a Screen Wick Condenser. Evaporator Power: 5300 W; Evaporator Heat Flux: 60 W/cm².

in the bench scale loops. Heat to the evaporator is supplied by two sources, an RF coil to supply 4-5 kW at the design high heat flux, and a lamp assembly to supply up to 20 kW at a lower heat flux. Heat is removed from the condenser with a gas-gap calorimeter.

CONCLUSIONS

Space power systems where capillary pumped loops can be used include the SP-100 system, the SDIO Multi-Megawatt system, and the Air Force Thermionic Nuclear Power System. The advantages of replacing an EM-pumped single phase loop with a CPL include a reduction in mass, and an increase in thermal efficiency. A CPL also has several advantages when compared with a heat pipe. Liquid can be subcooled in the vapor line, reducing the chance of boiling in the evaporator artery. A second major benefit is that only the evaporator and condenser require a wick. This simplifies fabrication, reduces mass, and allows the easy connection of multiple evaporators to a single vapor line. By replacing the heat pipes in a thermionic system with a multiple evaporator CPL, the number of vapor and liquid lines can be reduced, while still maintaining redundant cooling of each thermionic fuel element.

High temperature capillary pumped loops have been fabricated with sodium as the working fluid. The CPLs operated near 1000 K, with evaporator heat fluxes of up to 60 W/cm^2. With a 1.75-m long vapor line, a CPL carried 5300 W, giving a heat transport factor of 9300 W-m. The evaporator was elevated 15 cm above the condenser, which was the maximum possible due to mechanical limitations. A 20-kW CPL is currently being fabricated, with an vapor line length of 6 m.

Acknowledgments

The experimental work in this document was performed by Thermacore, Inc., Lancaster, PA, under contract F33615-89-C-2913 for the Aero Propulsion and Power Laboratory, Wright Laboratory, Aeronautical Systems Division, United States Air Force, Wright-Patterson AFB, Ohio. Funding was provided by the SDIO SBIR program. The technical monitor was A.S. Reyes. The capillary pumped loops were fabricated at Thermacore by Terry Duncan.

References

Gottschlich, J. M. (1989) "Capillary Pumped Loops for Aerospace Application," Society of Automotive Engineers Paper 892318, in Proc. Aerospace Technology Conference and Exhibition, Anaheim, CA, 25-28 September 1989.

Gottschlich, J. M. and R. Richter (1991) "Thermal Power Loops," Society of Automotive Engineers Paper 911188, in Proc. 1991 SAE Aerospace Atlantic, Dayton, OH, 22-26 April 1991.

Ku, J., E. J. Kroliczek, and R. McIntosh (1987) "Capillary Pumped Loop Technology Development," in Proc. Sixth International Heat Pipe Conference, Grenoble, France, 25-29 May 1987.

Maidanik, Y. F., et al. (1990) "Development, Analytical and Experimental Investigation of Loop Heat Pipes," in Proc. Seventh International Heat Pipe Conference, Minsk, U.S.S.R., 21-25 May 1990.

Rockwell (1991) "Heat Pipe Cooled Thermionic Fuel Element Technical Review", Phillips Laboratory, Albuquerque, NM, 9 April 1991.

Thermacore (1989) Vapor Resistant Arteries, U.S. Patent No. 4,815,528, 28 May 1989.

SULFUR HEAT PIPES FOR 600 K SPACE HEAT REJECTION SYSTEMS

John H. Rosenfeld, G. Yale Eastman, and James E. Lindemuth
Thermacore, Inc.
780 Eden Road
Lancaster, PA 17601
(717) 569-6551

Abstract

A preliminary investigation was performed to study the use of sulfur heat pipes in a lightweight space radiator for waste heat rejection at 600 K. Several space power concepts have a need for heat rejection at 600 K. Heat pipes have been shown in previous studies to be useful in reducing the mass of radiators; however, few high-performance, lightweight working fluids are available near 600 K. Sulfur has not been previously suggested as a heat pipe working fluid for this application because of its high liquid viscosity, which reduces heat pipe transport capability. However, it has been shown that the addition of several weight percent iodine to the sulfur has the effect of reducing dynamic viscosity by three to four orders of magnitude near 600 K. The addition of iodine significantly reduces the liquid pressure drop flow resistance in sulfur heat pipes. This appears to make sulfur heat pipes a viable approach for 600 K space heat rejection systems. Preliminary design calculations were performed to determine a preferred geometry and approximate weight of a sulfur heat pipe radiator operating at 600 K. Calculations were based on the requirements of SP-100 heat rejection. It was demonstrated that a radiator specific mass of 5 to 20 kg/m^2 is possible with a sulfur heat pipe radiator, depending on the required heat pipe lengths and heat transport requirement. Applications for several space power concepts are discussed, and compatibility with lightweight radiator materials is reviewed.

BACKGROUND

Several concepts for space power generation include a need to reject significant quantities of waste heat at temperatures near 600 K. These concepts include a nuclear powered Stirling engine system, a Brayton cycle concept, and a solar Rankine cycle. The current baseline heat rejection concept at 600 K for large space power systems is a NaK/Li liquid pumped loop. Although this approach is a low mass design, it is not a passive system, and the reliability for space systems has not been demonstrated.

Heat pipe radiators have been shown in previous studies to be useful in reducing the mass of radiators; however, few high performance lightweight heat pipe working fluids are available near 600 K. The known candidate fluids include mercury, cesium, rubidium, and sulfur.

Sulfur has historically been discounted as a potential heat pipe working fluid for three reasons: high liquid viscosity (preventing adequate condensate return), low thermal conductivity (producing undesirably high temperature gradients in the evaporator and condenser zones) and high chemical reaction rates (Lekae and Elkin 1960) with most potential wick and containment vessel materials (resulting in short operating life). It is believed that the technology now exists to overcome these adverse effects, so that heat pipes in the 600 K temperature range can now be considered seriously.

Of the four candidate fluids for operation at 600 K, cesium and rubidium have favorable liquid properties (such as viscosity, latent heat of vaporization, and surface tension), but have vapor pressures too low to be able to transport significant power. For heat pipes, this results in a low sonic vapor limit to heat pipe performance (Figure 1). The only solution to this problem is to design the heat pipe with a large diameter to accommodate the low vapor pressure, which significantly increases the mass of the radiator. Mercury also has many favorable

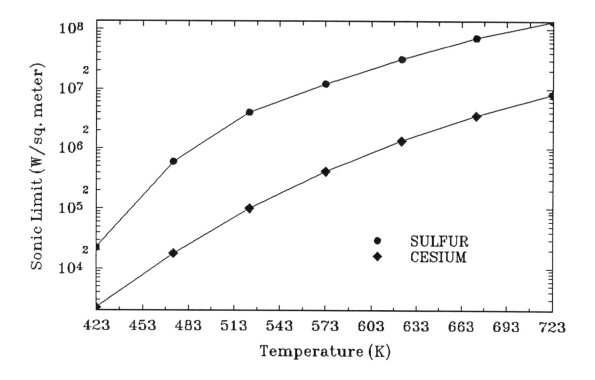

FIGURE 1. Sonic Limits for Sulfur and Cesium.

properties, including an appropriate vapor pressure (boiling point 357 C). However it is toxic, has a high liquid density that has tended to make it unacceptable for low mass structures like space radiators, and it has proven intractable in efforts to achieve the reproducibility of wetting angle that would make possible the use of its high surface tension. Additionally, mercury is apparently not compatible with low mass wall materials such as aluminum and titanium.

Although sulfur has previously been thought of unfavorably as a working fluid, it has recently been recognized that it may offer attractive performance at temperatures at which no other fluids can operate effectively. Earlier efforts by Ernst (1991) and Eastman (1991) showed sulfur to be compatible with aluminum in tests of 1000 hours related to its use in heat pipe heat exchangers for industrial heat recovery. However, due to a unique temperature-dependent polymerization property, sulfur has a liquid viscosity peak at about 475 K (200 C) that places it at a level about three orders of magnitude higher than the maximum level needed for useful heat pipe operation. Even at 575 K (300 C), the viscosity is still an order of magnitude too high (see Figure 2). Polasek (1975) and Polasek and Stulc (1976) showed that additions of small (3-10%) quantities of iodine will reduce the viscosity of sulfur by more than three orders of magnitude in the temperature range of interest, bringing it at a level where useful heat pipe operation can be expected. Timrot et al. (1978) confirmed the Polasek and Stulc (1976) findings on viscosity, showed some decrease in surface tension and indicated potential compatibility with chrome aluminum alloys and titanium as well as several non-metals. Figure 2, taken from Polasek and Stulc (1976), shows variation of liquid viscosity with temperature for pure sulfur and for sulfur/iodine mixtures. The role that iodine serves is to prevent formation of liquid-phase polymeric chains of sulfur, thereby causing the dramatic viscosity reduction.

Although Ernst (1991) and Eastman (1991) have shown short term compatibility of sulfur with aluminum, the compatibility of the sulfur/iodine mixture aluminum or other potential heat pipe wall/wick materials has not been established. Nickel-iron alloy was not recommended by Timrot, et al. (1978) in spite of earlier favorable reports by Fanelli (1946) and Polasek and Stulc (1976). Other materials that may be compatible include titanium, zirconium, aluminum oxide, silicon or silicon carbide and boron or silicon nitrides. It is thought that titanium and zirconium may form stable, thin sulfide layers that will prevent further attack, much as an aluminum oxide layer protects aluminum. The non-metals may be useful in the form of thin protective layers on metallic substrates.

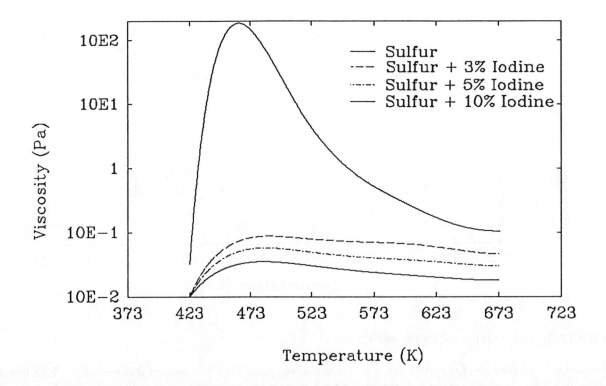

FIGURE 2. Effect of Iodine Additions to the Dynamic Viscosity of Sulfur.

With regard to thermal conductivity (Doi 1963), sulfur in its normal condition is generally considered an insulator. However, iodine (Horvath 1975) has a high thermal conductivity, comparable to silicon or tungsten and about one-third of copper and silver. No information is known to be available on the thermal conductivity of sulfur/iodine mixtures. However, if the conductivity were shown to be governed by the proportions present, a 10% iodine solution in sulfur would have a conductivity comparable to mercury, or about 12 W/m-K. Such a large thermal conductivity would greatly increase the usefulness of sulfur as a heat pipe fluid, since thermal interface dimensions could be made smaller without incurring excessive temperature losses. Scoping calculations have shown that the pool boiling critical heat flux for sulfur is over 100 W/cm^2 at 600 K. This further increases the potential for reducing thermal interface size and mass with sulfur heat pipes.

The density of sulfur is 1.2 g/cc at 600 K and iodine is 3.7 g/cc at 455 K. Therefore, the density of a sulfur/ 10% iodine mixture would be on the order of 1.45 g/cc, which is lower than cesium and is quite attractive for radiator service.

In summary, in spite of a limited data base (Karashaev et al. 1973; Ono and Matsushima 1957; Bacon and Fanelli 1943; Fanelli 1950; Niselson et al. 1974; and Raw et al. 1973), there is sufficient background information to indicate that it is worthwhile to investigate the sulfur/iodine system as a potential working fluid for spacecraft radiator heat pipes. Recent work by Rosenfeld and Lindemuth (1991) at Thermacore for SPI has collected the known data on the sulfur/iodine system and analyzed its potential for use in spacecraft radiators. No other work in the field is known in the past fifteen years, there being no apparent follow-on to the Czechslovakian (Polasek and Stulc 1976) or Soviet (Timrot et al. 1978).

Radiator Design Study Results

Scoping calculations were performed to compare the mass of a typical large SP-100 sulfur heat pipe radiator with that of one current alternative leading candidate, for example, a NaK/Li single-phase EM pumped loop designed by Space Power, Inc. (SPI) of San Jose, California. The NaK/Li pumped loop specific mass has been estimated at nearly 6 kg/m².

The scoping design of the sulfur heat pipe radiator was based on a geometry similar to that developed for an alkali metal heat pipe radiator at 600 K. In earlier work funded by the NASA Lewis Research Center, Thermacore worked with SPI to size a cesium heat pipe radiator for an SP-100 application (Singh and Rosenfeld 1987). The overall radiator configuration is shown in Figure 3. It consists of a series of long parallel heat pipes that dissipate waste heat via radiation to space. The heat pipes form a large hollow cylinder. Carbon/carbon radiating fins are attached to each heat pipe to reduce radiator mass by increasing heat pipe spacing. The earlier calculations showed that a cesium heat pipe radiator would have a larger mass then the EM pumped loop. The purpose of the recent calculations was to examine the usefulness of sulfur as the working fluid in a large heat pipe radiator operating at 600 K.

Due to sulfur's relatively large liquid viscosity, one important feature of a high-performance sulfur heat pipe will be a high permeability wick structure. In several earlier development programs at Thermacore, a wick structure that has been shown to have a low resistance to liquid flow is an open annulus wick formed by a screen spacer and the inside wall of the heat pipe (Figure 3). The scoping calculations were based on the open annulus wick for the radiating section of the heat pipes.

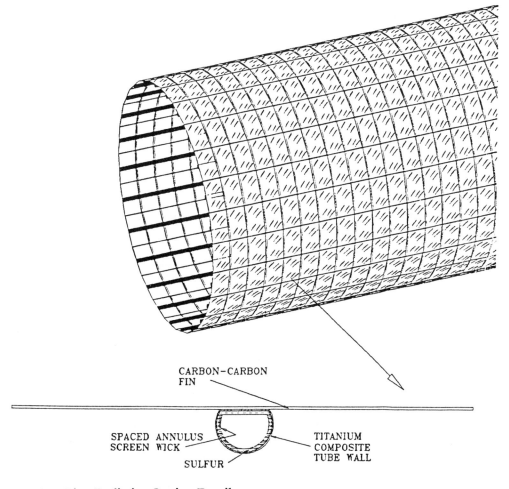

FIGURE 3. Heat Pipe Radiating Section-Detail.

1173

Another critical design area for sulfur heat pipes is the number and type of joints in a long segmented radiator. The large cylindrical radiator shown in Figure 3 has a required total length of nearly 30 meters. Fabrication limits and performance limits will both control the number and type of joints for a segmented radiator. Due to sulfur's unique combination of properties, a long sulfur heat pipe radiator may have many separate heat pipes joined in series to form the radiator. One typical joint design is shown in Figure 4. It consists of a series of parallel aluminum struts (fins) which provide extended surface for heat transfer and supports for the annulus wick structure. Aluminum is suggested for a strut material due to its high thermal conductivity.

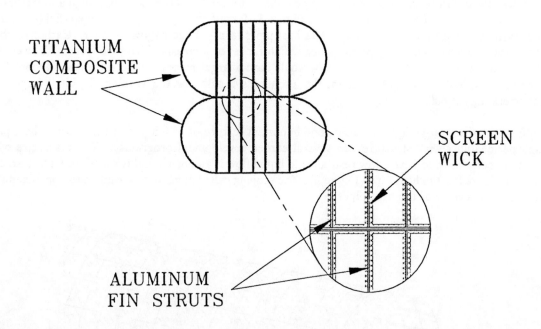

FIGURE 4. Joint Cross Section Detail.

One reason for dividing long radiators into segments is due to sulfur's relatively large liquid viscosity. Shorter segments result in lower liquid return drop for a fixed evaporator capillary pressure rise, which allows a thinner, lower-mass annulus for the liquid return wick.

Scoping calculations were performed to estimate the mass of a large sulfur heat pipe radiator using the design described above. The design was based on a radiator 30 m in length and 5 m in diameter, with each heat pipe carrying 20 kW of power away from the heat source. Mass was estimated for radiators with two, four, and ten segments. For a radiator with two segments, a specific mass of 10 to 20 kg/m² appears feasible. Calculations also showed that the mass can potentially be significantly reduced by increasing the number of segments; the mass could be reduced by up to 20% with four segments, and by up to 40% with ten segments. However, increasing the number of segments will also increase the thermal resistance due to interfaces, which can be strongly affected by a low thermal conductivity working fluid. Thus, an accurate radiator mass estimate will depend on accurate physical property data for the binary sulfur/iodine mixture.

The above scoping calculations demonstrate that a large cylindrical sulfur heat pipe radiator can potentially be produced with an areal specific mass similar to a NaK/Li single-phase pumped loop radiator. This result is significant because a heat pipe radiator is a passive heat rejection device. No auxiliary power, pumps, or mechanical parts are present which could potentially fail in a long-term use space application. Furthermore, smaller heat rejection systems such as those required for a 5 kWe to 50 kWe space power source are more likely to find a sulfur heat pipe radiator to be a favorable choice. This is true because smaller systems can feature shorter heat pipes with lower attendant mass.

In order to be able to predict the performance of sulfur/iodine heat pipes with acceptable accuracy, the physical properties of the working fluid must be quantified. In addition, suitable wall/wick materials must be identified and their compatibility confirmed. No data have been found for the thermal conductivity of sulfur/iodine solutions. Consequently, collection of data is recommended for the conductivity of representative sulfur/iodine solutions. Limited early work indicates that liquid sulfur is compatible with aluminum and possibly with titanium. No information is available in the literature regarding compatibility of sulfur/iodine solutions with these, or other, materials. Therefore, it is recommended that a useful data base be generated by conducting 1000 hour life tests followed by post mortem analysis.

The thermal conductivity of sulfur/iodine mixtures should be measured in the 600 K range. Mixtures of interest are expected to lie in the 5-20% iodine range. As a minimum, measurements should be carried out at 500, 600 and 700 K with solutions of 5, 10 and 20% iodine in sulfur.

Heat pipes to be life tested should include, but not be limited to, aluminum and titanium or their alloys.

The test vehicle could be wickless, gravity-aided heat pipes made of the candidate material to be tested. The heat pipes could be loaded with a selected sulfur/iodine working fluid in the range of 5-20% iodine, evacuated, sealed, tested and subjected to life testing at the design condition of 600 K. A failure of the tube wall or an increase with time in the thermal resistance across the heat pipe can serve as the diagnostic tools to indicate chemical incompatibility.

CONCLUSIONS

Sulfur heat pipe radiators appear to be a viable candidate for low-mass, reliable heat rejection systems near 600 K. The addition of 3% to 20% by weight has the effect of reducing sulfur liquid viscosity to acceptable levels. Depending on the liquid thermal conductivity and related transport properties of the sulfur/iodine binary mixture, a sulfur heat pipe radiator can potentially be designed with a specific mass of 5 to 20 kg/m^2 or lower, depending on the design requirements. Scoping calculations have shown that a sulfur heat pipe radiator can potentially be an attractive alternative to a NaK/Li pumped loop for heat rejection from small to moderately large space power systems. Accurate mass estimates for large radiators are not currently possible because complete transport property data are not available for the sulfur/iodine binary system, and because compatibility data with lightweight radiator wall materials has yet to be demonstrated. Collection of thermal conductivity data for sulfur/iodine mixtures and assembly of sulfur heat pipe compatibility life tests are recommended. Investigation into fluids other than iodine to reduce sulfur viscosity is also recommended. Short heat pipe section lengths and high permeability wick structures will also be important in minimizing sulfur heat pipe radiator mass. Development of innovative concepts in these areas will also help to increase the number of applications for sulfur heat pipes in space heat rejection systems.

Acknowledgments

The results reported herein resulted from a preliminary design study performed by Thermacore, Inc. for Space Power, Inc. The work was partially funded by the NASA Lewis Research Center, Cleveland, Ohio.

References

Bacon, R.F. and R. Fanelli (1943) "The Viscosity of Sulfur," J. Am. Chem. Soc., 65,: 639-648.

Doi, T. (1963) "Physico-Chemical Properties of Sulfur," Rev. of Phys. Chem. Jap., 33(2): 41-52.

Eastman, G.Y. (1991) Private communication regarding work conducted at Thermacore, Inc., Lancaster, PA, in 1976-77.

Ernst, D. M. (1991) Private communication regarding work conducted at Isothermics, Inc., Augusta, New Jersey, in 1975-77.

Fanelli, R. (1946) "Modifying the Viscosity of Sulfur," Ind. and Eng. Chem., 39(1): 39-43.

Fanelli, R. (1950) "The Surface Tension of Sulfur," J. Am. Chem. Soc., 72: 4016-4017.

Horvath, M.L. (1975) "Physical Properties of Inorganic Compounds," ed. S.A. Units, London.

Karashaev, A.A., S.N. Zadumkin, and M.P. Dokhov (1973) "The Surface Tension of Liquid Iodine," - In: Surface phenomena in Liquids and Liquid Solutions," ed. LGU, Vyp.2, p. 99-102.

Lekae, V.M. and L. N. Elkin (1960) "Materials for Industry Equipment Operating in the Medium of Elementary Sulfur at High Temperatures," In: Native Sulfur, Trudy GIGKhS, vyp.5, ed. Gosgortekhizdat, Moscow.

Niselson, L.A., Yu.P. Ballo, and K.V. Tretjyakova (1974) "Liquid-Vapour Equilibrium in Iodine-Sulfur and Iodine-Phosphorous Systems," Izvestiya AN SSSR, ser. Inorganic Materials 10, 5: 840-844.

Ono, K. and T. Matsushima (1957) "The Surface Tension of Liquid Sulfur, Sci. Res. (Tohoku Univ.) ser. A. 9: 309-318.

Polasek, F., "Heat Pipes with Sulfur Fillings and their Application," (1975) Fifth International Congress CHISA-75, Prague, August 25-29.

Polasek, F. and P. Stulc (1976) "Heat Pipe for the Temperature Range from 200 to 600°C," - In: Proc. Second International Heat Pipe Conference - Bologna, Italy, Vol. 2, p. 711.

Raw, H., T.R.N. Kutty, and J.R.F. Guedes de Carvalho (1973) "High Temperature Saturated Vapour Pressure of Sulfur and the Estimation of its Critical Quantities," J. Chem. Thermod. 5, pp. 291-302.

Rosenfeld, J.H. and J.E. Lindemuth (1991) Report - Sulfur Heat Pipes for 600 K Space Heat Rejection System, Thermacore, INc., Lancaster, PA, prepared for Space Power, Inc., July.

Singh, B.S. and J.H. Rosenfeld (1987) Final Report, Space Power Subcontract No. SUB-8076, Design and Analysis Support for Advanced Radiator Concept for Space Power Systems, Thermacore, Inc., Lancaster, PA, December 1987.

Timrot, D. L. et al., (1978) Thermophysical Properties of Sulfur-Iodine Binary System as a Promising Heat Transfer Medium for Heat Pipes, Advances in Heat Pipe Technology Proc. IV International Heat Pipe Conference, Pergeoman Press, Oxford, United Kingdom, 1981.

NUCLEAR ELECTRIC PROPULSION: AN INTEGRAL PART OF NASA'S NUCLEAR PROPULSION PROJECT

James R. Stone
Nuclear Propulsion Office
M.S. AAC-2
NASA Lewis Research Center
Cleveland, OH 44135
(216) 977-7120

Abstract

The National Aeronautics and Space Administration (NASA) has initiated a technology program to establish the readiness of nuclear propulsion technology for the Space Exploration Initiative (SEI). This program was initiated in the past fiscal year with a very modest effort identified with nuclear thermal propulsion (NTP); however, nuclear electric propulsion (NEP) is also an integral part of this program and builds upon NASA's Base Research and Technology Program in power and electric propulsion as well as the SP-100 space nuclear power program. Although the Synthesis Group On America's Space Exploration Initiative has identified NEP only as an option for cargo missions, recent studies conducted by NASA Lewis show that NEP offers the potential for the early manned Mars missions as well. Lower power NEP is also of current interest for outer planetary robotic missions. This paper reviews current plans for the overall nuclear propulsion project, with emphasis on NEP and those elements of NTP program which have synergism with NEP.

INTRODUCTION

Nuclear Propulsion has been identified as a key technology for human and robotic exploration of the solar system by the Advisory Committee On the Future of the U.S. Space Program (1990). The Synthesis Group On America's Space Exploration Initiative (1991) established Nuclear Thermal Propulsion (NTP) as the preferred approach for human missions to Mars, and also identified Nuclear Electric Propulsion (NEP) as a high-payoff technology which should be pursued. The National Aeronautics and Space Administration (NASA) initiated its Nuclear Propulsion Program in FY 1991 with emphasis on NTP. NEP activities have been underway at a modest funding level in NASA's Low Thrust Propulsion Technology Program for many years (Byers and Wasel 1987 and Stone and Bennett 1989), in the Power Technology Program, in the Civil Space Technology Initiative (CSTI) High Capacity Power Program (NASA 1988), and in the Department of Energy (DOE)/Department of Defense (DoD)/NASA SP-100 reactor program. NASA is providing funding for a new focused effort on NEP for the Space Exploration Iniative in FY 1992; this activity builds upon but does not relace the ongoing programs. The NEP Program is managed at the Office of Aeronautics, Exploration and Technology (OAET) by the Propulsion, Power, and Energy Division, and is closely coordinated with the NTP Program. The NEP Program is based on broadening the previously planned, but deferred Cargo Vehicle Propulsion Program (NASA 1989). Lewis Research Center (LeRC) acts as Lead Center for the program, with responsibility for technical integration and reporting; this function is performed by the Nuclear Propulsion Office. The work is performed by LeRC and the Jet Propulsion Laboratory (JPL). A preliminary version of this management structure is provided in Figure 1.

The NEP Program will establish the feasibility and practicality of electric propulsion for piloted and robotic solar system exploration. The performance objectives are high specific impulse ($2000 \leq I_{sp} \leq 10,000$ s), high efficiency (over 0.50), and low specific mass. Although the ultimate target is a multi-megawatt NEP system for piloted Mars vehicles, an evolutionary approach is planned; 100-kW class systems will be targeted for relatively near-term robotic missions, and longer term efforts will target megawatt-class systems applicable to cargo vehicles supporting human missions as well as to the piloted vehicles.

BACKGROUND

Prospects now appear good for broad acceptance and application of electric propulsion systems (Byers and Wasel 1987); however, current applications are limited to low-power stationkeeping and orbit control (Stone and Bennett 1989). The successful operation of these systems has led to increased user confidence, and further operational use of electric propulsion appears imminent.

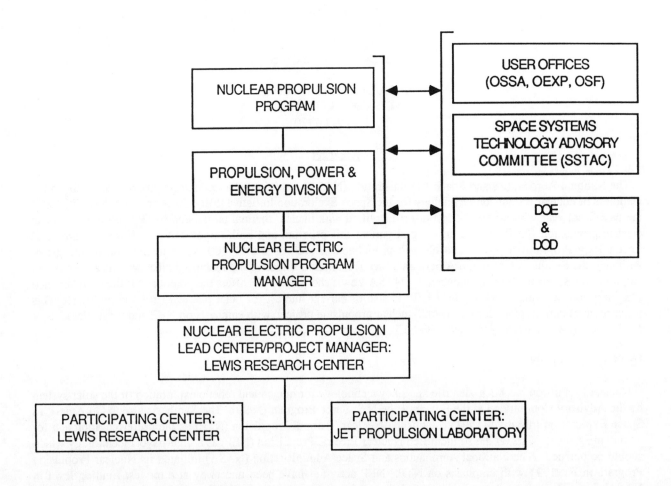

FIGURE 1.   Nuclear Electric Propulsion Program Management Structure.

Interest in high power electric propulsion has been building for several years. The National Commission on Space (1986) advocated a number of challenging missions, such as a return to the Moon, unmanned and manned exploration of Mars and its moons, and unmanned scientific exploration of the rest of the solar system. Many of these missions would be enhanced, and some would be enabled by high I_{sp} electric propulsion. To perform the challenging future missions, high power and high I_{sp} systems will be required. Candidate thrusters include electrostatic (ion) and magnetoplasmadynamic (MPD) engines, with electrodeless approaches representing a longer term possibility. Electric propulsion is also applicable to advanced robotic exploration missions, offering both decreased propellant mass (and/or increased payload) and reduced trip times for deep space missions. There are also many potential applications for high I_{sp} electric propulsion at the tens of kilowatts available from large solar power systems. High I_{sp} clearly offers propellant mass savings, but in order to practically exploit that benefit for space missions, it is necessary for the overall vehicle to exhibit acceleration levels sufficient to meet mission time lines. This requirement necessitates low specific mass and high efficiency propulsion and power systems in order to keep power system mass low.

The planning of the NEP Program builds on the earlier Cargo Vehicle Propulsion Program Plan (Stone 1988) augmented substantially by by a series of panel activities in support of the overall Nuclear Propulsion Project. These recent activities were initiated by workshops on NTP (Clark 1991) and NEP (Barnett 1991). These panels worked various planning issues and were coordinated through combined meetings and panel chairmen's meetings and were led by a DOE/DoD/NASA interagency steering group. Reports are being prepared by these panels, which were as follows:

- Nuclear Safety Policy Working Group,
- Mission Analysis Panel,
- NTP Technology Panel,
- NEP Technology Panel,
- Nuclear Fuels/Materials Panel, and
- Test Facilities Panel.

CURRENT PROJECT PLAN

The original Project Plan was prepared in Fiscal Year (FY) 1990 in parallel with Worshops on NEP and NTP (Barnett 1991 and Clark 1991, respectively) with the target of reaching Technology Readiness Level (TRL) of 6 (Table 1) for NTP and TRL-5 for NEP in 2006, in time to support the Space Exploration Initiative (SEI). The plan has been updated as a result of the panel activities described in the preceeding paragraph and a series of interagency and NASA intercenter meetings(Clark and Miller 1991 and Doherty 1991). Because of the slower than anticipated funding growth, the TRL targets for both NTP and NEP have been delayed to 2009.

TABLE 1.   NASA Technology Maturation Milestones for Space Application.

Level	Description

Technology Development

1. Basic Principles Observed and Reported.
2. Technology Concept/Application Formulated.
3. Analytical and Experimental Critical Function and/or Characteristic Proof-of-Concept.
4. Component and/or Breadboard Validation in Laboratory.
5. Component and/or Breadboard Demonstrated in Relevant Environment.
6. System Validation Model Demonstrated in Relevant/Simulated Environment.
7. System Validation Model Demonstrated in Actual Environment.

Advanced Development

8. Technology Applied to Construction of Component and/or Breadboard of Expected Flight Hardware Configuration.
9. Capability of Full Scale Subsystem Protoype Demonstrated in Ground Tests.
10. Capability of Full Scale Subsystem Prototype Demonstrated in Actual Use.

Flight Hardware Development

11. Full Scale System Prototype.
12. Capability Demonstrated in Flight Test of Flight Hardware.
13. Capability Demonstrated by Operational Flight Experience.

The top level work breakdown structure (WBS) for the project is give in Table 2, which also includes more detail where work will be conducted in FY 1992. The planned work is described in more detail in the following paragraphs.

Project Management

The Nuclear Propulsion Project Office coordinates the activities in various LeRC technology organizations, at the Marshall Space Flight Center (MSFC) (NTP) and at JPL (NEP). Coordination and communication with DoD and DOE laboratories are also important in management, particulary with regard to the nuclear sub-systems tasks, which are performed by DOE. The Idaho Engineering Laboratiry (INEL) has the lead in reactor systems, while the Los Alamos National Laboratory (LANL) has the lead in fuels development. This task also includes the development and implementation of a plan for informing the public of the rationale for SEI and the important role that nuclear propulsion will play in space exploration.

TABLE 2.   Nuclear Propulsion Project Work Breakdown Structure (Elements Planned for First-Year Activity).

PROJECT MANAGEMENT
    Project Management
    Information Systems
CONCEPT DEVELOPMENT AND SYSTEMS ENGINEERING
    System and Interface Requirements Definition
    System Level Design/Trade Studies
    Mission Analysis
    Engine System Modeling and Analysis
INNOVATIVE TECHNOLOGIES
    Studies/Modeling
    Proof of Concept Tests
NEP TECHNOLOGY
    Nuclear Sub-Systems
        Fuels
    Non-Nuclear Sub-Systems
        Thrusters
        Thermal management
        Power management, distribution & processing
NTP TECHNOLOGY
    Nuclear Sub-Systems
        Fuels
    Non-Nuclear Sub-Systems
        Materials
        Instrumentation and controls
        Turbopumps
        Nozzles
        Structures
FACILITIES
    Nuclear
    Non-Nuclear
        Space simulation vacuum chambers (NEP)
        High temperature facility (NTP)
SAFETY, QUALITY ASSURANCE, RELIABILITY AND ENVIRONMENTAL ISSUES
    Safety Planning and Integration
    Safety Analyses
    Reliability

Concept Development and Systes Engineering

This effort was initiated in FY 1991 for NTP in the areas of requirements definition and engine system modeling and analysis. These activities will be accellerated in FY 1992, and further design work will be initiated with significant industry participation. For NEP the effort will focus on system level design/trade studies. These NEP studies will identify combinations of reactor, power conversion, heat rejection and thruster systems which show significant enough promise in one or more of the mission categories to warrant subsequent conceptual design studies.

Innovative Technologies

High-risk/high-payoff technologies for performance enhancements beyond the 2009 target levels will be pursued on a paced basis. Efforts will be initiated with studies and modeling leading to proof of concept tests. For those technologies achieving successfull proof of concept, concetual designs will be developed and critical components designed and tested.

NEP Technology

Most of the fuels activity in the early phase of this program will be on NTP since near-term NEP fuels will probably be derived from the SP-100 Program. The thruster technology will be focused primarily on MPD initially since ion propulsion is funded by the NASA Low Thrust Technology Program. The MPD effort at LeRC and JPL will concentrate on increasing power level and efficiency while also advancing the theoretical understanding of electrode phenomena. Advanced heat pipe technologies will be addressed in the Thermal Management area, while high temperature electronics will be the first activity under Power Management, Distribution and Processing.

NTP Technology

The NTP technology developments by LeRC, MSFC and DOE may also benefit the NEP Program. Potential areas of synergy include fuels (at a basic level), instrumentation and controls (radiation-tolerant, high temperature), and health monitoring.

Facilities

Initial NTP facility activities will focus on high temperature hydrogen test capability. Early NEP facility activity will enhance existing capability of space simulation vacuum facilities by increasing the heat rejection and pumping capability resulting in increasing operating time at high power.

Safety and Environmental Issues

The initial activities in this area are largely common between NTP and NEP. Early activity will be primarily focused on planning and developing criteria appropriate for the ultimate piloted spacecraft application as well as ground testing issues. Preparation of Environmental Impact Statements will be initiated.

CURRENT STATUS

The NEP Program is underway at a significant pace, though not yet at a pace which assures our ability to meet the more aggressive proposed SEI timelines. The NEP effort is very much related to the ongoing SP-100 Program, but SP-100 currently addresses very general requirements, not specifically those of NEP. The status of advocacy for NEP is greatly improved over the recent past. NEP has often been proposed as enabling to many outer planetary missions, but received little serious attention or support from the space science community. Studies conducted in 1991 by JPL have done much to improve this situation and have aroused serious interest in the space science community (Yen 1991). Although Synthesis Group On America's Space Exploration Initiative (1991) selected NTP for piloted Mars missions, NEP was recommended for cargo applications. Recent LeRC studies (George, Hack and Dudzinski 1991) have shown that NEP may be quite competitive with NTP even for fast piloted missions to Mars.

CONCLUDING REMARKS

The Nuclear Electric Propulsion Program (NEP) is accelerating toward the capability to support robotic and human exploration of the solar system. Interagency and intercenter cooperation is enabling significant progress in spite of the fragmented funding for this program - from the CSTI High Capacity Power Program, SP-100, and NASA's Base Research and Technology Program. Potential users of this technology, in both the space science and human exploration communities are becoming incresingly aware of the enhancing and enabling capabilities of NEP.

Acknowledgments

I would like to acknowledge the considerable efforts of other members of the LeRC Nuclear Propulsion Office, LeRC technologists, and all those who participated in the intercenter/interagency planning and panel activities.

References

Advisory Committee on the Future of the U.S. Space Program (1990) *Report of the Advisory Committee on the Future of the U.S. Space Program*, U.S. Government Printing Office, Washington, DC.

Barnett, J.W. (1991) *Proceedings of the NASA/DOE/DOD Nuclear Electric Propulsion Workshop*, (JPL CP-____) Pasadena, CA, 19-22 June 1990.

Byers, D.C. and R.A. Wasel (1987) "The NASA Electric Propulsion Program," *AIAA Paper 87-1098*, May 1987.

Clark, J.S. (1991) *Proceedings of the NASA/DOE/DoD Nuclear Thermal Propulsion Workshop*, (NASA CP-____) Cleveland, OH, 10-12 June 1990.

Clark, J.S. and T.J. Miller (1991) "The NASA/DOE/DoD Nuclear Rocket Propulsion Project," (AIAA Paper 91-3413), *AIAA/NASA/OAI Advanced Technologies Conference*, Cleveland, OH, 4-6 September 1991.

Doherty, M.P. (1991) "Blazing the Trailway: NEP and Its Technology Program Plans," (AIAA Paper 91-3441), *AIAA/NASA/OAI Advanced Technologies Conference*, Cleveland, OH, 4-6 September 1991.

Hack, K.J., George, J.A. and L.A. Dudzinski (1991) "Nuclear Electric Propulsion Mission Performance for Fast Piloted Mars Missions," (AIAA Paper 91-3488), *AIAA/NASA/OAI Advanced Technologies Conference*, Cleveland, OH, 4-6 September 1991.

NASA (1988) *Technology for Future NASA Missions: Civil Space Technology Initiative (CSTI) and Pathfinder*, NASA CP-3016, Washington, DC, 12-13 September 1988.

NASA (1989) *Project Pathfinder -- Program Plan -- Fall 1988*, NASA Office of Aeronautics and Space Technology, Washington, DC, 27 January 1989.

National Commission on Space (1986) *Pioneering the Space Frontier*, Bantam Books, NY, June 1986.

Stone, J.R. (1988) *Pathfinder Cargo Vehicle Propulsion Program Plan*, NASA Office of Aeronautics and Space Technology, Washington, DC, November 1988.

Stone, J.R. and G.L. Bennett (1989) "The NASA Low Thrust Propulsion Program," *AIAA Paper 89-2492*, *AIAA/ASME/SAE/ASEE 25th Joint Propulsion Conference*, Monterey, CA, 10-14 July 1989.

Synthesis Group On America's Space Exploration Initiative (1991) *America at the Threshold. Report of the Synthesis Group On America's Space Exploration Initiative*, U.S. Government Printing Office, Washington, DC.

Yen, C.-W. (1991) "NEP Mission Study - Interim Report," Presented at *NEP Technology Panel Meeting*, NASA Lewis Research Center, Cleveland, OH, 18 July 1991.

NUCLEAR ELECTRIC PROPULSION TECHNOLOGY PANEL FINDINGS AND RECOMMENDATIONS

Michael P. Doherty
Nuclear Propulsion Office
NASA Lewis Research Center
Cleveland , OH 44135
(216) 977-7092

Abstract

Summarized in this paper are the findings and recommendations of a triagency (NASA/DOE/DOD) panel on Nuclear Electric Propulsion (NEP) Technology. NEP has been identified as a candidate nuclear propulsion technology for exploration of the Moon and Mars as part of the Space Exploration Initiative (SEI). The findings are stated in the areas of system and subsystem considerations, technology readiness, and ground test facilities. Recommendations made by the panel are summarized concerning (1) existing space nuclear power and propulsion programs and (2) the proposed multiagency NEP technology development program.

INTRODUCTION

The Stafford "Synthesis Committee" has suggested that nuclear propulsion will be required to safely, cheaply, and repeatedly conduct high performance Space Exploration Initiative (SEI) space missions (Synthesis Group 1991). The technology to perform these nuclear propulsion space missions with either Nuclear Electric Propulsion (NEP) or Nuclear Thermal Propulsion (NTP) must be developed and validated.

Six interagency technical panels comprised of National Aeronautics and Space Administration (NASA), Department of Energy (DOE), and Department of Defense (DOD) personnel were formed in FY91 to address mission analysis, safety, facilities, and technology development requirements for Nuclear Propulsion (NP) for advanced SEI space missions (Clark and Miller 1991). The panels met approximately monthly. Direction was provided and progress monitored by a Nuclear Propulsion steering committee consisting of program managers from the three organizations.

The Nuclear Electric Propulsion (NEP) Technology Panel was chartered to characterize NEP system and technology options in light of mission and safety considerations, and to initiate the planning for a NEP technology development program. The goals of the NEP Technology panel were to provide a basis for evaluating candidate power/propulsion systems, to identify nuclear and non-nuclear technology needs and plans, to define the major facility requirements for NEP, and to establish requirements for a NEP systems trade study. Twenty-one technical experts in nuclear reactor systems; power conversion, conditioning and processing; space systems engineering; and electric thruster technology were formed into working subpanels to achieve these goals. A summary final report is being prepared for publication (Doherty and Holcomb 1992).

Three broad classes of missions are being considered for NEP systems:

* Near Earth, such as orbit transfer, maneuvering, and station keeping;

* Planetary Exploration, such as robotic probes to the outer planets, comet nucleus sample return, asteroid exploration, and others; and

* Space Exploration Initiative, such as lunar cargo missions, and Mars cargo and piloted missions, including short trip time missions.

In addressing these three broad classes of missions, NEP subsystem technology options shown in Table 1 must be considered. This table includes all of the power and propulsion concepts discussed by Barnett (1991). To validate the required NEP technologies, programs must be initiated or focused in

TABLE 1.   NEP Subsystem Technology Options.

Reactor	Power Conversion	Thermal Management	Power Management & Distribution	Thruster
Liquid Metal Cooled	Dynamic	Heat Pipe	Silicon	Steady State Electrostatic
Growth SP-100	Rankine	Refractory Metal	Gallium Arsenide	Ion
Advanced Pin Cermet	Brayton	Carbon-carbon	Aluminum-Gallium Arsenide	
Boiling Potassium	Stirling	Ceramic Fabric	Silicon Carbide	Steady State Electromagnetic
	Static Thermoelectric	Pumped Loop		Magnetoplasma-dynamic (MPD)
Gas Cooled	Thermionic in core			Electron Cyclotron Resonance
NERVA Derived	ex core	Liquid Sheet/Droplet		Ion Cyclotron Resonance
Particle Bed	Electrochemical			Variable Specific Impulse
Pebble Bed	Magnetohydro-dynamic			
Cermet		Bubble Membrane		
				Pulsed Electromagnetic Deflagration
				Pulsed Plasmoid
Incore Thermionic				Pulsed Inductive
Vapor Core				Pulsed Electrothermal/ Electromagnetic
				Pulsed Electrothermal - MPD

reactor systems and fuels; power conversion, conditioning and processing; heat rejection and thermal management; and electric thrusters, with the goal of ground testing these subsystems prior to their use in outer space. Technology drivers are low mass, high reliability, and long life subsystems and systems.

PANEL FINDINGS & CONCLUSIONS

The NEP Technology Panel reached a number of findings and conclusions. These can be grouped under the headings of System and Subsystem Considerations, Technology Readiness, and Ground Test Facilities.

System and Subsystem Considerations

NEP has the potential to implement a broad range of mission applications, starting from near Earth and planetary exploration missions, and proceeding to SEI Lunar and Mars missions. Successfully addressing these missions will require evolutionary increases in net on-board electrical power for propulsion and measured decreases in system specific mass. System and subsystem considerations follow.

Systems

NEP systems which address the requirements for this broad range of missions will likely have many common technologies. For example, the SP-100 reactor and fuel technology currently being developed for space power applications potentially offers benefit to all the missions discussed. Also, dynamic power conversion technology, such as Potassium Rankine (K-Rankine) or Brayton, enabling for cargo missions, is the same technology necessary for piloted missions. Likewise, the Ion electric thruster, although not as physically compact as the Magnetoplasmadynamic (MPD) thruster, has a strong history of technology development, making it a candidate for the SEI missions, as well as for near Earth and planetary exploration missions.

For application to a propulsion system for planetary robotic missions, the SP-100 reactor and thermoelectric power conversion technologies, in combination with ion electric propulsion technology, seem well suited. Yen and Sauer (1991) show that timely development of these technologies offers a significant means to conduct rewarding robotic science missions in the early 21st century. Thermionic reactors offer promise for this application also. Such robotic missions are logical technology precursors to the higher power SEI missions.

Because NEP has the potential to implement a broad range of advanced space missions, and there is a host of candidate technologies, there is a need to perform a detailed systems/ subsystems trade study to determine optimum combinations of those technologies. This activity will provide a reference point for the initiation of detailed technology programs, and therefore is vital to conduct in the early program years for NEP.

Subsystems

The SP-100 reactor and fuel technology may be applicable to piloted missions to Mars. Recent mission and system studies (Hack et al. 1991) and (George 1991) have shown that this pin-type fuel, operating at higher temperatures in a larger core reactor, coupled to a K-Rankine power conversion system, with high temperature power management and distribution, lightweight heat pipe radiators, and ion electric thrusters, can propel a vehicle to Mars and back with trip times competitive with NTP systems at values of initial mass in low Earth orbit (IMLEO) somewhat less than NTP.

If greater performance is sought, more advanced reactor and power conversion technologies would be required (George 1991). Such an advanced system might be: 1) a 1500 K outlet temperature advanced pin-type liquid metal cooled reactor with K-Rankine power conversion, 2) a 2000 K outlet temperature gas cooled particle bed reactor with direct Brayton power conversion, or 3) an innovative design such as a gas (vapor) core reactor with Magnetohydrodynamic (MHD) conversion.

K-Rankine power conversion technology, in combination with SP-100 reactor technology, yields a lighter overall system mass than Brayton. Due to its higher cycle temperature ratio, Brayton cycle power conversion technology requires higher peak cycle temperatures than achievable with a 1375 K thermal source to result in a system mass competitive to K-Rankine. Therefore, given current SP-100 reactor technology, K-Rankine power conversion technology promises higher system performance.

Heat rejection subsystems for NEP must be lightweight, employing lightweight, high temperature materials. As the heat rejection subsystem mass is a dominant portion of the total NEP system mass, the performance of the heat rejection subsystem directly affects the overall performance of the system. Power Management and Distribution (PMAD) subsystems must be capable of high temperature, efficient operation, so that waste heat rejection can be accomplished using small radiator systems. High PMAD efficiency reduces both thermal and power system mass. As for electric thrusters, they must be efficient and be capable of long lifetimes.

Technology Readiness

The panel judged the NEP subsystem technology options according to their projected technology readiness. Table 2 displays the projected readiness of those technology options in Table 1 that would apply to the SEI missions. Within Table 2, any of the options listed in the middle column could be ground tested in a relevant environment - Technology Readiness Level 5 (TRL-5) - by the year 2005 (with adequate funding) and have been classified as

TABLE 2. Projected Technology Readiness of NEP Subsystem Technology Options Applicable to SEI Missions.

NEP Subsystem	Technology Options that could reach TRL[1]-5[2] by Year 2005 (with adequate funding)	Technologies Not Expected to reach TRL-5 by Year 2005
Reactor	Growth SP-100, Advanced Pin Cermet, NERVA[3] Derived Particle Bed, Pebble Bed	Boiling Potassium Vapor Core
Power Conversion	Rankine Brayton	Electrochemical Magnetohydrodynamic
Heat Rejection	Refractory Metal Heat Pipe Carbon-carbon Heat Pipe	Ceramic Fabric Heat Pipe Liquid Sheet Radiator Bubble Membrane
Power Management and Distribution	Silicon, Gallium Arsenide Aluminum Gallium Arsenide Silicon Carbide	
Thrusters	Ion Magnetoplasmadynamic (MPD)	Very high power MPD Electron Cyclotron Resonance Ion Cyclotron Resonance Variable Specific Impulse Deflagration Pulsed Plasmoid Pulsed Inductive

[1] TRL - Technology Readiness Level

[2] TRL-5 - subsystem technology ground test in a relevant environment

[3] NERVA - Nuclear Engine for Rocket Vehicle Applications

enabling for the SEI missions. Those options not expected to reach TRL-5 by 2005 are listed in the right-hand column of this table, and have been classified as "innovative". The year 2005 was chosen so that the technologies would be available in time to be considered for the SEI missions.

K-Rankine and Brayton based power systems are the recommended choices for SEI applications in the 2008-2020 timeframe, which require TRL-5 by 2001-2010. Other power system concepts are either suitable only for NEP applications requiring less power, or are presently deemed to have benefit-to-risk ratios too low as to expect their readiness in this timeframe (innovative). All of the reactor concepts listed in the middle column of Table 2 are relevant to Rankine and Brayton based power systems.

Rankine cycle space power conversion has many technological challenges to overcome. Challenges include turbine blade erosion due to vapor condensation, and condensate collection, management, and transport in a micro-gravity environment. Brayton cycle power conversion, on the other hand, is well understood and presents fewer technology issues.

Ion and MPD propulsion systems are the recommended choices for SEI applications in the 2008-2020 timeframe, requiring TRL-5 by 2001-2010. Ion propulsion is more mature than MPD propulsion, but has the disadvantage of being less power dense, requiring large thruster areas to accomplish the SEI missions. MPD, while being compact and having demonstrated high power operation, must show suitable efficiences. Both propulsion technologies must demonstrate acceptable life. Other propulsion system concepts are either suitable only for applications requiring less specific impulse, or are presently deemed to have benefit-to-risk ratios too low as to expect their readiness in this timeframe (innovative). If further studies indicate an advantage to developing any of these technologies, and feasibility issues have been resolved, then with adequate funding any of these technologies could be made available within the needed timeframe.

Pin-type fuels should be given the highest priority in the NEP fuels development program. The current SP-100 fuel pin design, Uranium Nitride (UN) pellets enclosed in PWC-11 clad tubing, should be irradiation tested at higher fuel temperatures (to 1500 K) to uncover fuel integrity issues associated with shorter life, higher temperature mission demands. Higher temperature cladding candidates for UN pin-type fuels should also be screened, and irradiation tests of these potential fuels performed. Fabrication and capsule irradiation tests of cermet and carbide fuels should be initiated if mission and system studies show significant benefit. Bhattacharyya et al. (1991) provide more discussion on the development of nuclear fuels for propulsion systems for SEI.

Ground Test Facilities

NEP technology will be validated at the subsystem level (TRL-5). There are two reasons for this. First, because there is only an electrical coupling between NEP power and propulsion subsystems, each subsystem should be capable of being adequately simulated in the testing of the other. That is, an electrical load simulating the electric thruster and power processing subsystems could be used in the NEP power subsystem test, and vice versa. Second, test environments for space power and space propulsion subsystems are very different from one another. Space power subsystems require a clean, static thermal vacuum environment, while the test environment for electric thrusters is characterized by the presence of an effluent. These requirements, in combination with the extreme distances between reactor and thruster subsystems to minimize radiation effects, place severe demands on a facility to test "everything under one roof". Flight system demonstrations from orbit about Earth, first using low power NEP systems for orbit raising missions, then larger NEP systems for planetary missions, and finally MW-class NEP systems for SEI cargo missions should provide the necessary system and flight experience to assure reliable performance of the piloted NEP system. The major facility requirements for validating NEP technology to TRL-5 are presented in Table 3.

TABLE 3.   Major Facility Requirements for Ground Testing NEP Subsystem Technologies in a
            Relevant Environment.

Subsystem	Requirement
Reactor	50 MWt[1] heat rejection, Vacuum vessel Reactor containment, Capability to test shielding Outlet temperature = 1500 - 2000 K Liquid metal handling facility[2] Control room; Maintenance, storage , decontamination and disposal facility Lifetime = 5-7 years
Power Conversion	2.5 MWe[3] (electric load), 12.5 MWt (heat dump) vacuum or inert gas insulation, support facilities Lifetime = 5-7 years Upgradable to 5 MWe
Thermal Management	10 MWe, Upgradable to 20 MWe
Power Management and Distribution	No major facility required
Thruster	2.5 grams per second effluent flowrate 5 MWe electric power 10 meter (m) diameter by 30 m long tank size

---

[1] MWt = megawatts thermal

[2] Assumes liquid metal cooled reactor

[3] MWe = megawatts electric

PANEL RECOMMENDATIONS

In view of the above panel findings and conclusions, the
following recommendations concerning NEP and a technology
development program in Nuclear Electric Propulsion are made:

*   NASA should plan and implement an evolutionary technology
development program in NEP, a program directed toward providing
the technologies for a piloted mission to Mars, while also
including interim program milestones which yield NEP
technologies for near Earth, interplanetary robotic, as well as
lunar and Mars cargo missions;

*   From the program outset, efforts should be initiated to (1)
determine performance and life limits of kW-class and MW-class
electric thrusters, (2) determine efficiencies, lifetimes, and
radiation tolerance of high-temperature power electronics, and
(3) address fundamental technology issues associated with
lightweight heat rejection systems;

* Accelerate the schedule for a ground test demonstration of the SP-100 space reactor and power conversion technologies in the late 1990s;

* Perform a systems/ subsystems trade study early in the program to clarify critically needed technology programs and to specify detailed technology requirements for system safety and performance;

* Demonstrate high power, dynamic power conversion technologies;

* If justified by systems trade studies, develop and demonstrate a new reactor technology;

* Demonstrate the current SP-100 fuels technology (UN with PWC-11 clad tubing) at higher temperature operating conditions to identify technological feasibility of the concept;

* Assess candidate facilities for power subsystem and propulsion subsystem testing for their suitability to meet ground testing requirements; and

* Provide a forum for the continued involvement of experts in all technology areas of NEP as the program is implemented.

CONCLUDING REMARKS

Work conducted in FY91 by an interagency panel in Nuclear Electric Propulsion technology has led to findings and conclusions regarding NEP systems, technologies, and technology demonstration. Recommendations by this panel concerning the existing SP-100 space nuclear reactor program, as well as for a focused program in NEP Technology are made. The panel provided a basis for evaluating candidate NEP power/propulsion systems, began to establish requirements for a NEP systems trade study, and defined the major facility requirements of NEP. The panel identified nuclear and nonnuclear technology needs, and proposed plans for a technology program to address those needs (Doherty 1991).

Acknowledgments

The author greatly wishes to acknowledge his indebtedness to the members of the Nuclear Electric Propulsion Technology Panel: Monte Parker and Kurt Schoenberg (LANL), Wayne Schmidt (AFAL), Hans Ludewig (BNL), Jim Lake and John Dearien (INEL), Jack Mondt, John Barnett, and John Brophy (JPL), Bick Hooper (LLNL), Bob Holcomb (ORNL), Dick Widrig (PNL), Don Gallup and Frank Thome (SNL), and Jeff George, Jim Sovey, Joe Nainiger, Jim Gilland, Al Juhasz, John Dickman, and Harvey Bloomfield (LeRC). Numerous other people followed the process, contributing to its final product. Their names are too many to list here. The panel members' personal dedication to meeting attendance, subpanel participation, action item resolution, and to the writing of the panel final report is greatly appreciated.

References

Barnett, J. W. (1991) "Nuclear Electric Propulsion Technologies: Overview of the NASA/DOE/DOD Nuclear Electric Propulsion Workshop," in <u>Proc. Eighth Symposium on Space Nuclear Power Systems</u>, CONF-910116, M.S. El-Genk and M.D. Hoover, eds., American Institute of Physics, New York, 1991.

Bhattacharyya, S.K., C.S. Olsen, and R.H. Titran (1991) "Development of Nuclear Fuels and Materials for Propulsion Systems for SEI," AIAA 91-3452, September 1991.

Clark, J. S. and T. J. Miller (1991) "The NASA/DOE/DOD Nuclear Rocket Propulsion Project: FY 1991 Status," AIAA 91-3413, September 1991.

Doherty, M. P. (1991) "Blazing the Trailway: Nuclear Electric Propulsion and Its Technology Program Plans," AIAA 91-3441, September 1991.

Doherty, M. P. and R. S. Holcomb (1992) "A Summary and Recommendations on Nuclear Electric Propulsion Technology for the Space Exploration Initiative," NASA CP-_____.

George, J. A. (1991) "Multimegawatt Nuclear Power Systems for Nuclear Electric Propulsion," AIAA 91-3607, September 1991.

Hack, K. J., J. A. George, and L. A. Dudzinski (1991) "Nuclear Electric Propulsion Mission Performance for Fast Piloted Mars Missions," AIAA 91-3488, September 1991.

Synthesis Group (1991) "America at the Threshold - America's Space Exploration Initiative," U.S. Government Printing Office, Washington D.C. 20402.

Yen, C. L. and C. G. Sauer (1991) "Nuclear Electric Propulsion for Future NASA Space Science Missions," IEPC-91-035, October 1991.

NEP MISSION SENSITIVITIES TO SYSTEM PERFORMANCE

James H. Gilland
Sverdrup Technology, Inc.
2001 Aerospace Parkway
Brook Park, OH 44142
(216) 977-7093

Abstract

Nuclear Electric Propulsion (NEP) mission performance is strongly affected by system performance. Power and propulsion system specific mass, specific impulse, and efficiency all combine to determine the performance limits for a given mission. Thruster technology determines the specific impulse and efficiency of the system. The effects of these parameters on the mission performance of NEP systems relative to other concepts has been analyzed to give guidance to thruster development goals for a range of missions: Lunar Cargo, Mars Cargo, and Mars Piloted. Mission sensitivities to system parameters are discussed, and technology requirements are identified for each mission.

INTRODUCTION

Trajectory calculations and mission analyses of electric propulsion vehicles are highly complex. The low vehicle acceleration characteristic of these vehicles require extended if not continuous acceleration throughout the trajectory. Low thrust mission analysis must therefore assess the full, integrated path of the vehicle. The exact optimum trajectory path for a given vehicle is determined by the system parameters of the vehicle, which determine the externally imposed accelerations. Changes in propulsion system performance, when integrated over the entire mission trajectory, can have dramatic impacts upon vehicle performance in terms of trip time or initial mass requirements. The propulsion system performance parameters are thus closely meshed with the mission performance results.

Propulsion system parameters can be defined as

System Specific Mass (α) - ratio of power/propulsion system mass to electric power input to
thrusters - kg/kWe,

Specific Impulse (I_{sp}) - Thrust per unit propellant weight flow, or exhaust velocity divided by
g_o - seconds, and

Propulsion System Efficiency (η) -   Conversion efficiency from electric to exhaust power - dimensionless.

Mission parameters that must be specified are

Destination - Target planets, Round trip or one way;
Trip Time; and
Payload.

To first order, these parameters can serve to characterize electric propulsion system performance capabilities (Jones and Monck 1984).

A typical approach to electric propulsion mission analysis has been to assume the performance capabilities of power systems and electric thrusters, and to optimize the mission based on these fixed parameters.   This approach gives results of interest to mission planners assessing the utility of certain technologies; however, technology developers receive little or no guidance as to what performance goals should be selected in their research. The study presented herein is aimed at showing the sensitivity of mission performance to each of the three systems variables described above, with no assumptions of specific power system or thruster system technology or performance. Instead, the results are intended to identify an envelope of performance within which electric propulsion technologies must operate for significant gains in mission performance.

MISSION SENSITIVITY ANALYSIS

Three missions have been considered for application of Nuclear Electric Propulsion (NEP) systems: Lunar Cargo, Mars Cargo, and Mars Piloted. These missions have been selected to span the range of mission requirements identified for the Space Exploration Initiative; some conclusions can be drawn pertaining to other NASA missions such as orbital transfer or robotic planetary probes. Mission requirements range from the mass efficient transportation of cargo with no emphasis on trip time (cargo missions), to the transportation of small payloads in minimum trip times (piloted missions). The differences in mission figures of merit and trajectory analysis among the three missions requires a different approach for each.

Lunar Cargo

The lunar cargo mission is essentially an Earth orbital transfer mission. As such, the NEP trajectory follows the familiar spiral path, gradually increasing orbital altitude with increasing vehicle tangential velocity. The mission assumptions used for this study are a round trip cargo mission to Low Lunar Orbit (LLO) from a Low Earth Orbit of ~500 km. Because of the relative magnitudes of the vehicle acceleration and the Earth's gravity, the trajectory requirements may be approximated by a constant ΔV value of 8 km/s each way (Palaszewski 1989). The NEP system may then be optimized analytically for maximum payload fraction (Auweter-Kurtz et al. 1985 and Gilland 1991). This optimization may be done parametrically for a fixed value of efficiency, with specific impulse and power level the parameters to be optimized. In addition, power level is normalized by payload mass, so that any payload may be assumed.

For mission performance comparisons, the results of the NASA 90-day Study (NASA 1989) for a hydrogen/oxygen (H/O) lunar cargo vehicle using an Earth return aerobrake with 20.5% mass fraction were used. Payload mass delivered to LLO was 58 Mg. The chemical/aerobrake vehicle was assumed to be reuseable for 5 missions, one per year. Previous studies of a comparable lunar NEP cargo vehicle have shown a 50% reduction in initial mass over the course of 5 missions (Hack et al. 1990). In these studies, the return propellant requirements of the NEP vehicle were found to be small relative to the payload and system mass. Performance comparisons in this study are on a single vehicle basis, which neglects the major benefit of NEP systems: reduced resupply mass.

Mars Cargo

NEP has often been considered for the Mars Cargo mission (Mason et al. 1989). The reference NEP cargo mission is a one way mission, with the vehicle departing from a Space Station Freedom Orbit of 407.5 km. Mars parking orbit is at Deimos altitude, 20077 km. In heliocentric space, the acceleration of the NEP vehicle is of the same order as the Sun's gravity; therefore, no approximation of constant ΔV can be made. Generally, the exact vehicle trajectory must be calculated numerically through a calculus of variations optimization process. A single invariant trajectory parameter has been identified which allows a mission to be approximated for use in system optimization. This approximation can be used to determine system performance with enough accuracy for parametric comparisons (Gilland, 1991).

In the system analysis, the optimum parameters power/payload mass (Pe/Ml) and payload fraction (μ) are calculated for a fixed trip time of 600 days, including spiral time, for a range of specific impulse and efficiency values. Mission performance is characterized primarily by payload fraction for the cargo mission, and the sensitivity of this parameter to Isp and efficiency is calculated. Power and initial mass requirements can also be derived based on representative payload masses.

Mars Piloted

The nature of a round trip piloted Mars mission introduces several additional degrees of freedom into the mission analysis, particularly the optimal balancing of outbound and inbound propulsion requirements. The emphasis upon minimum trip times for this type of mission also constrains the launch date availability compared to the less demanding cargo missions. In order to address the sensitivity of a round trip mission to system parameters, a reference opposition- class piloted mission was analyzed for a launch date of 2016, with a 30 day stay time at Mars. Total piloted trip time, defined as heliocentric transit time and Mars stay time, was fixed at 500 days. Earth departure orbit was 407.5 km, and Mars orbit was 20077 km. Payload masses were 124 Mg outbound, 40.3 Mg inbound. Propellant mass for orbital spirals at Earth and Mars are included in the vehicle initial mass; however, piloted trip time does not include spiral time at Earth, as the crew would not be aboard during this phase. A full numerical mission analysis and optimization was performed for a range of specific impulse and efficiency values for a fixed trip time. Two values of specific mass, 5 and 10 kg/kWe, were initially considered; however, the 10 kg/kWe case was not capable of achieving the specified trip time. Power was allowed to vary; A single case was run for a fixed power of 10 MWe, which results in a more limited mission capability.

Results

Lunar Cargo: The lunar cargo mission results are shown in Figures 1 through 6. Figures 1 through 3 show the optimized vehicle parameters (Isp, Power, and Initial Mass) for a 10 kg/kWe vehicle over a range of round trip times. Each curve represents a constant thruster efficiency. Figures 4 through 6 show similar results for a 20 kg/kWe total system specific mass. The initial mass of the comparable chemical/aerobrake system is shown as a horizontal line for comparison.

The lunar cargo mission studies assess the sensitivity of mission performance to α, Isp, and efficiency. System α plays a fundamental role in determining mission capability, as seen in Figures 1 and 4. Increasing α leads to lower values of Isp in order to perform a mission in a given trip time. Propellant and power system requirements increase accordingly, resulting in poorer vehicle performance. As an example, for a round trip time of 400 days and an efficiency of 0.4, the 10 kg/kWe system is found to operate best at an Isp of 4600 seconds, compared to a value of 2800 seconds for a 20 kg/kWe system. A 10 Mg penalty in vehicle mass is also imposed. This difference in performance increases with decreased trip time, until the 20 kg/kWe system reaches the chemical/aerobrake value at a trip time of ~220 days, compared to a value of 120 days for the 10 kg/kWe. Efficiency is found to play an important role in vehicle performance at values of 0.6 or less, while gains in efficiency to levels greater than 0.6 have a less drastic effect on performance. Even a thruster efficiency of 0.4 allows cargo mission performance superior to the chemical/aerobrake system. For reasonable efficiencies, optimum Isp values fall in the 2000 to 7000 second range, depending on specific mass and trip time. Corresponding power requirements range from 1 to 8 MWe. Efficiencies and Isp values in the preferred ranges could be accomplished using either MPD or ion thrusters, with propellant determining the suitable ion thruster choice. The MWe power requirements indicate the need for higher capacity nuclear power systems and thrusters.

Mars Cargo: Approximate Mars cargo system/mission analyses are shown in Figures 7 and 8. The mission analyzed is a 600-day total trip time mission, using a 10 kg/kWe NEP system operating at 4000 and 7000 seconds Isp over a range of efficiencies from 0.25 to 0.8. Results are shown in a normalized format; Figure 7 shows the vehicle initial mass variation with efficiency, normalized by the payload mass, with an estimated chemical/aerobrake payload fraction shown for comparison. The corresponding power requirements are shown in Figure 8, also normalized by payload mass. The dependence upon efficiency is reduced in this mission due to the relaxed constraint on trip time.

Mars cargo mission analyses focussed upon the effects of efficiency and Isp upon mission performance. Specific mass also plays a key role in NEP performance; however, this sensitivity has been explored amply in the literature (Hack et al. 1990). This study has focussed more on the propulsion system sensitivities which affect mission performance. For the Mars cargo mission, as with the lunar cargo, efficiencies of 0.4 or greater are desireable in order to obtain superior mission performance. The effects of Isp are less noticeable except at low efficiencies. The crossing of the two mass ratio lines at low efficiencies is an excellent example of the interchange of effects between Isp and efficiency. This interaction is important to model in system and mission optimization analyses. It is interesting to note that the desired Isp and efficiency values are similar for both the lunar and Mars cargo vehicles, indicating a possible commonality in thruster system design.

Mars Piloted: Mars Piloted results are shown in Figures 9 through 11. Only an α of 5 kg/kWe was able to achieve the 500-day round trip mission. Figure 7 shows the vehicle initial mass variation with efficiency and Isp, with power allowed to vary. Representative chemical/aerobrake and Nuclear Thermal Rocket mission performance are also shown. Figure 8 shows the corresponding power requirements. Figure 9 shows the mission sensitivity of NEP for a fixed power of 10 MWe, demonstrating the smaller performance space available.

Piloted Mars missions show a more dramatic sensitivity to efficiency due to the high energy requirements of the reduced trip time missions. The round trip nature of the mission further intensifies the effects of Isp and efficiency variations. Efficiencies greater than 0.6 are desired in order to significantly reduce system mass compared to other advanced propulsion concepts. At lower efficiencies, power system mass dominates the vehicle. The effects of changing Isp appear primarily in the power requirements, where there is a 50% - 60% increase in power requirements for the higher Isp value. By limiting power to 10 MWe, the role of effcency in the mission performance becomes more important in that the capability of the system to perform the desired trip time mission is very dependent upon efficiency. In general, the Mars and lunar missions both underscore the desirability for high Isp and efficiency, while identifying some representative operating regimes to serve as goals in technology development. It should also be noted that the α assumed for this study is 5 kg/kWe, a relatively optimistic value.

CONCLUSIONS

System and mission analysis results provide some guidance and insight into NEP system and technology needs. Conclusions can be made about efficiency, specific mass, specific impulse, and power requirements for SEI missions:

Efficiency: For both lunar and Mars cargo missions, efficiencies of 0.4 or higher are needed for NEP systems to be competitive with more conventional propulsion systems. The piloted Mars mission, with its greater energy requirements, needs efficiencies greater than 0.6 to compete with other propulsion options.

Specific Impulse: The lunar cargo mission specific impulse requirement depends on trip time as well as both specific mass and efficiency assumptions. Low efficiency and high specific mass values drive specific impulse values lower. Specific impulse values range from 2000 to 7000 seconds. The sensitivity to specific impulse for the Mars cargo mission is seen to be low, except at low efficiency values. Specific impulse values of 4000 or 7000 seconds provide significant reductions in vehicle mass until efficiency drops below 0.4.

Specific Mass: Sensitivity to specific mass is seen in the lunar cargo vehicle studies. Higher values of specific mass lead to increased sensitivity to efficiency, due to the greater mass impact from increased power. The higher specific mass system is not capable of achieving trip times as short as lower specific mass systems. For example, a specific mass of 10 kg/kWe could not be shown for the 500 day piloted Mars mission, as it was incapable of reaching this trip time with a reasonable value of initial mass or power. Lower specific mass values allow increased power and specific impulse to be used for a more efficient propulsion system.

Power: Power requirements are dependent upon mission trip time, payload mass, and difficulty; as well as specific mass, specific impulse, and efficiency. Power requirements for the lunar cargo mission were found to range from 1 MWe to 8 MWe, depending on specific mass and trip time. Mars cargo missions carrying hundreds of Mg of payload, are estimated to require 1 to 10 MWe, based on calculations for the 600 day cargo mission. The piloted Mars mission drives power requirements into the tens of MWe.

Technology implications: A possible progression of technology development can be seen over the range of SEI missions. The lunar cargo mission starts with modest efficiency, specific mass, and power requirements. Efficiencies of .4, specific masses of 20 kg/kWe, and powers of 1 - 2 MWe would be reasonable first steps in NEP system development. Mars cargo mission requirements show significant commonality with the lunar missions for all parameters, indicating the very real possibility for a common lunar/Mars cargo vehicle, operating at approximately 5 MWe, with common thruster technologies. Mars piloted vehicles drive technology to the next level, requiring more than 10 MWe, with lower specific masses and higher efficiencies, as well. An evolutionary technology development path is therefore evident for both power and propulsion systems.

Acknowledgement

This work was performed by Sverdrup Technology for the NASA Lewis Research Center Nuclear Propulsion Office under contract number NAS3-25266.

References

Auweter-Kurtz, M., H. L. Kurtz, and H.O. Schrade, "Optimization of Propulsion Systems for Orbital Transfer with Separate Power Supplies Considering Variable Thruster Efficiencies," AIAA Paper No. 85-1152, Presented at the 21st Joint Propulsion Conference, Monterey, CA, July, 1985.

Gilland, J. H., "Mission and System Optimization of NEP Vehicles for Lunar and Mars Missions," International Electric Propulsion Conference Paper No. 91-038, presented at the International Electric Propulsion Conference, Viareggio, Italy, October 1991.

Hack, K. J., J. A. George, J. P. Riehl, and J. H. Gilland, "Evolutionary Use of Nuclear Electric Propulsion," AIAA Paper No. 91-3821, Presented at the Space Programs and Technologies Conference, Huntsville, AL, September, 1990.

Jones, R. M. and J. Scott-Monck, "The Status of Power Supplies for Primary Electric Propulsion in the U.S.A.," in Proceedings of the 17th International Electric Propulsion Conference, International Electric Propulsion Conference Paper No. 84-83, held in Tokyo, Japan, May 1984, pp. 614-629.

Mason, L. S., K. H. Hack, and J. H. Gilland, "Nuclear Electric Propulsion for Mars Cargo Missions," in Proceedings of the Seventh Symposium on Space Nuclear Power Systems, Vol. 1, pp. 32-37., Albuquerque, NM, January, 1989.

NASA, Report of the 90-Day Study on Human Exploration of the Moon and Mars, National Aeronautics and Space Administration, Washington, D.C. November 1989.

Palaszewski, B.A., "Lunar Transfer Vehicle Design Issues with Electric Propulsion Systems," AIAA Paper No. 89-2375, Presented at the 25th Joint Propulsion Conference, Monterey, CA, July, 1989.

Priest, C., and G. Woodcock, "Space Transportation Systems Supporting a Lunar Base," AIAA Paper No. 90-0422, Presented at 29th Aerospace Sciences Conference, Reno, Nevada, January, 1990.

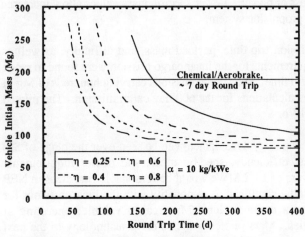

FIGURE 1. Sensitivity of NEP to Thruster Efficiency for the Lunar Cargo Mission, $\alpha = 10$ kg/kWe.

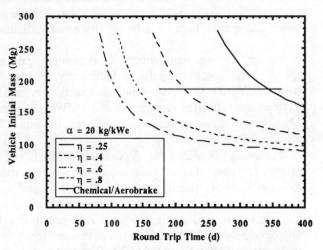

FIGURE 4. Sensitivity of NEP to Thruster Efficiency for the Lunar Cargo Mission, $\alpha = 20$ kg/kWe.

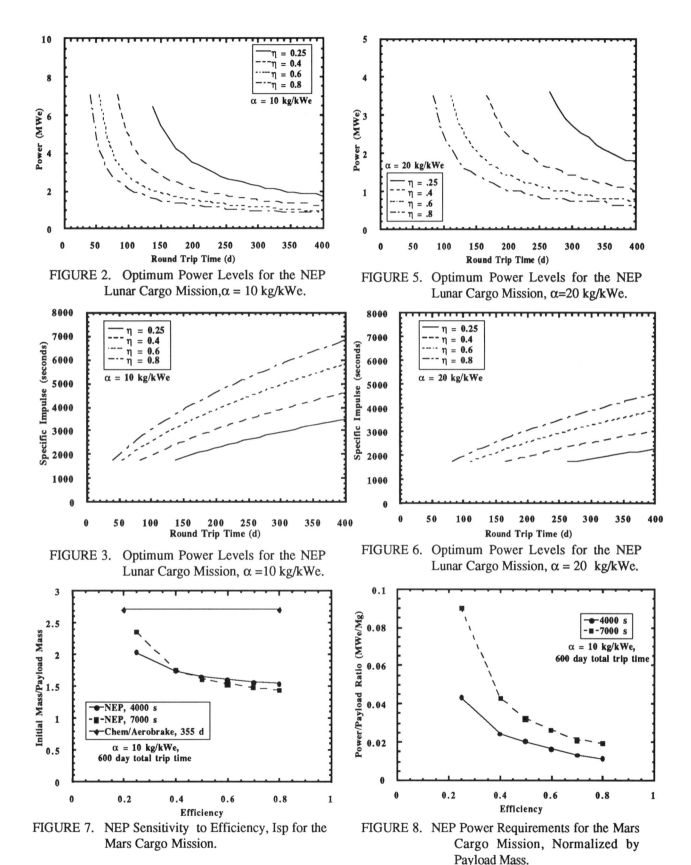

FIGURE 2. Optimum Power Levels for the NEP Lunar Cargo Mission, α = 10 kg/kWe.

FIGURE 5. Optimum Power Levels for the NEP Lunar Cargo Mission, α=20 kg/kWe.

FIGURE 3. Optimum Power Levels for the NEP Lunar Cargo Mission, α =10 kg/kWe.

FIGURE 6. Optimum Power Levels for the NEP Lunar Cargo Mission, α = 20 kg/kWe.

FIGURE 7. NEP Sensitivity to Efficiency, Isp for the Mars Cargo Mission.

FIGURE 8. NEP Power Requirements for the Mars Cargo Mission, Normalized by Payload Mass.

1197

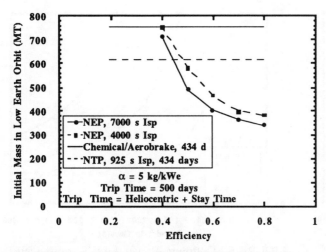

FIGURE 9. NEP sensitivity to Efficiency, Isp for the Mars Piloted Mission. Power varies.

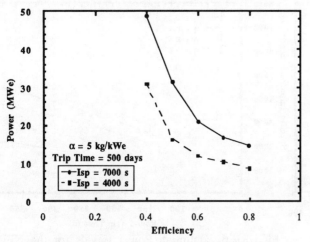

FIGURE 10. NEP Power Requirements for Mars Piloted Mission.

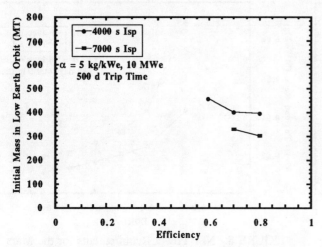

FIGURE 11. NEP sensitivity to Efficiency, Isp for the Mars Piloted Mission, Fixed Power.

1198

POTASSIUM–RANKINE NUCLEAR ELECTRIC PROPULSION FOR MARS CARGO MISSIONS

Richard D. Rovang, Gregory A. Johnson, and Joseph C. Mills
Rockwell International, Rocketdyne Division
6633 Canoga Avenue, HB21
Canoga Park, CA 91303
(818) 718–3318

Abstract

A system study was performed to determine the optimum initial mass to low Earth orbit (IMLEO), trip time, and power requirements for a Mars cargo mission utilizing a nuclear electric propulsion (NEP) spacecraft. The selected architecture consisted of parallel SP-100 type reactors, potassium–Rankine power conversion assemblies, and argon ion thrusters. Parametric potassium–Rankine cycle optimizations were performed to determine the operating state points, radiator area, and mass as a function of power level. NEP specific masses were determined over the anticipated power range, which then allowed mission optimizations to be completed, trading off IMLEO, trip time, and power level. These analyses resulted in a 5.5–MWe power system, 225 tonne IMLEO and a 480–day outbound trip to Mars.

INTRODUCTION

Under the Space Exploration Initiative (SEI), a number of propulsion systems are being considered to provide safe, reliable, and efficient transportation to Mars. Nuclear electric propulsion (NEP) has been identified as a major candidate for use on these SEI missions because it has a high specific impulse, it can provide electrical power on arrival, it is fully testable on Earth, and it has growth potential for missions beyond Mars.

The NEP option is attractive only if a high power, high performance, and low specific mass electric power system is available. A nuclear-powered, potassium–Rankine electrical generation system appears to be a promising approach for Mars missions in the 2016 timeframe. Significant potassium–Rankine development work was performed by the National Aeronautics and Space Administration (NASA), the national laboratories (in particular, Oak Ridge National Laboratory), and industry from 1960 through 1972. This work included corrosion loop, subsystem, and component tests in high temperature potassium environments over extended periods (Hoffman 1984, DeVan et al. 1984, and Anderson et al. 1983).

System studies were performed by Rockwell International in 1988 and 1989 under contract to Department of Energy-Idaho National Engineering Laboratory (DOE-INEL), NASA (Baumeister 1989), and with independent research and development (IR&D) funds. Results of these preliminary studies were presented at the NASA Department of Energy (DOE)/Department of Defense (DOD) Nuclear Electric Propulsion Workshop in Pasadena, California, on 19–22 June 1990. Subsequent to this work, which focused almost totally on the potassium–Rankine power generation system, additional studies have been performed. The purpose of these studies were to identify conceptual designs and to perform first order optimizations of potassium–Rankine-powered, ion-driven spacecrafts suitable for Mars missions. A piloted vehicle design was developed and previously reported (Rovang et al. 1991a and 1991b). This design was based on a potassium–Rankine power system utilizing UN-W/25Re cermet reactors as the heat source. Subsequently, a Mars cargo vehicle design based on a potassium–Rankine power system utilizing SP-100 type reactors as the heat source has been developed and is presented here.

OBJECTIVES

The primary objective of this spacecraft design is to develop a lightweight, reliable, and efficient transportation vehicle suitable for Mars cargo missions, either for a split sprint manned exploration mission or for a precursor mission.

CONCEPT

To meet the stated objective, a potassium–Rankine NEP spacecraft using an SP-100 type reactor as the heat source and ion thrusters is proposed. A gravity environment is induced on the spacecraft by imparting a rotational velocity around the central axis. This concept was previously used for the piloted vehicle design and has been carried through to the cargo vehicle design. This induced gravity simplifies operation of fluid systems in that all fluids can now flow downhill to collect in reservoirs or sumps and provide net positive suction head (NPSH) for pumping. With gravity assist, it is much easier to predict (and design) where all the fluids are in a given system. A gravity field also enables improved radiator performance by eliminating the need for wicking of the condensed heat pipe working fluid back to the evaporator reservoir. The heat pipes are now reflux condensers, and consideration of wettability, depriming, and dryout are minimized.

Some in-orbit assembly of the spacecraft may be required. The central truss is 68.5 m long but may be hinged or telescoped to facilitate stowage for launch into low Earth orbit (LEO) and subsequent deployment. Likewise, the radiator panels may be designed for stowage and subsequent deployment. If not, the modular radiator panels will then need to be attached to the central structure in orbit, and rigorous pre-start-up subsystem testing and verification will also be required before bringing the reactors online. Some pre-start-up testing will probably be required no matter which avenue is taken. Start-up activities, particularly melting of the lithium and potassium working fluids, will be facilitated by an onboard auxiliary power system in the range of 5 to 10 kWe. This auxiliary unit will be shut down after primary power start-up and will only be brought back online if an extended primary power shutdown is required. It is currently envisioned that once the potassium-Rankine power plant is started, it will run continuously throughout the Martian mission; however, shutdown and restart capabilities are included in the design.

DESIGN

The potassium-Rankine NEP system is designed as an autonomous vehicle to carry cargo to Mars and return empty for reuse. In order to determine a point design for the Mars Mission NEP vehicle, it was necessary to perform several mission analyses. Assumptions generic to each analysis are given in Table 1. This information was obtained from the current literature (Friedlander et al. 1990, and George 1991a and 1991b).

Propulsion system specific mass was determined over a wide range of powers. The results of this analysis are shown graphically in Figure 1. The specific mass in this figure accounts for the power subsystem, the thruster subsystem, structure, and power processing and controls. As shown in the graph, the specific mass decreases with increasing power. Thruster efficiency versus specific impulse was obtained from the current literature (Boeing 1990). This information is reproduced in Figure 2. The information in these two figures and Table 1 were used to generate the initial mass to low Earth orbit (IMLEO) as a function of trip time for the specified Mars mission.

Based on the results of this mission analysis, the 480-day outbound and 600-day return mission was selected for further study and propulsion system

TABLE 1. Mars Mission Assumptions.

Departure date	2016 opportunity
Payload to Mars (t)	144
Payload returned to Earth (t)	0
Earth parking orbit (km)	407 (Space Station orbit)
Mars parking orbit (km)	20,077 (Deimos orbit)
Thruster type	Ion
Propellant	Argon
Tankage fraction	0.05 (kg/kg propellant)
Propellant residuals	0.02 (kg/kg usable propellant)

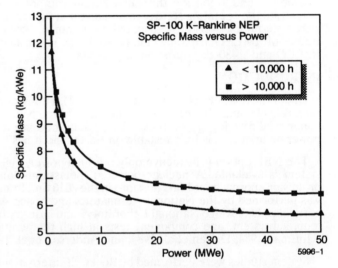

FIGURE 1. Potassium-Rankine/NEP Specific Mass versus Power.

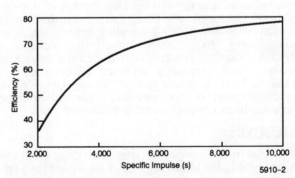

FIGURE 2. Ion Thruster Efficiency.

design. As shown in Figure 3, a "knee" in the curves occurs at the 480-day outbound missions, and the 600-day return mission provides significant mass savings over the 300-day return mission. The one-way mission provides some mass savings over the 600-day return mission but eliminates the option to reuse the Mars cargo vehicle. Hence, the 480-day outbound and 600-day return mission were selected for further study. The details for this mission are given in Table 2, which lists the duration, Δv, and propellant usage for each stage of the mission.

1200

This NEP spacecraft is composed of four major subsystems, including the potassium–Rankine electrical power generation subsystem, power management and distribution (PMAD), the ion thrusters, and tankage and payload. A description of these subsystems as well as a discussion of performance and system reliability follows.

Overall System Description

The selected NEP spacecraft configuration for the referenced Mars cargo mission is shown in Figure 4. This vessel is 68.5 m long, with a 60 m separation distance between the cargo and the nuclear reactors. A planar array

FIGURE 3. NEP Mars Cargo Missions.

TABLE 2. 480–Day Outbound 600–Day Return Mission Summary.

	Time (days)	ΔV (km/s)	Propellant (t)
Earth escape spiral	167.8	6.509	16.0
Helio acceleration	55.1	2.250	5.2
Helio coast	156.5	0.000	0.0
Helio deceleration	84.9	3.579	8.1
Mars capture spiral	15.7	0.680	1.5
Total Outbound	**480.0**	**13.018**	**30.8**
Mars stay	**90.0**	**0.000**	**0.0**
Mars escape spiral	2.4	0.,402	0.2
Helio acceleration	0.0	0.000	0.0
Helio coast	539.8	0.000	0.0
Helio deceleration	25.6	4.493	2.4
Earth capture spiral	32.2	5.992	3.1
Total Return	**600.0**	**10.887**	**5.7**
Total Mission	1170.0	23.905	36.5

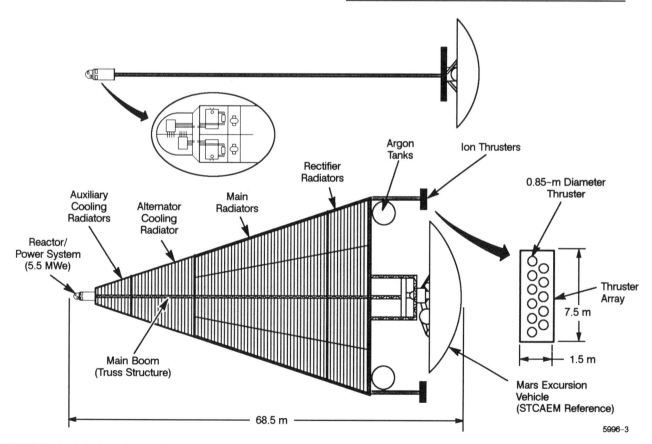

FIGURE 4. Nuclear Electric Propulsion Mars Transfer Vehicle.

of four radiators [the main power conversion assembly (PCA) potassium condensers, the auxiliary cooling loop, the rectifier coolers, and the alternator coolers] are the dominant physical feature of the spacecraft. These radiators are arranged within a 17.5 degree cone half–angle dictated by the reactor shield design. Any material outside this cone will result in neutron scattering and increased payload exposure. The two nuclear reactors and four PCAs are located at the narrow end of the spacecraft. The argon propellant tanks, payload module, and thruster arrays are located at the opposite end of the spacecraft, after the radiators. By spinning the spacecraft at 4 rpm, an induced gravity is produced on the spacecraft. Twenty–two ion thrusters are positioned in two arrays of 11 each, aft of the propellant tanks. The Mars excursion vehicle (MEV), based on the Space Transfer Concepts and Analysis for Exploration Missions (STCAEM) design (Boeing 1990), is placed along the spacecraft central axis beyond other structures. This placement provides minimum interference from the spacecraft during separation and docking, and allows departure and return maneuvers to be performed within the 17.5 degree shield cone.

An overall schematic diagram for the selected 5.5–MWe potassium–Rankine NEP system is shown in Figure 5. This diagram illustrates the subsystem and major component redundancies, relationships, and interfaces. Propulsion system performance characteristics and mass breakdown are given in Tables 3 and 4, respectively, while the vehicle mass summary is given in Table 5.

Potassium–Rankine Power System

The 5.5–MWe NEP system selected in this study has many similarities to previously reported designs (Baumeister 1990). Several design differences have been introduced, however, because of the presence of the induced gravity effects. Particularly, rotating fluid management devices (RFMDs) have been eliminated and the radiator designs have been modified to take advantage of gravity assist reflux action.

The power system consists of two lithium–cooled SP–100 type fast reactors, employing UN fuel pins, coupled to the potassium–Rankine power conversion system (PCS). A reactor outlet temperature of 1350 K was selected based on SP–100 temperature limits and, in turn, a turbine inlet temperature of 1300 K was selected based on trade–offs between system mass and efficiency. Major PCA components include the reactor/primary lithium loop, which extracts heat from the reactors, and parallel potassium power conversion loops, which pickup heat from the boiler producing potassium vapor. The vapor is then converted to electrical energy by a turboalternator. Auxiliary loops are provided to cool lithium and potassium loop components as well as the alternators. Dedicated heat pipe radiators are provided to remove waste heat.

The potassium–Rankine power system was optimized for minimum mass. This optimization consisted of computing system mass and radiator area for various

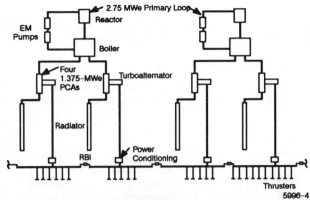

FIGURE 5. Overall Power System Schematic for 5.5–MWe Potassium–Rankine NEP.

TABLE 3. Propulsion System Performance Characteristics.

Specific mass (kg/kWe)	7.50
Specific area (m^2/kWe)	0.176
Gross cycle efficiency (%)	22.7
Net cycle efficiency (%)	21.6

TABLE 4. Propulsion System Mass Breakdown.

Subsystem	Mass (t)
Power system (two modules)	22.8
Total power module	11.4
Reactor	2.1
Shield	2.3
Primary heat transport	1.9
Power conversion assembly	3.0
Auxiliary cooling loops	0.3
Main heat rejection	1.0
Auxiliary heat rejection	0.7
PMAD	11.2
Ion thrusters	3.6
Structure	3.5
Total propulsion system	41.2

TABLE 5. Total Spacecraft Mass Breakdown.

Subsystem	Mass (t)
Payload	144.4
Propellant	37.2
Propellant tanks	1.9
Propulsion system	41.2
Total vehicle	224.7

condensing temperatures. SP-100 reactor and shield masses were obtained from the literature (Deane et al. 1989 and Newkirk et al. 1991). The results of this analysis are shown graphically in Figure 6. The minimum mass occurs at a condensing temperature of between 825 K to 850 K; 850 K was chosen to minimize radiator area.

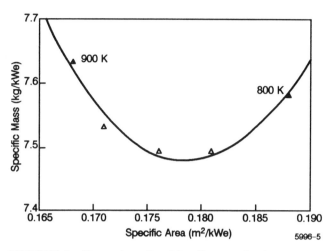

FIGURE 6. Potassium–Rankine System Optimization.

Power Management and Distribution

The PMAD consists of four parallel units. Each unit, shown in Figure 7, contains two remote bus isolators (RBIs) connected to an alternator output. One RBI is for the high voltage circuit, while the other is for the low voltage circuit. The high voltage circuit consists of cabling and rectifiers to convert the AC electrical power to DC power. Each high–voltage bus supplies power to the screen grids of one-fourth of the ion thrusters, while each low voltage bus supplies power to the low voltage loads of the same thrusters. Each thruster is connected to its associated high voltage bus with an RBI. Likewise, the low voltage portion of each thruster is also connected to its associated low voltage bus with an RBI. This arrangement allows for selecting any combination of thrusters throughout the mission. The PMAD configuration provides the capability of powering any thruster from any alternator, producing a high subsystem reliability.

Ion Thrusters

Performance characteristics for the thrusters envisioned for this mission are given in Table 6. Performance parameters were obtained from the literature (Friedlander 1990, and George 1991a and 1991b). The thrusters are arranged in two arrays of 11 each, 20 are required and 2 are redundant, thus ensuring that sufficient thrusters are available throughout the mission duration. The thruster array is divided into four subarrays. Each of these subarrays contains approximately one–fourth of the thrusters coupled to a high voltage bus and a low voltage bus, as previously described.

Reliability

The power system design philosophy and reliability philosophy are interwoven. The power system consists of two independent modules, each capable of providing 50% of the total electrical power. Both modules contain two power conversion units (PCUs), each providing 25% of the total power.

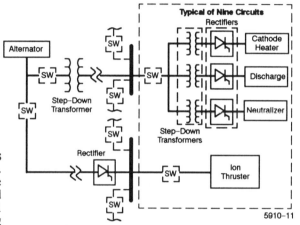

FIGURE 7. PMAD Unit (1 of 4) Electrical Schematic.

TABLE 6. Ion Thruster Performance.

Specific impulse (s)	9,000
Mass (kg)	165
Area (m²)	0.814
Efficiency (%)	78
Input power (kWe)	275
Beam current (A)	107.9
Thrust (N)	4.4

The design philosophy chosen does not utilize the concept of installed redundancy at the major subsystem level. That is, both reactors and all four turboalternators need to operate to produce the design electrical output of 5.5 MWe. However, the design does provide for a "graceful" degradation of electrical power. If, for example, one of the four turboalternators were to fail, the mission could still be continued but at a lengthened trip time, depending on the phase of the mission at the time of failure.

Redundant components, however, are used to enhance the reliability of any one major subsystem. For example, installed redundant heat pipes, primary electromagnetic pumps, and ion thrusters are used. Table 7 shows the probability of delivering the cargo to Mars orbit and of completing the mission at power levels from 100% to 25%. These values show reasonable probabilities of delivering cargo to Mars orbit and of completing the mission with 100% power at the end of mission. This reliability assessment was determined utilizing previously developed potassium–Rankine reliability information and techniques. The reliability analysis employed an iterative process

TABLE 7. Potassium–Ranking NEP System Reliability.

Power (%)	Reliability	
	Outbound	Total Mission
100	0.9332	0.8784
75	0.9785	0.9553
50	0.9994	0.9977
25	0.9999	0.9995

which created a system configuration reliability model, established a failure rate data base for components in the system, then assessed alternate configurations.

CONCLUSIONS

These conceptual design efforts show that a 5.5–MWe NEP spacecraft with a specific mass of 7.5 kWe is possible. At this power level and specific mass, a cargo mission to Mars will take 480 days. The vehicle could be left in Mars orbit or returned to Earth after a 90–day stay. A 600–day return trip requires less than 10 tonne of additional propellant and tankage. A 480–day Mars cargo mission plus 600–day return will require a total of 225 tonne of IMLEO. Additionally, the depth of contingency options made available by this design produces a high probability for mission success. These factors make the potassium–Rankine NEP cargo vehicle concept an attractive candidate for Mars cargo missions.

Acknowledgments

This work was performed at Rockwell International, Rocketdyne Division under independent research and development funding.

References

Anderson, R. V., et al. (1983), *Space Reactor Electric Systems – Subsystem Technology Assessment*, Rockwell International, Energy Systems Group, ESG–DOE–13398.

Baumeister, E. B., et al. (1989), *Ultra–High Power Space Nuclear Power System Design and Development*, NASA Contract NAS325808.

Baumeister, E. B., et al. (1990), "A Potassium–Rankine Multimegawatt Nuclear Electric Propulsion Concept," *Proceedings of the 25th Intersociety Energy Conversion Engineering Conference*, Vol. 1, pp. 121–124.

Boeing (1990), *Space Transfer Concepts and Analysis for Exploration Missions*. NASA Contract NAS8–37857.

Deane, N. A., et al. (1989a), *SP–100 Reactor Scalability Study Final Briefing Package*, DOE Contract DE–ACO3–89SF17787.

DeVan, J. H., et al. (1984), "Compatibility of Refractory Alloys with Space Reactor System Coolants and Working Fluids," *Proceedings of Symposium on Refractory Alloy Technology for Space Nuclear Power Applications*, CONF–8308130, Oak Ridge National Laboratory.

Friedlander, A., et al. (1990), *NEP Performance for 2016 Mars Opposition–Class Missions*, Final Report Presentation, Task Order No. 7, NASA–25809.

George, J. A. (1991a), *Technology and Mission Issues for Nuclear Electric Propulsion*, NEP Technology Meeting, NASA–LeRC.

George, J. A. (1991b), Personnel Telephone Conversation, NASA–LeRC.

Hoffman, E. E. (1984), "Refractory Alloy Components Accomplishments from 1963 to 1972, *Proceedings of Symposium on Refractory Alloy Technology for Space Nuclear Power Applications*, CONF–8308130, Oak Ridge National Laboratory.

Newkirk, D. W., et al. (1991), "SP–100 Multimegawatt Scaleup To Meet Electric Propulsion Mission Requirements," *Proceedings of the 26th Intersociety Energy Conversion Engineering Conference*, Vol. 1, pp. 467–470.

Rovang, R. D., et al. (1991a), "The Potassium–Rankine Nuclear Electric Propulsion Option," *Proceedings of the 26th Intersociety Energy Conversion Engineering Conference*, Vol. 1, pp. 397–402.

Rovang, R. D., et al. (1991b), "Potassium–Rankine Nuclear Electric Propulsion for SEI Applications," *Conference on Advanced SEI Technologies*, AIAA–91–3610.

MISSION ANALYSIS FOR THE POTASSIUM–RANKINE NEP OPTION

Elden H. Cross, Frederick W. Widman Jr., and D. Michael North
Rockwell International, Rocketdyne Division
6633 Canoga Avenue
Canoga Park, CA 91303
(818) 710-4729

Abstract

Mission analyses were conducted to select the design point of a nuclear electric propulsion (NEP) system for a manned mission to Mars. The propulsion system is comprised of ion thrusters with argon propellant and a potassium–Rankine cycle nuclear power plant. Mars parking orbits, departure dates, and outbound/return transfer times were varied to provide a minimum–mass system for a 390-day trip time. The study resulted in a power requirement of 46 MWe and an initial mass in low–Earth–orbit (IMLEO) of 700 tonnes.

INTRODUCTION

System studies of a multi-megawatt nuclear power system using the potassium–Rankine cycle were performed by Rockwell International under contract and internal funds in 1988 and 1989. This work has since been extended to apply the power system concepts developed in those studies to a manned Mars spacecraft with nuclear electric propulsion (Rovang et al. 1991). This paper describes the mission analyses that were conducted to select optimum operating conditions and mission parameters.

MISSION DESCRIPTION

The 2016 opportunity for an opposition-class mission was selected and the study goal was to minimize trip time within this mission opportunity while maintaining reasonable values of IMLEO and power level. A 390–day round trip (crew mission duration) which includes a 30-day stay at Mars was selected. Two mission characteristics which helped reduce the trip time were the use of a crew rendezvous vehicle and a high-altitude circular parking orbit at Mars. By sending the crew to meet the Mars spacecraft at near-escape velocity and retrieving them at the same point, the earth spirals, totaling 64 days, are not included in the total trip time and the crew exposure to Van Allen belt radiation is minimized. A Lunar Transfer Vehicle (LTV) was assumed to be available for crew transfer and would require approximately 100 tonnes of propellant for the two round trips to a high-energy rendezvous orbit. This mass is included in all IMLEO values. A 20,077 km parking orbit (same altitude as Deimos with a 30–hr period) was used at Mars. While requiring more capability from the Mars Excursion Vehicle (MEV) than a lower orbit, this orbit reduces the finite thrust losses during Mars capture and departure and also reduces the spiral time, making more time available for the heliocentric portions of the trip, and thus reducing the delta–v requirements. The mission is illustrated in Figure 1.

SPACECRAFT DESCRIPTION

The spacecraft design is shown in Figure 2 and is more fully described by Rovang et al. (1991). The reactor is located at the front of the vehicle, with the propellant tanks, thrusters, habitat, and MEV at the back, separated from the reactor by the radiators. The power system produces 46 MWe and the power/propulsion system specific mass (alpha) is 3.9 kg/kWe, including everything but the propellants and tankage. The vehicle rotates at 2–4 rpm about the longitudinal axis, providing artificial gravity, which not only benefits the crew but simplifies the fluid systems design in the power system.

The MEV uses an aerobrake during the descent into the Martian atmosphere and LO_2/LH_2 propulsion for deorbit, control, and landing. The delivered surface payload of 25 tonnes includes the surface habitat, supplies, and science equipment. The aerobrake and descent propulsion stage are expended on the Martian surface and the crew utilizes the MEV ascent stage (LO_2/LH_2) to return to the orbiting NEP Mars transfer vehicle (MTV). After crew transfer, the ascent stage is expended in the Mars parking orbit.

MISSION ANALYSIS

Mission analyses were conducted with the aid of the NASA Chebytop program. Launch dates and the split between outbound and inbound legs were optimized for each case. The earth departure orbit was the Space

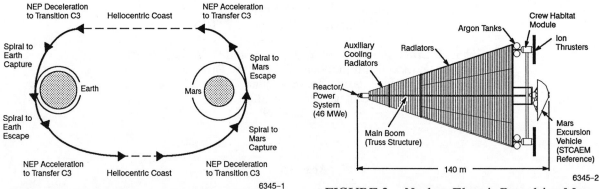

FIGURE 1. NEP Round Trip Mars Mission

FIGURE 2. Nuclear Electric Propulsion Mars Transfer Vehicle.

Station Freedom orbit at 407 km and the Mars parking orbit was at the Deimos altitude of 20,077 km. Payloads were taken from a report by Friedlander et al. (1990) and were 124 tonnes outbound and 40.3 tonnes return (the habitat mass in these payloads may be a little optimistic since the artificial gravity provided in the present design results in a two–part habitat with a connecting passageway). The outbound payload mass includes the MEV, surface payload, and the 40.3 tonnes return payload. NEP mass and performance data were taken from Rovang et al. (1991) and George (1991). Figure 3 shows the propulsion system alpha as a function of power level, and Figure 4 shows the ion thruster efficiency as a function of specific impulse. A tank mass fraction of five percent was used for the Argon propellant tankage with two percent allowed for residuals and boiloff. All missions include a 30–day stay at Mars and trip times do not include the earth spiral times.

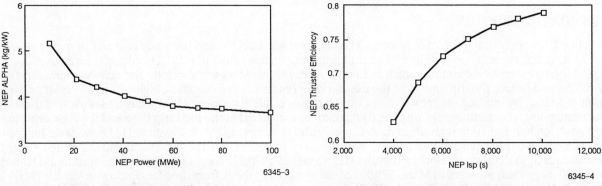

FIGURE 3. Ion NEP Specific Mass.

FIGURE 4. Ion Engine Thruster Efficiency.

The Chebytop low–thrust interplanetary trajectory analysis program was utilized for all mission calculations. The code models planet–vicinity as well as propulsive and coast periods for heliocentric transfer. A driver program was written and utilized to allow round–trip Earth–Mars mission/trajectory analysis for NEP transfer vehicles. The program was used to determine the optimum launch date, transfer time split between the inbound and outbound mission legs, power level, and thruster specific impulse to minimize vehicle IMLEO.

A comparison of several mission approaches is shown in Figure 5. The first option examined low circular orbits at both Earth and Mars with the crew aboard for the entire mission. Substantial reduction of IMLEO or trip time resulted from the use of a crew rendezvous at a high–energy Earth orbit. A representative LTV was used for the rendezvous and 100 tonnes were added to the IMLEO values to account for the propellants required. LTV characteristics for the rendezvous are summarized in Figure 6. The use of a Deimos–altitude orbit was also considered and showed further substantial gains. Finally, for the fixed values of specific mass and thruster efficiency listed, the specific impulse was optimized for each trip time and showed additional gains although much smaller than the other effects.

Circular Mars parking orbits from 1,000 km to 40,000 km were explored to determine the effect of the orbit altitude on vehicle mass. Figures 7 and 8 show the MEV and MTV masses as a function of parking orbit altitude. Scaling for the MEV to account for Mars parking orbit altitude is based on vehicle data in Woodcock (1990). The MEV mass increased substantially as altitude was increased due to the increased propellant required for descent and ascent (the aerobrake mass was held constant). Similarly, the MTV mass decreased due to the

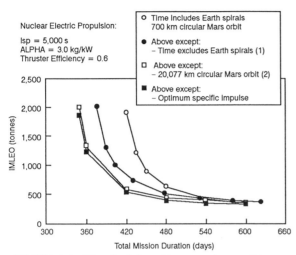

Nuclear Electric Propulsion:

Isp = 5,000 s
ALPHA = 3.0 kg/kW
Thruster Efficiency = 0.6

○ Time includes Earth spirals
  700 km circular Mars orbit
● Above except:
  – Time excludes Earth spirals (1)
□ Above except:
  – 20,077 km circular Mars orbit (2)
■ Above except:
  – Optimum specific impulse

(1) IMLEO includes 100 mt chemical propulsion crew rendezvous vehicle
    for Earth departure and capture
(2) Mars excursion vehicle scaled to perform mission to/from higher orbit

6345-5

FIGURE 5.   Optimization Trades to Minimize
            IMLEO.

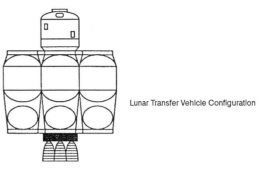

Lunar Transfer Vehicle Configuration

● Lunar transfer vehicle configuration

● Two round trips per Mars mission

  ● Transfer of crew from LEO to NEP vehicle at
    near–escape conditions and return to LEO
  ● Rendezvous with NEP vehicle upon return and
    return crew to LEO

Vehicle Characteristics:
(Single crew transfer, one round trip)

Round trip ΔV	=	8,000 m/s
Propellant	=	50,400 kg
Inert mass	=	6,600 kg
Crew module	=	5,000 kg
IMLEO	=	62,000 kg

Resupply mass for complete
departure and return mission =   100.8 tonnes

6345-11

FIGURE 6.   Earth Crew Rendezvous Vehicle.

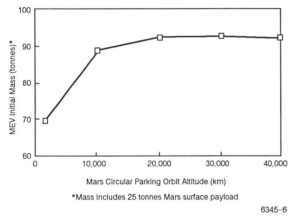

*Mass includes 25 tonnes Mars surface payload

6345-6

FIGURE 7.   Effect of Mars Parking Orbit Altitude
            on Mars Excursion Vehicle Mass.

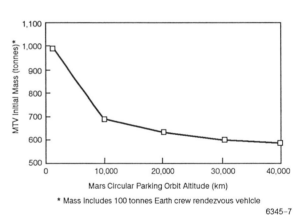

* Mass includes 100 tonnes Earth crew rendezvous vehicle

6345-7

FIGURE 8.   Effect of Mars Parking Orbit
            Altitude on Mars Transfer Vehicle
            Mass.

reduced delta v needed to capture into/escape from the higher orbit. The overall effect on IMLEO is shown in Figure 9. The value of the overall vehicle mass becomes asymptotic at altitudes near 40,000 km with a six percent IMLEO reduction over the Deimos altitude. Since other factors affecting the selection of a parking orbit have not been considered, the Deimos altitude was retained for the rest of the study.

One goal of the study was to minimize trip time and the results are shown in Figure 10. IMLEO is shown versus trip time and the knee of the curve is at about 390 days, with another 30–day reduction adding almost 70 percent to the IMLEO. The specific impulse was optimized for each point and the resulting values are noted on the curve. Electric power levels were also optimized and are shown in Figure 11. At 390 days the required power level is 46 MWe for minimum vehicle IMLEO. The mission is summarized in Table 1. The 390–day total includes 145 days for the outbound leg, a 30–day stay at Mars, and a 215–day return leg. The vehicle mass breakdown is included in Table 2, and Table 3 shows a breakdown of time, delta v, and propellant usage for each portion of the trip.

1207

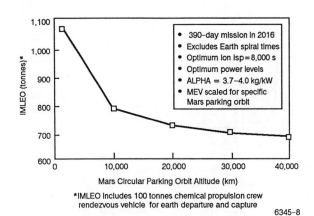

*IMLEO Includes 100 tonnes chemical propulsion crew
rendezvous vehicle for earth departure and capture

6345-8

FIGURE 9.   Effect of Mars Parking Orbit
Altitude on NEP Vehicle IMLEO.

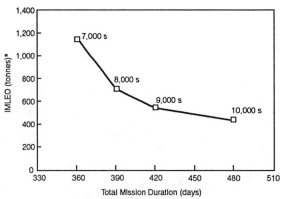

(Does not Include Earth departure or capture spiral times)

* IMLEO Includes 100 tonnes chemical propulsion crew rendezuous
vehicle for Earth departure and capture

6345-9

FIGURE 10.   Effect of Mission Duration on
NEP Vehicle IMLEO.

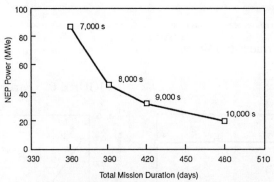

(Does not Include Earth departure or capture spiral times)

6345-10

FIGURE 11.   Effect of Mission Duration on
Optimum NEP Power.

TABLE 1.   390-Day NEP Mission
Summary.

Mission duration:	390 days*
Outbound transfer:	145 days*
Mars stay time:	30 days
Return Transfer:	215 days*
NEP Isp:	8,000 sec
NEP Power:	46.0 MWe
NEP ALPHA:	3.9 kg/kW

*Earth departure and capture spiral
times are not included.

D634-0005

TABLE 2.   390-Day NEP Mission
Vehicle Summary.

Usable Propellant:	275.5 tonnes
Residual propellant:	5.5 tonnes
Propellant tanks:	14.0 tonnes
Power/propulsion:	181.2 tonnes
Outbound payload:	124.0 tonnes
Rendezvous propellant:	100.0 tonnes
IMLEO:	700.2 tonnes

D634-0005

TABLE 3.   390–Day NEP Mission Characteristics.

	Time (days)	ΔV (km/s)	Propellant (tonnes)
Earth escape spiral:	45.2	6.085	44.8
Helio acceleration:	53.6	7.899	53.2
Helio coast:	23.3	0.000	0.0
Helio deceleration:	65.9	10.943	65.4
Mars capture spiral:	2.2	0.394	2.2
Total Outbound:	190.2	25.321	165.6
Mars stay:	30.0	0.000	0.0
Mars escape spiral:	1.5	0.336	1.5
Helio acceleration:	50.7	12.204	50.3
Helio coast:	122.8	0.000	0.0
Helio deceleration:	40.0	11.165	39.7
Earth capture spiral:	18.5	5.765	18.4
Total return:	233.5	29.470	109.9
Total mission:	453.7	54.791	275.5

D634–0005

CONCLUSIONS

Nuclear electric propulsion with the potassium–Rankine cycle power system offers the possibility of fast trip times with competitive values of IMLEO. The use of NEP for fast manned trips is only practical with lightweight systems; the alpha of 3.9 kg/kWe used in this design represents advanced technology and is toward the low end of values found in the literature. Further reductions may be possible with other thruster types, such as the pulsed inductive thruster, the MPD thruster, and others. The design shown herein also offers artificial gravity as a benefit to the design (due to reduction in power system mass), rather than a penalty, as in most concepts that have been considered.

Acknowledgments

This work was supported by Rocketdyne internal funds.

References

Friedlander, A., et al. (1990), "NEP performance for 2016 Mars Opposition-Class Missions," Final Report Presentation, Task Order No. 7, NAS3–25809.

George, J. (1991), "Technology and Mission Issues for Nuclear Electric Propulsion," NEP Technology Panel Meeting, Rosslyn, VA.

Rovang, R., G. Johnson, and J. Mills (1991) "Potassium–Rankine Nuclear Electric Propulsion Option," Preceedings of the 26th Intersociety Energy Conversion Engineering Conference, Vol. 1, August 4–9, 1991, Rockwell International, Rocketdyne Division, Canoga Park, CA.

Woodcock, G. R. (1990), "Space Transfer Concepts and Analysis for Exploration Missions," Third Quarterly Review, NASA Contract NAS8-37857, Boeing Aerospace and Electronics, Huntsville, AL.

FLOW BOILING IN LOW GRAVITY ENVIRONMENT

Basil N. Antar, and
Frank G. Collins
The University of Tennessee
Space Institute
Tullahoma, TN 37388
(615) 455 0631

Masahiro Kawaji
Department of Chemical Engineering
and Applied Chemistry
The University of Toronto
Toronto, Ontario M5S 1A4
(416) 978 3064

Abstract

An experimental proceedure for examining flow boiling in low gravity environment is presented. The proceedure involves both ground based and KC-135 flight experiments. Two experimental apparati were employed, one for studying subcooled liquid boiling and another for examining saturated liquid boiling. For the saturated experiments, liquid nitrogen was used while freon 113 was used for the subcooled experiments. The boiling phenomenon was investigated in both cases using flow visualization techniques as well as recording wall temperatures. The flow field in both cases was established by injecting cold liquid in a heated tube whose temperature was set above saturation values. The tubes were vertically supported with the liquid injected from the lower end of the tube. The results indicate substantial differences in the flow patterns established during boiling between the ground based, (1-g), experiments and the flight experiments, (low-g). These differences in the flow patterns will be discussed and some explanations will be offered.

INTRODUCTION

The phenomena of flow boiling has emerged as an essential technology issue for several space based applications. The understanding of this phenomena is necessary for the space operation of boilers and condensers, for evaluating heat pipe thermal effenciancy and for space based cryogen fluid transfer and storage, among others. Great deal of study and attention have been devoted to both pool boiling and flow boiling phenomen under terrestrial conditions. Both phenomena are exetremely complex and not fully understood in normal gravity conditions to date.

With the advent of space flight and exploration many system designs and operations require thorough understanding of fluid behaviour in low gravity environment. Low gravity fluid dynamics as a general topic is not yet fully understood, but in fact it is a field requiring a great deal of study. The cases of flows in which phase change takes place such as boiling, solidification, melting and evaporation, the flow of energy must be accounted for in addition to fluid dynamics. In the present study we confine the discussion to the flow boiling case only.

When cold liquid is injected into a hot pipe, the tube wall temperature will drop while the liquid temperature will rise. If the temperature of the tube is above the saturation temperature the liquid will begin to evaporate resulting in the establishment of a complex two-phase, liquid-vapor flow pattern. For the case of a vertical tube with liquid injected from the bottom end, the two-phase flow pattern consists of four distinct regions: single phase liquid, nucleate boiling region, saturated convective boiling and dispersed flow region. The flow pattern in the saturated convective boiling region was found by Kawaji et al. (1985) to depend on whether the injected liquid is saturated or subcooled. For the saturated case the flow pattern was annular in which the liquid flows along the tube wall with a vapor core. While in the subcooled case the flow took the form of inverted annular flow in which the there was a vapor film along the wall with a liquid core in the center of the tube.

In General, the heat flux in two-phase flows depends very strongly on the nature and the pattern of the flow of both both the liquid and the vapor. The heat transfer coefficient, h is a function of whether the fluid is liquid or vapor or a mixture of both and whether it is laminar or turbulent. Thus the first step in determining such essential parameters as h for a new flow situation a thorough understanding of the flow pattern. Also, the knowledge of the flow pattern is a mandated requirement for building detailed multi-field numerical models to describe a specific two phase flow.

EXPERIMENTAL APPARATUS

From numerous 1-g experimental observations it was concluded that the transient flow boiling process arising from injecting cold liquid in a hot tube is normally very short. However, the time for full quench is a strong function of the liquid injection rate. Experimenting with liquid nitrogen injected into a 12-mm OD tube a quench time of the order of 30 seconds was determined for tube lenghts of up to 1 m. The NASA/KC-135 airplane, through special flight maneuvers, can provide up to 30 seconds of approximately .01 g of laboratory conditions within its interior. Two classes of low gravity flight experiments were designed to be performed inside the NASA/KC-135 airplane. For both classes of experiments two different test sections were designed. One test section was made of quartz in order to allow for observing and recording the flow boiling phenomenon, and another made of stainless steel for collecting temperature data.

Saturated Liquid Boiling Experiment

One test section for this class of experiment was a quartz tube 12-mm OD, 10-mm ID and approximately 1 m in length. The quartz tube was housed within a plexiglass outer tube with the annular gap between the tubes evacuated. To prevent liquid condensation the quartz tube was wrapped with a heating wire. The liquid coolant used in the saturated boiling experiments was liquid nitrogen at atmospheric conditions. The flow loop was a once through in which the coolant was injected into the test section from an accumulator specifically designed to operate in the low gravity environment. The test section for these experiments was held vertically with respect to the aircraft floor. In all of these tests the liquid nitrogen was injected into the lower end of the test section with its upper end open to the atmosphere through a back pressure regulator. The flow patterns during boiling, were recorded using a high speed cine camera in which both close up and wide shots were used at different locations along the axis of the tube. In all of these tests film speeds ranging from 250 f/s up to 1000 f/s were used. A schematic of the flow loop for this set of test is shown in Figure 1.

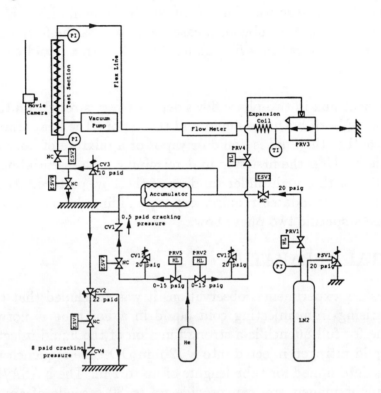

FIGURE 1. Low Gravity Liquid Nitrogen Flow Loop.

A second set of experiments in this class were conducted using the same apparatus with the quartz tube replaced with a stainless steel test section. The dimensions of this test section are the same as the quartz. Also, the same flow loop for these tests was used in order to repeat the conditions for the tests with the quartz tube. The objective of this set of experiments was for recording tube wall temperature data. Three copper-contantan, type-T thermocouples were attached to the outside surface of tube. The three thermocouples were placed 15 cm apart in the axial direction along the outside of the tube with the first was located at approximately 15 cm downstream of the entrance to the tube.

The saturated boiling experiments apparatus was flown twice on the NASA/KC-135 airplane. Once in April 1991 and another time in August 1991. Line quench tests were performed on a total of approximately 100 parabolas with each parabola offering a maximum of 25 seconds of .01 g. Each low gravity test commenced by opening the valve at the inlet to the test section at the start of the low gravity portion of the flight parabola. Great care was exercised to ensure that the tests were repeated under identical flow conditions in order to obtain a definitive picture for the flow patterns. Nevertheless minor variations in test conditions between runs were unavoidable.

For the case of a flow rate corresponding to approximately 1 bar of pressurization pressure a consistant flow pattern emerged when the flow was observed in slow motion. It was observed that, at a fixed location along the tube axis, the flow is initially gaseous nitrogen. This is then followed by a region of large liquid nitrogen droplets interspersed with liquid filaments which meandered between the tube walls along its flow path. Such flow presisted for a relatively long time. This flow regime was then followed by an annular region in which the liquid was observed to wet the tube wall. Figure 2 shows a schematic interpertation of the flow pattern observed. The flow pattern in these experiments was markedly different from those observed using the same appartus when the experiments were repeated in 1-g laboratory conditions. For the 1-g case the dispersed flow region was dominated by relatively small droplets uniformly flowing along the wall of the tube at much higher speeds than in low-g conditions. Nevertheless, this regime was followed by what appears to be an annular flow region in which the liquid was flowing along the wall of the tube.

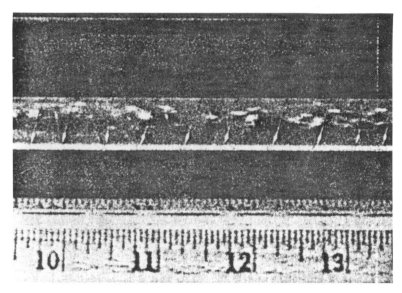

FIGURE 2. Typical Low-g Flow Pattern for Saturated Liquid Boiling.

Subcooled Liquid Boiling Expriments

In a second class of experiments, the flow pattern arising from injecting a subcooled liquid into a hot tube was investigated in low gravity environment. A test apparatus was constructed for that experiment in which the test section was a quartz tube 14 mm ID and 1.2 m long. The working fluid in these tests was freon 113 at approximately 30° C, and the test section was heated with a heating tape up to a temperature of 300° C. All of these tests were conducted under atmospheric pressure conditions whenever possible. The tube in these tests was aligned in the horizontal direction with respect to the floor of the aircraft. A schematic of the test loop for these experiments is shown in Figure 3. Details of the apparatus and of the tests performed are given in Kawaji et al. (1991).

This apparatus was flown on board the NASA/KC-135 in June 1990 and again in April 1991. Again as in the previous class of experiments, all the tests were conducted during the low gravity portion of each parabola. It was observed that the liquid immediately vaporized upon injection into the hot tube filling the test section with vapor. Subsequent to the initial full vapor region, a region of mixed liquid-vapor region appeared with the vapor forming a thick film between the liquid mass and the tube wall.

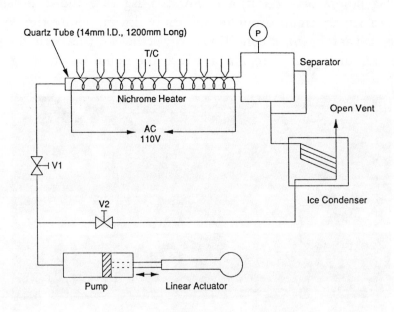

FIGURE 3. Flow Loop for Subcooled Liquid Boiling Experiments.

The liquid core in this region was often thin, more closely resembling a thick cylindrical liquid filament than that in the traditional inverted annular flow pattern. The thick liquid filament was observed to flow mostly in the middle of the tube with some meandering. The liquid filament was mostly smooth and continuous while sometimes bulging at places to nearly fill the tube. Figure 4 shows a trace of the flow as recorded by a still camera. Further downstream from the entrance the liquid filaments were seen to break up into large droplets. At no time during these tests did the tube become rewetted, although heat was being removed continuously from the tube by film boiling or forced convection to the vapor.

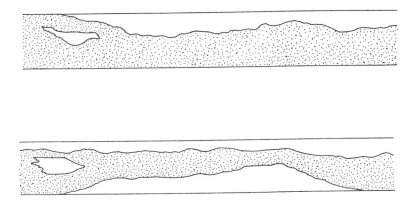

FIGURE 4. A Schematic of a Typical Flow Pattern for Subcooled Boiling in Low Gravity.

CONCLUSIONS

Although the experiments described above are not conclusive, the evidence from the experiments regarding the flow patterns during flow boiling in low gravity leads to the following. The dispersed flow region, the region between the fully vapor and the annular flow regions appear to be influenced the greatest by the gravity field. In low gravity, the liquid droplets are very large and nonuniform in shape. The droplet shape was found to vary from spherical to long meandering threads. In one-g environment, on the other hand, the droplets are spherical in shape and small flowing along the wall of the tube. This variation in the dispersed flow region may lead to diminishing heat flux in low gravity environment between the wall of the tube and the fluid.

Acknowledgments

All of the work described here was supported by the following organizations: Office of Space Commercialisation, NASA/Headquarters and the Boeing Aerospace corporation for B.N.A. and F.G.C., and the Canadian Space Agency for M.K.

References

Kawaji, M., Y. S. Ng, S. Banerjee, and G. Yadigaroglu (1985) "Reflooding with Steady and Oscillatory injection:Part I - flow regimes, void fraction and heat transfer," *J. Heat Transfer*, 107:670-678.

Kawaji, M., C. J. Westbye, and B. N. Antar, (1991) "Microgravity Experiments on Two-Phase Flow and Heat Transfer during Quenching of a Tube and Filling of a Vessel," *Trans. 27th National Heat Transfer Conf., City , ST, dates.*

DEFINITION OF CONDENSATION TWO PHASE FLOW BEHAVIORS FOR SPACECRAFT DESIGN

Thomas R. Reinarts and Frederick R. Best
Department of Nuclear Engineering
Texas A&M University
College Station, TX 77843-3133
(409) 845-4161

Wayne S. Hill
Foster-Miller, Inc.
350 Second Ave.
Waltham, MA 02154-1196
(617) 890-3200

Abstract

This paper presents an analysis of the condensation heat transfer data from 1-G and zero-G testing of a two-phase flow experiment employing R-12 (dichlorodifluoromethane) as the working fluid. The data is compared with condensation models. Condensation heat transfer coefficients from the working fluid to the inner wall of the condenser tube, calculated from the data by using an energy balance across the condenser cooling water, are presented for both zero- and 1-G conditions. The measured zero-G heat transfer is compared with Pohner's film model and Best's heat transfer coefficient. A comparison is made between the zero-G and 1-G heat transfer, and a discussion of the results follows.

INTRODUCTION

Future military and civilian space missions will benefit significantly from the energy-intensive phase change of liquid-vapor two phase flows. The advantages of incorporating two phase flows are well known, including reductions in system size, mass, and power consumption. Potential applications for two phase flow technology in future space missions are widespread, including thermal management systems, dynamic power systems, heat pumps, and storage and transfer of cryogenic propellants. Currently, relatively little is known with certainty about the behaviors of two phase flows in a microgravity environment. Improved modeling for zero-G conditions will allow for spacecraft designs that can take full advantage of the benefits of two phase flow.

A two phase experiment that used R12 (diclorodifluoromethane) as the working fluid was designed, constructed, and tested in zero-G aboard the NASA KC-135 plane. Ground testing was also performed. The experiment included adiabatic, condenser, and evaporator sections. The system also includes a two-phase pump, a proprietary device designed by Foster-Miller. The condensation data is presented here.

APPROACH

A schematic of the experimental apparatus is presented in Figure 1. Foster-Miller, Inc. was responsible for the design and fabrication of the experiment. Texas A&M University was responsible for equipment sizing calculations, two phase flow modeling, and data reduction and analysis.The experiment included two adiabatic sections, a condenser section, and an evaporator section. The two phase pump accepts an arbitrary mixture of liquid and vapor at its inlet, separates the phases centrifugally to high purity, and pumps them individually. This allows for direct measurement of the mass flow rate of each phase before the phases are remixed just upstream of the adiabatic inlet. In addition, it is not necessary to subcool the working fluid before the pump entrance. This also means that the evaporator power is not directly linked to the mass flow and outlet quality. Instead, energy can be added to or removed from the system as desired to change from one operating condition to the next. Consequently, a wide range of flow rates and qualities can be examined with relatively little evaporator power. A complete description of the experiment is given by Hill and Best (1991).

The condenser section analyzed here is the second of the two condenser sections in the working

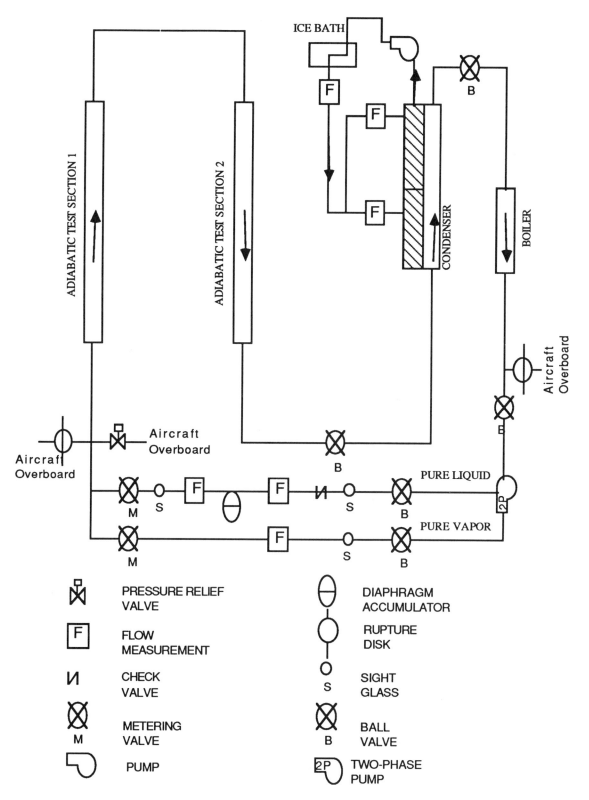

FIGURE 1. Two-phase Experiment Instrument Locations.

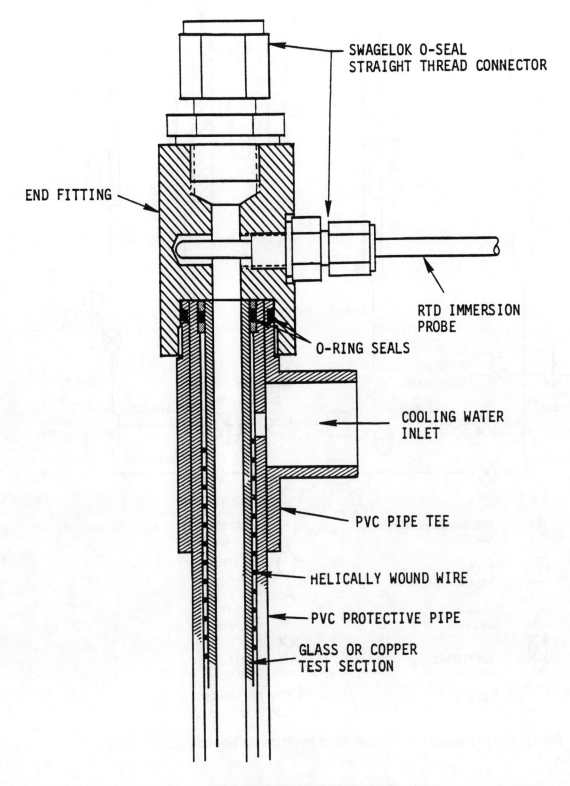

SWAGELOK O-SEAL
STRAIGHT THREAD CONNECTOR

END FITTING

RTD IMMERSION
PROBE

O-RING SEALS

COOLING WATER
INLET

PVC PIPE TEE

HELICALLY WOUND WIRE

PVC PROTECTIVE PIPE

GLASS OR COPPER
TEST SECTION

FIGURE 2. Condenser Detail.

fluid flow path. This section is a thin walled copper tube, with a length of 0.355 m, an outside diameter of 12.7 mm, and an inside diameter of 8.7 mm. The first condensing section is glass for flow visualization purposes. Since the thermal conductivity of glass is much lower than that of copper, the condensation heat transfer measurement is more accurately carried out in the copper section. A diagram of the condenser is shown in Figure 2. The cooling fluid is water flowing in an annular tube around the test section, counter-current to the working fluid. The water is cooled by running it through copper coils immersed in an ice water bath.

RESULTS

The data showed that the water coolant, R12 liquid, and R12 vapor flows were constant within 5 percent. The data for Figures 3 through 9 come from the afternoon flight of 9 April 1990, tenth parabola set, first parabola of the set.

Figure 3 is a plot of condenser cooling water inlet temperature versus time. This figure shows an unexplained oscillation in temperature of ±0.2 K around a slowly increasing mean. The slowly increasing average temperature is normal and is caused by the slow depletion of the ice bath. There is no ready explanation for the temperature fluctuation. However, the character of the flow is constant through the zero-G period and therefore does not contribute to a change in condensation heat transfer during zero-G. Having characterized the condenser's thermal hydraulic boundary conditions, the change in condensation due to zero-g can be analyzed.

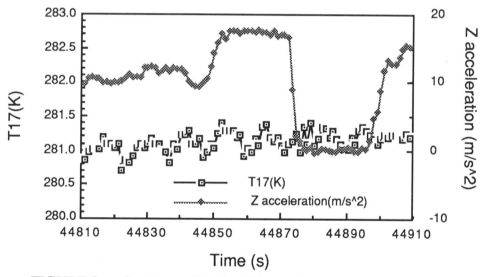

FIGURE 3.   Condenser Cooling Water Inlet Temperature versus Time.

Figure 4 is a plot of the temperature rise of the copper condenser's cooling water versus time. The accuracy of this differential temperature measurement is ±0.01 K. The cooling water ΔT rises slightly upon entering 2-G and decreases 31 percent (relative to the 1-G value) upon entering zero-G. This is a statistically significant reduction in cooling water temperature rise which indicates a decrease in condensation heat transfer. However, other factors, as described below, must be estimated or accounted for as competing effects before all of this temperature change can be ascribed to zero-G condensation effects.

Figure 4 shows that the condenser has not reached equilibrium by the end of the zero-G period. The temperature rise of the cooling water is still decreasing, although at a much reduced rate

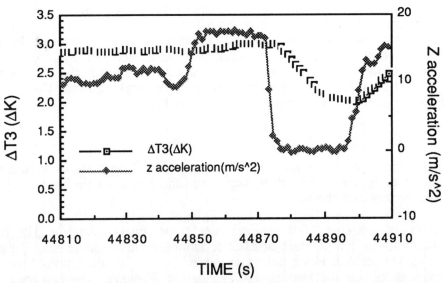

FIGURE 4. Copper Condensing Cooling Water Temperature Rise versus Time.

relative to the initial rate upon entering zero-G. Figure 5 shows the copper condenser cooling water outlet temperature versus time. Since the cooling water inlet temperature is constant, Figure 5 implies that the condenser is cooling down during zero-G. Thus, the cooling water is carrying away not only condensation energy but also some stored energy from the material of the condenser. The rate of temperature decrease at the end of the zero-G period is approximately 0.006 K/s which corresponds to a rate of less than 1 W of stored energy removal from all of the condenser material. The condensation heat transfer power is generally on the order of 100 W. Therefore, the temperatures at the end of the zero-G period, although not at equilibrium, are sufficiently close to contribute negligibly to experimental error. The condensation heat transfer coefficients determined from the data are calculated using the temperatures at the end of the zero-G period.

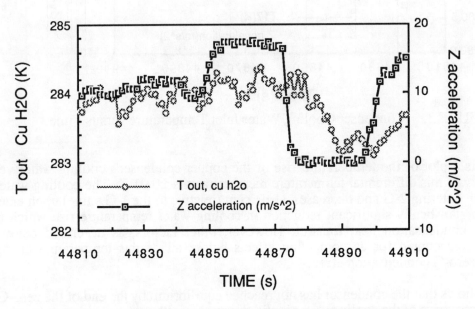

Figure 5. Cooling Water Outlet Temperature versus Time.

Another potential source of interference to experimentally determining the condensation heat transfer coefficient would be changes in system temperature and pressure. That is, if working fluid pressure and temperature decreased, this would also cause condensation heat transfer to decrease. Figures 6 and 7 show the working fluid mixture temperature and pressure at the inlet to the copper condenser section. Recall that there is a glass condenser section just before the copper section. These figures show that temperature and pressure rise slightly during zero-G. This can only contribute to an increase in the condensation heat transfer rather than the observed decrease. In fact, a temperature and pressure rise is expected for a system with a fixed rate of vapor and liquid injection which experiences a decrease in the ability to condense vapor. This temperature rise does

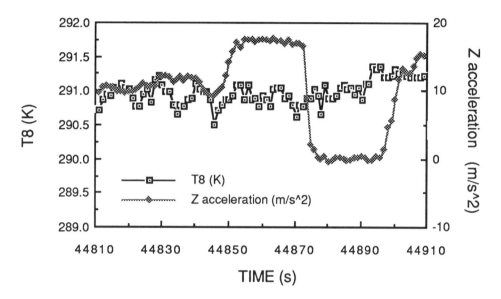

FIGURE 6. Copper Condenser Inlet Working Fluid Temperature versus Time.

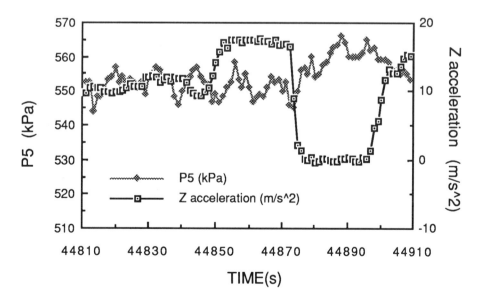

FIGURE 7. Condenser Inlet Working Fluid Pressure versus Time.

not contribute to an error in the measurement of condensation heat transfer because the actual temperatures are used in the calculation. Note that temperature and pressure seem to be leveling out at the end of the zero-G period.

Another factor which affects the time to reach equilibrium in the condenser is the time required for the flow regime to reorient itself to zero-G and for this new flow regime to propagate through the system. Visual flow observations show that the field reorients itself vertically within 1 s. Travel time from the mixing point to the condenser is on the order of a few seconds. The condenser, therefore, sees between 15 and 20 s of true zero-G flow. The data seem to show that the condenser is usually near equilibrium by this time. Having considered the major factors which affect condensation heat transfer in the experiment, the results of the calculation are presented.

The heat transfer coefficient from the working fluid to the inner copper tube wall is calculated by using an energy balance and the standard form of Newton's Law of cooling. First, an energy balance is used to calculate the overall energy transfer from the working fluid to the coolant. The heat transfer coefficient between the the outer copper wall and the water is essentially constant and equal to 7000 $W/m^2 \cdot K$. Ambient energy gains along the length of the condenser section must also be accounted for. For this experiment, a conservative estimate of ambient energy addition to the coolant of the the second condenser section is 5 W. For this and most other parabolas, 5 W is less than 5 percent (and often less than that) of the total energy addition to the coolant, and therefore can be neglected without significantly affecting the calculated heat transfer coefficients. Using this technique, the instantaneous condensation heat transfer coefficient was calculated and is shown in Figure 8.

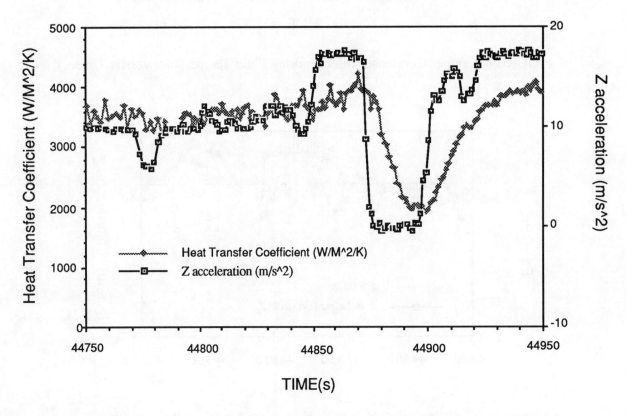

FIGURE 8. Measured Condensation Heat Transfer Coefficient versus Time.

For this parabola, the 1-G heat transfer coefficient from the working fluid to the inner copper wall is 3500 W/m²·K. The zero-G heat transfer coefficient is 2000 W/m²·K at the end of the parabola. The heat transfer coefficient is seen to be relatively constant in 1-G and to increase only slightly in 2-G, then decrease significantly in zero-G. The heat transfer coefficient is thought to be relatively insensitive to 2-G and very sensitive to zero-G because the flow is stratified in 1- and 2-G but annular in zero-G. Under stratified conditions, the increased "g" serves to pull condensed liquid off the inner, upper surface of the condensing tube more quickly than in 1-G, marginally improving condensation heat transfer. However, the annular flow conditions of zero-G effectively block condensation everywhere, degrading condensation heat transfer. This effect would not be as pronounced if the 1- and 2-G flows were annular because the flow regime reorientation would be from one annular condition to another.

Table 1 is a list of KC-135 flights, the measured condensation heat transfer coefficients for each parabola set for 1- and zero-G conditions, and the predicted zero-G condensation heat transfer coefficient to be discussed below. These results show that the measured zero-G condensation heat transfer coefficient is, with one exception, always less than the 1-G value. The average value of the measured zero-G to 1-G ratio is 74 percent with the lowest ratio being parabola set 2 of April 9, at 51 percent and the highest value parabola set 6 of May 30, at 99 percent. Stated another way, the measured condensation heat transfer coefficient decreased on average by 26 percent under zero-G conditions.

TABLE 1.   Condensation Heat Transfer Coefficients.

Flight (all 1990)	Parabola Set	Zero-G Heat Transfer Coefficient (W/m²/K) Predicted	Zero-G Heat Transfer Coefficient (W/m²/K) from Data	1-G Heat Transfer Coefficient (W/m²/K) from Data
April 6	3	935	1,930	3,300
April 9 a.m.	2	1,002	1,276	2,500
April 9 a.m.	3	890	1,863	2,854
April 9 a.m.	7	1,585	2,608	3,584
April 9 a.m.	8	1,340	1,789	3,123
April 9 a.m.	9	1,310	1,466	2,163
April 9 a.m.	10	1,870	2,000	3,500
April 9 p.m.	1	2,388	3,202	4,454
April 9 p.m.	4	2,269	3,251	4,158
April 9 p.m.	7	995	1,268	1,426
April 9 p.m.	8	2,190	3,773	4,158
April 9 p.m.	10	1,889	2,418	1,158
May 30	2	1,618	2,311	3,123
May 30	3	1,853	2,701	3,584
May 30	4	2,417	3,964	4,158
May 30	5	1,786	3,123	3,398
May 30	6	1,314	1,314	1,331

MODELING

Scoping design work for this experiment used Pohner's (1986) condensation model to predict heat transfer coefficients. This model performs a momentum and energy balance on a shear-controlled wall film. When this model was run for the test conditions actually achieved in the KC-135 flights, the heat transfer coefficients were found to be an order of magnitude lower than the measured values. The reason for the discrepancy is thought to be due to the difference between Pohner's film thickness and the film thickness in the experiment as recorded on the Digital Imager tapes. The tapes show that the liquid film is always wavy, frequently disturbed, and that there is significant entrainment. Pohner's model does not include entrainment nor allow for film waviness in heat transfer through the film. Entrainment removes liquid from the film, thinning it and increasing condensation. Waviness decreases the effective thickness of the film, also leading to increased heat transfer. Although Pohner's model is conceptually good, it was decided that a model specifically derived for low quality, thick films should be used rather than modifying Pohner's model.

Best (1986) proposed a heat transfer coefficient for low-quality two-phase flow conditions. The model assumes that the liquid film on the wall is so thick that the wall boundary layer is the same for the low quality flow as for a single phase flow with the same volumetric flow rate. Therefore, heat transfer is limited by energy transport through an all-liquid boundary layer at the tube wall, and the velocity profile in this boundary layer is characterized by the the superficial velocity of the two phase vapor-liquid core. The vapor and liquid velocities are assumed equal, and all properties are evaluated at bulk fluid saturation conditions. The local heat transfer coefficients, based on the Dittus-Boelter equation, is given by

$$\frac{hd}{k} = 0.023 \, Re^{0.8} \, Pr^{0.3} \left(1 + x \frac{v_{fg}}{v_f} \right)^{0.8}$$

(1)

where x is the quality, v's are specific volumes evaluated at saturation conditions, and all other terms are evaluated for the total mass flowing as saturated liquid.

Figure 9 is a plot of the heat transfer coefficient and the quality predicted using this equation as a function of position in the condenser for the same conditions as all the other figures in this section. The average heat transfer coefficient from Figure 9 is 1870 W/m^2·K as compared with the measured zero-G value of 2,000 W/m^2·K. The heat transfer coefficients predicted by this equation are shown in Table 1. On average, the predicted heat transfer coefficients are within 31 percent of the measured values, the largest discrepancy occurring for parabola set 3 of April 9, which is only 48 percent of the measured value. All of the zero-G predicted values are less than the measured values, but no obvious correlation was found between accuracy and flow regime.

CONCLUSIONS

The condensation heat transfer experiments demonstrated that the test article is suitable for carrying out such heat transfer experiments. Tests showed that there is a significant difference, a 26 percent reduction, in heat transfer between 1-G and zero-G for the flow conditions tested. The thick, wavy film produced in zero-G was unlike classical annular films and was not modeled well by the Pohner approach. A Dittus-Boelter type heat transfer correlation was found to match the results to within 31 percent on average. This correlation would be expected to be significantly improved if it were fitted specifically to the test conditions.

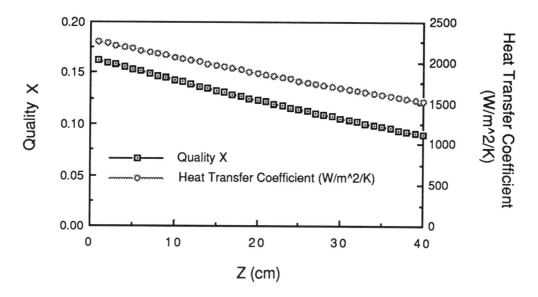

FIGURE 9. Condensation Heat Transfer Coefficient and QUality versus Position.

Acknowledgments

The work described here was performed under a Phase II Small Business Innovative Research (SBIR) program funded by the Air Force Phillips Laboratory. Additional support was provided by the Texas A&M Center for Space Power and the NASA Johnson Space Center.

References

Best, F. R., L. Kachnik, and D. J. Lee (1986) "Space Microgravity Two Phase Flow Safety Considerations," in *Trans. of the American Nuclear Society*, Vol. 52, pp 489-490.

Hill, W. S. and F. R. Best (1991) *Definition of Two-Phase Flow Behaviors for Spacecraft Design*, AFK-0062-FM-8933-418, Air Force Phillips Laboratory, Kirtland AFB, NM.

Pohner, J. A.(1986) "A Nusselt-Type Analysis of Steady, Condensing and Evaporating Annular Flows in Circular Tubes," in *Proceedings AIAA/ASME 4th Joint Thermophysics and Heat Transfer Conference*, Boston, MA, 2-4 June 1986.

TWO-PHASE FLOW CHARACTERIZATION
FOR FLUID COMPONENTS AND
VARIABLE GRAVITY CONDITIONS

John M. Dzenitis
NASA Johnson Space Center
EC/Crew and Thermal Systems Division
Houston, TX 77058
(713) 483-9147

Kathryn M. Miller
NASA Johnson Space Center
EP/Propulsion and Power Division
Houston, TX 77058
(713) 483-4546

Abstract

This paper describes a program initiated by the NASA Johnson Space Center to investigate vapor-liquid flow regimes and pressure drops in pipe components and variable gravity conditions. This program supports the Space Station Freedom External Active Thermal Control System design and future space missions, including the Space Exploration Initiative activities. The objectives for this program include studying two-phase flow behavior in fluid components (smooth pipes, bellows lines, quick-disconnect fittings), expanding the two-phase database for zero-g conditions, developing a database for low-g conditions (for example, Moon-g, Mars-g), and validating models for two-phase flow analyses. Zero-g and low-g data will be gathered using a Freon-12 flow loop during four test series on the KC-135 aircraft beginning in August 1991.

INTRODUCTION

Increasing waste thermal power levels caused an evolution in space system thermal control from conduction paths to single-phase working fluid loops. Space Station Freedom will employ the next step in that evolution. Its higher power levels and longer transport distances have led to use of a two-phase ammonia working fluid system, the External Active Thermal Control System (ATCS), providing a virtually isothermal energy sink at lower mass and pump power than possible with a single-phase system. Future space missions, such as the planetary voyages of the Space Exploration Initiative (SEI), will have even more demanding thermal management requirements. The full potential of two-phase systems will be used to make these missions more viable. Unfortunately, the highly complex phenomena of two-phase flow often requires empirical treatment, and at this time zero-g data is sparse, and there is no published low-g data.

The Johnson Space Center (JSC) has previous experience in the research and development of two-phase space systems, including: reduced-gravity flight experiments on the KC-135 aircraft (for example, Kachnik et al. 1987, Chen et al. 1988); ground and Shuttle flight experiments of heat pipes (for example, Rankin 1984, Kosson et al. 1990); and, most notably, ground evaluation and development of two-phase thermal control systems for Space Station Freedom (for example, Brady 1990).

The JSC project described in this text is a KC-135-based experimentation effort currently in progress. The following sections describe the program, experiment package, test objectives, testing status, and future work.

PROGRAM OVERVIEW

The program entitled "Two-Phase Flow Characterization for Fluid Components and Variable Gravity Conditions" was initiated in January 1991 to support both the Space Station Freedom External ATCS design and the SEI activities. The objectives for this program include

- studying two-phase flow phenomena in fluid components (smooth pipes, flexible bellows lines, quick-disconnects) at one-g and zero-g conditions;
- developing a database for two-phase flow in smooth pipes for low-g (Moon-g, Mars-g) conditions; and
- validating flow regime and pressure drop models for two-phase flow analyses.

This project is a cooperative effort between the Propulsion and Power Division and the Crew and Thermal Systems Division at NASA JSC. Both the divisions at JSC and McDonnell Douglas Space Systems Company are primary participants and responsible for managing the project, providing test support, and analyzing the test results for the Station portion of the program. A grant was awarded to Texas A&M University to provide test support and data reduction for the project, and analysis for the low-g portion of the project. Foster-Miller Inc. received a subcontract from Texas A&M University to provide additional technical and flight support to the program. In addition, the NASA Lewis Research Center was invited to participate in the program on a working level. The USAF Phillips Laboratory agreed to loan the experiment package to the JSC for the duration of the program.

The test plan and schedule is shown in Figure 1. The major parts of the project are modifications to the experiment package, two-phase flow model development, ground and flight testing, data reduction and analysis, and the final reports. Completion of the program is expected at the end of fiscal year 1992.

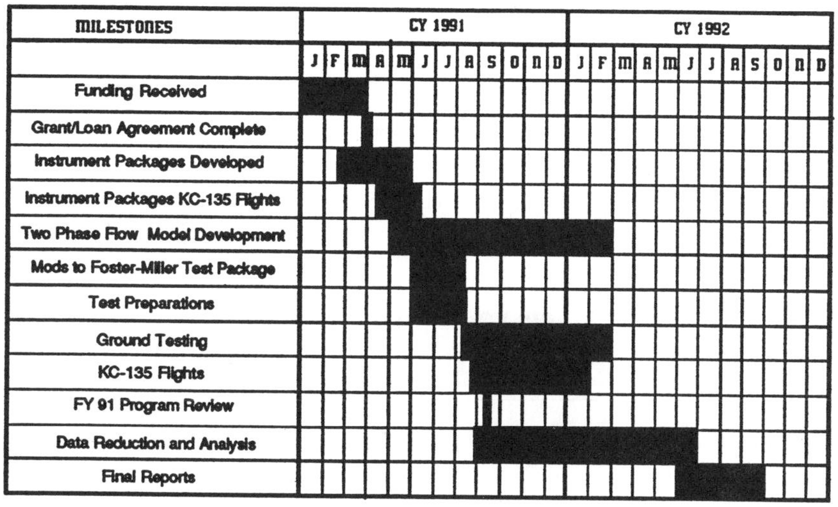

MILESTONES	CY 1991												CY 1992											
	J	F	m	A	m	J	J	A	S	O	N	D	J	F	m	A	m	J	J	A	S	O	N	D
Funding Received																								
Grant/Loan Agreement Complete																								
Instrument Packages Developed																								
Instrument Packages KC-135 Flights																								
Two Phase Flow Model Development																								
Mods to Foster-Miller Test Package																								
Test Preparations																								
Ground Testing																								
KC-135 Flights																								
FY 91 Program Review																								
Data Reduction and Analysis																								
Final Reports																								

FIGURE 1. Program Schedule.

EXPERIMENT PACKAGE DESCRIPTION

The experiment package (Figure 2) was developed by Foster-Miller Inc. as part of a Phase II Small Business Innovative Research contract performed for the United States Air Force (USAF) Phillips Laboratory at Kirtland Air Force Base. The Phillips Laboratory/Foster-Miller package was selected for this program due to its history of KC-135 flight testing during 1990 (Hill and Best 1991). The USAF agreed to loan the test package to the NASA JSC for the two-year project.

The experiment package was delivered to the Thermochemical Test Area at NASA JSC in April 1991 to begin modifications. Alterations were made to the package to enable different component testing and allow for a wider range of flowrates of both the liquid and vapor phases. The major changes to the equipment included: fabricating components (smooth pipes, bellows lines, and quick-disconnect/ bellows lines) to be used as the second adiabatic test section; adding ≈0.5 m (20 in) of inlet length to both adiabatic test sections; installing more accurate flowmeters for liquid and vapor measurements; and replacing the glass condenser section with copper for increased heat rejection capability.

In support of the flowmeter modifications, two small-scale tests on the KC-135 were conducted in April and June 1991. These tests verified that the low-flow liquid and vapor flowmeter systems would function accurately and reliably in the harsh environment of the KC-135.

In addition to being well-proven on the KC-135, this test loop offers a unique characteristic in employing a Foster-Miller two-phase pump which can process inlet mixtures of any quality. The phases are separated and then pumped individually, allowing independent control and measurement of their flowrates before they are combined. Two-phase flow of any quality can be produced by the pump with no evaporator load, and in this project the condenser is used only for removing the energy imparted by the pump. Freon-12 was selected for the working fluid because of its low toxicity, heat of vaporization, low surface tension, material compatibility properties, and high vapor density at acceptable pressures. Nominal operating conditions are 588.4 kPa (85 psia) and 294 K (70°F).

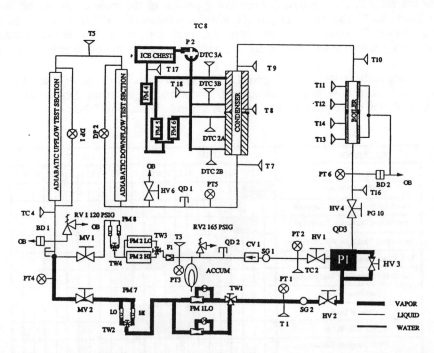

FIGURE 2. Two-Phase Experiment Loop Schematic, Post-Modifications (Best 1991).

The experiment package contains two adiabatic test sections having 1.2 m (48 in) length and a straight inlet length of ≈0.6 m (24 in). The smooth pipe test sections have 10.4 mm (0.41 in) and 4.6 mm (0.18 in) inner diameters. Vapor flow measurements are made with a low-range or high-range venturi meter. Liquid flow measurements are taken with a low-range or high-range turbine-type meter. Combining the minimum measurable flows with the pump upper limit allows roughly 0.001 → 0.090 kg/s (713 lbm/hr) of liquid and 0.001 → 0.010 kg/s (79 lbm/hr) of vapor.

For visual flow regime data, a Kodak Ektapro 1000 digital imaging system with two imager heads is used at rates up to 1000 frames per second. The two imager heads can be placed at different points along either of the test sections and the flows at these positions can be compared. In addition to qualitative flow regime identification, more quantitative analyses of film thickness, wave spectra, et cetera can be made by transferring the data to an image processing system.

TESTING OBJECTIVES

Flow Regime Testing

Flow regimes are the spatial distributions of the liquid and vapor phases flowing in a pipe. The prediction of other parameters, such as pressure drop and heat transfer coefficient, is specific to the flow regime experienced. The current project's goal for predicting flow regime transitions, and thus characterizing two-phase flows, is to develop a model which can be utilized for a wide range of fluids and environmental conditions; this includes the reduced gravity on the Moon or Mars and the zero gravity on orbit.

Part of the current testing program is an early effort in characterizing two-phase flow regime transitions in support of the SEI missions. The program will also provide additional information on the flow conditions of the Space Station External ATCS. For example, the model developed by Taitel and Dukler (1976) can be used to predict the trend of the transitions. Figure 3 shows this model for varying gravity fields.

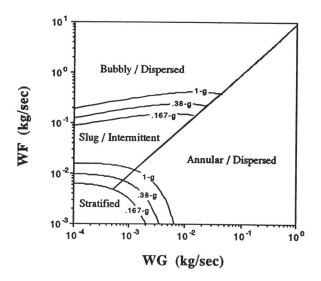

FIGURE 3. Taitel and Dukler (1976) Flowmap Predictions for Different Gravity Levels.

The predicted transitions are shown to occur at lower liquid and vapor flowrates as the gravity level is decreased, with the exception of the slug to annular transition which is predicted to depend only on the quality of the flow. Using such models, test points will be selected close to the predicted transition zones. As the program progresses, the test matrix will be refined to provide a majority of data in the demonstrated transition regions. The data collected for the different gravity levels will be used to assess a variety of two-phase flow regime models.

Pressure Drop Testing

Pressure drop can have profound effects on pipe and pump sizing, and hence mass and power, of two-phase thermal control systems. These effects are particularly critical for space systems, where penalties for under- or over-designing are significant. The pressure drop testing of this project has two thrusts: to provide one- and zero-g two-phase pressure drop data through fluid components to verify design procedures for Space Station External ATCS, and to provide pressure drop data for smooth pipes at different gravity levels.

To verify sizing procedures and improve Space Station External ATCS models, three major components that experience two-phase flow were selected for testing: smooth pipes, flexible bellows lines, and quick-disconnect fittings. The components were roughly ranked by relative predicted pressure drop (based on mass flowrate, diameter, and an effective length). The net results attributed approximately half of the total pressure drop to smooth pipes and a quarter each to flex lines and quick-disconnects. Next, a scaling procedure was developed to map Station conditions to the experiment package; the Martinelli parameter, Reynolds number, and Weber number were incorporated. When the mapping procedure was completed, it was found that the diversity of diameters, mass flowrates, and qualities that could exist on the Station covered much of the operating range of the package. This range was then divided into a grid of constant total mass flow and constant quality lines, and the test points were taken from the intersections.

Once reduced data is available, comparisons will be made involving the component pressure drops measured in one-g and zero-g for each test section and those predicted by common models. It will also be determined whether there is any systematic variation of prediction errors with the scaling groups. Later in the project, some pressure drop correlations based on the data will be developed.

TEST STATUS

The first KC-135 test series using the experiment package was during the week of 19 August 1991. This series consisted of four flight days of zero-g parabolas, with two days each testing smooth pipes and bellows lines. Post-flight calibrations, ground testing, and data reduction are in progress at Texas A&M at the time of submission of this paper.

FUTURE WORK

The next major goal is completion of pressure drop data reduction for Flight Series I. Comparison of measurements and predictions will then begin.

The remaining flight series are scheduled as follows:

- Series II, 11-15 November 1991: 4 days zero-g;
- Series III, 16-20 December 1991: 2 days zero-g, 2 days Moon-g; and
- Series IV, 27-31 January 1992: 1 day Moon-g, 3 days Mars-g.

The data reduction and analysis cycle will be undertaken again for the data from these flights, and new correlations and models will be developed as the project continues. Data from the flight and ground testing will be presented in the final reports to be published in September 1992.

Acknowledgements

This work is being performed at the NASA Johnson Space Center as a cooperative effort between the Propulsion and Power Division, the Crew and Thermal Systems Division, and the McDonnell Douglas Space Systems Company. Space Station funding is being provided through the divisions for the components studies, and by the JSC Center Directors Discretionary Funds for the advanced low-g studies.

The authors wish to acknowledge Dr. Frederick Best and his staff at Texas A&M University and Dr. Wayne Hill of Foster-Miller Inc., subcontracted to Texas A&M, for providing expertise in both two-phase flow analyses and KC-135 testing. The authors would also like to thank the USAF Phillips Laboratory on behalf of NASA for loaning the equipment package for the duration of this program.

References

Best, F. R. (1991) "NASA JSC Microgravity Two Phase Flow Project," in *Test Equipment Data Package July 1991*, Texas A&M University, College Station, Texas.

Brady, T. (1990) "Summary of Results from the Testing of Three Prototype Thermal Bus Systems for Space Station Freedom," in *AIAA/ASME 5th Joint Thermophysics and Heat Transfer Conference*, AIAA-90-1740.

Chen, I-Y., R. S. Downing, R. Parish, and E. Keshock (1988) "A Reduced Gravity Flight Experiment: Observed Flow Regimes and Pressure Drops of Vapor and Liquid Flow in Adiabatic Piping," *AIChE Symposium Series*, 84(263): 203-216.

Hill, W. S. and F. R. Best (1991) "Definition of Two-Phase Flow Behaviors for Spacecraft Design," in *Final Report*, Contract No. F29601-88-C-0062, Air Force Phillips Laboratory, Kirtland AFB, New Mexico.

Kachnik, L., D. Lee, F. R. Best, and N.Faget (1987) "A Microgravity Boiling and Convective Condensation Experiment," in *Trans. ASME Winter Annual Meeting*, ASME Paper 87-WA/HT-12, held in Boston, MA, 13-18 December 1987.

Kosson, R., R. Brown, and E. Ungar (1990) "Space Station Heat Pipe Advanced Radiator Element (SHARE) Flight Test Results and Analysis," in *28th Aerospace Sciences Meeting*, AIAA-90-0059.

Rankin, J.G. (1984) "Integration and Flight Demonstration of a High-Capacity Monogroove Heat-Pipe Radiator," in *AIAA 19th Thermophysics Conference*, AIAA-84-1716.

Taitel, Y., and A.E. Dukler (1976) "A Model for Predicting Flow Regime Transitions in Horizontal and Near-Horizontal Gas-Liquid Flow," *AIChE Journal*, 22(47): 47-55.

ANALYSIS OF A CYLINDRICAL RESERVOIR CONCEPT FOR CONTROL OF A SPACECRAFT CAPILLARY PUMPED THERMAL MANAGEMENT SYSTEM

William J. Krotiuk
General Electric Astro Space
P. 0. Box 800
Princeton, NJ 08543-0800
(609) 951-7481

Abstract

The SINDA85/FLUINT computer program is used to predict performance of a cylindrical reservoir intended to provide temperature control for a spacecraft capillary pumped thermal management system. The reservoir is constructed of stainless steel, and employs a stainless steel sintered metal wick structure to control the evaporation process and eliminate exiting vapor flow. Internal vanes are provided to position vapor in a 0–G environment. Recommendations for 0 and 1–G testing are also provided.

INTRODUCTION

A capillary pumped loop (CPL) thermal management system is intended to be used on the Earth Observing System (EOS) orbital platform to maintain instruments within their operating temperature range. The CPL system, which uses anhydrous ammonia as the working fluid, consists of cold plates, condensers and a reservoir. The reservoir is attached to the liquid line connecting the condensers and cold plates, and maintains system operating conditions during heat load variations by expelling or accepting liquid at constant pressure. This paper provides the predicted performance and testing recommendations for a cylindrical reservoir designed at GE Astro Space. This prototype design has been ground tested at NASA Goddard Spaceflight Center (GSFC). This or one of the other considered reservoir designs will be included in the Capillary Pumped Loop Flight Experiment (CAPL) which will be flown on the Space Shuttle to test components being considered for the CPL thermal management system of EOS.

DESIGN AND PERFORMANCE OBJECTIVES AND REQUIREMENTS

The CAPL prototype reservoir is designed to meet the following performance objectives for both 0–G and 1–G operation: elimination of vapor expulsion under normal CPL operating modes, minimization of inlet port pressure drop to less than 1 psi (6895 Pa) with 100–watt heater load, minimization of expulsion time to less than 1 hour with 100–watt heater load, and the capability of discharging a 150 in^3 (2.458x10^{-3} m^3) liquid ammonia inventory at –20 C. The reservoir is designed to operate between –20 C and + 50 C, and to survive a temperature range between –60 C and 80 C. The reservoir design must also provide saturation temperature control within ±0.5 C. In order to insure acceptable operation of the GE reservoir under all operating conditions, the reservoir has been designed to control pressure, temperature and exit flow to promote liquid evaporation at a constant saturation condition.

DESIGN CONCEPT

The GE CAPL reservoir consists of a cylindrical tank with interior bubble positioning vanes, a porous sintered stainless steel exit plug, and a porous sintered stainless steel inner liner. A port near the exit plug connects the reservoir to the CPL loop to provide system control. Electrical resistance heaters are externally mounted on the surface of the reservoir opposite the reservoir outlet; insulation is located at the end near the exit port. Liquid expulsion is accomplished by using the strip heaters to evaporate liquid. The volume change resulting from the density change of the evaporation process expels liquid from the reservoir with a small pressure drop. Incoming liquid flow from the CPL system is accommodated by condensing the vapor within the reservoir, thus increasing density.

TRANSIENT CALCULATIONS

The SINDA85/FLUINT computer code (Cullimore et al. 1990) has been used to analyze the GE CAPL reservoir for both 0–G and 1–G conditions. These analyses are important to verify the design calculations, and to predict reservoir performance. Design calculations consider each design parameter separately and do not account for their interaction. A FLUINT analysis is capable of simultaneously solving for the heat transfer, fluid flow, and capillary flow phenomena. Although the FLUINT code provides analytical insight, the code's predictions should be interpreted recognizing its analytical and modeling limitations. The assumptions of one–dimensional flow, and homogeneous, isothermal fluid–thermal conditions within each control volume are the two most important analytical limitations. Despite these limitations, the FLUINT analyses of the reservoir are considered appropriate because the reservoir has been divided into a number of control volumes (lumps) for analysis in FLUINT. The flow in the reservoir is primarily one–dimensional, axially out or into the exit port, and the control volume boundaries have been chosen to define volumes where axial flow dominates. The reservoir is designed to possess capillary flow in the sintered metal wicks which is not coincident with flow in the dominant direction, and local mixing flow within a control volume will not be in the dominant flow direction. However, in each control volume, these flows are smaller in magnitude than the flow in the dominant axial direction; thus, the code's predictions are considered acceptable. The number of control volumes also approximately bound areas where homogeneous, isothermal conditions are expected to exist. Since the reservoir is designed to operate isothermally, the isothermal assumption in each control volume is consistent. Additional vertical control volumes were added to be able to predict the buoyancy effects in a 1–G field and to account for the nonhomogeneous nature of two phase flow in the presence of gravity. In order to further justify this simple modeling approach, a separate task to develop a multidimensional, transient computer code to analyze a CPL reservoir is proceeding at GE Astro Space (Shyy and Gingrich 1991).

0–G Transient Analysis

The FLUINT model used to analyze the GE CAPL reservoir is shown on Figure 1. Strip heaters are distributed over the outer wall surface adjacent to control volume 41. The outer wall surface adjacent to control volumes 41 and 31 is covered with a layer of 10 mil (2.54×10^{-4} m) AL–Teflon tape possessing an emissivity of 0.88. Multi–layer insulation (MLI) blankets, with a through emissivity of 0.05, cover the outer wall surface adjacent to volumes 21 and 11. Heat transfer between the fluid volumes and the inner wall surface accounts for the presence of the sintered stainless steel wall wick and exit plug. A typical convective heat transfer coefficient of 1 BTU/hr/ft²/F (5.7 W/m²/K) is assumed between the wall and control volume 11. The exit line, control volume 1, is assumed to be thermally insulated.

The 0–G liquid expulsion case was analyzed using FLUINT. The reservoir was initially assumed filled with liquid at temperatures of –20 C, 10 C, 20 C, 25 C and 50 C. The exit line, control volume 1, is connected at one end to volume 11 and, at the other end, to a boundary maintained at the saturation pressure corresponding to the initially assumed temperature. The reservoir transient response was calculated for the case where 100 watts of heater power is added to the reservoir. Hot and cold case environmental temperatures of 5 C and –15 C were used. The environmental temperature bounds were determined by NASA GSFC from orbital analyses of the CAPL experiment in the Shuttle bay for earth and space facing configurations (Hoang 1990). The results of the FLUINT analyses using an environmental temperature of 5 C are shown in Figure 2. Figure 3 shows the results using a –15 C environmental temperature. For the cases analyzed, calculations show that the required mass expulsion of 150 in³ (2.458×10^{-3} m³) of liquid at –20 C is reached before two phase flow exits the reservoir. The pressure drop for the exiting liquid flow is always below the required pressure drop of 1 psi (6895 Pa). Reservoir fluid transient temperatures are calculated to remain within 0.1 C of the initial temperature. In conclusion, the design calculations and FLUINT analyses show that the GE CAPL reservoir will meet all fluid–thermal design and operational requirements in 0–G.

TESTING RECOMMENDATIONS

Two means for testing the GE CAPL reservoir in normal gravity have been identified. For simplicity, initial testing in a thermal vacuum chamber is not proposed. Therefore, the two considered performance testing methods must minimize convective heat loss to simulate a space environment. For the first considered test approach, the reservoir is placed in a cooled enclosure shroud to simulate thermal radiation in space. The enclosure shroud uses externally mounted cooling coils to maintain a shroud "environment" temperature. The enclosure volume is filled with a non-condensible gas to eliminate condensing heat transfer to an air environment; but convective heat transfer will still exist at the uninsulated surfaces of the reservoir. The other reservoir testing option uses cooling coils, mounted directly to a portion of the reservoir wall, to provide cooling. The cooling coils are located on the reservoir surface between the heaters and the internal exit plug. For this testing method the entire surface of the reservoir is insulated to minimize convection heat transfer. FLUINT analyses of the GE CAPL reservoir have been performed for the two configurations to determine which 1–G testing configuration better represents the actual orbital conditions.. In both cases the reservoir is horizontally oriented with the exit line parallel to the earth's surface.

1–G Transient Analysis using a Enclosure Shroud

The FLUINT model used to analyze the GE CAPL reservoir is shown on Figure 4. The reservoir has been further divided to model 1–G buoyancy effects. Frictional flow junctions are provided between volumes 41–42, 42–43, 11–12 and 12–13. Capillary flow is specified between volumes 21–22 and 22–23. Strip heaters are distributed over the wall surfaces adjacent to volumes 41, 42 and 43. Insulation covers the walls adjacent to volumes 11, 12, 13, 21, 22 and 23. The outer surface of the walls adjacent to volumes 31, 32, 33, 41, 42 and 43 are covered with 10 mils (2.54×10^{-4} m) of Al–Teflon tape. The enclosure shroud is modeled as a boundary temperature of 5 C. A convective heat transfer coefficient of 1 BTU/hr/ft²/F (5.7 W/m²/K) is assumed on all wall surfaces to transmit heat to the air space inside the enclosure shroud. The enclosure air space temperature is assumed equal to the enclosure shroud temperature. As for the 0–G transient analysis, the liquid expulsion case has been analyzed. The transient response for four initial fluid temperatures, –20 C, 10 C, 20 C and 25 C, have been calculated. The 100 watts from the heaters was assumed evenly divided between wall nodes 141, 142 and 143. The other initial assumptions are identical to those described for the 0–G analysis. Transient temperatures are calculated to remain within 0.1 C of the initial temperature for the duration of the analyzed transient. Results indicate that 136, 131, 122 and 116 in³ (2.23×10^{-3}, 2.15×10^{-3}, 2.00×10^{-3} and 1.90×10^{-3} m³) of liquid is expelled at fluid temperatures of –20 C, 10 C, 20 C and 25 C before an exit void fraction of 10% is reached. (It should be noted that a void fraction of 10% corresponds to a mass quality of less than 1% over the entire range of analyzed reservoir temperatures.) The limit in liquid expulsion capability occurs because the upper portions of the sintered exit plug become depleted of liquid, probably due to the inability of the capillary action of the sintered exit wick to supply liquid flow at a large enough rate to balance the exiting liquid mass flow. Calculations

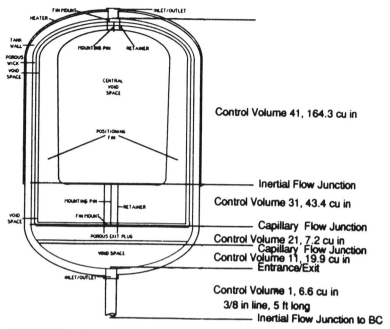

FIGURE 1. GE CAPL Reservoir 0-G Control Volume Model.

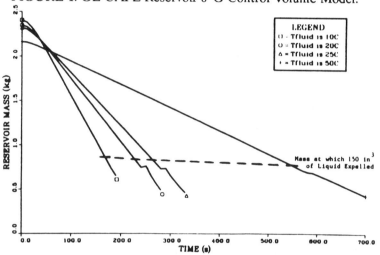

FIGURE 2. CAPL Reservoir Mass in 0-G with 100-Watt Heaters and $T_{env} = 5$ C.

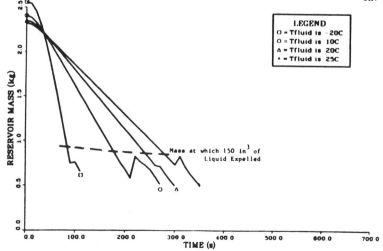

FIGURE 3. CAPL Reservoir Mass for 0-G with 100-Watt Heaters and $T_{env} = -15$ C.

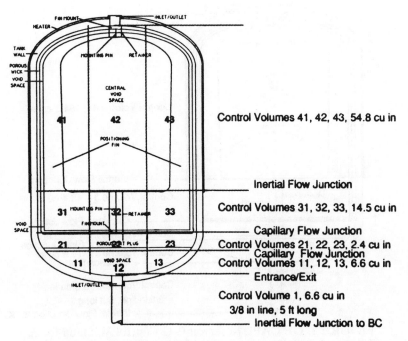

FIGURE 4. GE CAPL Reservoir 1-G Control Volume Model.

indicate that the exit flow pressure drop remains substantially below 1 psi (6895 Pa).

1-G Transient Analysis using Wall Mounted Cooling Coils

The FLUINT model used to analyze the GE CAPL reservoir for the condition where the outer surface of the reservoir is cooled by coils is very similar to the 1-G model with an enclosure. The only difference is the heat transfer connections to the external environment. For this situation, the entire surface of the reservoir is insulated. Convective heat transfer is simulated by assuming a 1 BTU/hr/ft²/F (5.7 W/m²/K) heat transfer coefficient on the outer surface of the insulation. The external air temperature is assumed to be 20 C (68 F). The cooling coils are mounted on the outer wall surface adjacent to control volumes 31, 32 and 33. Horizontal, liquid expulsion transients were performed with a 100-watt heater input for fluid temperatures of -20 C, 10 C, 20 C, 25 C and 50 C. The results from the FLUINT analysis of the reservoir with cooling coils are similar to those obtained from the 1-G horizontal analysis with an enclosure. Figure 5 indicates that, at -20 C, only about 143 in³ (2.34×10^{-3} m⁻³) of liquid mass will be expelled before the exit flow reaches 10% void fraction. At 10 C, 20 C and 25 C, the analyses predict that 131, 126 and 122 in³ (2.15×10^{-3}, 2.06×10^{-3} and 2.00×10^{-3} m³) of liquid will exit before the void fraction reaches 10%. Calculations show that the 1 psi (6895 Pa) exit pressure drop requirement is met with substantial margin. Reservoir temperatures are calculated to remain within 0.1 C of the initial temperature throughout the transient.

Comparison of 0-G and 1-G FLUINT Predictions

The 1-G FLUINT transient analyses illustrate the interrelated effects that could not be accounted for in simplified design calculations; but the code's accuracy, especially for expelled liquid mass, must be verified by testing. However, the FLUINT analysis does provide insight into the sensitivity of the reservoir design features. 0-G and 1-G predictions for liquid expulsion mode at 20 C are compared on Figures 6 and 7. Figure 6 shows that the 1-G predictions for reservoir mass using cooling coils more closely match the 0-G predictions with a 5 C environment for about 200 seconds of the calculated transient. After that time, 1-G buoyancy effects promote vapor flow through the top of the exit plug and the exit flow becomes two phase. The results for the case with a cooling shroud at 5 C diverge from the 0-G predictions throughout the calculated transient. The differences between the two 1-G predictions are primarily due to the differences in convective losses. The convective losses for the 1-G case with an enclosure are large enough to cause the results to diverge. Figure 7 shows that neither 1-G method adequately predicts 0-G exit line void fraction. For analysis, a typical free convection heat transfer coefficient was used to calculate convection from the reservoir or insulation surfaces. The actual value may be as low as 1/2 to 1/3 of the assumed number. The determination of surface heat loss is dependent on a number of variables including surface properties, surface area, reservoir temperature, enclosure temperature, and enclosure gas temperature. Due to the uncertainties and difficulties in specifying insulated and uninsulated reservoir surface areas for testing in an enclosure, and the fact that the cooling coil approach more closely matches the 0-G results, the cooling coil approach is recommended for 1-G testing. The apparatus for the cooling coil approach also provides simpler, reproducible temperature control.

Testing in a 1-G Environment

The GE CAPL reservoir has undergone 1-G testing at NASA GSFC using the cooling coil method. Liquid expulsion and system thermal control tests were performed. The GE CAPL reservoir was tested in normal gravity with the reservoir's major axis configured at various angles relative to the earth's surface. Because of the desire to perform testing at adverse tilt conditions, FLUINT calculations have been performed to assess the impact of operating the

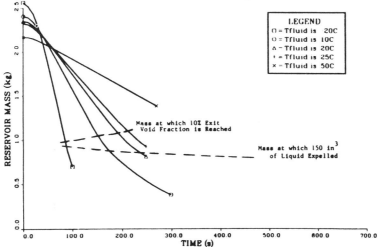

FIGURE 5. CAPL Reservoir Mass in 1-G with 100-Watt Heaters and Wall Mounted Cooling Coils.

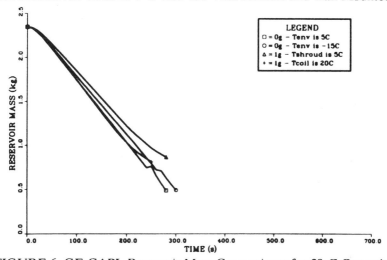

FIGURE 6. GE CAPL Reservoir Mass Comparisons for 20 C Operation.

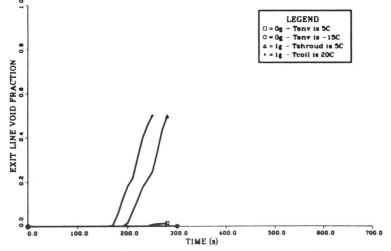

FIGURE 7. GE CAPL Reservoir Void Fraction Comparisons for 20 C Operation.

reservoir in an adverse tilt. FLUINT liquid expulsion calculations were performed for an initial temperature of 20 C, and various reservoir orientations: horizontal, vertical with the exit down, 1 in (2.54 cm) adverse tilt, 2 in (5.08 cm) adverse tilt and 5 in (12.7 cm) adverse tilt. Tilt height is defined as the difference in elevation between the reservoir end cap radial center line at the operational exit line and at the closed end. For the liquid expulsion tests, the reservoir was filled with liquid ammonia and cooled to an initial temperature. The outlet line from the reservoir was connected to the test CPL loop which acted as a pressure sink. After the reservoir temperature stabilized, the heaters were turned on. These reservoir tests were performed in horizontal, vertical and titled orientations. The system thermal control tests are intended to determine the reservoir's capability to control system saturation temperature. In these tests the reservoir was connected to a thermal capillary pumped loop (CPL), and oriented in horizontal, vertical or tilted orientations. The CPL was operated for various operational transients, including a start–up, to determine the ability of the reservoir to control temperature.

Liquid expulsion analysis results indicate that the behavior of the GE CAPL reservoir in a horizontal orientation differs substantially from 0–G conditions. Further, it is expected that 1–G testing with adverse inclinations, with the exit line pointing up, results in fluid behavior more dissimilar from 0–G conditions. To approximate 0–G fluid behavior in 1–G, calculations and tests were performed with the exit line pointing down. This orientation more closely approximates the 0–G fluid behavior of the bubble positioning vanes and minimizes the severity of 1–G operation. FLUINT calculations illustrate the closer similarity between 0–G and 1–G exit line down orientations during liquid expulsion. The predicted exit line void fractions for the 0–G and the 1–G vertical, exit down orientation are almost identical; while, the exit void fraction plots for 1–G horizontal and 5 in (12.7 cm) adverse tilt orientations differ dramatically from the 0–G predictions. The pore radius of the exit plug is capable of supporting an ammonia liquid head greater than the length of the reservoir, and the wall wick pore size and flow area are capable of supplying liquid flow to the heater area without dryout occurring. Testing in a 1–G vertical, exit down orientation verified the exit plug performance and the capillary capability of the wall wick. The GE CAPL reservoir performed well during testing. The results of these tests will be discussed in a future paper.

Testing in a 0–G Environment

As with 1–G testing, 0–G testing of a CAPL reservoir should consist of liquid expulsion and system thermal control tests. The testing should cover the range of operating temperatures, –20 C to 50 C. Analyses predicting 0–G operation of the reservoir should be performed before testing using the analytical model verified in the 1–G tests. The pre–test analyses would provide insight into the operation of the reservoir. After testing, analyses should be redone using the actual test conditions which could differ from those originally planned. The post–test analyses would provide insight into the actual performance of the reservoir, and correlation and verification of the analytical model.

CONCLUSIONS

The calculations and FLUINT analyses performed for the GE CAPL reservoir indicate that the fabricated unit will be capable of meeting all design requirements for 0–G operation. Since the GE CAPL reservoir has been designed for 0–G operation, reservoir performance is not optimal in 1–G. This condition is primarily due to the inability of the bubble positioning vanes to operate in 1–G. In a horizontal configuration, calculations show that the reservoir will meet all design requirements except, possibly, the vapor free liquid expulsion requirement at –20 C. With adverse tilts of 1, 2 or 5 in (2.54, 5.08 and 12.7 cm), the reservoir was predicted to perform almost as effectively as the horizontal orientation. In a vertical orientation, with the exit down, the reservoir should exceed all design requirements because this configuration closely matches 0–G liquid–vapor positions. 1–G testing results meet or exceed the conservative analytical predictions and verify the reservoir design. The test results will be discussed in a future paper.

Acknowledgments

This work was performed as part of the technical activities supporting the design and fabrication of the EOS orbiting platform under contract to NASA GSFC. The effort could not have been completed without the continuing support from the personnel in the Thermal Group and the EOS Program at GE Astro Space. Additionally, support from the Advanced Thermal Management Group and the EOS Program at NASA GSFC is greatly appreciated. Finally, I wish to acknowledge the fabrication expertise supplied by Thermacore, Inc. Their knowledge of sintered metal technology was integral in building the prototype reservoir.

References

Cullimore, B., R. G. Goble, C. L. Jensen, and S. G. Ring (1990) *SINDA '85/FLUINT, Systems Improved Numerical Differencing Analyzer and Fluid Integrator, User's Manual, Version 2.3*, MCR–90–512 Revision 4, Martin Marietta Corporation, Denver, Colorado, NASA Contract NAS9-17448, 1990.

Hoang, T. (1990) "Temperature on Smaller Reservoir," Internal Memo, NSI, Goddard Spaceflight Center, Greenbelt, Maryland, 20 September 1990.

Shyy, W., and W. K. Gingrich (1991) "Modeling of Two–Phase Thermocapillary Flow in a Spacecraft Thermal Control Loop", in *Microgravity Flows – 1991*, FED–Vol. 111, American Society of Mechanical Engineers, 1991.

HEAT PIPE COOLED THERMIONIC
REACTOR CORE FABRICATION

M. Harlan Horner and Thomas H. Van Hagan
3550 General Atomics Court
San Diego, CA 92121-1194
(619) 455-2472

William R. Determan
Rockwell international,
Rocketdyne Division
6633 Canoga Ave.
Canoga Park, CA 91303
(818) 718-3375

Abstract

Thermionic and driver fuel elements in an in-core heat pipe cooled reactor will reject heat to a surrounding array of redundant heat pipes. Such structures present a formidable fabrication problem if approached conventionally through assembly of tubular structures.

A reasonable method for fabrication of such reactor cores is described in this study. The technique selected involves the use of hot isostatic processing to minimize the amount of material in heat pipe walls, maximize available vapor flow cross sectional area and provide accurate location of fuel elements.

INTRODUCTION

Initial conceptualization of the configuration of a heat pipe cooled thermionic reactor (HPTI) envisioned an array of individual heat pipes thermally bonded to·the thermionic fuel elements (TFEs) making up the reactor critical assembly (Mills et al. 1992; and Determan 1992). The plan included redundant cooling of TFEs by surrounding each with six individual heat pipes. At the power level of this study, 40 kW$_e$, additional driver elements in positions surrounding the the TFE array are required to provide enough ^{235}U to achieve criticality. Waste heat from TFEs, and all driver element heat, is transported with the surrounding heat pipes to an additional array of heat pipes at one end of the reactor. That array, coaxially located with both types of elements, subsequently fans out in a conical configuration to create the power system heat rejection radiator. The general reactor layout is shown in Figure 1. Plane sections through seven TFEs within the core are shown in Figure 2 for two configurations evaluated for the heat pipe geometry. The Bowtie geometry, (Figure 2a), is preferred since more redundancy in cooling results with just two TFEs serviced by each heat pipe rather than three for the Tricusp design, (Figure 2b). Figure 3 illustrates the heat pipe and tube structure of the reactor core by way of an isometric section. The Bowtie heat pipes are shown as a structure integral with the TFE location tubes and sharing the same walls. The single heat pipe walls have wicks bonded to all surfaces accounting for the apparent thickness of the walls in the illustration. Not shown are the arteries within each heat pipe visible in the section of Figure 2a. Even though the structural walls are thin, the structure has sufficient rigidity for the application.

Preliminary design work on the reactor system led to a set of requirements for the selected core assembly design and fabrication method. These are summarized below:

A. The volume of metal comprising heat pipe walls within the core should be minimized in order to maximize vapor flow volume. A thickness of 0.5 mm was considered a minimum to allow reasonable control of tolerances during forming operations.

B. Tolerance buildup within the reactor core should be uniformly distributed to assure accurate indexing for TFE and driver fuel elements in order that a critical array with predictable behavior be created.

C. The method of core assembly should not expose completed TFEs to isothermal temperatures in excess of 1500 K to avoid damage caused by chemical interaction between graphite integral cesium reservoir material and the refractory metals of construction.

D. The fabrication method should provide access for non-destructive examination of core structural and heat transfer bonds, and provide for thermal performance testing of the completed core heat pipe structure before installation of TFEs and driver elements.

FABRICATION PROCESS SELECTION

A review of methods and technologies for fabrication of the HPTI core was conducted. The methods considered were:

ELECTRO-CHEMICAL MACHINING (ECM) or ELECTRICAL DISCHARGE MACHINING (EDM) - A solid block of Nb-1%Zr alloy is shaped by drilling and ECM or EDM methods to produce TFE positions and heat pipe cavities or holes. Procurement of a large block of material with a suitably refined grain structure required to provide liquid metal corrosion resistance was found not to be currently possible without unreasonable developmental investment. Computer numerical control wire EDM techniques are only marginally state-of-the-art for this structure and more conventional cavity sinking methods, required to create end structures for the heat pipes, would be prohibitively expensive.

STACKED TUBE METHOD - An assembly of Nb-1% Zr tubes shaped to form heat pipes and TFE sockets is brazed together to form the core structure. Although intuitively simple, this method has several difficulties. Each interface between heat pipes or heat pipe and TFE socket consists of three layers of metal (a wall for each structure, and a layer of braze) creating an effective wall thickness more than double the minimum required to fabricate accurate heat pipe and TFE socket tubes, adding unacceptable mass to the core structure. Slight variations from size or configuration ideality for each tube are unavoidable and will add to create a tolerance buildup condition difficult to deal with in a real situation. Installation of wicks and attachment of end caps and loading tubes to heat pipes could exacerbate the tolerance buildup problem through welding shrinkage.

HOT ISOSTATIC PROCESSING (HIP) - Niobium-1% Zirconium cylindrical tubes, tube sheets, heat pipe wicks and thin sheet sections are combined with molybdenum removeable inserts by electron beam welding, then HIP'ed. Removal of the molybdenum inserts is accomplished by chemical etching once the HIP'ed assembly is accurately machined to the reactor core configuration.

The HIP method appears to offer a more complete solution to the fabrication challenge since the total heat pipe geometry with TFE positions is formed in one bonding step, and heat transfer interfaces and wall thicknesses are minimized by single layer construction. The following sections provide a more complete explanation of the selected processing technique and the manufacturing experience base, and discuss the intended fabrication procedure.

HOT ISOSTATIC PROCESSING (HIP) METHODOLOGY

The HIP process has been used by General Atomics to fabricate niobium-alumina trilayer collectors and niobium-molybdenum composite electrical lead structures for thermionic fuel elements for a number of years (Gulf General Atomic Company 1969; and 1973). The HIP sequence applies high (69-103 MPa) inert gas pressure at temperatures of 1670 to 1823 K to cause intimate contact and bonding through creep and interdiffusion. Since temperature, pressure and time are all controllable, the bonding process can be tailored in a repeatable way to control the amount of creep and interdiffusion experienced. A schematic of a typical HIP chamber and support equipment is shown in Figure 4. Figure 5 depicts steps used for bonding and/or densifying a composite structure of two components. A more complex structure demonstrating niobium-to-niobium bonding and use of molybdenum filler pieces is illustrated by Figure 6. Conical and small insert pieces are niobium, the massive pieces are molybdenum. The tube to the rear of the illustration is niobium. The bonding package was made up by sliding the niobium cylinder over the five pieces in the center and electron beam welding on the conical end pieces. The assembly was bonded at 1823 K for three hours under 69 MPa argon pressure. A portion of the resulting structure was cut from the bonded assembly and split in two pieces as shown in Figure 7. Intimate niobium-niobium and niobium-to-molybdenum bonding is shown on the right half of the figure. On the left half, preferential acid etching was used to remove a portion of the molybdenum and reveal the intimate diffusion bonding between all portions of the assembly. The small niobium insert is seen to be completely bonded to the outer niobium cylinder.

Suitably sized facilities for bonding the reactor core structure were sought by reviewing the literature of three companies currently vending HIP services and through conversations with one representative. Eight facilities with .61 m diameter or greater working zones from 1.22 m to 3.05 m in length are available at the these companies. Bonding temperatures from 1477 K to 1723 K at pressures <69 MPa are offered. A significant application for such equipment is densification of large titanium investment castings for the turbine industry. Processing such components is similar to reactor core assembly not only with respect to size and and part complexity, but the impurity levels within the bonding chamber in each case must be carefully controlled to to low values to prevent modification of properties through contamination.

Reactor Core Fabrication Sequence

The general approach planned for the assembly process involves positioning and electron beam welding of the reactor core components, with heat pipe cavity inserts installed, into a monolithic array before bonding as described above. Tube sheets at each end of the core, the outer cylinder of the core structure, and tubes that establish TFE and driver element locations within the core will be considerably thicker at this stage than when the structure is complete. Tolerance buildup occurring as a result of welding shrinkage and creep during the HIP process will appear as small, evenly distributed dimensional changes within the heat pipe cavities and not as shifts in TFE or driver element positions. The core assembly and inspection procedure is given in the steps outlined below:

Step 1. Fabricate reactor core subassembly details to include:
  - Reactor core assembly outer shell with control and reflector attachment points rough machined into position,
  - Driver and TFE tubes,
  - Heat pipe septums (0.5 mm thick),
  - Heat pipe wicks and arteries,
  - Heat pipe inserts (molybdenum),
  - Core structure face sheets, and
  - Beryllium oxide core reflector segments.

Step 2. Assemble reactor components on an accurate EB welding fixture to correctly position them with respect to the TFE and driver element positions. These steps are illustrated by the next several figures. In Figure 8, a heavy tube sheet is lowered onto the baseplate of the welding fixture. Bullet nosed cylindrical guides, located at TFE or driver element positions, are shown mounted to the baseplate. Figure 9 shows the lower base plate of the fixture with the tubesheet in position, and a thick wall TFE tube being placed into a hole in the tubesheet where it will be centered by the locator pin. The monolithic structure generated by assembling the reactor core components is shown in Figure 10. Several characteristics of this assembly should be pointed out. TFE tubes are counter-bored near each end so that they may be expanded into the tube sheet to provide a closely fitting joint for electron beam welding. All of the electron beam joints illustrated are planned as shallow penetration welds with the majority of the unwelded faying surfaces shown joined by the subsequent HIP step. The heat pipe wall inserts are sized to be closely fitting to the tubes so that bonding pressure is first applied to the tube-insert interfaces during the HIP process. Molybdenum filler pieces with wicking and central arteries are relatively loosely fitting components so that tolerance buildup during the assembly process is primarily accommodated in heat pipe structures during bonding.

Step 3. Electron beam weld the structure at all tube sheet-to-tube and tube sheet-to-outer cylinder joints, then leak check the welded assembly by the helium mass spectrometry technique.

Step 4. HIP the reactor core assembly to bond all faying niobium alloy surfaces and the wicks and arteries to heat pipe walls. The gas pressure load will be controlled to a level suitable to accomplish the bond without damage to the wick and artery structures. Once the structure is bonded, it will be inspected by ultrasonic test methods to verify adequate bonding between the heat pipe septum array and the tubes, and between the tubes, face sheets and outer cylinder. Access for this testing can be observed in Figures 3 and 10. Methods for correcting bonding deficiencies are straightforward. The monolithic rigid nature of the bonded structure, illustrated by these Figures, allows accurate machining of the core detail as described in the next step.

Step 5. Machine and mechanically inspect the bonded reactor core subassembly using TFE and driver element tubes as locating datums. The following detailed machining is to be included in this work:

- Core structure face sheets will be machined to a final thickness of 2 mm (from about 5 mm) and heat pipe fill-tube weld preparations will be created. Weld bosses, mounting pads and screw fastener locations required on either face of the core will also be machined at this time.

- TFE and driver tubes will be machined on the internal diameter to final wall thickness and include preparations for TFE and driver element weld installation.on both faces of the core.

- Mounting pads and brackets for fixed radial reflector elements, nuclear and thermal sensors, and control drums will be finish machined on the core structure outer diameter.

- Drilling of holes into each heat pipe location will be performed to facilitate removal of molybdenum inserts.

Step 6. Repeat ultrasonic testing for all niobium-niobium joints to assure the machining process caused no damage.

Step 7. Chemically etch the molybdenum heat pipe inserts from the machined core structure and visually inspect the heat pipe cavities to verify acceptable molybdenum removal, and acceptable wick and artery condition.

Step 8. Vacuum degas assembly, weld in heat pipe loading tubulation, leak check bonds and welds, and charge heat pipes with sodium using a distillation and tubulation weld back process. This step, and subsequent processing and testing requiring heating of the reactor core assembly will be performed in a vacuum chamber capable of maintaining a pressure of $<1 \times 10^{-10}$ bar during the high temperature process.

Step 9. Perform fully instrumented core-heat pipe thermal performance test using electrical heaters and calorimeters in TFE and driver element positions in order to verify adequate heat transfer into and out of each heat pipe.

Step 10 Mount nuclear safe shutdown poison and control elements to the core-heat pipe structure and install the TFEs and driver elements by electron beam welding joints at either end of the elements. At this point, load the sodium heat transfer medium into the heat transfer gaps in the same manner heat pipes were loaded, and electron beam weld TFE interconnects at the radiator heat pipe end of the core structure to create series connected TFE pairs. It should be made clear that installation of TFEs and drivers, and loading of heat transfer gaps will take place in stages proceeding from groups of TFEs near the center of the core. Loading tubes near the core center will be inaccessible once TFEs and drivers further out the core radius are installed. Figure 11 illustrates the configuration of the TFE-core interface at the shield end of the reactor. Loading tubulation for the heat pipes and for the TFE sodium bond are shown in position as they would appear prior to sodium loading operations. Weld joints machined integrally with the reactor tube sheet are shown for TFE and core heat pipe loading tube installation welds.

Step 11. <u>Perform electrical discharge testing</u> of each TFE pair (twelve converters in series) to verify that the installation and processing operations had no deleterious effects on TFE function. The nature of this testing is described in more detail in a companion paper of this conference (Durand and Horner 1992).

CONCLUSIONS

A method of assembly for a heat pipe cooled thermionic reactor core structure that is based on hot isostatic processing was devised because of the advantages the technique offers in comparison with other methods. All of the fabrication procedures apply current demonstrated technology-no breakthroughs are required. Accurate TFE and driver element positioning are provided through machining locations for the elements into tubes that were initially accurately located by rugged fixturing. Processing of the critical array for thermal bonding of the TFEs and driver elements, and for electrical testing of the TFEs is performed at modest temperatures. Incremental testing of the bonded core structure and of installed TFEs allows quality appraisal at appropriate times (at times when repair or removal and replacement are possible). Hot isostatic processing facilities of the size and capacity required for performing this work are commercially available in this country and other required facilities are available or well within current technology.

Acknowledgments

The work described herein was performed under Air Force Phillips Laboratory contract F29601-90-C-0059 under the direction of Project Officer Captain Thomas Wuchte.

References

Determan, W. R., (1991) "Thermionic In-Core Heat Pipe Design and Performance," *9th Symposium on Space Nuclear Power Systems*, Albuquerque, NM, 12-16 January 1992.

Durand, R. and M. H. Horner, (1991) "Testability of a Heat Pipe Cooled Thermionic Reactor," *9th Symposium on Space Nuclear Power Systems*, Albuquerque, NM, 12-16 January 1992.

Gulf General Atomic Company, January 6, 1969, *"Research and Development on Fission-Heated Thermionic Cells for Application to Nuclear Space Power Systems,"* Summary Report for Period July 1, 1968 through October 31, 1968, USAEC Report GA-8965.

Gulf General Atomic Company, June 30, 1973, *"Development of a Thermionic Reactor Space Power System,"* Final Summary Report, USAEC Report Gulf-GA-A12608.

Mills, J. C., W. R. Determan and T. H. Van Hagan, (1991) "A Fast Spectrum Heat Pipe Cooled Thermionic Power System," *9th Symposium on Space Nuclear Power Systems,* Albuquerque, NM, 12-16 January 1992.

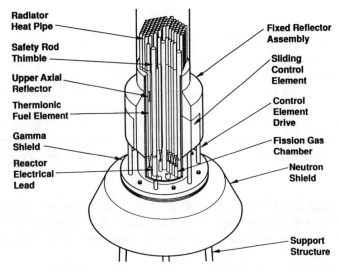

FIGURE 1. Heat Pipe Thermionic Reactor Layout

2a. Bowtie Core Design   2b. Tricusp Core Design

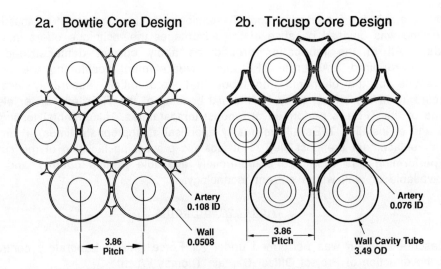

FIGURE 2. Possible Reactor Heat Pipe Configurations

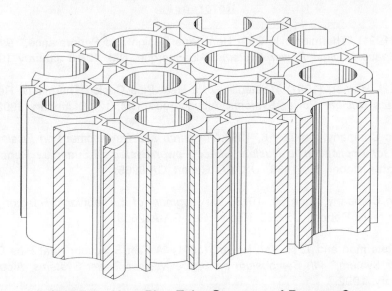

FIGURE 3. Heat Pipe-Tube Structure of Reactor Core

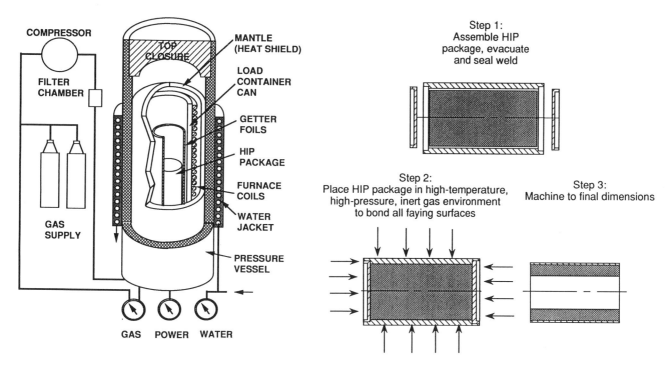

COMPRESSOR

FILTER
CHAMBER

GAS
SUPPLY

TOP CLOSURE

MANTLE
(HEAT SHIELD)

LOAD
CONTAINER
CAN

GETTER
FOILS

HIP
PACKAGE

FURNACE
COILS

WATER
JACKET

PRESSURE
VESSEL

GAS   POWER   WATER

FIGURE 4. Hot Isostatic Processing Facility

Step 1:
Assemble HIP
package, evacuate
and seal weld

Step 2:
Place HIP package in high-temperature,
high-pressure, inert gas environment
to bond all faying surfaces

Step 3:
Machine to final dimensions

FIGURE 5. Example of Hot Isostatic Processing Steps

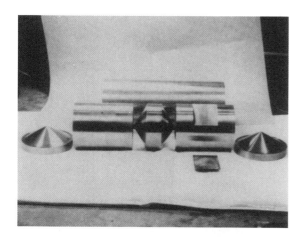

FIGURE 6. Collector Current Lead Assembly
for Gas Pressure Bonding

Note: The tube in back is 152 mm long.

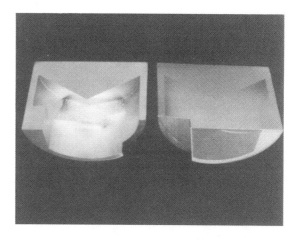

FIGURE 7. HIP'ed Lead Structure Illustrating
Core Assembly Process

Note: Intimate Mo to Nb bonding is shown on right,
preferential removal of Mo by acid etching is shown
on left.

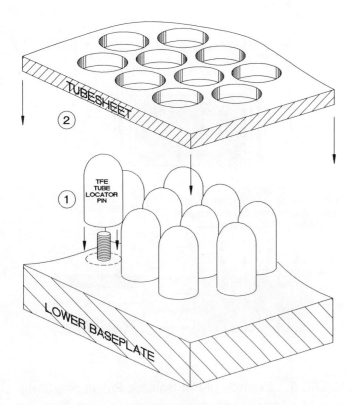

FIGURE 8. Nb-1% Zr Tubesheet is Mounted to Welding Fixture

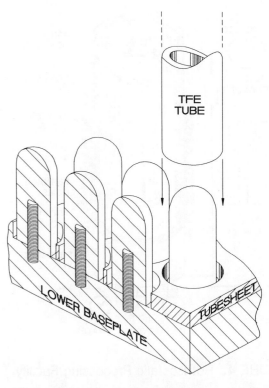

FIGURE 9. Tubes are Positioned into the Tubesheet

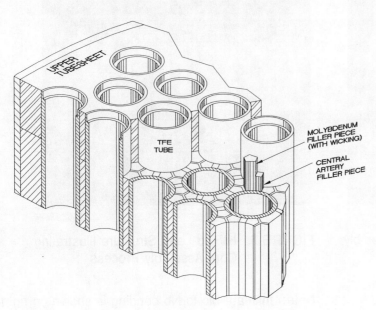

FIGURE 10. Core Structure Assembly Ready for Welding

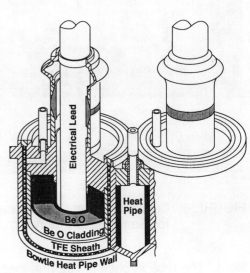

FIGURE 11. Sodium Bonded TFE Configuration

1244

PERFORMANCE OPTIMIZATION CONSIDERATIONS FOR THERMIONIC FUEL ELEMENTS IN A HEAT PIPE COOLED THERMIONIC REACTOR

Elizabeth A. Bellis
General Atomics
P.O. Box 85608
San Diego, CA 92186-9784
(619) 455-4316

Abstract

A heat pipe-cooled, in-core thermionic (HPTI) reactor design has been proposed in support of the Air Force Thermionic Space Nuclear Power Program. As part of this design, the performance of the power conversion system has been characterized. This paper focuses on the performance optimization studies carried out for a thermionic fuel element (TFE) which will be used in a reactor design capable of producing 40 kWe over a 10 year operating life. The technical approach to the optimization studies closely couples converter lifetime constraints with converter performance to produce the best possible design choice.

INTRODUCTION

This paper addresses the performance optimization of the TFE for a HPTI space power reactor. Figure 1 provides an overview of the concept. The design couples TFE power converters with in-core sodium heat transfer assemblies to provide a highly survivable, redundant, passively cooled (no pumped loop, no valves) reactor power system with no single point failures. Waste heat is transferred axially from the core using in-core transport heat pipes to ex-core sodium radiator heat pipes.

The modular design concept is scalable over the entire power range from 10 to 100-kWe, and the creative use of redundant core heat transfer assemblies offers a significant reduction in system complexity, with outstanding reliability and survivability. The same TFE technology being developed and tested in the TFE Verification Program (Dahlberg and Samuelson 1990) is used over the entire 10 to 100-kWe range.

SYSTEM DESCRIPTION

The key component of the power conversion system is the TFE. The proposed TFE is based on the F-series thermionic converter which is supported by a significant data base developed in prior thermionic technology programs (GA Technologies 1985). The TFE is comprised of six F-series thermionic converters, or cells, of varying length stacked vertically and connected electrically in series as shown in Figure 2. TFE electrical leads protrude from the bottom and top of the TFE string. The bottom lead is connected electrically in series with an adjacent TFE to produce a TFE pair with 12 series-connected thermionic cells. This pairing arrangement confines the electrical penetrations to one end of the reactor and minimizes the complexity of the power transmission circuit. The top TFE leads can then be connected in a series/parallel arrangement to produce the desired combination of reactor output voltage and current.

The thermionic cells within the TFE are roughly the size of a D-size flashlight battery and are static power conversion devices which directly convert thermal power from nuclear fuel into DC electrical power. Electric power is generated thermionically in the space between a tungsten emitter and niobium collector. Current is conducted from one cell to another through a tungsten emitter stem and tantalum transition piece. To enhance the electron transport properties in the interelectrode space, low pressure cesium vapor is introduced from a graphite cesium reservoir coupled with each individual TFE.

TECHNICAL APPROACH

Optimizing the electrical output of an in-core thermionic reactor must satisfy both the nuclear and thermionic requirements of the TFE, as well as system requirements for lifetime and reliability. This typically has been a highly iterative process, involving a strong analytical interaction between core neutronics, thermionic performance and TFE lifetime calculation. These affect the TFE optimization in the following ways:

Core Neutronics The collection of TFEs that comprise the reactor core must be in a configuration capable of achieving nuclear criticality. This requirement imposes restrictions on TFE length, the number of TFEs and the geometry and type of TFE materials chosen for candidate core configurations. The optimum TFE geometry for electrical performance may not be consistent with the results of criticality calculations. In such cases, appropriate adjustments are made until both criticality and power requirements are met.

Thermionic Performance Operating parameters that characterize thermionic performance of a converter include emitter temperature, collector temperature, interelectrode gap spacing, cesium pressure and electrode surface work function. The collector temperature for fast reactor systems of this type is generally set by the minimum heat rejection temperature necessary to meet radiator sizing requirements. Maximizing the emitter temperature and reducing the interelectrode gap spacing tends to improve the thermionic performance. These parameters are limited, however, by the lifetime considerations discussed below.

TFE Lifetime The lifetime of a fueled converter is limited by thermal distortion of the emitter and ultimate shorting to the collector once the interelectrode space is closed. To extend the lifetime of the converter it is necessary to reduce the emitter temperature, increase the electrode spacing and/or increase the cladding thickness. In view of the system lifetime requirement of 10 years, the optimum combination of emitter temperature and electrode spacing is chosen to meet electrical requirements while the emitter thickness is maximized up to the criticality limits of the core.

The HPTI TFE optimization accommodated these considerations with three state-of-the-art computational tools:

• TECMDL (Rasor Associates, Inc. 1990) - TECMDL is a thermionic performance code developed by Rasor Associates, Inc. that estimates the performance and current-voltage characteristics of a converter in the ignited mode for varying electrode temperatures, electrode materials, cesium pressures and interelectrode spacings. TECMDL results have been compared with experimental results from various converters and produce output voltages within 0.1 volts of the experiment and current densities within 1 A/cm^2.

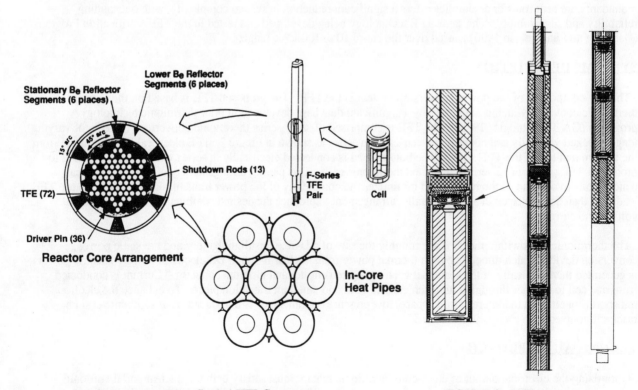

FIGURE 1. Heat Pipe Cooled TFE Concept.   FIGURE 2. TFE Arrangement.

TECMDL has been integrated into an overall thermal/electrical performance code which calculates the required thermal input to the emitter based on the anticipated thermal losses and electrical power production. The thermal losses from the emitter include radiation across the interelectrode gap, conduction through the cesium vapor, thermal conduction down the emitter lead stem and electron transfer to the adjacent collector. Losses in the electrodes themselves are also accounted for in the model. Input parameters can be varied over a wide range of conditions to produce parametric surveys allowing the selection of optimum operating points.

- LIFE-4 (Westinghouse Electric Corporation 1982) - LIFE-4 was originally developed to calculate the thermal and mechanical behavior of light-water reactor fuel elements during steady-state operation and design-basis transients. It has since been modified under the leadership of Los Alamos National Laboratory to better suit the high cladding temperatures typical of thermionic converters. LIFE-4 has a broad range of analytical capabilities and has been shown to exhibit excellent agreement with experimental data. In the context of the HPTI design, it has been used to assess emitter deformation resulting from fuel creep, swelling, sintering, restructuring and fission gas release.

- MCNP (Los Alamos National Laboratory 1986) - MCNP is a general-purpose Monte Carlo code that can be used for neutron, photon or coupled neutron/photon transport. The code treats an arbitrary three-dimensional configuration of materials in geometric cells. One of the strongest features of MCNP is its use of continuous-energy data. Geometry and particle transport in complex problems can be modeled nearly exactly. For the HPTI design it is used to establish critical core configurations and calculate core radial and axial thermal profiles.

PARAMETRIC ANALYSES

For the HPTI fueled converter, the LIFE-4 code was used to predict the maximum allowable emitter temperature based on closure of an initial .0381-cm interelectrode gap at the end of a 10 year system life. The materials of construction were uranium dioxide fuel with a 5% central void and tungsten cladding which serves as the emitter structure. Parametric studies were performed using an F-series emitter with an outer tungsten cladding diameter of 2.794 cm. The studies resulted in maximum allowable emitter temperature being expressed as a function of fuel linear power density for varying clad thicknesses. The results are shown graphically in Figure 3. It should be noted that the key parameter linking converter lifetime to thermionic performance is the linear power density of the fuel. A functional dependence exists which allows one to calculate a maximum allowable thermionic current density corresponding to the linear power density illustrated in Figure 3 for a specific emitter operating temperature.

Figure 4 illustrates the parametric analyses which led to the selection of the optimum emitter operating temperature of 1750 K. The ten year lifetime limit represents the point at which the shortest length converter would operate. For a constant current to be maintained throughout the TFE string, the intermediate and long cells would move to the left of this point and operate at lower current densities. As one moves to the left of the ten year lifetime limit on the 1750-K curve, a relatively high TFE efficiency is maintained and margin exists for subsequent emitter graduation.

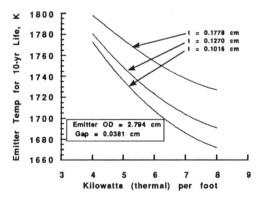

FIGURE 3. F-Series Emitter Characteristics.

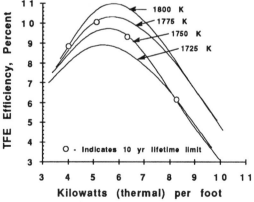

FIGURE 4. Emitter Temperature Optimization.

ADDITIONAL TFE DESIGN CONSIDERATIONS

Emitter Temperature Management

If left uncorrected, the axial thermal power distribution within a TFE for a fast reactor follows a cosine shape, which results from the variation of the neutron population within the core. This produces a corresponding temperature profile in the TFE emitters, reducing the electrical output of the outboard, lower-temperature converters. Two steps are taken in the design process to correct this problem: (a) fuel zoning to flatten the thermal profile, and (b) graduating the lengths of the emitters to flatten the emitter temperature profile. The latter technique takes advantage of the electron cooling afforded by the thermionic emission process to control emitter temperature: the shorter converters are located at the center of the TFE, placing the smallest emitter areas (highest current densities) where the thermal power is highest, and vice versa. For the purpose of graduating the 6-cell emitter string, an average TFE with a representative power distribution was chosen. Table 1 characterizes the resulting emitter graduation scheme. Margin was allowed in the final design selection to accommodate emitter temperature variations in other TFE strings with slightly different power profiles. The lengths of the thermionic cells are symmetric around the centerline with the shortest cell exhibiting the highest allowable current density of 5.3 A/cm^2 which corresponds to the 6.2 kW/ft fuel linear power density indicated in Figure 4 at an emitter temperature of 1750 K.

TABLE 1. Graduated Emitter Parameters.

	Cell 1	Cell 2	Cell 3
Fueled Length, cm	8.05	5.65	5.19
Linear Power Density, kW/ft	4.64	5.84	6.20
Power Peaking Factor	0.79	1.00	1.06
Current Density, A/cm^2	3.42	4.87	5.30

Using the emitter lengths given in Table 1 as nominal values, the TFE can be lengthened or shortened to vary power production and TFE efficiency. Figure 5 illustrates the results of this trade study performed for a fixed core configuration of 72 TFEs (36 TFE pairs). The design point chosen for the HPTI reactor concept corresponds to a gross reactor output power of 49.2 kWe (40 kWe net) at a TFE efficiency of 10.2%. The core height at this point is about 49 cm which includes the fueled and intercell regions. A reactor core with 72 TFEs of this height is able to achieve nuclear criticality and meet excess reactivity requirements. A design margin of 9 kWe has been allowed to compensate for TFE failures over the lifetime of the system, power conditioning and transmission losses, radial zoning and local temperature variations. As one moves to the right on the figure, the efficiency drops off sharply due to the excessive power losses within the converter electrodes. A point is reached at about 60 cm in height where the losses surpass the increase in electricity production and the power curve begins to drop off.

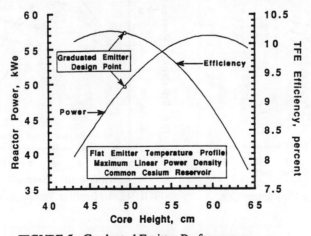

FIGURE 5. Graduated Emitter Performance.

1248

Cesium Management

One consequence of using graduated emitters that will be addressed here is the effect of using an integral graphite cesium reservoir. Ideally, there is an optimum cesium pressure which results in maximum power production for each cell length within the TFE string. Each TFE, however, has a single, thermally coupled cesium reservoir which provides the same pressure cesium to all cells in the string. With other conditions remaining equal, the optimum cesium pressure is based on cell current density. Figure 6 depicts the dependence of each cell's electrical power production capability on cesium reservoir graphite temperature. The chosen design point corresponds to a pressure of 2.86 torr. The design cesium pressure is not far off optimum for two of the three cell lengths and on a relatively flat portion of the curve for the longest cell length operating at 3.4 A/cm². Figure 7 further illustrates the minimal losses realized when providing a common cesium pressure to all cells as opposed to individual cell optimums.

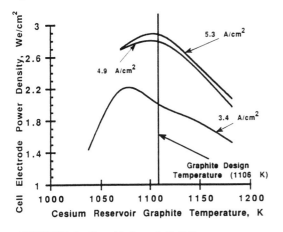

FIGURE 6. Sensitivity of Cell Power to Graphite Temperature.

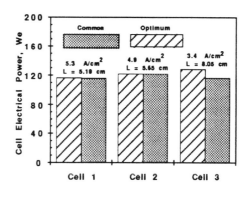

FIGURE 7. Effect of Common TFE Reservoir.

Thermionic Technology Improvement

Potential improvements to performance have been investigated to a limited extent as part of the design optimization process. One of the most attractive options is the use of a high strength emitter material. The purpose of the high strength material would be to significantly reduce the potential of emitter-to-collector shorting due to fuel swelling and subsequent emitter deformation. The fuel linear power density (and hence the operating current density) could be raised until fuel temperature limits were approached. In addition, the interelectrode gap spacing could be reduced to enhance thermionic performance. Potential benefits in output power resulting from the use of a high strength emitter material are illustrated in Figures 8 and 9.

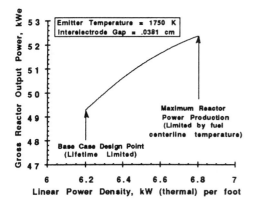

FIGURE 8. Performance Increase due to High Strength Emitter.

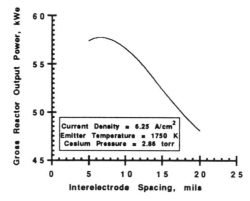

FIGURE 9. Performance Increase due to Interelectrode Gap Reduction.

TFE PERFORMANCE

The results of the TFE optimization process for the HPTI reactor design are summarized in Table 2. Included are the thermal and electrical performance characteristics of the TFE as well as pertinent cell dimensions. The tabular data are for a 72 TFE reactor core producing a net electrical output of 40 kWe over the lifetime of the power system.

TABLE 2. TFE Optimization Summary.

Emitter Outer Diameter	2.794 cm
Fueled Emitter Lengths	5.19, 5.65, 8.05 cm
Interelectrode Spacing	0.0381 cm
Nominal Emitter Temperature	1750 K
Nominal Collector Temperature	1085 K
Cesium Reservoir Graphite Temperature	1106 K
Cesium Pressure	2.86 torr (common)
Electrode Current Densities	5.3,4.9,3.4 A/cm^2
TFE Voltage	2.84 volts
Reactor Voltage	34.1 volts
TFE Current	241.4 A
TFE Efficiency	10.2%
Heat Rejected to Radiator (per TFE)	6.05 kWt
Thermal Power Produced (per TFE)	6.73 kWt
Electrical Power (per TFE)	685 We

CONCLUSIONS

Performance optimization studies for the HPTI reactor TFE have yielded a viable design capable of meeting the power requirements set forth in the Air Force Thermionic Space Nuclear Power Program. Results of neutronic studies, electrical performance calculations and lifetime surveys have been closely coupled to produce the optimum TFE configuration for this program. This methodology can be applied to similar TFE design programs where lifetime, electrical performance and nuclear concerns cannot be considered independently.

Acknowledgments

This work has been supported by the U.S. Air Force Phillips Laboratory, Kirtland Air Force Base, New Mexico under a General Atomics subcontract to the Rocketdyne Division of Rockwell International.

References

Dahlberg, R. C. and S. L. Samuelson (1990) "Thermionic Fuel Element Verification Program - Summary of Results," in *Proc. Seventh Symposium on Space Nuclear Power Systems*, CONF-900109, held in Albuquerque, NM, 7-10 January, 1990.

GA Technologies (1985) *SP-100 Thermionic Technology Program Annual Integrated Technical Progress Report for the Period Ending September 30, 1985*, GA-A18182, GA Technologies, San Diego, CA, November 1985.

Los Alamos National Laboratory (1986) *MCNP--A General Monte Carlo Code for Neutron and Photon Transport*, LA-7396-M, Rev. 2, Los Alamos National Laboratory, Los Alamos, NM.

Rasor Assosicates, Inc. (1990) *TECMDL Ignited Mode Planar Thermionic Converter Model*, E-563-004-C-082988 NSR-31/90-0775, Rasor Associates, Inc., Sunnyvale, CA, August 1990.

Westinghouse Electric Corporation (1982) *LIFE-4 (Rev. 0) Users' Manual*, Advanced Reactors Division Report WARD-0X-94000-12, Vol. 1, Westinghouse Electric Corp, Novemver 1982.

RADIATION SHIELD REQUIREMENTS
FOR MANNED NUCLEAR PROPULSION SPACE VEHICLES

Paul H. Sager
General Dynamics Space Systems Division
PO Box 85990
San Diego, CA 92186-5990
(619) 547-7298

Abstract

Manned nuclear propulsion space vehicles require radiation shielding to protect the crew from a number of diverse radiation sources: the propulsion system reactor, the Earth's Van Allen radiation belts, anomalously large solar proton events (ALSPEs), and galactic cosmic radiation (GCR). The sources are characterized not only in terms of species and energy spectrum, but also by frequency, duration, and probability of occurrence. Such factors as effectiveness of available vehicle materials (such as propellants) in providing shielding and operational strategies (such as multiple periapsis burns) must be factored into the design and mission planning for the vehicle. The optimum distribution of the shielding to limit exposure to the crew and meet established dose limits with minimum vehicle mass was determined for a typical Mars transfer vehicle using a NERVA-derivative nuclear rocket engine. For this case, the optimum shielding for a 434-day mission was also adequate to limit the exposure of crew to short-term exposure to historical ALSPEs.

INTRODUCTION

Nuclear thermal propulsion (NTP) space vehicles require shielding of the crew compartment, propellant, and critical electronic components from both reactor and naturally occurring space radiation. Space radiation sources include Van Allen radiation belts, solar particle events, and galactic cosmic radiation (GCR).

In designing manned nuclear-powered space vehicles, the sources must be defined for a specific mission and the distributions of the shielding must be optimized to conform to accepted standards and to provide an optimum vehicle design. The recommended crew blood-forming organ (BFO) ground dose limit is specified as 0.50 Sv (50 rem) by the National Council on Radiation Protection and Measurements (NCRP 1989). The optimum distribution of shielding is a function of the space vehicle design, as well as the dose limits to the crew and critical components.

In the 1960s and 1970s, extensive studies were carried out to define the space radiation environment and develop shielding requirements for manned space vehicles using nuclear thermal propulsion, as well as isotopic and nuclear reactor electric power systems. The shielding studies of that era are summarized in proceedings of the symposium held in Las Vegas in early 1971 (Warman 1972).

The recent consideration of the Space Exploration Initiative (SEI) has prompted renewed interest in nuclear propulsion systems for a manned mission to Mars. In its report, the Stafford Synthesis Group (1991) recommended consideration of nuclear thermal propulsion to reduce the transit time and thereby the crew exposure to GCR and consideration of nuclear electric propulsion to limit the mass in low Earth orbit of supporting cargo space vehicles.

RADIATION SOURCES

Natural Radiation Environments

The primary sources of natural radiation in space are trapped particle radiation (Van Allen belts), galactic cosmic radiation (GCR), and solar particle events (solar flares).

The Van Allen belts include trapped protons, electrons, and heavy ions. High-energy protons (10 to many hundred MeV) have a maximum intensity at about 1.5 Earth radii with an integral flux above 74 MeV of 2×10^4 $cm^{-2}s^{-1}$ (Holley 1972). Low-energy protons (0.1 to 10 MeV) are found in layers about the Earth, with increasing energy toward the Earth. Electrons are found in two belts of high intensity with a wide separation at lower and varying intensity electrons. The inner belt of high intensity occurs at 1.4 Earth radii and has an integrated intensity above 0.5 Mev of greater than 10^6 $cm^{-2}s^{-1}$ peaking at 4.5–5.0 Earth radii. Heavy ions (helium nucleii) are also trapped in the geomagnetic field out to about 4 Earth radii. Above 0.5 MeV/nucleon energy, the helium-to-proton flux ratio is approximately 2×10^{-4} at 3.1 Earth radii.

GCR is comprised of protons, helium ions, and heavier ions (High -Z and High-Energy-HZE-particles). The maximum cosmic particle fluence rate has been estimated to be about 4 $cm^{-2}s^{-1}$. Protons make up approximately 87% of GCR and helium ions about 12% (NCRP 1989). The HZE particles make up only 1% of GCRs, but they are highly penetrating. Iron is the most important of GCR dose because of its high linear energy transfer (LET). It has been pointed out that a cosmic ray iron nucleus is 13,500 times as dangerous as a cosmic ray proton (Letaw).

Solar particle events include emissions of protons, helium, and sometimes heavier ions. Anomalously large solar particle events (ALSPEs) can have intensities as high as 10^4 particles $cm^{-2}s^{-1}$ per steradian and fluences as high as 10^{10} protons cm^{-2} with energies greater than 10 MeV. Long-range predictions of ALSPEs are not possible, but once a solar flare is observed, the arrival of the significant particle fluence rates is reasonably predictable and crew members in a space vehicle can take cover. It has been observed, for example, that the average delay time from maximum flare mission to the first significant particle fluence rates of protons with energies of about 10 MeV is about 15 hours with a range of 15 minutes to 60 hours.

Among the recorded ALSPEs, the February 1956 event has been defined as the most severe one (Townsend). The intensity of this ALSPE, however, was measured indirectly (Smart) and there is an uncertainty of 3 in the data (Townsend). Other ALSPs that have been used for shield design studies include November 1960 and August 1972.

Reactor Radiation

The reactor in a nuclear thermal propulsion source is an intense source of neutrons and gammas. Some of the neutrons produced in the fission process leak from the reactor, but most of these are attenuated and captured by the hydrogen propellant. Only at the depletion of the propellant at the end of the mission do the neutrons present a significant dose to the crew.

The gammas, which constitute the major crew dose from the reactor, originate from the fission process, fission product decay, and radiative capture and inelastic scattering. The radioactive capture and inelastic scattering occurs not only in the reactor core, but also in other engine and vehicle structure and components, including the shield itself.

SHIELDING REQUIREMENTS

Crew Compartment Shielding

In manned space vehicles with extended-duration missions, shielding of the sleeping compartment is critical regardless of the propulsion system used. The vehicle is subject to continuous exposure to GCR and — because long-range prediction of ALSPEs is impossible — random exposure to an ALSPE. For nuclear space vehicles, allowance must also be made for exposure to radiation from the nuclear reactor.

For purposes of this study, it was assumed that the total crew exposure should be limited to the guidelines published by the National Council on Radiation Protection and Measurements (NCRP 1989). These guidelines permit a 30-day exposure of 25 rem and an annual exposure of 50 rem to blood-forming organs.

Assuming effective solar flare monitoring systems can be deployed either on-board or located at ground stations or space platforms where warning communication systems can be maintained, the crew compartment can be equipped with a "storm cellar" to which the crew can evacuate in the event of an impending ALSPE. Further, if the bunks are installed in the storm cellar, the crew can use this feature to reduce the exposure to GCR, particularly that due to secondary particles produced by the penetration of HZE particles.

Reactor Shielding

The reactor shielding requirements are based on dose limitations of the crew and avionics and on heating limitations of the propellant and system components.

The crew is at least partially protected by the crew compartment shielding. Design optimization, however, leads to additional shielding at the reactor, because shadow shielding at reactor is generally smaller in frontal area than the base of the crew compartment. For a given reactor radiation attenuation required, the mass of shielding located at the reactor is much less than shielding located at the base of the crew compartment.

The radiation exposure of the crew to the nuclear reactor can also be reduced by increasing the separation distance between the reactor and the crew compartment.

In addition, if the propellant tank used to contain the propellant for the final maneuver is mounted in-line, the propellant provides substantial attenuation for all but the final phase of the final maneuver.

Exposure of the propellant to reactor radiation in the propellant tank and feedline of an NTP vehicle results in an increase in vapor pressure on boiling of the propellant if not suppressed. This can require higher operating pressures and heavier tank weights if boiling is suppressed, or interuption of operation and possible damage to the rocket engine turbopump if not suppressed. Accordingly, shadow shielding must be included in the engine design to limit heat deposition in the propellant.

The avionics and the components situated in the vicinity of the reactor must be protected from natural and reactor radiation. Hardening techniques are available for electronics, but very large integrated circuits and other sophisticated electronics are very vulnerable to radiation damage. Elastomeric seals are also susceptible to damage. Radiation heating can also present a problem for thick-wall components near reactors.

CASE STUDY

The shielding required for a typical piloted Mars NTP transfer vehicle (Figure 1) provides insight into the impact of the specific source elements. By using a triple-perigee burn, the mission can be accomplished with a single NERVA-type nuclear rocket engine with a thrust of 75,000 lbf (Schnitzler 1991). The engine has a specific impulse of 925 s. The vehicle includes jettisonable propellant tanks for the trans-Mars injection and Mars orbit capture maneuvers. An in-line tank contains the propellant for trans-Earth injection and Earth orbit capture. An extended separation distance from the mission module to the engine is achieved by use of an open truss structure. The overall mass in low Earth orbit (IMLEO) is 613 Mg.

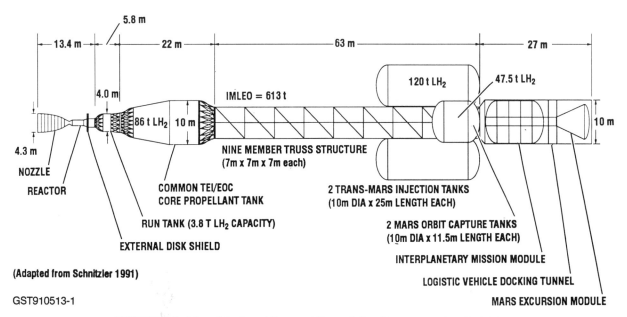

FIGURE 1. Mars Nuclear Thermal Propulsion Transportation Vehicle.

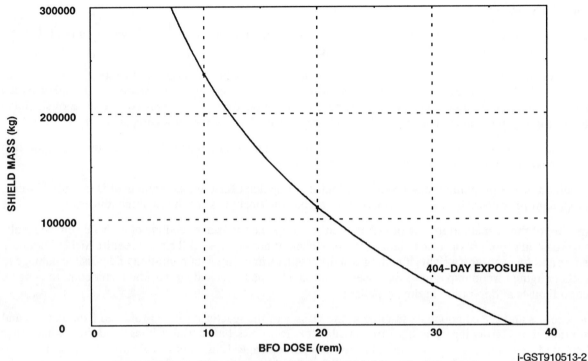
FIGURE 2. Galactic Cosmic Radiation Shielding Requirements.

i-GST910513-2

In the design illustrated, an external shield with a mass of 4.5 Mg is provided in addition to the internal shield of 1.5 Mg (3,300 lb). The results of this study indicate that it may be possible to eliminate this component by making a modest increase in the engine internal shield.

Radiation Dose Limits

The mission for which this vehicle is designed is a reference NASA 2016 opposition mission with a 30-day surface stay and an inbound Venus swingby (Borowski 1991). Because the total mission time is 434 days, the allowable crew exposure is:

$$\frac{434}{365}(50 \text{ rem}) = 59.5 \text{ rem}.$$

Budgeting 1.5 rem for transit through the Earth's radiation belts and 1 rem for the 30-day stay at Mars (Nealy et al. 1991), the allowable dose for the interplanetary flight is 57 rem.

GCR Shielding

The annual dose for the average of the galactic cosmic ray fluxes between solar maximum and and solar minimum is presented by Nealy et al. (1991). Assuming a mission module with a total surface area ($5.34 \times 10^6 \text{ cm}^2$) corresponding to the vehicle presented above, the shield mass (Figure 2) was derived as a function of crew dose for water or water-equivalent (such as polyethylene).

Solar Flare Shielding

It was shown by Nealy et al. (1991) that water is also effective as a shielding for solar flares. Because solar activity can be monitored, it is assumed that a warning system can be provided so that the crew can take shelter in a "storm cellar" in the event of a large solar particle event. To limit disruption and provide additional protection from the reactor and GCRs, the sleeping compartment could be shielded and serve as a storm cellar.

Assuming a sleeping compartment with six bunks, the total surface area would be approximately $2.7 \times 10^5 \text{cm}^2$. Scaling the data presented by Nealy et al (1991) for the 1989 flare fluences, the shield mass as a function of crew dose for a 404-day exposure was derived (Figure 3).

Reactor Shielding

The NERVA nuclear rocket engine was designed with a 1500 kg (3,300 lb) internal shield to limit the radiation heating in components immediately above the engine and heating of the hydrogen propellant. As reflected schematically in Figure 1, the installation of an external disk shield was envisioned for manned vehicles requiring additional shielding.

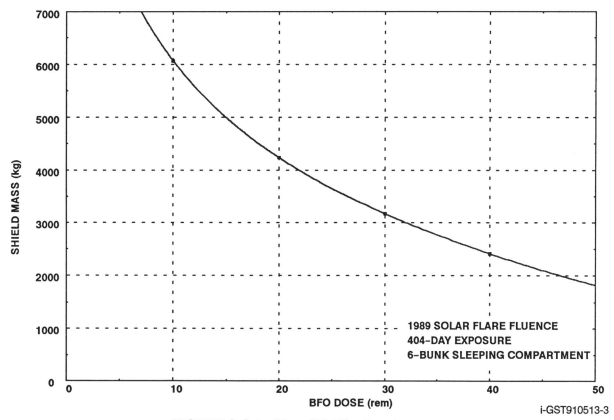

FIGURE 3. Solar Flare Shielding Requirements.

To maintain a given shielding shadow angle, it is necessary to design the external shield with a much larger diameter than that of the internal shield. The mass of the shield varies as the square of its diameter; therefore, a substantial penalty is involved in adoption of the external shield.

Because the design of a new nuclear rocket engine is not limited to previously adopted parameters, the internal shielding included in the thrust chamber should be considered a variable.

The crew dose for the vehicle presented in Figure 1 with the NERVA engine was determined to be 37.4 rem for the reference Mars mission (Schnitzler 1991). Scaling this data point exponentially, the crew dose was determined as a function of shield mass (Figure 4).

Shield Distribution Optimization

The optimum distribution of shield mass was determined with a pattern search. It was assumed that the crew would spend one-third of the time in the sleeping compartment, and the doses received from the GCR and reactor were adjusted accordingly. Also, the GCR and reactor doses were scaled to account for the mission module hull, which was assumed to be aluminum with an effective thickness of 2 cm.

It was determined that, due to the extensive area of the mission module, the optimum design is obtained for no GCR shielding. As indicated in Figure 5, the optimum distribution is obtained with a storm cellar shielding thickness of 13 cm (3550 kg) and a reactor shield mass of 2250 kg.

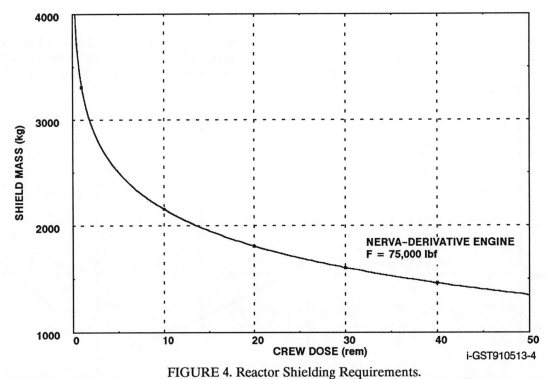

FIGURE 4. Reactor Shielding Requirements.

The corresponding crew exposures for this shielding system are shown in Table 1.

TABLE 1. Crew Exposures for Storm Cellar Shielding Thickness
of 13 cm (3550 kg) and a Reaction Shield Mass of 2250 kg.

Source	Dose (rem)
GCR	31
Solar Flare	21
Reactor	5
Total	57

A tissue thickness of 5 cm was included in the BFO dose calculations for the GCR and solar flares. The reactor doses, on the other hand, were scaled from data on whole-body dose. Accounting for the tissue penetration of the reactor gammas results in a shield savings of only about 1% and was therefore ignored for this study.

Short-Term Exposure

The optimum design was checked to determine whether the dose would exceed the NCRP-98 limit of 25 rem for a 30-day exposure. Scaling the doses reported for three major flares for the optimum design (Schnitzler 1991), the 30-day exposures shown in Table 2 were obtained.

TABLE 2. ALSPE Crew Exposures for Storm Cellar Shielding Thickness at 13 cm (3550 kg).

ALSPE	2/56	11/60	8/72
ALSPE Dose (rem)	23	22	13
GCR (rem)	2	2	2
Total (rem)	25	24	15

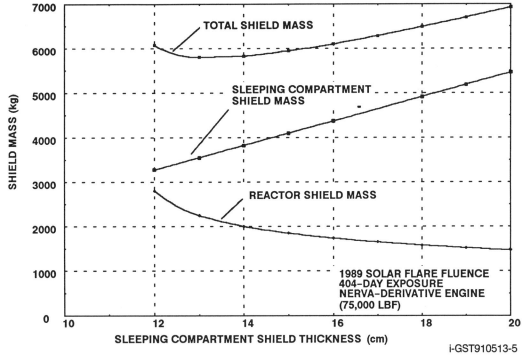

i-GST910513-5

FIGURE 5. Total Shielding Requirements.

The design appears to be marginal for the February 1956 ALSPE. However, because the measurements for this flare were made indirectly and have an uncertainty factor of 3, they are not considered reliable data upon which to base the design.

Also, it should be noted that the reactor dose is not included in the 30-day dose. Since the major portion of the reactor dose is received during the final maneuver at Earth capture, and the probability of the ALSPE occurring within the 30 days prior to this maneuver is very small, the expected dose from this source is negligible.

CONCLUSIONS

As a result of this study, it is concluded that an acceptable shielding system can be designed for the reference NTP mass transfer vehicle using rocket engine internal shielding and shielding of the sleeping compartment. For this particular case, the optimum shield system is comprised of 2250 kg of engine shielding and 13 cm (3550 kg) of water or a hydrogenous material for the sleeping compartment.

The optimum design was based on the sum of the fluences of the 1989 solar flares and is believed to be a reasonable basis for purposes of this study. A check of the exposures due to the three ALSPEs observed in February 1956, November 1970, and August 1972, indicates that the optimum design satisfies the 30-day limits of NCRP-98.

Future studies should account for the reduced shield requirements resulting from shorter mission transit times. Since the dose due to both solar particle events and GCR could be reduced, not only would the crew sleeping compartment shielding requirements be reduced, but the allowance for reactor dose could be increased with a resulting decrease in reactor shielding requirements.

A more comprehensive study would also include the separation distance between the mission module and the reactor as an independent variable for the optimization. With a longer vehicle spine, the shielding mass would be less, but the vehicle structural mass would be greater.

Finally, the composition of the reactor engine shielding should be examined. In the NERVA engine, the material used for the internal shield was a composite mixture of boron carbide, aluminum, and titanium hydride. By using tungsten as the gamma shield material, it should be possible to increase the shield effectiveness without necessarily increasing its thickness.

Acknowledgments

This study was supported by internal funding at General Dynamcs Space Systems Division. The author wishes to thank Bruce Schnitzler of Idaho National Engineering Laboratory and John Nealy of NASA Langley Research Center for their helpful comments on the space radiation environment and space vehicle shielding requirements.

References

"America at the Threshold," Report of the Synthesis Group of America's Space Exploration Initiative, U.S. Government Printing Office, Washington, D.C., May 1991.

Borowski, S.K., "Nuclear Thermal Rocket Workshop Reference System – Rover/NERVA," *Proceedings of the NASA/DOE/POD Nuclear Thermal Propulsion Workshop*, edited by T.S. Clark, NASA CP-10079, Lewis Research Center, Cleveland, OH, August 1991.

"Guidance on Radiation Required in Space Activities," National Council on Radiation Protection and Measurements, NCRP Report No. 98, Bethesda, MD, July 31, 1989.

Holly, F. Eugene, "The Geomagnetically Trapped Radiation Environment," *Proceedings of the National Symposium on Natural and Manmade Radiation in Space*, NASA Technical Memorandum NASA TMX-2440, Washington, D.C., January 1972.

Letaw, John R., et al., "Radiation Hazards on Space Missions," *Nature*, 330 (24/31): 709–710.

Nealy, J.E. et al. (1991) "Radiation Exposure and Dose Estimates for a Nuclear-powered Manned Mars Sprint Mission," in *Proceedings of Eighth Symposium on Space Nuclear Power Systems*, Conf-910116, M.S. El-Genk and M.D. Hoover, eds., American Institute of Physics, New York, pp. 531–536.

Schnitzler, B.G. and S.K. Barowski, "Radiation Dose Estimates for Typical Piloted NTR Lunar and More Mission Engine Operations," Paper AIAA 91-3407, presented at the AIAA/NASA/OSI Conference on Advanced SEI Technologies, Cleveland, OH, September 4–6, 1991.

Smart, D.F. and M.A. Shea, "Solar Proton Events During the Past Three Solar Cycles," *Journal of Spacecraft*, 26(2): 403–415.

Townsend, L.W., et al., "Large Solar Flare Radiation Shielding Requirements for Manned Interplanetary Missions," *Journal of Spacecraft*, 26(2): 126–128.

Warman, E.A., *Proceedings of the National Symposium on Natural and Manmade Radiation in Space*, NASA Technical Memorandum, NASA TMX-2440, Washington, D.C, January 1972.

THE INTEGRATED POWER AND PROPULSION STAGE:
A MISSION DRIVEN SOLUTION UTILIZING THERMIONIC TECHNOLOGY

Robert M. Zubrin and Tal K. Sulmeisters
Martin Marietta Astronautics
PO Box 179
Denver, CO 80201
303-977-1925

Michael G. Jacox and Ken Watts
EG&G Idaho, Inc
Idaho National Engineering Lab
PO Box 1625
Idaho Falls, ID 83415-3515
208-526-0973

Abstract

The design of an Integrated Power and Propulsion Stage (IPAPS) and mission analysis demonstrating its utility is presented. Power for the IPAPS is provided by a modified SEHPTR out of core 40-kWe thermionic reactor which has been modified to allow direct thermal heating of hydrogen propellant to 2200 K. The reactor can thus function either as a 40-kWe power source or as a 400-kWt nuclear thermal rocket engine with a specific impulse of 736 s and a thrust of about 111 N. The IPAPS itself then consists of this reactor system, radiators, shield, hydrogen tank, a set of hydrogen arcjets with a specific impulse of 1400 s and 2.2-N thrust, avionics, and a set of power connects to support payload power utilization. The IPAPS becomes the bus for the mission instruments, providing flight avionics, guidance, telemetry, and command and control systems, thus eliminating the need for an independent upper stage for mission payload delivery. Since the arcjets and the SEHPTR thermal thruster utilize the same propellant, division of propellant between high priority high thrust maneuvers using the thermal thruster and for slow maneuvers using the arcjets does not have to be preallocated, hence contingency operations can be readily accommodated. The great increase in thrust made possible using a thermal system allows the IPAPS to transfer payloads from low Earth orbit (LEO) into highly elliptical destination orbits using a perigee kick strategy with greatly reduced gravity losses compared to electric propulsion (EP) spiral and reverse apogee kick transfers. The net result is that the IPAPS can achieve the same mass savings as EP (relative to chemical propulsion) while reducing transfer times in missions of interest from the 6 to 12 months typical of EP down to 2 to 3 weeks. It is concluded that the IPAPS concept holds large performance benefits and cost savings for missions requiring significant amounts of electric power and significant delta-V beyond an initial LEO orbit.

INTRODUCTION

Many space missions require or could be enhanced by large amounts of electric power, and it has been observed on numerous occasions that such missions could potentially be enhanced further by utilizing the on-board electric power source to drive advanced propulsion to deliver the spacecraft to its destination orbit, as well as to provide the power for the mission's payload. The most obvious way to attempt to use the electric power source in such a dual role is to drive an electric propulsion (EP) system, as the high specific impulse offered by such devices frequently makes possible a significant mission mass benefit. Unfortunately, the extremely low thrust limitations inherent in electric propulsion almost always causes the transfer times it produces to be unacceptably long, especially precisely those large ΔV missions for which the EP system offers a significant mass saving relative to chemical propulsion. However, if the electrical power source is nuclear, it is also a source of thermal power at levels an order of magnitude greater than its electrical output. This makes possible a different form of dual use for the power source, with direct thermal power generating rocket thrust without the intermediate step of electrical conversion. While the specific impulse offered by such direct thermal thrusters (700 to 800 s) is well below that offered by E.P., it is still much greater than that possible with chemical propulsion, and for missions of interest offers the lion's share of the mass benefits available with EP Moreover, the level of thrust such systems make possible may be 2 orders of magnitude greater than that of the same reactor's EP capability, with a corresponding drastic reduction in the required trip time. Furthermore, since the thermal thruster uses the same propellant as one of the main EP options (the hydrogen arcjet), both systems can be integrated on the same stage, so that the EP option can be taken advantage of for those operations where time is not of the essence. The electric power source also enables active refrigeration, thus making long term storage on-orbit of the hydrogen propellant a viable option. Thus there is good reason to believe that provided an electric power reactor design is available which is amenable to the production of direct thermal thrust as well, an integrated stage incorporating the symbiotic utilization of electric power, direct thrust, hydrogen arcjets, and active refrigeration offers a superior option for accomplishing many space missions. In this paper we shall discuss such a concept, which we call the Integrated Power and Propulsion Stage (Sulmeisters et al. 1991), and propose its use as a bus for instruments.

THE IPAPS STAGE

The Integrated Power and Propulsion Stage (IPAPS) is depicted in figure 1. It consists of a 40-kWe/400-kWt Small Exocore Heat Pipe Thermionic Reactor (SEHPTR) (Jacox et al. 1990), with a mass of 2600 kg, 400 kg of hydrogen arcjet equipment, 9150 kg of liquid hydrogen, and 1350 kg of tankage, structure, and avionics, for a total wet mass of 13500 kg. The SEHPTR reactor can be used to either drive the arcjet system with 40 kWe, generating 2.2 N (0.5 lbs) of thrust at a specific impulse of 1400 s, or it can be used to directly heat hydrogen propellant to 2200 K with 400 kWt in a pressure fed mode to produce 111 N (25 lbs) of thrust at 736 s specific impulse. Since both the arcjet and the thermal thruster use a common propellant, no preallocation of propellant between high thrust and low thrust maneuvering systems is needed, and either propulsion system can be used as contingency requires. The large hydrogen tank replaces the boom ordinarily required for payload separation from reactor in nuclear systems, with the hydrogen contained therein providing extra shielding beyond the minimum which is included within the SEHPTR's integral shield. The hydrogen tank is covered with 8 cm of double aluminized mylar/double silk net multi layer insulation with 20 layers per cm, which allows a heat leak from the ambient environment of 19 W. In addition about 2 W is deposited in the tank due to neutron and gamma transport. Using turbo-Brayton refrigerators which attain 10% of the ideal Carnot efficiency at 20 K, and allowing 45% extra margin on heat leak, 3.8 kWe is required to actively refrigerate the hydrogen tank to zero boiloff levels.

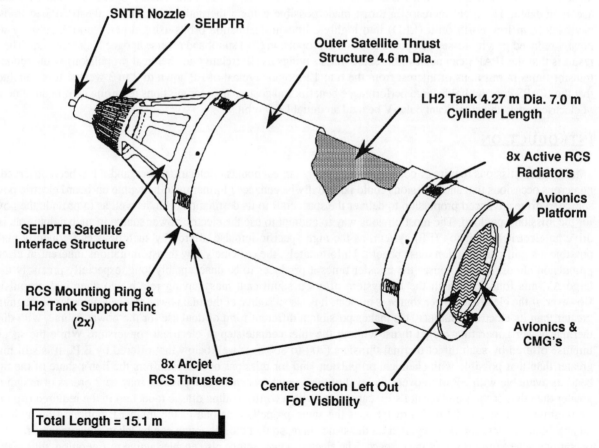

Figure. 1 The Integrated Power and Propulsion Stage (IPAPS).

The IPAPS point design described above is 15 m long. When properly placed in a Titan IV fairing it allows a 10 m barrel section (10 m length by 4.5 m diameter) to accommodate the payload instruments. If geosynchronous orbit (GEO) delivery is required, this payload instrument group can have a mass of up to 6500 kg. By comparison, a Titan IV equipped with a Centaur upper stage can deliver a 40 kWe thermionic reactor plus a payload satellite of only 2.2 tonnes.

THE SEHPTR REACTOR

The SEHPTR reactor concept was initially developed as a response to potential user concerns with space nuclear power systems. Studies performed by the Air Force have revealed that spacecraft developers would be reluctant to use a primary power subsystem with mission ending single point failure modes. The ability to perform adequate ground testing prior to launch is critical for both DoD and NASA systems and may be difficult to accomplish for many nuclear power concepts. Being able to safely launch and deploy a nuclear system is critical to user acceptance and the launch approval process, The SEHPTR concept addresses these concerns in several innovative ways.

Heat is generated within hexagonal shaped fuel elements of UO_2 clad in tungsten. The outer surfaces of the fuel elements radiate the heat to the thermionic heat pipe module (THPM) energy conversion devices. Hence there are no pumps of circulating core coolant loops associated with this design and the fuel is not in physical contact with the power conversion units. This radiative coupling between the core and the balance-of-plant is the key to allowing for complete ground testing prior to launch and also enables the concept to provide direct thermal propulsion by passing hydrogen between the fuel elements and the THPMs. Figure 2 provides a schematic illustration of the hydrogen flow through the SEHPTR reactor.

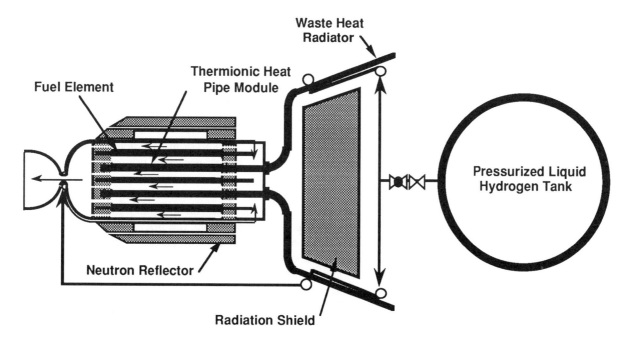

Figure 2. Schematic of Propellant Flow for Dual Mode SEHPTR Operation.

The baseline THPM concept is a single cell core-length thermionic energy converter configured with a cylindrical emitter sleeve on the outside, surrounding a central collector on the inside. The external cylindrical emitter is an annular heat pipe which receives heat from the tungsten fuel surface and distributes it isothermally both axially (down the length of the module) and circumferentially (around the circumference of the cylindrical module.) The inner surface of the heat pipe emitter forms the converter cathode surface. The converter anode is an internal cylindrical sleeve bonded to a central collector heat pipe through an electrically insulating material. The insulator is provided to electrically isolate the thermionic converter cell from the central collector heat pipe. The collector heat pipe receives waste heat from the convertor anode sleeve and transports it to the radiator. The heat pipe emitter and the collector heat pipe are separated by the interelectrode gap, which is maintained by ceramic centering spacers. The spacers are split rings which fit into slots cut in the anode sleeve of the collector heat pipe. Cesium pressure is provided by integral, solid sorption type reservoirs. Electrical connections are made at each end of the converter for both the emitter and the collector so that the device operates as a double ended cylindrical diode. Sixty two thermionic heat pipe units are employed to provide 40 kWe.

Neutronically, the core is reflected top and bottom by BeO reflectors and controlled by two sets of windowed radial reflectors. Reactor control is achieved by allowing neutrons to leak or be reflected by this dual rotating windowed reflector. The reactor is shut down when the inside and outside windows or voids in the reflector are aligned, and the reactor becomes operational as the windows close, reflecting neutrons back into the core region. Either the inside or outside reflector and drive mechanism are capable of independently controlling the reactor through its full range from startup to shutdown throughout its life.

The SEHPTR concept offers some significant advantages over competing space nuclear power concepts. The radiative coupling of the core to the balance-of-plant allows the SEHPTR concept to be acceptance tested at the system level with an electrical core simulator to validate the flight worthiness of the system prior to launch. Additionally, the separation of the energy conversion devices from the fuel allows acceleration of the development program and broader participation from potential thermionic manufacturers. The reliability data base required to gain user acceptance can be developed at significantly reduced cost if if the fuel and emitter interactions are minimized. The SEHPTR concept offers inherent safety features not found in all thermionic concepts. The presence of significant amounts of tungsten in the core guarantees subcriticality in all launch accident scenarios, and the material expansion and Doppler broadening provide a negative temperature coefficient of reactivity. Redundancy is provided in removing heat from the core, converting heat to electricity, and controlling the reactor throughout life in both start-up and shutdown sequences. The THPM design offers several approaches for scaling the SEHPTR concept to higher power levels including using a multi-cell approach while maintaining full testability.

ORBIT TRANSFER TRADES

A detailed series of trades were run for using the IPAPS stage and other forms of advanced propulsion to transfer a 1500 kg satellite from a properly inclined 300-km altitude low Earth orbit (LEO) to its destination orbit. The three destination orbits chosen were a 4 hour circular orbit, a circular geosynchronous orbit, a 24 hour elliptical orbit with a 300 km altitude perigee, and an "oblate" 24 hour orbit with a 10519 km perigee. Such orbits are useful for many satellite applications.

The four propulsion systems analyzed were cryogenic chemical propulsion with an Isp of 450 s, hydrogen arcjets with an Isp of 1400 s, ion thrusters (xenon or argon) with an Isp of 5000 s, and the IPAPS stage using the SNTR thruster. The SNTR uses the thermal energy of the SEHPTR nuclear reactor to produce thrust with direct heating of hydrogen propellant to 2200 K, yielding an Isp of 736 s at about 111 N of thrust. All stages were assumed to have a tankage fraction of 0.15, and in addition the arcjet and IPAPS/SNTR propulsion systems incurred a further mass penalty of 400 kg for thrusters, power processing unit, etc, while for the ion system the mass penalty was taken as 550 kg. The arcjet was taken to have an efficiency of 0.4, and the ion thruster an efficiency of 0.7. 40 kWe of power was assumed available for electric propulsion, and 400 kW thermal for the IPAPS/SNTR. The SEHPTR reactor system was assumed to mass 2600 kg.

The cryogenic option used the normal high thrust Hohmann transfer maneuver to move from its initial LEO orbit to its destination orbit, while the IPAPS employed a series of perigee kicks to enable it to use Hohmann transfers as well. The EP systems moved to their elliptical destination orbits in two different ways. In Mode A, electrical propulsion was used to spiral the spacecraft out to a circular orbit whose radius was equal to the apogee radius of the elliptical orbit. Then a high thrust reverse apogee kick using cryo was applied to immediately transform the circular orbit into the appropriate elliptical orbit. In Mode B, the electrical propulsion system was used to spiral out from LEO to the circular orbit, after which the EP system was used in a sequence of reverse apogee thrust arcs to gradually lower the perigee and thus transform the circular orbit to the appropriate elliptical orbit. Mode B is more propellant efficient than Mode A, as it avoids the use of chemical propulsion, but its employment will increase the total time from LEO departure to initiation of payload satellite operations by about 25%.

The Edelbaum equation was used to compute the spiral EP ΔVs. The ΔVs of the reverse thrust arcs were calculated as being the same as their high thrust values. Both the cryo and the EP. systems were assumed to be single stages with no drop tanks. The ΔVs calculated for both the cryo and EP transfers from LEO to the destination orbits are given in table 1.

Table 1. Ideal ΔV for Satellite transfers from LEO to Destination Orbits.

Destination	Cryo ΔVs (km/s)	EP ΔVs (km/s)
24 hr elliptical	V1 = 2.759	V1 = 5.463
	V2 = 0	V2 = 1.363
24 hr oblate	V1 = 2.697	V1 = 5.298
	V2 = 0.506	V2 = 0.892
4 hr circular	V1 = 1.130	V1 = 2.140
	V2 = 0.963	V2 = 0
GEO circular	V1 = 2.467	V1 = 5.20
	V2 = 1.797	V2 = 0

For calculation purposes, a margin of 5% on all ΔVs was assumed. That is, the ΔVs used were those given above multiplied by 1.05. No cosine losses due to off-optimum thrust direction were included in the analysis. Table 2 gives the total initial mass in LEO (IMLEO) of the satellite transfer mission under all options. No additional propellant mass for orbit changes has been included.

Table 2. IMLEO and Transfer Time of Satellite Mission for Alternative Modes.

Destination	Cryo		Arcjet		Ion		IPAPS	
	Mass	Time	Mass	Time	Mass	Time	Mass	Time
24 hour elliptical, Mode A	9188	0.5	10998	256	7905	434	7263	16
24 hour elliptical, Mode B	9188	0.5	8449	327	5514	522	7263	16
24 hour oblate, Mode A	10597	0.5	9349	213	6911	366	7875	16
24 hour oblate, Mode B	10597	0.5	7962	265	5427	431	7875	16
4 hour circular	7469	0.1	5460	57	4905	110	6478	1
GEO circular	16422	0.2	7062	152	5259	262	9477	12

In Table 2. all masses are given in kg, and all transfer times are given in days. It can be seen that the Arcjet offers little mass advantage over cryo propulsion in reaching the 24 hour elliptical or oblate orbits. The ion thruster offers mass advantages against both arcjets and cryo propulsion in all cases. The IPAPS always has a significant mass advantage compared to the cryogenic system, and is better than the arcjet in reaching elliptical orbits. The mass saving of the pure EP systems may also be somewhat overstated, moreover, if it turns out that an operational requirement dictates the emplacement on the satellite of a high thrust (NTO/MMH biprop, Isp=330 s) system to meet the requirements of rapid redeployment. Since for practical purposes the SNTR system is a high thrust system, the IPAPS's SNTR/arcjet combination would not need a contingency NTO/MMH biprop system to meet such a requirement. Furthermore, while the EP systems offer mass advantages in many cases, the time they require to attain their desired operating orbit may be excessive, especially for the 24 hour elliptical orbit options (200-500 days.) It can be seen, however, that the IPAPS can deliver its payload without excessive time penalties, while providing comparable mass savings as the EP systems.

The mass advantages of the advanced propulsion options over cryo could be used to lower IMLEO, as shown above, or it could be used to enable either a larger instrument compliment or a much larger amount of maneuvering propellant to be delivered while keeping IMLEO constant. In Table 3 we show the results for such an analysis, with the IMLEO of all options constrained to the 15000 kg limit of the Titan IV launch to high inclination or a Titan III launch to 28 degree inclination. Maneuverability is assessed in delta-V capability (in km/s) of the satellite system after its destination orbit has been achieved. In the IPAPS column, the entries marked with a * are those in which the SNTR is used to achieve the design orbit quickly, after which arcjets are used for on-orbit maneuvers. Since the SNTR and the arcjet both use a common propellant, hydrogen, such an operational mode is highly attractive.

Table 3. On Orbit Repositioning Capability of the Satellite (km/s) and Transfer Time (days).

Destination	Cryo		Arcjet		Ion		IPAPS	
	ΔV	Time	ΔV	Time	ΔV	Time	ΔV	Time
24 hour elliptical, Mode A	1.51	0.5	2.58	350	22.92	823	3.90	35
24 hour elliptical, Mode B	1.51	0.5	5.72	482	37.78	1210	7.43*	35
24 hour oblate, Mode A	1.05	0.5	4.29	340	28.57	797	3.44	33
24 hour oblate, Mode B	1.05	0.5	6.39	430	38.44	923	6.54*	33
4 hour circular	2.21	0.1	10.64	155	42.70	334	4.60	1.7
	----	----	-----	----	-----	----	8.75*	1.7
GEO circular	- 0.32	0.2	7.47	337	40.73	783	2.32	23
	----	----	------	-----	-------	----	4.43*	23

It can be seen that the use of ion EP can result in an enormous increase in satellite repositioning ability, expanding the satellites' on-orbit delta-V capability by more than an order of magnitude. The trip time for the pure EP stages are, however, hopelessly long. The IPAPS combination of SNTR and arcjet, on the other hand, is more than sufficient to expand the orbital maneuverability by a factor of 6 relative to cryo, while avoiding the very large transfer time penalties of the pure EP system.

For all of these reasons, it appears that the IPAPS combination of SNTR/arcjet systems is far and away the best propulsion option for a large variety of satellite missions.

IPAPS Perigee Kick Strategy

In order to effectively employ the 25 lb thrust IPAPS/SNTR system for satellite transfer, a burn strategy known as "perigee kicks" must be employed. In this strategy, the SNTR engine is fired only for a short time while the spacecraft is flying near the periapsis of its orbit, and then shut off as the spacecraft coasts out to apogee and back to perigee, when the kick is repeated. In this way, the effect of a series of modest thrust burns at perigee may be superimposed to create the equivalent (to first approximation) of a single high thrust burn, and gravity losses kept close to a minimum.

A fourth-order Runge-Kutta code, TRANSFER, was used which integrates the primitive equation of motion of a rocket vehicle operating in a planetary gravitational field to simulate the trajectory. The perigee kick strategy was defined by the maximum angle between the the radius vector drawn from the center of the Earth and the spacecraft and the vector from the center of the Earth and the orbit's perigee within which the engine would be fired. Thus, if the thrust arc angle was set at 30 degrees, the SNTR would fire from an orbital position of -30 degrees, through perigee, and cut off at the +30 degree position. The results of an investigation of thrust arc strategy are shown in Table 4.

Table 4. IPAPS Thrust Arc Options.

Arc (degrees)	No. of Kicks	Time (h)	IMLEO	Gravity Loss (%ΔV)
15	338	1689	7104	0.5
30	176	865	7151	2.2
45	116	561	7232	4.2
60	84	389	7343	7.6
75	65	305	7522	12.8
90	52	228	7798	18.4
105	42	162	8140	29.8
120	36	130	8631	39.3

It can be seen that the narrower the perigee kick angle, the lower the IMLEO but the greater the time required to arrive at the design (elliptical) orbit. It was decided that the best compromise between the desired objectives of short trip time and low IMLEO could be met if the thrust arc was set at about 60 degrees, and this was therefore baselined. If operational considerations dictated an increased importance to either quick transfer or low IMLEO, the IPAPS stage could accommodate by altering its thrust arc angle accordingly. The history of the IPAPS spacecraft orbit during its series of perigee kicks is given in Table 5.

Table 5. IPAPS Perigee Kick History.

Kick No.	Time(hrs)	Perigee	Apogee	Max. Velocity
0	0	290 km	310 km	7.74 km/s
10	15.3	382	1272	7.93
20	33.4	463	2519	8.13
30	54.2	554	4191	8.35
40	79.1	646	6503	8.58
50	109.9	736	9849	8.84
60	150.4	825	15096	9.12
70	208.8	912	24449	9.42
80	309.2	997	45680	9.74
84	378.5	1009	64234	9.91

Because the perigee kick is not a perfect high impulse burn at perigee itself, the perigee is raised slightly with each successive burn. Thus after the 84st burn above, a small trim burn at apogee is required to either lower or raise the perigee if it is desired to make the orbit correspond exactly to either the "elliptical" (300 km perigee) or "oblate" (10519 km perigee) 24 hour orbits defined earlier.

Solar Electric Performance Comparisons for Generic Missions

In figures 3a and 3b we show a comparison of the performance of the IPAPS stage with solar electric transfer vehicles for missions to both 24-hour elliptical orbits and also to a circular geosynchronous orbit. In this study the mass to power ratio "alpha" of the solar electric power system was regarded as a free variable. Examining the figures we see that if survivable solar power systems are employed, the IPAPS outclasses the solar electric systems on both missions. If light weight solar electric systems are employed, the IPAPS is comparable to the pure arcjet on both missions. Only if ultralight weight solar electric systems without batteries are employed does pure electric propulsion show a significant mass advantage over the IPAPS, however even in these cases the IPAPS has a trip time which is at least an order of magnitude less than the electric propulsion options. It may thus be noted that if such light weight solar electric systems were ever made available a thermionic reactor which could only produce electric power would essentially be obsolete, while one that had the IPAPS/SNTR capability for both thermal propulsion and electric power would still offer many mission advantages.

ORBITAL DECAY AND RADIOLOGICAL HAZARDS TO EARTH

Data generated by the Idaho National Engineering Lab (INEL) has shown that the SEHPTR reactor builds up most of its radioactivity within 1 day of operation, and that most of this has decayed away by 100 days after shutdown. These are predominantly short term radio-nuclides. Long term (30 year half-life) radionuclides are built up more slowly, accumulating in an approximately linear fashion with operation time, however, as we have seen, the IPAPS only requires about 16 days to raise itself to an 1000 by 67000 km elliptical orbit, after which orbital re-entry via natural decay is a threat to be dealt with several millennia in the future (at which time all the significant longer term radio nuclides will have decayed away.) Thus the prime threat to be dealt with is the case where the IPAPS operates for such a short time that it has not yet substantially raised itself above its initial orbit of 300 km.

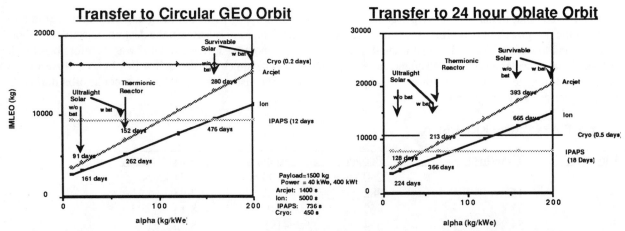

Figure 3. Regardless of hypothetical future improvements in electric propulsion power/mass ratios, IPAPS preserves an order of magnitude advantage over EP in trip time while garnering most of the mass benefits.

If we examine Table 5 we see that at the end of 10 perigee kicks, the IPAPS has been raised from its initial 300 km circular orbit to a 385 by 1300-km elliptical orbit. While the total time elapsed is 15 hours, the total burn time since initiation or orbital operations will have been only about 5 hours. The very large majority of time on this orbit is spent at high altitudes where aerodynamic drag is close to zero, but to be conservative, we shall assume that a full 1/3 of the orbital time is spent at the 385 km altitude (and assume zero drag for the rest of the trajectory). As an alternative case, we shall also examine the case where a single perigee kick is performed (0.5 hour burn time) with no effect in increasing altitude (total engine failure to perform), after which the orbit is allowed to decay. The 10 kick case we call "Case X," while the single kick case we call "Case Y." The two cases are summarized in Table 6.

Table 6. IPAPS Orbital Decay and Radiological Hazard.

	Case X	Case Y
Orbit (km)	385 x 1300	300 x 300
Burn time	5 hours	0.5 hours
Initial Radioactivity	1640000 Ci	1310000 Ci
Orbit Decay Time	600 days	100 days
Re-Entry Radioactivity	10 Ci	18 Ci

The above analysis assumed Solar Max conditions and also that the IPAPS oriented itself perpendicular to the direction of motion (maximum drag, minimum time to re-entry.) Since the failure was assumed to occur near the beginning of operational life, the propellant tank was assumed to be full, which gives the spacecraft a ballistic coefficient of 250 kg/m^2.

It can be seen that the maximum amount of radioactivity returned to Earth after an early IPAPS re-entry is in the neighborhood of 20 Ci. This is too small to form a significant plume, groundshine, skyshine, or inhalation dose hazard to the public. The primary hazard that it would represent would be as the cause of an ingestion dose due to crop contamination. While standards for condemnation of crops due to radiological contamination vary from state to state, nation to nation, and by isotope, a good typical working figure for embargo of crops is about 1 Ci/km^2. Thus, if upon re-entry, the SNTR was to disperse itself homogeneously so as to contaminate the maximum acreage possible, the total area of embargoed crops would be about 20 km^2. Even in the unlikely event that this were to occur over prime farm land, the damages would not be excessive. For example, wheat currently is selling for about $2.50 a bushel. if we assume 40 bushels to the acre, the total monetary value of the crop over 20 km^2 is about $500,000.

The low radiological hazard presented by the IPAPS is a result of the presence of the thermal thruster, which allows it to generate 40 times the thrust at a given power level as would be possible using the arcjet for orbit raising. A pure nuclear electric stage operating at the same power level and performing the same mission would thus present approximately 40 times the radiological hazard presented by the IPAPS.

CONCLUSIONS

We find that the IPAPS concept offers numerous advantages for space missions requiring or enhanced by significant amounts of on-board electric power. The SEHPTR thermionic reactor provides a compact, survivable source of electric power in the 10 to 100 kWe range, while simultaneously providing ten times this power as a heat source for a hydrogen thermal thruster with a specific impulse over 730 seconds. The system thrust is thus increased by a factor of 50 over that which would be available from an arcjet, leading to an order of magnitude decrease in trip-time while preserving most of the mass saving advantages of electric propulsion systems. In addition to removing the traditional show-stopper to the implementation of advanced propulsion on satellite transfer missions, this sharp reduction in trip time made possible by the thermal thruster makes the SEHPTR thermionic reactor a superior power system for most missions regardless of how light the solar photovoltaic systems weight may become in the future. The addition of on-board arcjet thrusters using a common hydrogen propellant with the thermal thruster enables the stage to take full advantage of high impulse EP when time permits, thus enabling a very large on-orbit maneuver capability, while requiring no pre-allocation of propellant between high-thrust and low thrust systems. The high power available from the thermionic reactor makes long-term storage of the hydrogen on orbit using active refrigeration feasible and attractive. The cost of missions will be reduced because no separate orbit transfer stage will be needed, and this will also reduce mission risk because the possibility of the failure of an independent upper stage will be eliminated. The IPAPS stage itself is highly reliable, incorporating three independent modes of propulsion: thermal, arcjet, and cold hydrogen gas. The IPAPS's medium thrust perigee-kick mode of orbit transfer allows reasonable trip times while improving the accuracy of guidance of orbit transfer, as thrust vector errors occurring on one perigee kick can be corrected on the next. The incorporation of the thermal thruster on the reactor also reduces by a factor of 50 the radiological hazard posed by the reactor to the terrestrial environment in the event of an early mission failure. For all of these reasons it is clear that the Integrated Power and Propulsion Stage represents the proper form of implementation of thermionic technology for space missions, and we recommend that it be pursued vigorously.

Acknowledgments

Robert Spencer drew the CAD drawing of the IPAPS in figure 1, and together with Jim Greenwood, did most of the overall layout design of the IPAPS vehicle. Sid Earley made many contributions to the mission analysis. John Malloy and his team at the Babcox and Wilcox company have contributed many ideas leading to improvement in the original SEHPTR design. The IPAPS stage concept was developed and studied under a contract from the Air Force Phillips Laboratory directed by Dr. Michael Schuller.

References

Jacox, J. G., R. G. Bennett, L. B. Lundberg, B. G. Miller, and R. L. Drexler, (1990) "Small Ex-Core Heat Pipe Thermionic Reactor Concept (SEHPTR)," Proc. IECEC, Monterey, CA, June, 1990.

Sulmeisters, T. K., R. M. Zubrin, and S. M. Earley, (1991) "Advanced Propulsion System Applications to Upper Stages," AIAA 91-1988, AIAA/SAE/ASME 27th Joint Propulsion Conference, Sacramento, CA, June 24-26, 1991.

CESIUM VAPOR SUPPLY SYSTEMS FOR LONG LIFETIME THERMIONIC SPACE NUCLEAR POWER SYSTEMS

Georgy M. Gryaznov, Evgeny E. Zhabotinsky, Victor I. Serbin,
Boris V. Slivkin, Yuri L. Trukhanov, Leonid M. Sheftel

SPU "Krasnaya Zvezda"
USSR, Moscow 115230

Abstract

For long lifetime space nuclear power system (SNPS) of second generation the cesium vapor supply systems must be applied which differ on principle from system used for "Topaz" SNPS. The main differences are multiple usage of cesium, weak dependence of system mass and dimensions upon SNPS lifetime, essential increase of effectiveness for impurity gas removal from interelectrode gap (IEG) of thermionic fuel element (TFE). Possible schemes of cesium vapor supply systems with its multiple usage are considered. These schemes include one-side and two-side diffusion of impurity gases from IEG, combination of diffusion with vapor pumping through IEG. The estimates of supply system effectiveness relatively to removal of impurity gases from IEG and its coomparison by that criterion are given. It is shown that for long lifetime SNPS it is necessary to use regenerative supply systems with cesium vapor flowrates up to several hundreds grams per day. The selection of the most expedient supply system scheme and its parameters are determined by TFE type and gas discharge into IEG.

INTRODUCTION

For thermionic SNPS with lifetime of several years the application of systems for cesium vapor supply to in-core thermionic reactor (ITR) which are different principally from cesium vapor supply system (CVSS) for SNPS "Topaz" is necessary. The main ones of these differencies are next:

1. Cesium multiple usage under which the value of system vapor flow rate don't be connected with cesium stocking in vapor generator;
2. Weak dependence of cesium mass in vapor generator from SNPS lifetime stipulated sufficiently small leakages of cesium from CVSS only; and
3. Considerable increase of effectiveness for gaseous impurity removal from IEG as a result of vapor flow rate in CVSS.

The CVSS posssible schemes for various TFE types for long lifetime thermionic SNPS are considered below; the estimates of its efficiency with reference to impurity gas removal are adduced.

THE CVSS SCHEMES

TFE with non-separated and separated cavities of IEG and fuel pins can be applied in ITR. In first case the gas removal from fuel pin are performed to IEG, in the second the special duct is used, the tightness degree of which with regard to IEG can be various. TFEs can have three- or five-layer collector assembly, blind or through cesium duct. The CVSS schemes for TFE with non-separated cavities of IEG and fuel pins are shown on figure 1. These schemes include:

1. One-side diffusion of impurity gases from IEG to cesium chamber (TFE with blind cesium duct);
2. Two-side diffusion of impurity gases from IEG to cesium chambers arranged on top and bottom of ITR (TFE with through cesium duct); and
3. Combinations of one- or two-side scheme with vapor flow through IEG.

The CVSS schemes for TFE with separated cavities are shown on figure 2. In that case gas removal from IEG can be organized by analogy with previous. The gas removal ducts of fuel pins can be connected to either cesium chambers or lowered pressure zone behind the throttle. The successive cesium vapor pumping through IEG and gas removal ducts of fuel pins is possible in principle. The connection of fuel pin ducts with lowered pressure zone is possible for TFE with three-layer collector assembly. For TFE with five-layer collector assembly the connection of IEG and fuel pin ducts is possible in zone of TFE leads. For CVSS a vapor generator of regenerative type ensuring multiple usage of cesium is applied. The level of cesium vapor pressure is determined by temperature of generator evaporation zone. In condensation zone the cesium vapor is condensed and supplied to evaporation zone again with the help of capillary structure and noncondensible gases are removed to space. The temperature of condensation zone has to be in range from 150 to 200 C for ensuring sufficiently small cesium leakages through gas-removal ducts and for exclusion of opportunity to form the solid cesium-impurity combinations which could cork up the vapor generator capillary structure. With that the possibility for system operation with any expedient vapor flow rate defined by throttle conduction at ITR outlet is ensured. On principle a few independent CVSSs can be applied each of which attends to own TFEs group.

THE EFFECTIVENESS OF VARIOUS CVSS SCHEMES

The average pressure of impurity gases in IEG can be described by next expression for CVSS scheme with one-side diffusion under condition of constant gas-discharge density stant along the length of all TFEs (Gryaznov et al. 1990):

$$\bar{P}_g = P_{cs} \left[\frac{Q_g}{Q} + \Delta P_d (Q, L_d) + \frac{Q_g L}{3\pi \cdot A_{IEG} \cdot d_e \cdot D \cdot n \cdot P_{cs}} \cdot \frac{T_{IEG}}{T_{thr}} \right] . \qquad (1)$$

For scheme with two-side diffusion under the same conditions of impurity gas removal to each cesium chambers the analogous equation is:

$$\overline{P}_g = P_{Cs}\left[\frac{Q_g}{Q} + \frac{1}{2}\Delta P_d(Q,L_d) + \frac{Q_g \cdot L}{12\,\pi \cdot \Delta_{IEG} \cdot d_e \cdot D \cdot n \cdot P_{Cs}} \cdot \frac{T_{IEG}}{T_{thr}}\right]. \quad (2)$$

Here P_{Cs} - vapor pressure in cesium chambers, Q - vapor volume flow rate in CVSS, Q_g - total gas discharge for n TFEs which are served by given CVSS, d_e and Δ_{IEG} - TFE emitter diameter and IEG width, D - diffusion coefficient for impurity gas in cesium vapor, T_{thr} and T_{IEG} - average vapor temperatures in the throttle and IEG, L and L_d - lengths of ITR core and TFE outlet duct, ΔP_d - impurity gas pressure drop for TFE removal duct which is conditioned by its diffusion resistance.

In equations (1) and (2) the influence of magnetic forces is not taken into account so far as that influence is sufficiently small under specific power density up to 3 W/cm². The passage cross-sections of removal ducts from TFE must be selected so that the contribution of second term in equations (1) and (2) would be lesser considerably (in 5 to 10 times) than the contribution of third term. The rise of CVSS vapor flow rate can be expedient up to values under which the contribution of first term in equations (1) and (2) becomes lesser considerably than total contribution of second and third terms. With that two-side diffusion scheme can be efficient in four times as compared with one-side scheme under other equal conditions. The performed estimates show that for thermal neutron ITR with L=0,5 to 0,7 m, Δ_{IEG} =4 10⁻⁴ m, d_e=(10-20)·10⁻³ m, T_{IEG} =1400 to 1600 C, T_{thr} =800 to 900 K, n=60 to 100 and for hydrogen and gases with molecular weight 28 the expedient rates of flow is in range of 200-500 g/day for one-side diffusion scheme. For two-side diffusion scheme the expedient rates of flow rise up to 1000 g/day and more. The increase vapor flow rates above these values will influence upon P_g feebly under given gas discharge to IEG. The further drop of P_g can be achieved by decreasing of gas discharge to IEG. With that purpose the use of TFE with separated TFE and fuel pin cavities is imagined as efficient. Under connection of fuel pin ducts with cesium chambers the value $\alpha \cdot Q_g$ will enter to second and third terms of equations (1) and (2), where α - the portion of total gas discharge directly to IEG which depends on tightness degree for sealing devices between cavities. If α<1 then first term in equations (1) and (2) will be predominated up to values of flow rates which is more in $1/\alpha$ times than the above-mentioned ones. If gas-removal ducts of fuel pins is connected not with cesium chambers but with condensation zone of vapor generator then and first term in equations (1) and (2) will be decreased in $1/\alpha$ times under given vapor flow rate. For scheme with vapor pumping through IEG the restrictions by vapor pressure drop along TFE length must be observed; this drop must not exceed the value of about 133 Pa under considerations for permissible deviations from optimal cesium pressure for all electrogenerating cells in TFE. It restricts the possible values of vapor flow rates through IEG and, consequently, decreases the influence of vapor pumping on removal of impurity gases from TFE.

The calculations with account of possible IEG narrowing because of lifetime form-changing of TFE emitters in permissible limits show that for one-side diffusion scheme the vapor pumping can decrease $\bar{P}_q$ for hydrogen till 15% approximately in comparison with the case of pumping absence. For two-side diffusion scheme and organized vapor pumping through IEG the possible decrease of $\bar{P}_q$ don't exceed a few per cents. As an example it may be pointed out that for SNPS "Topaz" under gas discharge from TFE (5 to 7)·10^{-4}W (by hydrogen mainly) and vapor flow rate about 10 g/day the IEG pressure had been determined by pressure in outlet commutation chamber practically which could reach 133 Pa for hydrogen. Under these vapor flow rates the contribution of diffusion terms in equation (1) turned out to be lesser in 1,5 order than the commutation chamber pressure. In conditions of SNPS "Topaz" the contributions of these terms would be equal approximately under vapor flow rates 300 to 400 g/day. With that the IEG average pressure wouldn't exceed the value of 4 Pa. Such level of hydrogen pressure must not influence neither plasma losses nor IEG heat conductivity, i.e. must not cause degradation of TFE electrical performances.

CONCLUSION

For long lifetime thermionic SNPS it is expedient to use multiple-usage cesium vapor supply systems under flow rates up to several hundreds grams per day. The similar systems can provide the high effectiveness of impurity gas removal from TFE IEG. The choice of the most expedient supply system and its parameters is determined by TFE type, gas discharge into IEG and permissible level of impurity gases in IEG.

R e f e r e n c e s

Gryaznov, G. M., "Multiple-usage cesium vapor system for SNPS of the second generation", in Anniversary specialist conference on nuclear power engineering in space. Transactions. Part 1. Obninsk, USSR, 1990.

Acknowledgments

This work was carried out of the SPU "Krasnaya Zvezda" (USSR, Moscow).

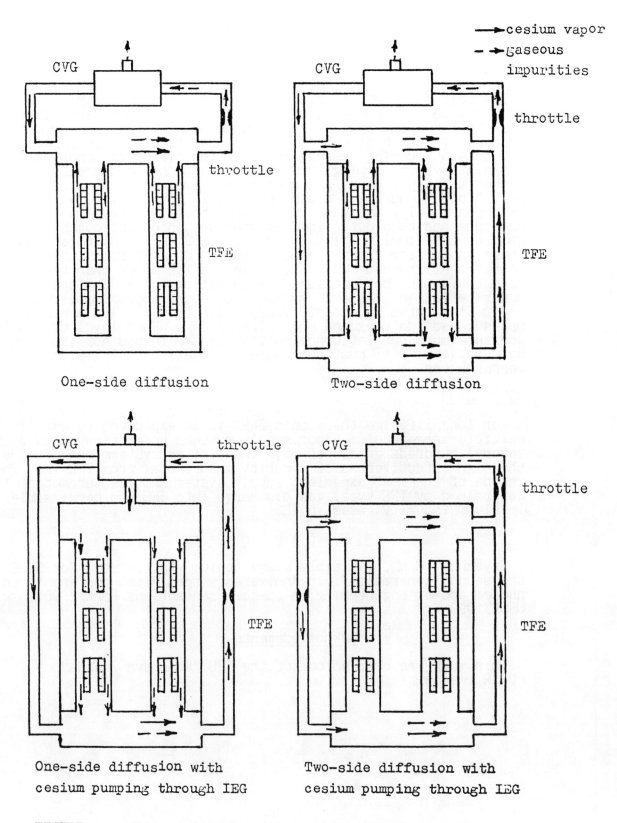

One-side diffusion

Two-side diffusion

One-side diffusion with
cesium pumping through IEG

Two-side diffusion with
cesium pumping through IEG

FIGURE 1. Schemes of Cesium Vapor Supply System for TFE with
Non-separated Cavities of IEG and Fuel Pins

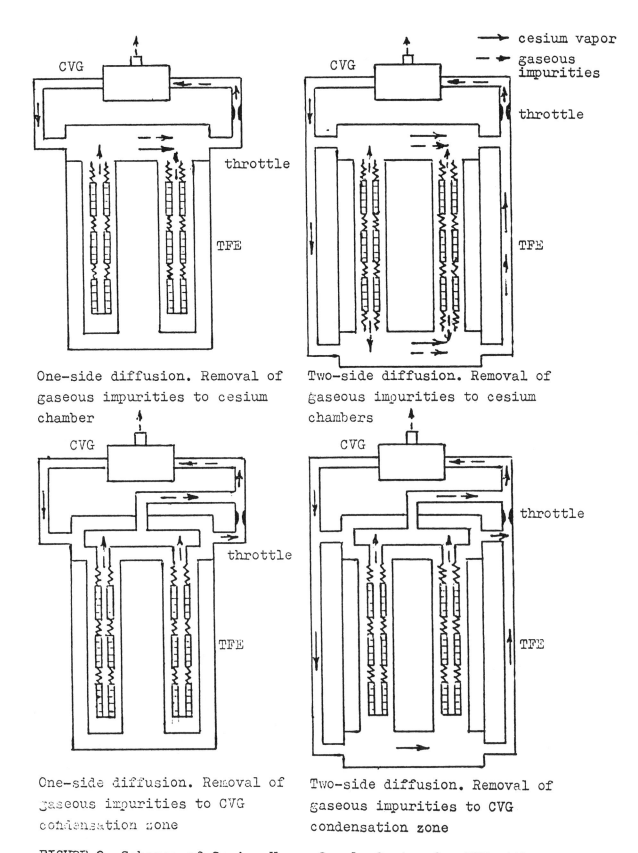

One-side diffusion. Removal of gaseous impurities to cesium chamber

Two-side diffusion. Removal of gaseous impurities to cesium chambers

One-side diffusion. Removal of gaseous impurities to CVG condensation zone

Two-side diffusion. Removal of gaseous impurities to CVG condensation zone

FIGURE 2. Schemes of Cesium Vapor Supply System for TFE with Separated Cavities of IEG and Fuel Pins

CONTROL ALGORITHMS FOR "TOPAZ " TYPE THERMIONIC SPACE NUCLEAR POWER SYSTEMS OF SECOND GENERATION

Irma V. Afanasyeva, Georgy M. Gryaznov, Evgeny E. Zhabotinsky, Gennady A. Zaritzky, Victor I. Serbin

SPU "Krasnaya Zvezda"
USSR, Moscow 115230

Abstract

Control algorithms for "Topaz" type thermionic space nuclear power systems (SNPS) of second generation with electrical power up to 100 kW are considered. These algorithms are based on results obtained under creation of "Topaz"SNPS in ground and flight conditions and develop directly control principles of that SNPS. The most general demands for SNPS under start-up and nominal modes taking into account of limitations necessary for security of prolonged lifetime are given. These demands must include the security of nuclear and radiation safety under possible emergency situations with SNPS and spacecraft. Control algorithms answering to these demands to a high degree are formulated. Therefore these algorithms can be considered as the most expedient for thermionic SNPS. In particular it is shown that duration of start-up mode for thermionic SNPS will be equal from 1,5 to 2 hours. The specific values of parameters for these algorithms must be optimized in each concrete case.

INTRODUCTION

The choice and elaboration of control algorithms for thermionic SNPS is one of the most difficult tasks of SNPS creation. This problem had been solved sucessfully for "Topaz" SNPS. The results of "Topaz" SNPS ground and flight tests have allowed to conclude that base principles of developed and realized algorithms is expedient to use for that type SNPS of second generation and also to outline the ways of further control algorithm improvement directed towards an increase in SNPS operation performances, lifetime and safety. The possible control algorithms for thermionic SNPS of "Topaz" type with thermal neutron incore thermionic reactor (ITR) under electrical power up to 100 kW are considered below; these algorithms satisfy a totality of demands for these SNPS in the most degree.

OPTIMAL CONNTROL ALGORITMS

Start-up Mode

For start-up mode with bringing up to electrical power the main factor are next:

1. Time for SNPS start-up to switching of all spacecraft user feeding from start-up battery on SNPS;
2. Observance of restrictions on thermal power, rate of coolant temperature rise, temperature difference along moderator discs and other which are necessary for ensuring long SNPS workability; and
3. Capacity of start-up battery.

The criterion for optimal start-up mode will be the minimum time of SNPS bringing to electrical power under observance of all restrictions. This time is defined by next processes:

1. By heating-up of cesium vapor supply ducts to level about 600 K excluding from possibility of cesium vapor condensation under its entering to ITR; and
2. By filling the ITR cavities with cesium vapor up to normal pressure.

For one-side diffusion cesium vapor supply system which is one of the most expedient for long lifetime SNPS the temperature state of supply ducts will depend upon coolant outlet temperature and character of its time change mainly. For time minimization of filling the ITR cavities with cesium vapor the temperature in vapor generator evaporation zone under its unsealing must be maintained by regulator on given level which may exceed an icipated optimal one on start-up mode.

The beginning of start-up mode is ITR bringing up to minimally controlled level (MCL) of thermal power (5-10% from nominal value). The ITR bringing up to MCL can be performed under program of two-rate start-up analogous for "Topaz" SNPS, according to which reactivity input with maximally possible rate for existing control drum drives is fulfilled until to approximately critical state, and then the rate of reactivity input decreases sharply (down to 10 times from maximum). For "Topaz" SNPS flight prototypes the time of bringing up to MCL presented about 4,5 min. The time of two-rate start-up can be minimized with account of the scatterings for reactor initial subcriticality and control drum efficiency under the condition of given limitation on reactor period in process of MCL attainment. But and by this optimization it is unlikely that the time of bringing up to MCL can be less than 3-3,5 min. It is possible to use the programs for ITR start-up to given reactivity or period level with the maintenance of these values for MCL attainment. However with account of sufficiently small time for MCL attainment such programs can't ensure any seriuos advantages in comparison with two-rate start-up program even under application of precise means for period or reactivity measure. After MCL attainment the thermal power channel with programly changing setting is switched on. In heating-up regime beginning when coolant temperature is low still the control is carried out by misalignment between measured thermal power and its program value which is formed as independent function of time, for example, under linear law:

$$N_{giv}(t) = N_{MCL} + k_N \cdot t. \tag{1}$$

When coolant temperature at ITR outlet achieves the are given value T_{giv} (500 to 600 K) at the moment t_0 the temperature channel with programly changing setting for outlet coolant temperature is switched on. With that the setting for power is formed as function of misalignment between measured temperature and its program value at $t > t_c$:

$$N_{giv}(t) = N_{giv}(t_0) + k_1 \psi_1(\Delta T) + \psi_2 \left[k_2 \int_{t_0}^{t} \psi_3(\Delta T) dt \right], \tag{2}$$

where $\psi_1(x), \psi_2(x), \psi_3(x)$ - functions determining the characteristics of corresponding channels which can be either linear without or with insensitivity zone or relay ones; k_1 and k_2 - amplification coefficient; $\Delta T = T_c(t) - T_c^{giv}(t)$, T_c - coolant temperature.

The conditions of optimization for this part of start-up algorithm can be formulated by next way:

1. Minimal time t_f until cesium vapor feeding to ITR;
2. Temperature of "cold" point at the ducts for cesium vapor feeding to ITR $T_{c.p.}(t) > T_{min}$ by $t \leq t_f$;
3. Coolant temperature $T_c(t) \leq T_{lim}$, $t > t_0$ and its time derivative $\partial T_c / \partial t \leq (\partial T_c / \partial t)_{lim}$;
4. ITR thermal power which determines emitter temperature of thermionic fuel element in the absence of electron cooling, $N(t) < N_{per}$, $t \leq t_f$; and
5. Absence of non-permissible power excursions under switching of temperature channel.

It can be noted that for "Topaz" SNPS the time t_f exceeded the time of bringing on MCL by order approximately. In that way the time of heating up is one of the main component for start-up mode.

The setting for coolant temperature can be selected as:

$$T_c^{giv}(t) = T_c(t_0) + \left[T_{lim} - T_c(t_0) \right] \left[1 - exp\left(\frac{t_0 - t}{\tau} \right) \right].$$

The value T_{lim} must be selected from the condition that the value of thermal power corresponding to T_{lim} didn't exceed the permissible value N_{per} with given probability under all possible scatterings of parameters for heat removal system units. The value τ is selected so that the restriction for increase rate of coolant temperature can be satisfied. The coefficients k_N, k_1, k_2, in equations (1) and (2) are selected so that non-permissible excursions of thermal power should be excluded in moment of temperature channel switching on. The described algorithm is sensitive to inexactitudes of thermal power measuring at a least degree as calculated comparisons with other possible algorithms have shown.

During cesium vapor feeding the control lasts according to temperature channel. As experimental investigations have shown, for guaranteed exclusion from discharge processes under low vapor pressures in course of cesium feedind to ITR the voltage at ITR terminals mustn't exceed a few volts, i.e. ITR must be in mode near to short circuit one. If with that ITR current exeeds the permissible level then transition is expedient from temperature channel of control to current channel which maintains permissible short circuit current by means of thermal power correction. ITR must be on short circuit mode until attainment of given cesium vapour pressure in interelectrode gap of thermionic fuel element. It can be noted that for "Topaz" SNPS this time could be in order of 20 min. The switching over of ITR to load can be performed according to either time signal which meets wittingly a demand for achievement of vapor pressure distribution near to steady-state readings of pressure sensor, or ITR current stabilization. The conducted investigations demonstrate that start-up mode time for thermionic SNPS with thermal neutron ITR will be 1,5-2 hours.

Nominal Mode

The control algorithm for nominal mode includes:

1. Maintenance a given current by means of thermal power correction;
2. Restriction of coolant temperature at ITR outlet by value about 600 C if thermal power increase for compensation of degradation process under given ITR current results in attainment of this temperature;
3. Maintenance a given voltage by means of fast-acting regulator redistributing the ITR current of given permanent value between users and ballast load; and
4. Maintenance a temperature of vapor generator condensation zone on a few setting in range 575 to 625 K what ensures a possibility of adjustment to optimal pressure.

The setting for thermal power under control according to ITR current is formed as:

$$N_{giv}(t) = N_{giv}(t_0) + K_4 \varphi_1(\Delta I) + \varphi_2 \left[K_5 \int \psi_3(\Delta I) dt \right], \qquad (3)$$

where $\Delta I = I(t) - I_{giv}$.

On attainment of top coolant temperature limit ΔI in equation (3) replaced by $\Delta T(t) = T(t) - T_{per}$. The amplification coefficient in equation (3) are selected so that total working time of control drum drives would answered to its lifetime characteristics with taking into consideration of clearance possibility in course of SNPS exploitation. The conducted investigations had showed that amongst possible linear controls the control (3) ensures a minimal drive working time under given dynamic exactness of ITR current maintaining in range up to 3 to 5% including and under clearance at kinematic chain "drive-rotating drum" in range of 2 to 3%. If SNPS must be exploited on essentially different levels of electrical power it is expedient to foresee a several current settings at mode regulator; the switching of these settings which can be performed either automatically according to signals of on-board program device or according to radiocommands from Earth, ensures SNPS transition from one working mode to another.

ENSURING OF NUCLEAR AND RADIATION SAFETY

Under emergency situation ITR must be shut down with the help of all existing independent groups of drives. By that the indications of emergency state which demand immediate SNPS shutdown are in general case:

1. Overshoot of on-board network voltage beyond permissible limits both from below and from above taking into consideration the least possible power consumption of on-board apparatus;
2. Decreasing in coolant pressure and flow rate; and
3. Violation of thermal modes of exploitation for automatic control system (ACS) apparatus.

To these indications the exceeding of limit thermal power, period or reactivity under start-up, failures of control drum drives or thermal power sensors and others may be added.

Application of enumerated emergency state SNPS indications to control algorithms under limited rate of reactivity input at maximally possible failures ensures the impossibility of nuclear fuel melting under any accidents including total failure of heat removal system. It is conditioned by thermal decomposition process for hydride-zirconium moderator which requires the great heat expenditures and results in ITR transition to subcritical state due hydrogen losses from moderator in range of few per cents. These circumstances will exclude the fuel melting and under hypotetical situations with total automatic control system failure apparently.

CONCLUSION

Above described control algorithms for thermionic SNPS of second generation with thermal neutron ITR meet to the most general requirements under SNPS application for power supply of various spacecrafts at all working modes. Thus these algorithms can be regarded as the most expedient ones for thermionic SNPS with long lifetime. The concrete the values of parameters used for control algorithms must be optimized for each specific case.

Acknowledgments

This work was carried out at the SPU " Krasnaya Zvezda" (USSR, Moscow).

MULTIMEGAWATT MPD THRUSTER DESIGN CONSIDERATIONS

Roger M. Myers and James E. Parkes
Sverdrup Technology, Inc.
M.S. SPTD - 1
NASA Lewis Research Center Group
Brook Park, OH 44142
(216) 433-8548 and (216) 433-2603

Maris A. Mantenieks
NASA Lewis Research Center
M.S. SPTD - 1
Cleveland, OH 44135
(216) 433-2423

Abstract

Performance and lifetime requirements for multimegawatt magnetoplasmadynamic (MPD) thrusters were used to establish a baseline 2.5 MW thruster design. The chamber surface power deposition resulting from current conduction, plasma and surface radiation, and conduction from the hot plasma was then evaluated to establish the feasibility of thruster operation. State-of-art lithium heat pipes were found adequate to cool the anode electrode, and the nominal liquid hydrogen propellant flow rates were sufficient to cool the applied-field magnet, the cathode, and backplate. Unresolved issues having an impact on thruster design are discussed to help focus future research.

Introduction

Magnetoplasmadynamic (MPD) thrusters are currently being considered for application to primary propulsion systems on robotic and piloted interplanetary missions. The high specific impulses offered by MPD thrusters, between 3000 and 7000 seconds, can reduce launch mass requirements by over a factor of two compared to chemical/aerobraked systems if the power-to-thrust conversion efficiency is over 50 percent. To satisfy the total impulse requirements for these missions, the thruster system must process between 0.5 and 10 MW of power while lasting over 10,000 hours (Hack 1991, Gilland 1990 and 1991).

MPD thrusters processing over 500 kW of input power have been operated continuously for several hours (Wegmann 1990), and other designs have been pulsed at powers between 5 and 20 MW to study performance scaling (Gilland 1987, Schoenberg 1991). While specific impulses meeting performance requirements have been demonstrated with several thruster configurations (Sovey and Mantenieks 1988), the thruster efficiencies are still substantially below those required for missions of interest. However, data exist indicating this problem can be overcome (Myers et al. 1991a). MPD thruster research between 1970 and 1987 focussed on self-field accelerators, in which the electromagnetic body forces result from the interaction of the discharge current and the self-induced azimuthal magnetic field. Recent work (Myers 1991) has shown that improved performance results from applying an external magnetic field, for which additional propellant acceleration terms result from the presence of the axial and radial magnetic field. The voltage and thrust of these engines have been shown to increase monotonically with applied-field strength at constant discharge current, indicating a lower discharge current requirement than would be necessary in a self-field accelerator to achieve the same power and thrust levels. This not only reduces the cathode current density requirements, but also will increase the propulsion system efficiency via reduced power processor and transmission line losses. At the system level, preliminary designs for both quasi-steady (King and Rudolph 1982, Rudolph and Ogg 1985) and steady-state (Coomes 1986, Gilland et al. 1990) MPD thruster systems have been presented. However, none of these considered in detail the implications of the thermal loading and materials issues confronting long term thruster operation, and all but one (Gilland et al. 1990) considered only self-field MPD thrusters. The purpose of this work is to incorporate recent experimental results and model predictions into a multimegawatt steady-state thruster design, in an attempt to demonstrate the feasibility of long life thrusters operating at high power. In addition, by presenting a preliminary thruster design we focus attention on unresolved issues requiring further research. While large uncertainties remain in the design of operational multimegawatt MPD thrusters, the potential advantages of using these devices, including simplicity, robustness, and relative ease of testing (Sovey et al. 1991), indicate that an assessment of the issues controlling design of operational thrusters is needed.

In this report the impact of applied-field thruster geometry on performance and life is first discussed in order to establish the overall design of a 2.5 MW thruster and illustrate the uncertainties in the geometry. The baseline design is then used with the plasma properties and known electrode and insulator phenomena to establish surface heat loads and potential lifetime limiters. Finally, a summary of the study results and unresolved issues is presented.

Thruster Performance Scaling

Applied-field MPD thruster data have been recently obtained at power levels between 50 and 220 kW, and included direct measurements of thruster performance, calorimetric studies of electrode power deposition, and a preliminary characterization of the thruster plasma characteristics (Myers 1991, Myers et al. 1991b, Mantenieks and Myers 1991, Gallimore et al. 1991). The laboratory MPD thrusters, shown schematically in Figure 1, consisted of a central, 2% ThO_2 tungsten cathode with a coaxial water-cooled copper anode. Propellant was injected through the insulating backplate at the rear of the chamber. Measurements were made over a wide range of thruster propellant flow rates, discharge currents, and applied magnetic field strengths with argon, hydrogen, and argon/hydrogen mixture propellants. The performance measurements clearly showed that, for the operating conditions studied, the thrust scaled as:

$$T \ \alpha \ \frac{R_a^2}{k_1 L_c R_c} B + k_2 f(L_a) \tag{1}$$

where R_a and R_c are the anode and cathode radii, L_a and L_c are the anode and cathode lengths, B is the magnetic field strength, and k_1 and k_2 are constants. The second term indicates that while the anode length influenced the magnitude of the thrust, it did not affect the rate at which the thrust increased with magnetic field strength. The thruster discharge voltage scaled as:

$$V \ \alpha \ \frac{R_a^2}{k_3 R_c} B + k_4 f(L_a) \tag{2}$$

where k_3 and k_4 are constants and the other symbols are defined above. The data used to establish these scaling relationships were also used to study the efficiency behavior. For the operating condition examined, the rate of efficiency increase with magnetic field strength increased quadratically with anode radius. However, the absolute magnitude of the efficiency decreased with increasing anode radius. This result is illustrated in Figure 2, which shows thruster efficiency as a function of applied magnetic field strength for four different anode radii. The field strengths for the larger thrusters were limited by the capabilities of the anode cooling system. Note that, presuming the trends remain constant, for applied-field strengths greater than 0.25 T the efficiency of the 5.1 cm radius thruster will reach the 50% required.

The calorimetric studies of electrode power deposition showed that between 50 and 80% of the input power was deposited into the anode for the selected operating conditions . While the total power deposited into the anode increased quadratically with anode radius and linearly with applied-field strength, the fraction of the input power deposited into the anode decreased with both increasing anode radius and applied-field strength. This effect was due to the more rapid increase in thruster power than anode power. Typical results are shown in Figure 3, where the anode power fraction is plotted as a function of applied-field strength for four anode radii. The anode power measurements indicate that the dominant mechanism of anode power deposition was current conduction across a 10 - 40 V anode fall voltage. Evaluation of the thruster plasma properties showed that the electrons in the andoe fall region were magnetized, with electron Hall parameters ranging from 10 - 500. For these conditions electron transport into the anode is severely restricted, which leads directly to the high anode fall voltages and associated high anode power deposition. These results suggest that the anode power fraction can be reduced via appropriate tailoring of the propellant injection and anode and applied-magnetic field shapes. Shaping the anode/applied-field such that the magnetic field lines cross its surface reduces the importance of cross-field transport and promises to reduce the dependence of the anode power on the applied field strength.

These results have several implications. First, the solenoidal coil for the applied magnetic field should have the capability of producing a field strength of approximately 0.5 T or greater in the thruster chamber. As discussed below, heat flux considerations dictate a minimum magnet radius of approximately 20 cm due to the requirement for multiple heat shields. A numerical code written by LaPointe (1989) was used to design the required number of turns and current for the magnet. For a magnet radius of 20 cm, 60 turns of 1 cm square conductor require 2300 A to produce the required field strenth. This increases to 4800 A if the coil radius is increased to 40 cm. Second, the anode radius and length will be approximatley 15 and 30 cm, respectively, and it will most likely be a straight cylinder or have a converging-diverging geometry to minimimize anode losses. The anode size is determined not only by performance requirements, but also by the need to keep the deposited power density below limits set by the heat removal techniques. Third, the cathode should be as short and have as small a radius as possible, though these parameters will be strongly dependent on the lifetime requirements.

Lifetime Considerations

The system lifetime required for most missions being considered is 10,000 hours. Achieving this lifetime using several thrusters with short lifetimes significantly increases the system complexity. The approach adopted in this work is to maximize the probability of success with a single thruster.

The principle life limiter in MPD thrusters is cathode erosion. Recent work has identified evaporation and oxidation as the principal erosion mechanisms, with the latter playing a significant role only if the net evaporation rate is reduced to less than 30 ng/C (Myers et al. 1988, Polk et al. 1990, Myers et al. 1991a). These experimental results also show that backscattering of evaporated cathode material plays a dominant role in establishing the net mass loss rate. For certain steady-state thruster geometries with high pressure chambers, erosion rates as low as 2-4 ng/C can be achieved with high purity propellant (Auweter-Kurtz, 1990). Such values may satisfy the mission lifetime requirements .

The exponential dependence of evaporation rate on surface temperature indicates that the best method of reducing the cathode erosion rate is to reduce its operating temperature. For a thermionic emitter, the surface temperature is approximately linearly dependent on work function, and only varies logarithmically with current density. Thus, the best way to minimize the operating temperature is to minimize the cathode work function. A promising material for this purpose is barium-calcium-aluminate impregnated tungsten, which is currently used in ion thruster hollow cathodes. The operating temperature for these cathodes in ion thrusters is approximately 1400 K at emission currents of 10 - 20 A. The low temperature is required to ensure that the

1280

barium evaporative loss rate does not exceed the barium diffusion rate to the surface. The hollow cathode enclosure raises the surface pressure on the emitting surface, which enhances backscattering and increases the residence time of the low work function material. Recent testing with hollow cathodes in MPD thrusters (Mantenieks and Myers 1991) indicates that hollow cathodes can be built to operate in the 2000 A current range. While not conclusive, these results indicate it will be possible to succesfully build low work function cathodes for MPD thrusters, and that either rod-shaped or hollow cathode geometries could be successfully used without significantly compromising thruster performance.

The impact of sputtering on anode and cathode material loss rates was evaluated by estimating the sputtering threshold energies for hydrogen using the method of Bohdansky (1980). The resulting threshold energies were 406 and 130 eV for tungsten and molybdenum, respectively. While these values may be in error as a result of model inaccuracies, it appears that little or no sputtering will occur on these surfaces with the 13 eV proton energy corresponding to a specific impulse of 5000 seconds.

2.5-MW Thruster Design

The proposed 2.5 MW MPD thruster designed to operate at a discharge voltage of 250 V and a current of 10,000 A, is shown in Figure 4. It consists of a 15 cm radius, 30 cm long, cylindrical molybdenum anode and a 5 cm diameter, 10 cm long, rod-shaped barium-calcium-aluminate impregnated porous tungsten cathode. Propellant is injected into the chamber through small holes in the boron nitride backplate. While a cylindrical anode is shown, changing to a converging or a converging-diverging anode, should that improve performance, would not significantly change the thermal analysis issues discussed below. Molybdenum was selected for the anode to reduce the thruster mass, though this may have to be changed to tungsten due to potential recrystalization of molybdenum at the 1400 K anode operating temperature (see below). A hollow cathode was not chosen due to the uncertainties in their operation at high currents, though again the design could easily be modified to accomodate this change. Use of a large radius cathode was forced by the lifetime requirements: Impregnated cathode current densities should be below 20 A/cm^2 to ensure adequate lifetime. To satisfy this requirement the cathode surface will have to be textured (grooved or pitted) to provide for a factor of four increase in surface area.

Fifty 1 cm diameter lithium heat pipes transport heat deposited into the anode to a pyrolitic graphite radiator. The lithium heat pipes, operating at a temperature of 1400 K, are capable of removing 400 kW from the anode while remaining over an order-of-magnitude below the sonic heat transfer limit (Dunn and Reay 1982). Lifetests indicate that with a TZM wall, lithium heat pipes will last over 9000 hours. The heat fluxes that must be accomodated by the heat pipes and MPD thruster chamber surfaces are discussed below.

Heat Flux Estimates

Thruster

Heat loads to MPD thruster surfaces arise from current conduction across plasma/electrode interfaces; the presence of the 1 - 2 eV, high velocity plasma; and radiative exchange between the inner chamber surfaces. The largest heat load is to the anode, for which the power deposition can be estimated from (Gallimore 1991):

$$ P_{an} = J\left(V_a + \frac{5kT_e}{2e} + \Phi\right) + P_{rad} + P_{flow} \qquad (3) $$

where the first term represents the power deposited by the current carrying electrons, the second represents the power radiated from the cathode and the plasma, and the third is the flow energy transported to the anode. Previous work (Myers 1991, Gallimore et al. 1991) has shown that the first term is the dominant power loss mechanism, and that the anode fall voltage is the largest source of electron energy. The magnitude of the first term was estimated assuming an anode fall voltage of 15 V, an electron temperature of 1.5 eV, and an anode work function of 4.5V. While the latter two values are easily justified based on measurements in high power quasi-steady and 100 kW class steady-state thrusters, the assumed anode fall voltage is a factor of 2 - 3 lower than that estimated from calorimetric measurements of low power, high specific impulse thrusters operated at high applied-field strengths (Myers 1991). However, the data indicate that the large anode fall may be greatly reduced by shaping the anode and magnetic field so as to reduce the requirement for cross-field current conduction. If this reduction is not achieved, it is doubtful whether MPD thrusters will be viable candidates for propulsion applications. For the assumed values, the first term in Eqn. 1 yields a power deposition of 2.33 x 10^5 W for a 10,000 A discharge current, corresponding to 82 W/cm^2 for the proposed geometry. This heat flux is well within the radial heat flux capabilities of the lithium heat pipes. The anode power drops to 2.14 x 10^5 W if the anode material is barium impregnated tungsten with a work function of 2.6. This result illustrates the premium placed on reduction of the anode fall voltage.

It will be shown below that the other terms in the anode power balance are very small, so that the lithium heat pipe/pyrolitic graphite radiator can be sized using the 233 kW found above. Assuming an emissivity of 0.9 and a temperature of 1400 K for the radiator surface, the outer diameter of the radiator must be 1.2 m.

The effect of internal chamber surface radiation was estimated by calculating the cathode - anode, anode - anode, cathode - backplate, and anode - backplate view factors and assuming grey body radiation at a uniform surface temperature. The view factors were estimated using relations for straight, finite length, cylinders (Siegel and Howell 1972). The electrode surface temperatures were set to 1400 K based on use of lithium heat pipes for the anode and a barium-calcium-aluminate impregnated tungsten cathode. The electrode emissivities were set to 0.4, and the insulating backplate emissivity was set to 0.1, typical values for the materials used. Three equations for the surface heat fluxes, in terms of the emitted power and the radiosity, were developed and solved simultaneously. The backplate temperature was left as a free-parameter as there was no independent method of establishing its temperature. The resulting equations are:

$$Q_B = 3.8 \times 10^{-10} T_B^4 - 1490 \text{ , W}$$

$$Q_C = -5.6 \times 10^{-11} T_B^4 + 221 \text{, W}$$

$$Q_A = -3.3 \times 10^{-10} T_B^4 + 1218 \text{, W} \qquad (4)$$

where Q_A, Q_B, and Q_C are the heat fluxes from the anode, backplate and cathode (a negative value indicating net heat input), and T_B is the backplate temperature. These equations do not depend on thruster power level, but rather on the geometry, materials, and surface temperatures. The largest radiated heat flux is to the backplate, reaching almost 1 kW if T_B is 1100 K. Heat fluxes to the cathode and anode are negligibly small since their temperatures are equal. An important issue not yet addressed in detail are the emissivity changes which will occur over the lifetime of the thruster. In current testing of MPD thrusters the backplate is typically coated with contaminants within the first few minutes of operation.

The contribution of the plasma to the chamber surface heat flux consists of thermal particle flux, free-free, free-bound, and line radiation. While the scaling of the plasma density and temperature are poorly understood, order-of-magnitude estimates can be obtained from simple kinetic theory and plume plasma measurements in applied-field thrusters (Myers et al. 1991b).

The thermal loads on the cathode and backplate resulting from particle flux to the surfaces were estimated from the product of the particle flux and the energy deposition per particle. The particle heat flux to the anode was included in the first term of Eqn. 3. Assuming equal electron and ion temperatures, the surface power fluxes are given by:

$$P_s = \frac{NC}{4} E_p \qquad (5)$$

for an attracting surface potential, or:

$$P_s = \frac{NC}{4} \exp\left(-\frac{V_s}{kT}\right) E_p \qquad (6)$$

for a retarding potential. In these equations, c is the mean thermal speed, N the particle density, V_s the difference between the plasma and surface potential, and E_p the energy deposited into the surface by the incoming particle. The ion energy term must include the recombination energy (13.6 eV/atom for hydrogen) if the surface is conducting. These relations do not include thermalization in the boundary layer, so that the ion energy to the cathode includes the kinetic energy of the flow and that acquired across the cathode fall, recombination, and thermal energies. Assuming a 1 eV temperature, and setting Vs to - 5V for the cathode and utilizing the floating potential for the insulating backplate, the electron and ion thermal fluxes were estimated to be approximately 2.2 kW onto the cathode and 8.9 kW onto the backplate for the chosen geometry. If the ions recombine on the backplate surface this value increases to 17 kW. In addition, removing the cathode from the high speed flow will reduce the ion heat flux to its surface to 1.3 kW.

Free-free, free-bound, and line radiation contributions were estimated using the formalisms described in Lochte-Holtgraven (1968) and Griem (1968). Assuming a Maxwellian distribution function for the particles, which may be inappropriate in a highly magnetized plasma, the power emitted via free-free radiation is:

$$P_{ff} = 1.13 \times 10^{-41} \Omega V T_e^{1/2} N_e N_i \qquad (7)$$

and the free - bound radiation is:

$$P_{fb} = 5.4 \times 10^{-52} T_e^{-1/2} N_e N_i \Omega V \left(\frac{E_H}{hn^2} + \frac{kT_e}{h} \left(\exp\left(-\frac{E_H}{kT_e n^2}\right) - 1 \right) \right) \qquad (8)$$

where T_e is the electron temperature, is the solid angle, n is the lowest quantum level into which recombination occurs, V is the emitting volume, and N_e and N_i are the electron and ion density. Line radiation was estimated by multiplying the resonance transition energy by the electron excitation rate, and assuming the plasma is optically thin to obtain a worst-case value. The result was:

$$P_l = 2.1 \times 10^{-32} N_e N_a \Omega V \exp\left(-\frac{E_2}{kT_e}\right) \qquad (9)$$

where E_2 is the resonance line energy, N_a the neutral atom density, and other terms are as defined above. The contributions of these terms with an electron temperature and density of 1.5 eV and 1×10^{14} cm^{-3}, respectively, and an ionization fraction of 50% are 1.9, 1.9 and 9.0 kW for the free - free, free - bound, and line radiation, respectively. A fraction of this power will go to each of the thruster surfaces.

In addition to the power deposited into the electrodes and backplate there will be ohmic losses in the electrode current leads. These losses must be minimized without increasing the heat conduction from the electrodes to the low temperature current conductors. Depending on the design, these conduction losses can reach 10 kW, power which must be removed from the system either by radiation or regenerative cooling.

Regenerative cooling with cold gaseous hydrogen may be beneficial. Assuming that the required 2.5 g/s of hydrogen propellant is vaporized at 25 K near the magnet, and that the hydrogen is in equilibrium with the 1400 K boron nitride backplate when it enters the thruster, a total of 48 kW of power is absorbed by the hydrogen. This potential heat sink can absorb all the power from plasma radiation and particle flux, and most likely will be able to cool the electrical conductors attached to the anode and cathode.

Applied-Field Magnet

The principal problem with applied-field thruster design is to minimize the power loss in the applied-field coil without sacrificing the simplicity and robustness that makes MPD thrusters attractive. Thus, while it may be possible to build a superconducting coil with the desired field strenghts, the complexity associated with the helium liquefaction system makes this option unattractive. While developments in high-temperature superconductivity may alleviate this problem, the authors are not aware of a demonstrated high-temperature superconductor that will satisfy the field requirements. The optimal applied-field coil appears to be an aluminum coil kept at 20 K by a liquid hydrogen bath. The resistivity of aluminum at liquid hydrogen temperatures, 1.2×10^{-9} ohm-cm, is over an order-of-magnitude below that of oxygen-free copper. Several multi-Tesla field magnets have been designed using this concept (Gibson et al. 1990, Cope et al. 1990). For the coil described above the total ohmic power dissipated in the coil would be 60 or 479 W for the 20 and 40 cm radius coils, respectively. Using the heat of vaporization of hydrogen, 433 kJ/kg at 101 kPa, this would vaporize 0.14 or 1.11 g/s of hydrogen if the full 0.5 T field strength were required. This is less than the approximately 2.5g/s hydrogen required for thruster operation, indicating that the mass flow rate might be controlled using a small heat source (< 500W) to vaporize the needed hydrogen.

In addition to the ohmic losses in the applied-field coil the magnet cooling system must process the heat radiated from the thruster anode and the back surface of the 1.2 m diameter radiator to the coils (Figure 4). Reducing the emissivity of the anode/heat pipe surface to 0.14 by either polishing or coating the surface yields a heat transfer of 5.2 kW from the anode and 0.87 kW from the radiator for the dimensions given above. Radiation shields are thus necessary to reduce this heat flux to a level commensurate with vaporizing the required mass flow rate of liquid hydrogen. The minumum number of radiation shields required was estimated using standard radiation shielding equations from Siegel and Howell (1972). Appropriate emissivities were used for each shield material, which ranged from polished tantalum (0.11) near the anode to polished copper (0.06) near the magnet. Shield materials were selected which could withstand the temperatures to which they would be subjected. Providing for a factor of two margin on the boil-off, and using the 20 cm diameter magnet, the radiated heat flux to the magnet must be reduced to 480 W, or a factor of 12 below that emitted from the anode. This can be accomplished using ten 0.08 cm thick heat shields inserted between the anode and magnet and the radiator and magnet. Thermal expansion estimates for the heat shields show that a maximum of sixteen shields may be accomodated in the 5 cm gap between the anode and magnet, though many more could be accomodated behind the radiator.

A potential problem with using liquid hydrogen vaporization to absorb the heat generated in, and transferred to, the magnet is localized boiling on the magnet coils leading to thermal runaway. The resistivity of aluminum increases rapidly above 25 K, which could potentially result in catastrophic failure of the magnet. While two phase flow of liquid hydrogen is poorly understood, results given by Baker (1985), indicate that the vapor mass flow rate should not exceed between 0.1% and 1 % of the total mass flow rate to ensure that the flow remains in a "bubble" mode. While this requirement is not firm, it is clear that the heat leaks to the magnet should be minimized, as even the 0.14 kW generated by ohmic heating of the 20 cm radius magnet will cause the gas fraction to exceed 5 percent. This indicates that the smaller magnet may prove better due to its lower ohmic losses. However, the problem might be solved by forcing a large recirculating flow of liquid hydrogen through the magnet to keep localized boiling from becoming a problem, and providing a mechanism for extracting only the needed propellant from this stream.

Summary

Performance and lifetime requirements were used to establish a baseline 2.5 MW applied-field MPD thruster designed to operate at a discharge current of 10,000 A, a voltage of 250 V, and a hydrogen propellant flow rate of 2.5 g/s. The thruster consists of a 15 cm radius, 30 cm long, molybdenum anode, and a 2.5 cm radius, 10 cm long impregnated tungsten cathode. Lithium heat pipes are used to transport the ~ 233 kW of power deposited onto the anode to a 1.2 m diameter pyrolytic graphite radiator. The applied-field magnet is made of aluminum conductor cooled to 20 K using a liquid hydrogen bath. Power dissipated by the coil when generating a 0.5 T magnetic field is less than 0.02% of the thruster power, and the magnet could be kept at a constant temperature by vaporizing a fraction of the required 2.5 g /s hydrogen mass flow rate. Chamber wall heat flux estimates based on the plasma characteristics, materials and geometry showed that it should be possible to regeneratively cool all thruster surfaces other than the anode using 25 K gaseous hydrogen coming out of the magnet.

Acknowledgements

This work was done at the NASA Lewis Research Center under Contract NAS3-25266.

References

Auweter-Kurtz, M., et al. (1990) "Cathode Phenomena in Plasma Thrusters," AIAA Paper 90-2662.

Baker, R. (1985) <u>Cryogenic Systems</u>, Oxford Press, NY.

Bohdansky, J., Roth, J., and Bay, H.L. (1980) "An analytical formula and important parameters for low-energy ion sputtering," <u>J. Appl. Phys.</u>, Vol. 51, No. 5., pp. 2861-2865.

Cope, D.B., Snyder, M.D., and Weissman, R.M. (1990) "The Magnetic and Structural Design of a Liquid Hydrogen-Cooled Magnet Utilizing High Purity Aluminum Conductors and a Boron/Epoxy Composite as Structural Reinforcement," in <u>Advances in Cryogenic Engineering</u>, Vol. 35, Plenum Press, N.Y.

Coomes, E.P., King, D.Q., Cuta, J.M., and Webb, B.J. (1986) "PEGASUS: A Multi-Megawatt Nuclear Electric Propulsion System," AIAA Paper 86-1583, June 1986.

Curran, F. M., Bennett, G.L., Watkins, M.A., Byers, D.C., Brophy, J.R., and Sercel, J.C. (1991) "An Overview of the NASA Electric Propulsion Program," IEPC 91-002, Presented at the 22nd International Electric Propulsion Conference, Vaireggo, Italy, October 1991.

Dunn, P.D. and Reay, D.A. <u>Heat Pipes</u>, 3rd ed., Pergamon Press, 1982.

Gallimore, A.D., Myers, R.M., Kelly, A.J., and Jahn, R.G. (1991) "Anode Power Deposition in an Applied-Field Segmented Anode MPD Thruster," AIAA Paper 91-2343, June 1991.

Gerwin, R. (1991) "Estimates of Orbit-Raising and Mars Mission Capabilities of Nozzle-Based Coaxial Plasma Thrusters," AIAA Paper 91-3494, Sept. 1991.

Gibson, C.R., Dew, M.W., and Johnson, E.R. (1990) "Design and Analysis of a High-Purity Aluminum Inductor," in <u>Advances in Cryogenic Engineering</u>, Vol. 35, R.W. Fast, editor, Plenum Press, N.Y.

Gilland, J.E., Kelly, A.J., and Jahn, R.G. (1987) "MPD Thruster Scaling," AIAA Paper 87-0997, May 1987.

Gilland, J.E., Myers, R.M., and Patterson, M.J. (1990) "Multimegawatt Electric Propulsion System Design Considerations," AIAA 90-2552, July 1990.

Gilland, J.E. (1991) "Nuclear Electric Propulsion Mission Sensitivities," in <u>Magnetoplasmadynamic Thruster Workshop</u>, Proceedings, R. M. Myers, ed., NASA CP 10084, 16 May 1991.

Griem, H. R. (1968) <u>Plasma Spectroscopy</u>, McGraw-Hill Book Co., 1964.

Hack, K.J., George, J.A., and Dudzinski, L.A. (1991) "Nuclear Electric Propulsion Mission Performance for Fast Piloted Mars Missions," AIAA Paper 91-3488, Sept. 1991.

King, D.Q. and Rudolph, L.K. (1982) "100 kW MPD Thruster System Design," AIAA Paper 82-1897, Nov. 1982.

LaPointe, M.R. (1989) Personal Communication, Sverdrup Technology, NASA Lewis Research Center Group, Brook Park, OH, 44142.

Lochte-Holtgraven, W. ed. (1965) <u>Plasma Diagnostics</u>, North Holland Publ. Co., Amsterdam, 1968.

Mantenieks, M.A. and Myers, R.M. (1991) "Hollow Cathode MPD Thruster," IEPC 91-076, presented at the 22nd International Electric Propulsion Conference, Varieggo, Italy, October 1991.

Myers, R. M., Suzuki, N., Kelly, A.J. and Jahn, R.G. (1988) "Cathode Phenomena in a Low-Power Magnetoplasmadynamic Thruster," AIAA Paper 88 - 3206, July 1988, see also <u>Journal of Propulsion and Power</u>, Vol. 7, No. 5, Sept.-Oct. 1991, pp. 760-767.

Myers, R.M. (1991) "Applied-Field MPD Thruster Geometry Effects," AIAA Paper 91-2342, June, 1991, see also NASA CR 187163, August, 1991.

Myers, R. M., Mantenieks, M.A., and LaPointe, M.R. (1991a) "MPD Thruster Technology," AIAA Paper 91-3568, Sept. 1991, see also NASA TM 105242.

Myers, R. M., Werhle, D., Vernyi, M., Biaglow, J., and Reese, S. (1991b) "A Preliminary Characterization of Applied-Field MPD Thruster Plumes," AIAA Paper 91-2339, June 1991, see also NASA CR 187165, August 1991.

Polk, J.E., Kelly, A.J., Jahn, R.G., Kurtz, H., Auweter-Kurtz, M., and Schrade, H.O. (1990) "Mechanisms of Hot Cathode Erosion in Plasma Thrusters," AIAA Paper 90-2659, July 1990.

Rudolph, L.K and Ogg, G.M. (1985) "Orbit Transfer Using High Power MPD Thrusters," AIAA Paper 85-1478, July 1985.

Siegel, R. and Howell, D. (1972) Thermal Radiation Heat Transfer, McGraw-Hill, NY 1972.

Schoenberg, K., Gerwin, R., Barnes, C., Henins, I., Mayo, R., Moses, R., Scarberry, R., and Wurden, G. (1991) "Coaxial Plasma Thrusters for High Specific Impulse Propulsion," AIAA Paper 91-3570, Sept. 1991.

Sovey, J. S. and Mantenieks, M.A. (1988) "Performance and Lifetime Assessment of MPD Arc Thruster Technology," AIAA Paper 88-3211, July 1988.

Sovey, J. S., Vetrone, R.H., Grisnik, S.P., Myers, R.M., and Parkes, J.E. (1991) "Test Facilities for High Power Electric Propulsion," AIAA 91-3499, Sept. 1991.

Wegmann, T., Auweter-Kurtz, M., Kurtz, H., Merke, W., Loesener, O., and Schrade, H.O. (1990) "Steady State High Power MPD Thrusters," AIAA Paper 90-2555, July 1990.

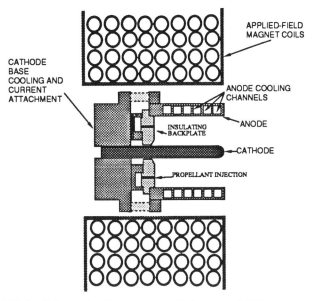

FIGURE 1 - Schematic of water-cooled laboratory MPD thrusters.

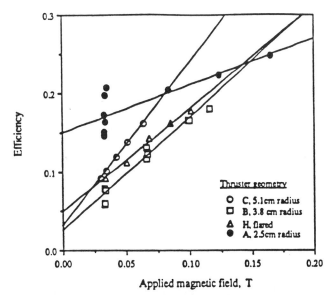

FIGURE 2 - Efficiency versus Applied-Field Strength for Four Anode Radii Operated on 0.1 g/s Argon Propellant. Discharge Current was 1000A, Anode Length was 7.6cm (Myers 1991)

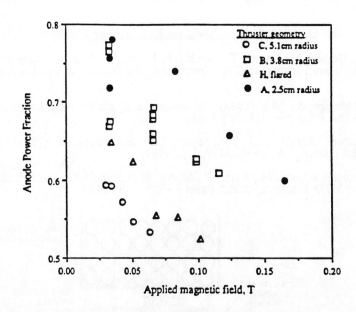

FIGURE 3 - Anode Power Fraction versus Applied Magnetic Field Strength for Four Anode Radii Operating on 0.1 g/s Argon Propellant. Discharge Current was 1000A, Anode Length was 7.6 cm (Myers 1991).

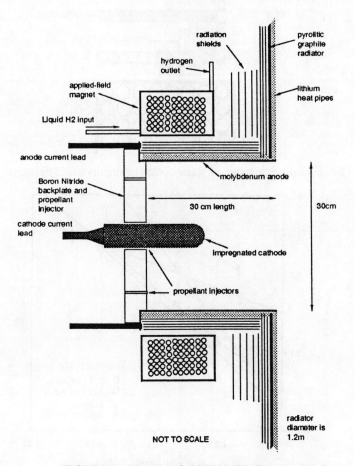

FIGURE 4 - Proposed 2.5 MW MPD Thruster Design.

A MULTI-MEGAWATT ELECTRIC THRUSTER TEST FACILITY

Keith I. Thomassen and E. Bickford Hooper
Lawrence Livermore National Laboratory
P.O. Box 808, L-637
Livermore, CA 94550
(510) 422-9815

Abstract

A Multi-Megawatt Test Facility (MTF) for electric thrusters is described, based on modifications of the Lawrence Livermore National Laboratory (LLNL) Magnetic Fusion Test Facility-B (MFTF-B). MTF would use the large (66 m long, 11 m maximum radius) vacuum vessel, the cryopumping system including 1000 m^2 of cryopanels and 11.5 kW cooling at LHe temperature, 250 MVA dedicated power line, and other existing facilities. As a result, significant cost and time savings would be realized over construction of a completely new test facility.

INTRODUCTION

One of the propulsion options being considered in the Space Exploration Initiative (SEI) is Nuclear Electric Propulsion (NEP), wherein nuclear power is converted to dc or rf power to drive an electric (ion or plasma) thruster. (Doherty, et al. 1991) Several propulsion devices have been investigated during the last several decades, most notably the ion thruster and the magneto plasma dynamic (MPD) thruster. The former has been operated at steady state (but pulsed) power levels up to 200 kW and at relatively high efficiency, while the latter has operated to 600-kW levels but at low efficiency. Short pulsed MPD thrusters have operated at levels up to 10 MW. Other concepts are being explored but are not as well advanced.

None of the concepts are yet viable for the SEI applications, and the thrust level, efficiency, and lifetimes for the thruster must be improved. Fundamental to this development is a rather large facility, or several facilities, (Sovey, et al. 1991) to do the component development and lifetests (10,000 hours, typically) necessary to qualify them for space use. We describe here a Lawrence Livermore National Laboratory (LLNL) facility based on the former Mirror Fusion Test Facility-B, (MFTF-B), (Karpenko 1983, Coensgen 1986) which would become a Multi-Megawatt Test Facility (MTF). The requirements driving the design of MTF will also be discussed. Techniques to safely and effectively handle large amounts of gas, including hydrogen or deuterium, are suggested. MTF would be a minor modification of the existing MFTF-B.

REQUIREMENTS

The primary defining requirements for the facility are its size and pumping speed for the various gases used in thrusters, typically hydrogen, hydrazine, ammonia, lithium, argon, or xenon. The MPD thruster obtains a specific impulse in the desired range with low atomic number (Z) gases, whereas ion engines use higher Z to allow operation at higher voltage. Other requirements, such as heat removal and facility power are important but not difficult to provide. The cryopanels and refrigeration systems associated with the pumping speeds (tens of millions of liters per second) are a major cost element, as is the large vacuum tank (tens of meters, characteristic distance).

Large dimensions are needed to prevent the walls from interferring with the thruster plume, either through wall collisions whose effects are transmitted back through the gas, or through plasma currents that are interrupted by the wall. Also, to contain the necessary amount of cryopanel surface area the chamber must be of sufficient size. For MW class thrusters the vacuum tanks need to be of the order of 10 m radius and 20 m long.

Depending on the gas, pumping must be accomplished by cryopanels at liquid helium (LHe) temperatures (4 K) or with gaseous helium (20 K). The pumping speed of a cryopanel is constant in the molecular flow regime, at a value of the capture probability, η_c, times the pumping speed of a circular aperture separating two regions at different pressures. The resulting pumping speed is $\eta_c A v_t/4$, where the thermal velocity is $v_t = (8kT/\pi m)^{1/2}$, A the area of the aperture (or cryopanel), T the gas temperature, k Boltzmann's constant, and m the mass of the gas molecule. For example, the "black-hole" deuterium pumping speed is 3.12×10^5 L s^{-1} m^2. The capture probability for the MFTF "Z"-array cryopanels is 0.26 (Hood and Bond 1980, Valby and Pittenger 1980), yielding a deuterium pumping speed of 8×10^4 L s^{-1} m^2 (Margolies and Valby 1983, Krause and Kozman 1986, Coensgen 1986). Note that the pumping speed varies as $(T/m)^{1/2}$ so the pumping speed is increased by $2^{1/2}$ for hydrogen but reduced for heavier gases.

A 1 MW thruster will have a typical gas flow of roughly 1 g s^{-1}, and (for example) 1 g of deuterium occupies 5.6 L of space at S.T.P., so the gas flow is 4.3×10^3 Torr L s^{-1}. With the measured deuterium pumping speed we would require a panel area of 5,300 m^2 to reach 10^{-5} Torr (1.3 mPa). Clearly, it is desirable to pump the gas at a higher pressure to minimize panel area. However, thruster performance characterizations are compromised if the chamber pressure is much higher than 10^{-5} Torr, so one must have a differential pressure between the thruster exhaust region and the cryopanel surfaces which provide the bulk of the pumping.

Baffles or conductance limiting paths that take advantage of the free streaming flow from the thruster must be employed. These will permit a large pressure differential between the thruster and the panel surfaces to provide significant enhancement. An arrangement of this type was used in the Tandem Mirror Experiment Upgrade (TMX-U) together with titanium-gettering pumping to handle gas loads from neutral beams (Turner, et al. 1984). Neutral beam pumping on MFTF also utilized optimized baffling (Stone and Duffy 1983). In addition, the speed of cryopumping can be enhanced by arranging cryopumps appropriately; in MFTF, for example, cryopumps for the flowing plasma exiting the magnetic confinement region were arranged in a "V" configuration, yielding an effective capture probability of 0.67 (Hood and Bond 1980, Valby and Pittenger 1980). Finally, the use of liquid nitrogen (LN) cooled baffles may provide a radiation temperature of 77 K in the pumping region, eliminating or reducing the need for local LN-shrouding of the LHe cryopumps and further increasing the capture probability. Calculations using Monte Carlo codes will be required to design an optimized system.

The MFTF-B FACILITY

In Figure 1 we show the MFTF-B facility, with its 66 m long vacuum vessel (Gerich 1986). The end tanks and center cell are of comparable length, but the ends are 11 m diameter whereas the center section is 8 m in diameter, resulting in a total volume of 4,300 m³. Closed loop LN and LHe refrigeration systems of 500 kW and

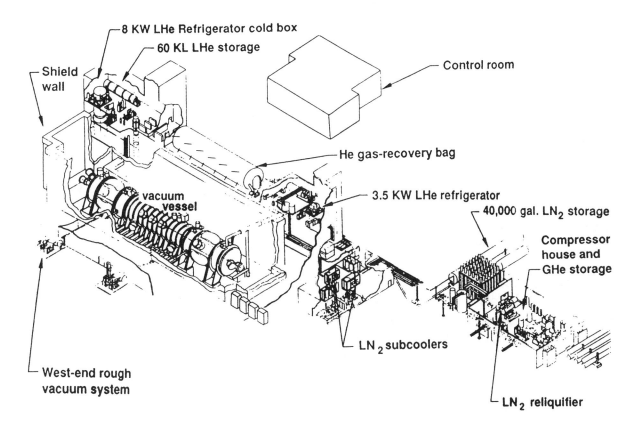

Figure 1.   The MFTF-B Facility, featuring its Closed Loop Helium and Nitrogen Cryogenic Facilities.

11.5 kW respectively constitute the main features of the cryogenic plant. These include the recovery and purification systems, in addition to the cold boxes, low and high pressure storage tanks, and double jacketed, vacuum insulated cryogenic transfer lines. Inside the vessel there are cryopanels totaling 1000 m^2 in area. These are a combination of helium cooled and nitrogen shrouded panels. One end tank, or one end with the central section, could provide the necessary test space for electric thrusters.

There are several pumping systems to bring the vessel from atmospheric pressure to base pressures in the 10^{-8} Torr range. These include the Roots blowers shown in the figure for rough pumping, several 5000 L s^{-1} turbo-molecular pumps, and cryosorption and cryocondensation pumps installed on the vessel through 36 inch gate valves. Finally, the primary high vacuum pumps are the cryopanels with their 80 million L s^{-1} pumping speed for D_2 (corresponding to 1000 m^2 area).

There is an electrical substation on site, converting 230 kV power to 13.8 kV, with 100 MW steady state capacity, more than adequate for the refrigerators that constitute the primary load (about 15 MW) and the thruster power. Fibre-optic links from the control room connect to the Camac-based local controls of the MFTF-B subsystems and the diagnostics on the machine.

CONVERSION TO MTF

To convert this facility to the MTF the superconducting magnets currently in the vessel must be removed, baffles designed and installed, and the facility controls consolidated for the specific operations needed for thruster testing. Diagnostics specific to thruster testing and lifetesting must be added.

The design of the optimal placement of baffles and cryopumping surfaces will need careful analysis. Experience with MFTF and TMX-U showed that significant enhancement of pumping capability is possible with careful arrangement of the pumping components. Water cooled surfaces will be required to handle the direct energy of the thruster exhaust so as to minimize heat loads on the cryosystems.

A major problem to be addressed is the regeneration of the cryopanel surfaces, which will load up with frozen gases after an extended period of operation. This is normally done by warming the surfaces slightly and pumping the gas that evolves (at a fairly high pressure, so that the throughput is high and the "downtime" is minimized). Options are being explored for continuously pumped panels, for example using a "Venetian blind" or "window shade" arrangement that allows surfaces to be isolated from the impinging gas and sealed sufficiently from the pumping volume that the temperature and pressure near the pumping surface can be raised and the gases pumped. This will allow degassing single panels while the rest of the system continues to pump the thruster gas.

If hydrogen or deuterium are used in the thruster, the issue of safety in an up-to-air accident must be considered. The amount of gas used in thruster testing is large, and may yield an explosive mixture of hydrogen and air when the cryopanels rapidly degas (Pittenger 1976). If the partial pressure following degassing is less than 13 Torr, combustion will not occur. A realistic maximum hydrogen partial pressure, corresponding to a pressure rise of 1 atmosphere in the event of an explosion, is 50 Torr. This corresponds to 25 kg in the MFTF-B vacuum vessel. A thruster operating at 1 g/s will consume 25 kg in 7 hours. Consequently, unless single panel degassing is fast enough to complete the recycling of the pumping system in 7 hours, innovative solutions will be required for safety. One possibility is the rapid filling of the vacuum vessel with an inert gas in the event of a significant vaccum accident.

Thruster specific equipment, including a thrust stand, will have to be integrated into the MTF. Extensive computerized data handling and analysis capability is already available at LLNL (Butner, et al. 1990) and will simplify both facility diagnostics and those associated with particular thruster tests.

This facility was a major investment for the magnetic fusion program, one not now in use. It offers a very cost effective way to do thruster development for the SEI program.

ACKNOWLEDGEMENTS

Work performed by LLNL for USDOE under Contract W-7405-ENG-48.

References

Butner, D. N., et al. (1990), "MTX Data Acquisition and Analysis Computer Network", *Rev. Sci. Instrum.*, 61(10): 3277-3279.

Coensgen, F. H. et al. (1986), "MFTF-B Acceptance Tests and Operation," Plasma Physics and Controlled Fusion Research 1986, International Atomic Energy Agency, Vienna, 1987, Vol. 3, pp 333-341.

Doherty, M. P., et al. (1991), "Nuclear Electric Propulsion Technology Panel Final Report", NASA Lewis Research Center, Cleveland OH, to be published.

Gerich, J. W. (1986) "The Design, Construction, and Testing of the Vessel for the Tandem Mirror Fusion Test Facility," *J. Vac. Sci. Technol. A*, 4(3): 1742-1748.

Hood, C. B. and J. W. Bond (1980), "Large Scale Cryopumps for Fusion Power Systems" in *Cryogenic Processes and Equipment in Energy Systems*, W. M. Toscano, R. C. Longsworth, F. J. Zimmerman, and C. F. Gottzmann, eds., Am. Soc. Mech. Eng., New York, NY, 1980, pp 155-161.

Karpenko, V. N. (1983), "The Mirror Fusion Test Facility: An Intermediate Device to a Mirror Fusion Reactor," *Nucl. Technol./Fusion*, 4(2), Part 2: 308-315.

Krause, K. H. and T. A. Kozman, Tech. Editors (1986), "MFTF-B PACE Tests," UCID-20819, pg. 5-35, Lawrence Livermore National Laboratory, Livermore, CA 94550.

Margolies, D. and L. Valby (1983), "The (Changing) MFTF Vacuum Environment", *J. Vac. Sci. Technol. A* 1(2): 1308-14.

Pittenger, L. C. (1976), "Vacuum Engineering for Fusion Research and Fusion Reactors", UCRL-78501, Lawrence Livermore National Laboratory, Livermore, CA 94550.

Sovey, J. S., R. H. Vetrone, S. P. Grisnik, R. M. Myers, and J. E. Parkes (1991), "Test Facilities for High Power Electric Propulsion," AIAA Paper No. 91-3499, presented at the AIAA/NASA/OAI Conf. Advanced SEI Tech., Sept. 4-6, 1991, Cleveland, OH.

Stone, R. and T. Duffy (1983), "Optimized Baffle and Aperature Placement in Neutral Beamlines,", in *Proc. 10th Symp. Fusion Engineering*, IEEE Cat. No. 83CH1916-6 NPS, Vol. 2, pp 1451-54, held in Philadelphia, PA, IEEE New York, NY (1983).

Turner, W. C., et al. (1984), "Gas Pressure in the End Plug Regions of the TMX-U Thermal Barrier Experiment," *J. Vac. Sci. Technol. A* 2(4): 1576-1582.

Valby, L. E. and L. C. Pittenger (1980), "The Mirror Fusion Test Facility: An Intermediate Device to a Mirror Fusion Reactor," in *Proc. 4th Topical Meeting on Tech. of Controlled Fusion*, CONF 801011, held in King of Prussia, PA, U. S. Dept. of Energy, Washington, DC, July 1981, Vol 2, pp 970-975.

RESISTIVE PLASMA DETACHMENT
IN
NOZZLE BASED COAXIAL THRUSTERS

Ronald W. Moses, Jr., Richard A. Gerwin, and Kurt F. Schoenberg
Los Alamos National Laboratory
P. O. Box 1663
Los Alamos, NM 87545
(505) 667 5622

Abstract

Nozzle based coaxial plasma thrusters constitute potentially attractive electric propulsion engines that are both compact (with high thrust density) and robust. The coaxial plasma thruster can be viewed as an evolutionary state of magnetoplasmadynamic (MPD) thruster development that may satisfy the demanding performance requirements of Space Exploration Initiative (SEI) relevant cargo or piloted missions. Previous work (Schoenberg *et al.*, AIAA Technical Report 91-3570) has shown that ideal magnetohydrodynamics (MHD) plays a major role in multi-megawatt coaxial plasma thruster dynamics, particularly in the behavior of high-grade plasma acceleration by a converging-diverging magnetic nozzle. In this paper, we examine the detachment of high-grade MHD plasma from a magnetic nozzle by resistive diffusion. A quantitative description of the resistive detachment process is derived. Included is a discussion of non-ideal MHD effects, including classical and anomalous resistivity, that can enhance the detachment. This analysis supports the hypothesis that magnetic nozzle design for high-performance thruster operation requires an optimization between the conflicting requirements of efficient plasma acceleration and plasma detachment. A qualitative prescription for such an optimized magnetic nozzle design is discussed.

INTRODUCTION

Scoping studies (Gerwin *et al.* 1991) have shown that nozzle-based coaxial plasma guns constitute potentially attractive, high performance electric thruster systems. One can view the coaxial plasma gun as an evolutionary state of magnetoplasmadynamic (MPD) thruster development that may satisfy the demanding, high-power performance requirements of SEI-relevant cargo or piloted missions. In addition, an understanding of the quasi-steady-state dynamics of coaxial plasma guns provides an important perspective for optimizing the performance of present generation MPD thrusters.

Multi-megawatt coaxial plasma thrusters benefit from the utilization of high-grade, energetic plasma to produce thrust (Schoenberg *et al.* 1991). Energetic plasma generation is necessary in order to meet the high specific impulse requirements of advanced space missions. Furthermore, efficient thruster operation requires good coupling of the energetic plasma with the self and applied magnetic fields. Within the context of magnetohydrodynamics (MHD), these requirements naturally lead to an economy of scale in operating efficiency with respect to power and spatial dimension.

Multi-megawatt coaxial plasma thrusters can access ideal-like MHD operation where plasma acceleration is produced by a combination of $\mathbf{E} \times \mathbf{B}$ drift and magnetoplasma flow through a converging-diverging magnetic nozzle (Schoenberg *et al.* 1991). This acceleration mechanism potentially allows for efficient thrust generation with minimal electrode losses. However, once accelerated, plasma must also efficiently detach from the thruster with minimal drag. In this paper, we examine the detachment of high-grade MHD plasma from a magnetic nozzle by resistive diffusion. A quantitative description of the resistive detachment is derived, and includes a discussion of non-classical effects, like anomalous resistivity, that can enhance the resistive detachment process. This analysis supports the hypothesis that magnetic nozzle design for high-performance thruster operation requires an optimization between the conflicting requirements of efficient plasma acceleration and plasma detachment. We conclude with a qualitative prescription for an optimized magnetic nozzle design.

RESISTIVE PLASMA DETACHMENT

The introduction of an applied poloidal field, $\mathbf{B}_{rz}$, into a magnetic self-field plasma thruster is intended to provide a magnetic cushion between the wall and the accelerating plasma. This magnetic nozzle ideally would direct the plasma flow as in a conventional nozzle, while providing magnetic insulation to reduce transverse losses from the plasma to the electrodes. The "tensile strength" of the $\mathbf{B}_{rz}$ field also resists the swirling motion induced by a "no slip" boundary condition that would occur if only the self field, B_θ, were present. Such vortex formation is detrimental, not only due to the loss of thrust power into turbulent motion, but also due to enhanced transport.

Although the $\mathbf{B}_{rz}$ field can provide significant improvements of internal thruster dynamics, it also leads to the fundamental problem of plasma detachment. In the case of a self-field thruster, there is no topological linkage between B_θ and any mechanical part of the thruster. As a ring of plasma moves out of the thruster, it carries the trapped B_θ with it. Toroidal flux carried into the plume represents a loss of energy but not a plasma deceleration mechanism. By contrast, a field line of $\mathbf{B}_{rz}$ in the throat of a thruster will connect back to the chamber walls and ultimately link a J_θ coil. If the plasma were ideal, it could exit the thruster by motion parallel to the $\mathbf{B}_{rz}$ line. However, it remains trapped on that $\mathbf{B}_{rz}$ line. Efficient thruster operation therefore requires, in addition to efficient plasma acceleration, an efficient mechanism for plasma detachment.

Plasma detachment can be mediated by several fundamental mechanisms, all of which are non-ideal within the context of MHD. Hooper (1991) showed that for collisionless plasma, "··separation from the field is determined by the hybrid (electron-ion) Larmor radius··," and plasma-field separation could be significantly constrained. Here, we consider the effect of finite resistivity. That is, we examine the case where plasma becomes sufficiently resistive outside of the thruster to diffuse across the $\mathbf{B}_{rz}$ field.

A simple analytic model of resistive plasma detachment may be constructed in which the assumption of classical Coulomb resistivity presents a "worst case" as regards the difficulty of detachment. The possibility of a micro-turbulent plume and the associated anomalous resistivity is also discussed with regard to plasma detachment.

Consider a ring of δN ions moving out of the thruster as in Figure 1. The ring is taken to be small as compared to the thruster, and it has a velocity in the r-z plane $\mathbf{V}_{rz}$. We consider motion from the nozzle throat, where $\mathbf{B}_{rz}$ is parallel to $\hat{z}$, on to a point far from the thruster where the cluster of particles, δN, is effectively detached from $\mathbf{B}_{rz}$.

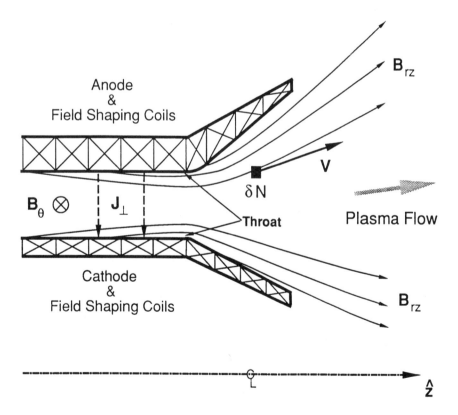

FIGURE 1. Schematic of a Coaxial Plasma Thruster with Resistive Detachment of Plasma from the $\mathbf{B}_{rz}$ Field.

The resistive drag for plasma crossing a field is $\mathbf{J} \times \mathbf{B}_{rz}$, where $\mathbf{J}$ is the plasma current density. A simple expression for the gross magnitude of velocity change due to resistive drag is

$$\Delta V_{rz} \equiv \int |\dot{V}_{rz}| dt . \tag{1}$$

The scalar ΔV_{rz} does not give the direction of total velocity change. However, a value of ΔV_{rz} comparable to the propellant exhaust velocity, V_e, indicates a major loss of thrust. The objective of this analysis is to evaluate Equation (1) and to compare it to the exhaust velocity.

The magnitude of momentum change for the cluster of δN ions of mass m_i is simply

$$\Delta p_{rz} \equiv m_i \delta N \Delta V_{rz}. \tag{2}$$

Neglecting gradients in plasma pressure, the momentum change, Δp_{rz}, is given by

$$\Delta p_{rz} = \int |\mathbf{J} \times \mathbf{B}_{rz}| d^3 x dt \tag{3}$$

where the volume integration is only over the considered ring of particles δN.

Assuming a singly ionized plasma of density n, the Ohm's law, including Hall terms, is

$$\mathbf{E} + \left(\mathbf{V} - \frac{1}{en}\mathbf{J} \right) \times \mathbf{B} + \frac{1}{en}\nabla P_e = \eta_\| \mathbf{J}_\| + \eta_\perp \mathbf{J}_\perp. \tag{4}$$

where $\mathbf{J}_\|$ and $\mathbf{J}_\perp$ are current components parallel and perpendicular to $\mathbf{B}$ respectively, and $\eta_\|$ and $\eta_\perp$ are the corresponding resistivities. In the general solution for thruster dynamics, the complete Ohm's law must be considered. However, in this discussion we are primarily looking for the effect of detaching the plasma from the $\mathbf{B}_{rz}$ field. Consequently, the θ-component of Equation (4) is emphasized. By azimuthal symmetry, there is no θ-component of the Hall term involving ∇P_e. Furthermore, the azimuthal component of the Hall term, $\mathbf{J}_{rz} \times \mathbf{B}_{rz}$, is neglected in the plume since $\mathbf{J}_{rz}$ is largely disconnected from the electrodes, and B_θ is small compared to B_{rz}. Since the resistive drag term arises through currents perpendicular to $\mathbf{B}$, it is also assumed that $\eta_\perp \mathbf{J}_\perp$ is the dominant term of interest on the right-hand side (RHS) of Equation (4). For steady state flow, Faraday's law requires $E_\theta = 0$ which, when combined with the other conditions described above, leads to

$$\eta_\perp J_\theta = - V_{rz\perp} B_{rz}. \tag{5}$$

Here, the scalar $V_{rz\perp}$ is defined by $V_{rz\perp} = \mathbf{V}_{rz} \cdot \hat{\theta} \times \mathbf{B}/B$. Entering Equation (5) into Equation (3), yields

$$\Delta p_{rz} = - \int \frac{V_{rz\perp}}{\eta_\perp} B_{rz}^2 d^3x dt$$

$$= - \int \frac{B_{rz}}{\eta_\perp} (2\pi r B_{rz} V_{rz\perp}) dt dr dz$$

$$= - \int \frac{B_{rz}}{\eta_\perp} d\Phi dr dz \tag{6}$$

where Φ is the poloidal magnetic flux passing through the ring of δN particles.

As the plasma moves from the throat of the thruster out into the plume. it undergoes an adiabatic expansion with $T \propto n^{\gamma-1}$. Taking $\gamma = 5/3$ leads to the relation

$$\left(\frac{T}{T_o}\right)^{3/2} = \frac{n}{n_o} \tag{7}$$

where the subscript "o" references any radial position in the throat of the thruster. We note that classical resistivity has the simple form

$$\eta_\perp \approx 1.04 \times 10^{-4} Z_{eff} ln\Lambda\, T^{-3/2}(eV) \equiv \eta_{\perp o} \left(\frac{T_o}{T}\right)^{3/2} \tag{8}$$

where everything is in SI units except for T in eV. Combining Equations (2) and (6)-(8) leads to

$$\Delta V_{rz} = \frac{1}{\eta_{\perp o} m_i n_o} \int_0^{\Phi_o} \frac{B_{rz}}{2\pi r} d\Phi. \tag{9}$$

The poloidal flux linking the ring of δN particles in the throat of the thruster is Φ_o, while the flux linkage for the detached ring is 0. The integration of Equation (9) simply follows the r-z trajectory of the ring between the throat and detachment, using Φ as the variable of integration.

The Alfvén speeds associated with B_θ and B_{rz} are defined as $V_{A\theta} \equiv B_\theta / \sqrt{\mu_0 m_i n}$ and $V_{Az} \equiv B_z / \sqrt{\mu_0 m_i n}$ respectively. Also, the exhaust velocity without considering the detachment issue is $V_e \approx \sqrt{3} V_{A\theta 0}$ (Gerwin et al. 1990, Hooper 1991, Schoenberg et al. 1991). For simplicity, we assume that the poloidal field is approximately constant in radius and in the z-direction at the throat, leading to $\Phi_0 = \pi r_0^2 B_{rz0}$. The condition for having resistive detachment minimally affect the exhaust velocity is,

$$\Delta V_{rz} \ll V_e, \tag{10}$$

which is equivalent to

$$\int_0^1 \left(\frac{B_{rz}}{B_{rz0}}\right)\left(\frac{r_0}{r}\right) d\left(\frac{\Phi}{\Phi_0}\right) \ll \frac{2\sqrt{3}\eta_{\perp 0}}{\mu_0 r_0} \frac{V_{A\theta 0}}{V_{Az0}^2} = \frac{2\sqrt{3}\Delta_0}{R_M r_0}\left(\frac{V_{A\theta 0}}{V_{Az0}}\right)^2, \tag{11}$$

where $R_M \equiv \mu_0 V_{A\theta 0}\Delta_0/\eta_{\perp 0}$ is the magnetic Reynolds number and Δ_0 is the channel width at the throat.

Two conflicting requirements for designing an optimal magnetic nozzle arise here. First, inside the thruster, the $\mathbf{B}_{rz}$ field should be strong enough to guide particles moving at the Alfvén speed along specified flow lines, preventing particles from leaving the annular self-field region and concomitantly incurring substantial diffusion across the $\mathbf{B}_{rz}$ flux, Φ. Second, after exiting the thruster, the plasma must become sufficiently resistive to diffuse through the $\mathbf{B}_{rz}$ flux without incurring a significant velocity reduction, ΔV_{rz}.

Previous work has shown that the condition for high-grade plasma-magnetic-field coupling is met by having the RHS of Equation (11) be < 1. Just how small this term should be is the objective of ongoing research. In the work of Schoenberg et al. (1991) it was conjectured that R_M should be $\geq$ 100 for effective magnetic insulation dominated by B_θ in the presence of a weak $\mathbf{B}_{rz}$. In the experimental work of Schoenberg et al. (1991), the radial dimensions were $r_0 \sim 25\,cm$ and $\Delta_0 \sim 10\,cm$. Unoptimized thruster operation was demonstrated, with plasma exhaust velocity of order of the Alfvén speed, $V_e \sim 10^5\ m/s$. This result was interpreted as evidence for magnetoplasma acceleration by a converging-diverging magnetic nozzle. For nozzle accelerated plasma, we expect the RHS of Equation (11) to range from 0.01 to 1.

The left-hand side (LHS) of Equation (11) compares the momentum imparted to plasma, if it were moved across a flux Φ_0 in the throat of the gun, to the momentum change when plasma is disconnected from Φ_0 in the exhaust plume.

A rapidly diverging field in the exhaust plume leads to the LHS of Equation (11) being ~1. Hence, a consequence of having field lines turn quickly away from the plume is the violation of Equation (11). This is consistent with the result obtained by Gerwin *et al.* (1990), where the time rate of change of momentum lost in the plume is equivalent to the thrust.

Alternatively, a plasma entering a nozzle of straight diverging field lines would behave as if it came from a point source, where $B_{rz}/B_{rzo} \approx (r_0/r)^2$. For this case, where field lines are essentially in the axial direction out to a radius r_D before bending back to the sides of the thruster, the LHS of Equation (11) will be $\leq (r_0/r_D)^3$. If, for example, plasma left the throat region with $r_0 = 25cm$ on a field line diverging at $0.2rad$ from the axis, then the LHS of Equation (11) would reach 0.01 for plasma at ~5m from the throat. Doubling the length of this controlled field divergence would cut the LHS of Equation (11) by an additional factor of 6. Thus, if the RHS of Equation (11) were held to the most severe constraint noted above of 0.01, plasma detachment could be achieved with ~18% momentum loss by a 10m long shaped field. Trim coils necessary for this field shaping are expected to be small. If, on the other hand, the RHS of Equation (11) is closer to unity, the shaping coils could make detachment an insignificant issue in 2 or 3 meters.

THERMAL CONDUCTION IN THE PLUME

This model, based on adiabatic cooling of the exhaust plume, is invalid if electron thermal conduction from the throat of the thruster sustains a significantly higher temperature in the plume. Applying our notation to the previous work of Gerwin (1990), we get a parallel thermal diffusivity of

$$D_{\parallel}^{e} = \frac{3T^{5/2}}{ne\eta_{\perp o}T_0^{3/2}}$$

and a heating time constant for plasma a distance l_z from the source of

$$t_{e\parallel} \sim l_z^2/D_{\parallel}^{e}.$$

Comparing the heating time constant to the time for the escaping plasma to traverse l_z in the plume, $t_z = l_z/(\sqrt{3}V_{A\theta o})$, defines a figure of merit, τ, for the validity of the resistive detachment model:

$$\tau \equiv \frac{t_{e\parallel}}{t_z} \sim \frac{e\eta_{\perp o}l_z n V_{A\theta o}}{\sqrt{3}T_0}. \tag{12}$$

Model validity requires $\tau > 1$. As an example, for the experimental conditions discussed by Schoenberg (1991) ($ln\Lambda = 10$, $T_0 = 10eV$, $n_0 = 10^{20}m^{-3}$, $V_{A\theta o} = 10^5$, and $l_z = 1m$), τ is $\gtrsim 1$. In this case the plume is effectively thermally disconnected from the thruster. This example indicates that, for relevant conditions, adiabatic expansion of a plasma can be considered without explicit treatment of axial thermal conduction.

THE EFFECT OF ANOMALOUS RESISTIVITY

The preceding discussion of resistive detachment can be substantially altered by microturbulence driven anomalous resistivity. The role of plasma microturbulence in MPD thrusters has recently been extensively reviewed by the Princeton group (Choueiri 1991). The essential conclusion of this work is that the Lower Hybrid Drift Instability and its various generalizations provide a robust mechanism for the generation of microturbulence in coaxial thrusters when the cross-field current drift velocity, V_d, exceeds the ion thermal velocity V_{thi}. This occurs when the velocity ratio (V_d/V_{thi}), called the "drift parameter," exceeds unity.

We shall base our estimate of the drift parameter on the MHD axial momentum equation for cold steady flow, which we write schematically as $\rho V(\Delta V/\Delta_z) = JB$ (Schoenberg *et al.* 1991). Here, $\rho \equiv m_i n$ is the plasma density, $J = neV_d$ is the current density, and Δ_z is the characteristic axial length over which the velocity increment (ΔV) is produced.

In the body of the thruster, one can envision an annular "self-field" plasma region dominated by an azimuthal magnetic field B_θ and a radial current density J_r, surrounded by an annular magnetic nozzle comprised of a poloidal magnetic field, B_{rz}. For a plasma well represented by the ideal MHD model, plasma flow in the throat of the annular nozzle is at the Alfvén speed associated with B_θ. If one then utilizes this upper bound $(V_{A\theta o})$ for the velocity increment (ΔV), one can arrive at an upper bound for the velocity ratio of interest by means of the axial momentum equation:

$$\frac{V_d}{V_{thi}} \lesssim \left(\frac{c}{\omega_{pi} \, \Delta_z}\right) \left(\frac{V_{A\theta o}}{V_{thi}}\right), \tag{13}$$

where c is the speed of light and $\omega_{pi} = (ne^2/\varepsilon_0 m_i)^{1/2}$ is the ion plasma frequency.

Because of ion-slip heating effects associated with plasma production (Gerwin 1991, Schoenberg *et al.* 1991), the flow velocity and ion thermal speed are roughly comparable. Consequently, (V_d/V_{thi}) is small ($\ll 1$) in the body of the thruster when the "ion skin depth" (c/ω_{pi}) is small compared with macroscopic dimensions. The latter condition has been shown to be closely related to the smallness of the "Morozov Hall parameter" $\Xi \equiv (m_i/e)(I/\dot{M})$ where I is the current and $\dot{M}$ is the axial mass flow rate (Schoenberg *et al.* 1991).

In the plume, the plasma is dominated by the applied field B_{rz}. Consequently, it is the azimuthal current density that characterizes the magnetic body force in the axial momentum equation. Again, the axial momentum balance equation yields

$$\frac{V_d}{V_{thi}} \sim \frac{c}{\omega_{pi}} \frac{V_{A\theta o}}{V_{Arz}} \frac{\Delta V}{\Delta_z} \frac{1}{V_{thi}} \tag{14}$$

Here, $V_{A\theta o}$ is the Alfvén speed within the throat of the thruster, whereas V_{Arz} indicates a representative Alfvén speed associated with the B_{rz} field out in the plume. Noting that the poloidal flux, Φ, can be roughly represented as

$$\Phi \sim (2\pi r \Delta_z)(B_{rz} \sin \alpha), \tag{15}$$

where α is the angle of B_{rz} with the axis of symmetry, one can obtain from Eq. (6) the estimate

$$\frac{\Delta V}{\Delta_z} \sim \frac{(V_{Arz} \sin \alpha)^2}{D_\eta}, \tag{16}$$

where $D_\eta \equiv (\eta/\mu_o)$ is the resistive diffusivity in the plume. (In deriving this expression, the simplifying assumption has been made that the plasma escape velocity is practically axial.)

The use of Equation (16) in Equation (14) yields the following estimate for the drift parameter

$$\frac{V_d}{V_{thi}} \sim \left[\frac{(c/\omega_{pi}) V_{Arz}}{D_\eta} \right] \left(\frac{V_{A\theta o}}{V_{thi}} \right) \sin^2 \alpha = \left(\frac{\omega_{ce}}{\nu_e} \right) \left(\frac{V_{A\theta o}}{V_{thi}} \right) \sin^2 \alpha, \tag{17}$$

where ω_{ce} is the electron gyro-frequency in the plume and ν_e is the electron-ion momentum-transfer collision frequency. For classical resistivity and reasonable values for plasma parameters in the plume, ($T \sim$ a few *ev*, hydrogen ion density $n \sim 10^{19} \ m^{-3}$, $B_{rz} \sim 0.05$ Tesla) one finds that the electron magnetization (the electron Hall parameter) is large, $(\omega_{ce}/\nu_e) \sim 10$ to 100. Consequently, it is likely that the drift parameter in some part of the plume is large (based upon classical resistivity), and that at least part of the plume will likely become microturbulent unless the applied field lines are only weakly divergent ($\sin^2 \alpha \ll 1$).

Equation (17) qualitatively implies that a weakly-diverging $\mathbf{B}_{rz}$ field engenders a subcritical drift parameter and hence a more "classical" plume. Although classical resistivity nominally represents a worst case with regard to the difficulty of detachment, it has the considerable merit of well-known physical processes whose actions can be calculated and utilized to produce optimized engineering designs for the detachment region. In contrast, the subject of microturbulence and anomalous transport in plasmas is still an area of basic research, with questionable value for presently contributing to knowledgeable engineering design.

CONCLUSION

A delicate balance exists between thruster efficiency, exemplified by making the right-hand side of Equation (11) < 1, and plasma detachment, achieved by making the left-hand side of Equation (11) much smaller than the right-hand side. We have shown that a simple adiabatic expansion of the escaping plasma can lower the temperature and increase the classical plasma resistivity enough to allow for effective plasma detachment from the magnetic nozzle of the thruster. We have also shown that electron thermal conduction from the thruster to the plume need not inhibit this process. Having indicated the conceptual possibility of efficient resistive detachment, the design of the field shaping coils required to achieve this result must be determined quantitatively by more detailed experimental and theoretical work.

Non-classical effects can also affect the detachment process. Ion drift, produced by cross-field currents, is believed to be important for the generation of Lower Hybrid Drift micro-turbulence and associated anomalous resistivity. In the body of the thruster, the drift parameter should be limited to a subcritical value (and hence the maintenance of high-grade plasma with a desirably large magnetic Reynolds number), by operating the thruster with a small ion skin depth (c/ω_{pi}). In a macroscopic context, this is equivalent to operation with a small "Morozov Hall parameter", $\Xi \equiv (m_i/e)(I/\dot{M})$. However, in the plume, a large drift parameter is expected due to the large electron Hall parameter, (ω_{ce}/ν_e), unless the applied field lines are only weakly divergent. This weakly divergent configuration would tend to produce a more quiescent plume, which might be manipulated so as to minimize resistive drag. In a more strongly divergent magnetic field, one would expect to have a more resistive plume due to microturbulence, and the possibility of exerting deliberate control over the detachment process becomes problematic.

Finally, magnetic field line tearing and reconnection may be an alternative to both classical and anomalous resistive diffusion for plasma detachment. For this to occur, the z-components of the poloidal field would be in opposite directions at the cathode and anode to allow formation of a $\mathbf{B}_{rz}$ field null between the electrodes. Although non-ideal processes occur at the X-line during reconnection, the bulk of the exhaust plasma is disconnected from the thruster without resistive diffusion. The physics of this process is a subject for future reports.

Acknowledgments

This work was performed at Los Alamos National Laboratory and supported by the U. S. Department of Energy.

References

Choueiri, E. Y., Ph.D. thesis, *Electron-Ion Streaming Instabilities of an Electromagnetically Accelerated Plasma*, Princeton University, October 1991.

Gerwin, R. A., G. J. Marklin, A. G. Sgro, and A. H. Glasser, *Characterization of Plasma Flow Through Magnetic Nozzles*, Technical Report AL-TR-89-092, Air Force Astronautics Laboratory, 1990.

Gerwin, R. A., K. F. Schoenberg, and D. J. Rej, "Preliminary Scoping Studies for Nozzle-Based Coaxial Plasma Thrusters," in *Proceedings of the Eighth Symposium on Space Nuclear Power Systems,* page 493, University of New Mexico, Albuquerque, NM, 1991.

Hooper, E. B., "Plasma Detachment from a Magnetic Nozzle," AIAA/SAE/ASME 27th Joint Propulsion Conference, AIAA-91-2590, Sacramento, CA, June 24-26, 1991.

Schoenberg, K., R. Gerwin, C. Barnes, I. Henins, R. Mayo, R. Moses, R. Scarberry, and G. Wurden, "Coaxial Plasma Thrusters for High Specific Impulse Propulsion," Technical Report AIAA 91-3770, American Institute of Aeronautics and Astronautics, 1991, submitted for publication in the *AIAA Journal of Power and Propulsion.*

THE CONCEPT OF A NUCLEAR POWER SYSTEM
WITH ELECTRIC POWER OF ABOUT 2 MW
FOR A POWER TRANSFER SPACECRAFT

Nikolay N. Ponomarev-Stepnoi, Vladimir A. Pavshoock, and
Michael A. Diachenko
Kurchatov Institute of Atomic Energy
Moscow 123182, USSR
(095) 196-9595

INTRODUCTION

High profitableness as well as knowledge-capacity of rocket-space technics the development of which allows to inculcate advanced technologies and to consolidate scientific-technic potential favour the fact that all new firms rush to take part in the space exploration intensifying the competition at the market of space paying loads and making the space exploration leaders - USA ans USSR - to advance the existed arsenal of powerfull spacecrafts and to create qualitatively new ones.

The proposed nuclear power system (NPS) with electric power of about 2 MW for a power transfer spacecraft (PTS) with electic jet propulsion may favour to achieve the significant progress in the field of Solar System planets exploration, the lunar base creation, space industry realization, the conduction of some pure scientific high-power experiments in space, and, at last, it may be possible to reduce several times the prime cost of 1 kg of pay load delivery to the geostationary orbit and other high ones.

PTS is a space shuttle (an interplanetary tug) which transfer cargoes from the low Earth orbit of 600-800 km hight to high orbits and back. Cargoes supplied with the units of supplimentary delivery are delivered to the orbit by transfer vehicles "Proton", "Energia" and others. PTS docks with cargoes at the low orbit. PTS is delivered to the low orbit by the transfer vehicle "Proton".

TABLE 1. PTS Extravehicular Abilities.

Characteristics	"Proton"+BU[a]	"Energia"+BU
Delivery orbit parameters	H = 600-800 km	H = 600-800 km
Pay load (waiting orbit)	i = 51°	i = 51°
Cargo mass delivered to the waiting orbit	17 t	90 t
Cargo mass delivered by PTS to the GEO	9 t	70 t
Duration of the direct trip to the GEO	35 days	120 days
Cargo mass delivered by PTS to the lunar orbit	10 t	73 t
Duration of the direct trip to Moon	45 days	125 days
Duration of the back trip	25 days	25 days
Cargo mass delivered by PTS to Martian orbit	7 t	60 t
Duration of the direct trip to Mars	320 days	450 days
Cargo mass delivered by PTS to Jupiter orbit	1 t	35 t
Duration of the trip to Jupiter	500 days	700 days

[a] Booster unit.

It is evident that the programs of far planets exploration, lunar bases creation, piloted expeditions to Mars, and, at last, the projects of global space telephone communications require the intensive increase of vehicular abilities of space technics which predetermines the sharp extension of spacecraft powerfulness.

POWER SYSTEM SELECTION

Accomplished in the USA (Khanetal 1989) and USSR studies on determination of application fields of different energy types for spacecraft supply testify that from the power level of 300 kW and with the spacecraft operation life more than 1 year nuclear power has no alternatives.

Glushko et al. (1988) considered the thermionic NPS option with power 7.5 kW and specific mass 6 kg/kW. However, due to the absence of experimental data regarding NPS of the similar power, it is impossible to give confident preference to the thermionic option. Moreover, as Adamov et al. (1990) and Polukhin et al. (1990) pointed out, NPS with turbomachine energy conversion (TMEC) may have advantages in comparison with thermionic and thermoelectric options due to the following features:

- Considerably less weight of the reactor system with equal electric power;
- Higher efficiency (Caveny 1988);
- Higher quality of electric energy;
- Higher technological ability due to considerably lower coolant temperature;
- Principal possibility of separate power loop running tests;
- Higher TMEC reliability due to the absence of element backup limits outside the reactor.

Therefore, this paper suggests the NPS concept with TMEC. It should be pointed out that the USSR has accumulated great experience in NRE development, that there is a base of reactor critical test facilities and specialists of high qualification. At the same time, the USA possess vast scientific-technical experience achieved during the accomplishment of NERVA program. With the selected electric power level (2 MW), the design of the reactor and radiation shield is quite close to the optimum one according to specific masses, configuration, and fuel loading, and specific masses of TMEC units are reduced to the level of 2-4 kg/kW (Caveny 1988).

The conducted project-ballistic analisys determined the required parameters of electric power as well as the characteristics of the electrical jet propulsion system (EJPS). The main limits adopted during calculations are as follows:

- System mass and dimensions must not exceed the ability limits of "Proton" transfer vehicle;
- Radiation dose accumulated by the pay load during radiation belts passing must not exceed $5 \cdot 10^4$ rad;
- Circular orbit of 600-800 km hight is considered to be radiation-safe;
- PTS on-board systems operation life is 1-2 years at the first stage of operation, and then, after the further running tests, it will be brought to 5-7 years;
- Number of PTS trips during the operation life - up to 10;
- Total dose of radiation exposure in the instrument compartment form the reactor and Earth radiation belts is
  - gamma-ray - not more than 10^6 rad,
  - fast neutron fluence - not more than 10^{13} $1/cm^2$.

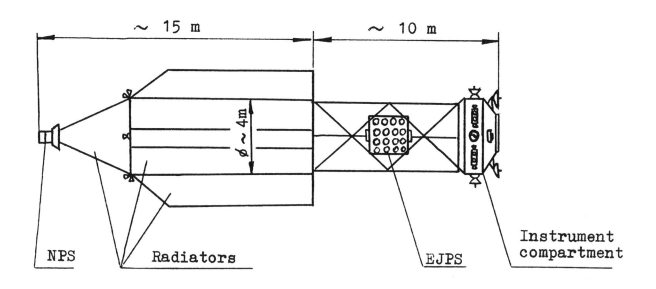

FIGURE 1. Schematic of the Major Components of
the Power Transfer Spacecraft.

NUCLEAR POWER SYSTEM WITH TMEC

The results of NPS with TMEC studies conducted at Kurchatov Institute of Atomic Energy are given below as an example. The final selection of the power system type can be made only after special theoretical and experimental studies are performed considering two type options. The main NPS calculated parameters are

- Thermal power - up to 10 kW;
- Electric power - about 2 MW;
- Energy conversion system - turbomachine (Brayton cycle);
- Total operation time - not less than 10^4 hours;
- Number of start-ups per operation life - up to 30;
- Maximum coolant temperature - up to 1500 K.

The conducted studies result in determination of the following main NPS design characteristics

- Mass of a gas-cooled reactor - 1000 kg;
- Fuel - UC, $(U,Zr)C$, UN with U-235 of 90% enrichment, fuel cladding - Zr, W^{184}, reflector - Be;
- Mass of radiation shielding - 1000 kg;
- Mass of energy converter (turbine, compressor, and unipolar generator) - 3500 kg;
- Coolant - helium-xenon mixture (1-3% Xe);
- Mass of heat pipe radiators with mean temperature of about 700 K - 3000 kg;
- Radiator area - about 300 m^2;
- Mass of automatic control system and energy supply system - 1000 kg;
- Mass of NPS construction - 1500 kg;
- Total NPS mass - 11,000 kg;
- Specific mass - 5.5 kg/kW.

POWER TRANSFER SPACECRAFT

PTS consists of a power system module with a nuclear reactor and shielding (Fig.1), TMEC located in the radiation shielding cone, a heat pipe radiator consisting of four deploying half-cylindric sections as well as a pulling out truss located inside the radiator. The units located on the pulling out truss are

- Instrument compartment with a docking system, devices of orientation, navigation, and communication and the additional propulsion system;
- Main electric-jet propulsion system (I_{sp} = 4600 c);
- Xenon fuel tank.

PTS main mass characteristics are

- NPS - 11,000 kg;
- EJPS - 5000 kg;
- Pulling out truss, fuel tank - 1000 kg;
- Instrument compartment, docking system - 2000 kg;
- Additional propulsion system, unconsidered elements - 1000kg;
- Fuel (xenon) - 8000 kg;
- Total PTS dry mass - 20,000 kg.

CONCLUSIONS

The conduction of extensive space studies, the lunar base creation, the accomplishment of a manned Mars mission will undoubtedly result in high economic and scientific-technical effects, will open new markets and give new opportunities. One of the main elements in realization of the given problems is the creation of powerful power propulsion complices on the base of nuclear reactors. The atomic industry branch has gained great experience in development of the similar systems which open good prospects for beneficial application of international division of labour in terms of SEI Program.

Acknowledgments

This research was conducted in Kurchatov Institute of Atomic Energy under internal funding support.

References

Khanetal, E. U.(1989) "Overview of Key Issues for Space Nuclear Power Systems," presentation to the Sixth Symposium on Space Nuclear Power Systems, held in Albuquerque, NM, 8-12 January 1989.

Glushko, B. et al. (1988) "Way to Mars", Pravda newspaper, May, 24, 1988.

Adamov, E. O. et al. (1990) "Status of Works in the USSR on the NPPP Development for Interplanetary Space Flights", Presentation to the Anniversary Specialist Conference on Nuclear Engineering in Space, held in Obninsk, USSR, 15-19 May 1990.

Polukhin, D. A. et al. (1990) "Power Propulsion System for Spacecrafts", Presentation to the Anniversary Specialist Conference on Nuclear Engineering in Space, held in Obninsk, USSR, 15-19 May 1990.

Caveny, L. (1988) "Space Propulsion : Status and Prospects", Moscow, Mir, 1988, p.p. 283, 308.

HYTEC A HIGH EFFICIENCY THERMALLY REGENERATIVE
FUEL CELL FOR SPACE APPLICATIONS

Douglas N. Rodgers
General Electric Company
175 Curtner Ave., M/C 575
San Jose, CA 95125-1088
(408) 925-1929

Prodyot Roy
General Electric Company
6835 Via Del Oro, M/C S76
San Jose, CA 95153-5354
(408) 365-6464

Samir A. Salamah
General Electric Company
6835 Via Del Oro, M/C S54
San Jose, CA 95153-5354
(408) 365-6361

Abstract

The Hydrogen Thermo-Electrochemical Converter (HYTEC) is an improved means of thermal-to-electrical energy conversion for space or other remote locations. The system is based on a fuel cell where the fuel element, hydrogen gas, is regenerated from the fuel cell product, metal hydrides. The HYTEC work at General Electric has resulted in a system concept design and supporting experimental data. Unique features in the system include the use of a niobium-titanium alloy for the electrodes and a lithium-sodium mixture to transport the hydride reaction product. Experimental data to date has shown the feasibility of the concept.

INTRODUCTION

Conversion of heat to electricity without moving parts is highly desirable for space power and other remote systems applications. The Hydrogen Thermo-Electrochemical Converter (HYTEC) system (Roy 1987) has the potential for providing this capability with higher efficiency and lower total system weight than currently available technologies. In this paper a brief description is presented along with a summary of the development progress to date.

System Principles

HYTEC is a thermally regenerated fuel cell system based on hydrogen gas fuel, a molten salt electrolyte, and an alkali metal to react with the hydrogen and then transport it to be regenerated thermally.

The HYTEC system started with the knowledge that some molten salts, such as alkali metal halides will act as hydride ion conductors when a small amount of lithium hydride is added to the melt. Some of these mixtures are nearly 100% ionic conductors and have a low resistance to the hydride ion transport.

A thermally regenerative fuel cell system was designed around this concept by using hydrogen gas at several atmospheres pressure on one side of the cell and a lithium/lithium hydride mixture to provide a low pressure sink on the other side. A pair of thin metal membrane electrodes contain the electrolyte, separate the reactant from the product, and collect the current.

The lithium hydride product is removed from the fuel cell and heated in another part of the system where the hydride is decomposed to yield high pressure hydrogen gas and lithium which are recycled. A thin metal membrane is used as a diffusion barrier to provide absolute separation of the regenerated gas from the liquid metal. A continuous cycle system was then designed from these concepts. A schematic diagram of the system is shown in Figure 1.

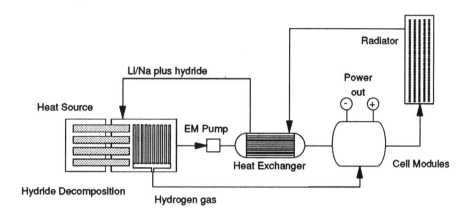

FIGURE 1. Schematic of the HYTEC System.

HYTEC FEATURES

- By using hydrogen gas and metal diffusion membranes HYTEC solves the problem of separating the fuel cell reactants and products. This problem has defeated previous thermally regenerative fuel cell system concepts. Without a complete separation, the fuel cell becomes fouled with products carried over to the reactant side of the cell. A set of diffusion membranes at the decomposition stage guarantees 100% separation since only hydrogen will diffuse through the membrane.

  Fouling of the fuel cell electrode by corrosion products carried into the reactant side of the cell from the cycle is also eliminated by the metal diffusion membranes, since pure hydrogen gas is produced by the separator;

- A new feature of HYTEC is the use of a lithium-sodium mixture to carry the hydrogen. The mixture gives the ability to select system operating temperatures and pressures independently by selecting the relative amounts of lithium and sodium. Better overall system optimization is possible compared to using lithium or sodium alone;

  and

- Due to the high mobility of hydride ions in the electrolyte, and the rapid diffusion of hydrogen gas through certain metals, HYTEC is intrinsically a high specific power device. Since the hydrogen fuel and the lithium hydride reaction products are both light elements, the weight of HYTEC system is very low.

INCENTIVES FOR HYTEC DEVELOPMENT

In order to evaluate the potential benefits and the feasibility of developing the HYTEC system concept, both system design and experimental work were initiated. The system design was done on the basis of a cell performance model derived from the fundamental properties of the cell materials. The model and the results of the systems studies have been described by Salamah et al. (1991a and 1991b).

The studies show that power systems using a HYTEC conversion cycle have a significantly greater power density than both existing systems and other developmental system projections. Studies were done for small (<10 kWe) and large (10 to 100 kWe) systems.

One reason for the high power density is the flexibility of the HYTEC system design which allows close matching of the heat source characteristics to the conversion system. For example, with isotopic heat sources, which have a relatively high fuel weight (and cost), the HYTEC system can be sized with a higher thermal efficiency. Although the HYTEC mass increases the fuel mass decreases (less heat is needed to drive a more efficient conversion system) resulting in a lower overall mass.

For a reactor system such as SP-100 the fuel weight is relatively small, so the optimum mass for the total power system has a HYTEC system designed with a lower thermal efficiency and smaller mass.

EXPERIMENTAL WORK

Low Current and Open Circuit Cells

Initial experimental work was directed toward proving that a real cell would behave as predicted by the HYTEC analysis. Single cells containing a reservoir of lithium as a hydrogen sink and using hydrogen gas from a cylinder were built and tested. Electrode membranes were made from iron, nickel and tantalum. Figure 2 is a diagram of the test cells.

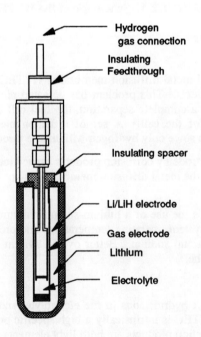

FIGURE 2. HYTEC Test Cell Diagram.

Results of the early cell experiments showed that open circuit voltages were accurately predicted by the hydrogen gas pressure and the decomposition pressure of lithium hydride. Cells were successfully operated at temperatures from 550 to 800 K.

Current production from the early cells was small due to the choice of electrode material and the thickness of the electrode diffusion membranes. However, the cell currents were found to be proportional to the calculated electrode permeation, as expected. Cells with iron membrane were operated in the open circuit condition for several thousand hours at 775 K.

Material Choices for Cell Construction

One of the environments in the cell which dictate material choices is the liquid lithium side of the cell where materials such as vanadium, niobium, and tantalum can be used. Of these, vanadium has the highest hydrogen permeability.

Recently, a 55% niobium/titanium alloy, called beta-titanium, has been found to have a significantly greater permeability than vanadium. Thin wall (0.13 mm) tubing was fabricated from beta-titanium bars. Preliminary testing indicated a factor of 8 increase in hydrogen permeability compared to vanadium. The measurements were made electrochemically in an apparatus which is shown in Figure 3 along with the results. The experiment consisted of passing current from one disk electrode to the other while measuring the voltage between each disk electrode and the molybdenum electrode. The slope of the voltage versus current curve is a relative measure of the hydrogen permeability of the disk material.

The gas side electrode was originally chosen to be Pd/25%Ag. This commercially available alloy is commonly used for hydrogen purification because of its high permeability for hydrogen. The alloy, as well as pure palladium, was found to be attacked by the electrolyte, however. This was not expected since chloride salts are compatible with palladium. Some basic compatibility testing showed that the addition of a few percent LiH caused rapid attack of the Pd. Since beta-titanium is compatible with the electrolyte, it is used for the hydrogen gas electrode as well as the lithium electrode.

Cell Performance at Higher Currents

After testing the initial series of cells at open circuit and low currents, more advanced cells were fabricated with vanadium electrodes. These cells showed improved performance since the electrodes were produced in a thin wall (0.13 mm) tube configuration, as compared to 0.26-mm thick membranes.

Flow Dependence of Cell Current

Short circuit currents in static cells typically decrease after a few minutes, due to formation of LiH reaction product on the lithium side of the anode electrode. A dynamic system test was needed to determine cell performance with a stream of liquid metal removing the hydride reaction product.

A test was run using vanadium electrode cells in a pumped sodium loop. The results showed stable cell currents for several hours at even the lowest measurable flow velocities. Further testing is needed to determine the minimum flow required to remove the LiH reaction product from the electrode surface.

Maximum Current Density

The flow dependence test produced the maximum current density of 0.018 A/cm^2 at approximately 50% of open circuit voltage. This low value was due at least in part to the lack of wetting of the electrode surface by the electrolyte as shown in X-ray photographs of the cells. Further experiments are planned with substitution of beta-titanium for vanadium and flowing lithium which is expected to bring the current densities closer to 0.1 to 0.2 A/cm^2. This is the range of current density that leads to optimum overall power generation system mass.

Sodium-Lithium Mixtures

Both lithium and sodium have been considered as the liquid metal for the HYTEC system. New hydrogen partial pressure data (Werich et al. 1988) on lithium-sodium alloys, however, show some advantages of the mixture over either sodium or lithium at very high temperature. The most important advantage is that by adding composition as a

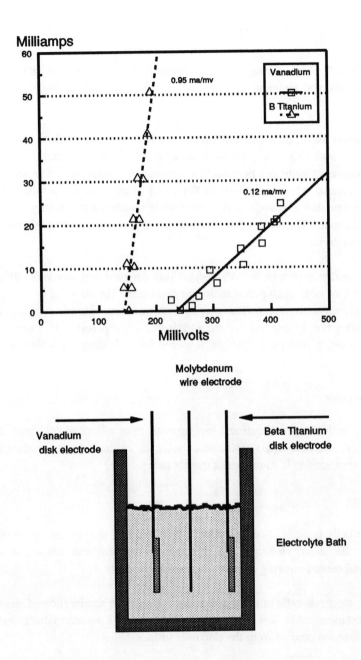

FIGURE 3. Relative Permeability of Beta-Titanium and Vanadium.

variable, the hydrogen decomposition pressure and temperature can be selected independently. Use of the Li/Na alloy will permit the use of heat sources with outlet temperatures between 1000 and 1400 K.

The data from Werich et al. (1988) did not cover the temperature range of interest to HYTEC. We predicted the data on the basis that the logarithm of the pressure versus the reciprocal of the temperature is a straight line. We did an experiment which verified that a 50% Li/Na mixture does indeed behave at lower temperatures as we predicted. We also linearly interpolated the mixture data to yield the pressure data required by the system analysis calculations. The interpolation must still be verified by experiment, especially in the area of low lithium concentrations.

FUTURE DEVELOPMENT

Further development of HYTEC will address the following areas:

- Demonstrate prototypic current densities for single cells;

- Obtain additional data for the lithium-sodium-hydrogen system and hydrogen permeation through beta-titanium;

- Improve the cell performance model to better represent actual measurements;

- Develop methods to assemble the unit cells which have been constructed into cell arrays; and

- Design and test a prototype HYTEC system.

Acknowledgments

HYTEC has been developed with General Electric Internal Research and Development funding. Studies of power systems employing HYTEC were performed with KAPL funding.

References

Roy P. (1987) Hydrogen Thermo-Electrochemical Converter, U.S. Patent No. 4,692,390, 8 September 1987.

Salamah, S. A., D. N. Rodgers, R. S. Narkiewicz, and P. Roy (1991a) "HYTEC, High Efficiency Thermally Regenerative Power Conversion for SEI Missions," in *Proc. of the AIAA Conference on Advanced SEI Technologies*, 91-3524, held in Cleveland, OH, 4-6 September 1991.

Salamah, S. A., D. N. Rodgers, D. G. Hoover, and P. Roy (1991b) "SP-100 Reactor/HYTEC - A High Efficiency Static Conversion Power System," in *Proc. 26th Intersociety Energy Conversion Engineering Conference*, 910283, held in Boston, MA, 4-9 August 1991.

Werich T. W., et al. (1988) "Zersetzungsdrucke in Lithium/Natrium/Wasserstoff-Systemem," Z. Phy. Chem., 158: 69-79

AMTEC/SHE FOR SPACE NUCLEAR POWER APPLICATIONS

Thomas K. Hunt, Robert K. Sievers
and
Joseph T. Kummer
Advanced Modular Power Systems, Inc.
4667 Freedom Drive
Ann Arbor, MI 48108
(313) 677-2113

Jan E. Pantolin and David A. Butkiewicz
Environmental Research Institute of Michigan
4667 Freedom Drive
Ann Arbor, MI 48108
(313) 677-1170

Abstract

The Alkali Metal Thermoelectric Converter (AMTEC) is a high efficiency device for the direct conversion of heat to electricity. Applications for this converter range from space power to remote terrestrial missions. Significant progress has been made on both component and converter cell development, with particular interest focused here on high efficiency and zero-G operation. Experiments were performed that demonstrate that electromagnetic pumping will operate without a gravity induced head, allowing pumped cell design to operate in zero-G. A cell has also been developed that will operate in zero-G without an EM pump. Wick return cells, without pumps, were also tested and found to be feasible for zero-G operation. Design calculations, using test validated models, indicate that these cells can have efficiencies greater than 25%. These results will strongly influence state of the art and advanced design development.

INTRODUCTION

The Alkali Metal Thermoelectric Converter (AMTEC), also known as the sodium heat engine (SHE), was invented at Ford by Kummer and Weber (1968). An AMTEC operates as a thermally regenerative electrochemical concentration cell using sodium as the working fluid (Weber 1974 and Hunt et al. 1978, 1981). Its operation relies on the high sodium ion conductivity and low electronic conductivity of a sodium beta"-alumina solid electrolyte (BASE). The static nature of AMTEC plus the high efficiency and modularity characteristics have made this technology attractive for space and remote terrestrial applications.

AMTEC/SHE development programs are underway at the Environmental Research Institute of Michigan (ERIM), the Jet Propulsion Laboratory (JPL), the Advanced Modular Power Systems (AMPS), and Beta Power in the United States, at Asea Brown Boveri in Germany, and at the Electrotechnical Laboratory and at Kyushu University in Japan. The AMTEC/SHE program at ERIM is a continuation of the program formerly conducted at Ford and which had an historical focus on terrestrial applications, but is now focused on space applications. The other major U.S. efforts have always been directed primarily toward space applications of 1 Kwe scale devices. A wide variety of development work has been described in the literature (see Cole 1983, Bankston et al. 1984, Hunt et al 1987). The current status of the United States development effort is described in this paper, as it relates to key technical objectives of high efficiency, high power density, reliability and zero-G operation.

A large development effort has been concentrated on electrode power density. Significant progress has been made in this area by both the JPL and ERIM programs, and efforts are now beginning to shift to other issues. Zero-g operation, which will be the particular focus of this paper, is one of the key developmental issues. Very little development work, however, had been performed in this area. No work had been done, for example, to develop an EM pump, which is found on almost all cell design concepts, for operation with a zero inlet pressure head under superheated conditions. Only a limited effort had been mounted to develop a cell that did not need an EM pump (Sherrit et al. 1988). This paper will describe the significant advances that have now been made on both of these subjects.

Development Status

A wide variety of continuously operating laboratory test cells have been built and tested. Through these tests all of the technical principles of AMTEC operation have been verified experimentally and models describing AMTEC operation have been refined to allow rational prediction of the performance of new systems. Highlights of the experimental programs include the following.

1. A high efficiency 23 watt AMTEC cell utilizing a single beta"-alumina tube, in a specially designed furnace environment that simulated high temperature series connection by eliminating all lead thermal conduction losses, was built and tested at temperatures up to 805°C (Hunt et al. 1981). That cell was heated internally with an electric heater and the conversion efficiency, measured simply as the electric power output divided by the heater power input, reached 19% at 805°C. In the intervening development period, improved current collection methods together with more durable electrodes have allowed power densities to increase by a factor of 2, up to .45 watts/cm^2. It may be anticipated that a similar test run with current state-of-the-art electrodes and current collection bus systems could yield thermal to electric conversion efficiencies approaching 25% if all thermal conduction losses, including lead and EM pump losses, are eliminated and the condenser achieves a specular emissivity of 0.02.

2. A simple, single-tube cell was operated under continuous load at approximately 6 watts for over 10,000 hours. During this entire period the output power was essentially constant. Operation of this cell was halted due to a leaking valve used for the initial evacuation of the condenser chamber. Following replacement of the valve, the cell was restarted and operated continuously for a further 4,000 hours at reduced output.

3. TiN electrodes for AMTEC systems have been developed with demonstrated lifetimes above 1600 hours at specific power densities of 0.45 watts/cm^2 at 800°C (Novak et al. 1988). The power density for these electrodes increases with temperature at a rate of approximately 0.002 W/cm^2/K. Electrodes in the tungsten-rhodium group, have been developed and tested at 850°C for 1400 hours (Ryan et al. 1991, Underwood 1991) with durable power densities in the same range as TiN electrodes.

4. A detailed understanding of electrode performance has been achieved leading to improved electrodes and permitting accurate model predictions to be made for new system designs. (Williams et al. 1990a,b and Underwood et al. 1991)

5. Photodeposited molybdenum microgrids with molybdenum mesh overlay current collectors (Ryan et al. 1991) and copper microscreen grids with molybdenum wire current collectors (Hunt et al. 1978) have been developed and have been shown to reduce contact resistance to about 0.05 ohm-cm^2.

6. Series and mixed series-parallel connection of multiple tubes at the high operating temperature has been achieved offering the potential for reduced thermal conduction losses (Novak et al. 1989). Improved management of thermal radiation losses using microscreen radiation shields has been successful in laboratory test cells (McBride et al., 1988).

7. A multi-tube AMTEC device reached a power output above 500 watts operating at 750°C.

8. Operation of an AMTEC cell in zero-G, has been simulated by running a small, 1 watt, cell at 600°C in both conventional and inverted orientations with respect to gravity. EM pumping in simulated zero-G (inverted orientation) has also been demonstrated.

Substantial progress has been made on the development of AMTEC components and cell designs, and many of the most critical issues have been resolved. Material lifetime demonstration, structural durability, reliable high temperature feedthroughs, and electrolyte to metal sealing remain as key development tasks.

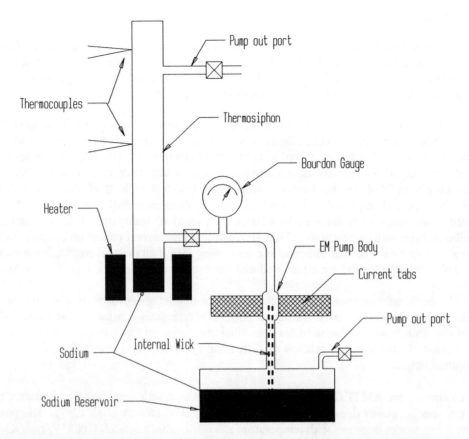

Figure 1 Schematic Diagram of Test Apparatus for Wick-Fed Electromagnetic Pump.

EXPERIMENTAL METHODS AND RESULTS

Zero-G EM Pump

The pump test was carried out using the apparatus shown schematically in Figure 1. The thermosyphon/-receiver chamber, shown on the left, was fabricated from 5-cm diameter, 304 SS tubing with .63 cm diameter SS tubing connecting the pump out line to its SS Nupro bellows valve at the top and to the similar valve and remainder of the system at the lower section. During EM pump operation in these experiments, this chamber serves as a stand-in for the BASE tube to which sodium would be delivered in an operating AMTEC system. As shown, beyond the lower valve, a stainless steel Bourdon Tube pressure gauge was fitted to allow measurement of the pump pressure reached during operation. The source reservoir chamber was also constructed of 5-cm O.D., 304 SS tubing and a pump out port with valve provided as shown in the figure. Following its fabrication by crushing a thin-wall stainless tube containing a fine ss felt tube as the wick, and the attachment of leads by vacuum brazing, the EM-pump tube was inserted into a fitting in the top of the source reservoir and sealed in place. The lower end of the pump tube was placed in the source chamber so that the tube of felt wick material in the pump tube extended beyond the lower end of the tube by approximately 0.5 cm and reached the bottom of the chamber, well below the sodium level. The upper end was connected as shown to the pressure gauge. All of the critical pump tube components were baked out under vacuum at temperatures above 700°C in order to achieve a high level of cleanliness so as to reduce the time required to achieve sodium wetting as well as the lowest possible contact resistance between the sodium and the SS pump channel walls. The receiver chamber was heated with a "clam-shell" heater and the remaining plumbing and the source chamber were brought to operating temperature with heating tapes. The pump, gauge and connecting tubing were operated at approximately 250°C.

With the system hot and the source chamber evacuated, a D.C. current was applied to the pump leads with the mid-valve closed. In this mode, the static pressure produced by the pump was determined. Measurements were made over a period of several days as sodium wetting of the EM pump channel walls improved. Figure 2 shows the rise in pump pressure as a function of time at various current levels.

The results of these initial experiments, indicate that it is possible to both prime and operate a small electromagnetic pump in opposition to gravity using a simple wick structure and presently available commercial wick materials. We believe that successful pump operation 'up hill', as demonstrated in this test provides a reasonable worst case test for zero-G operation of the pump related portion of an AMTEC sodium return system.

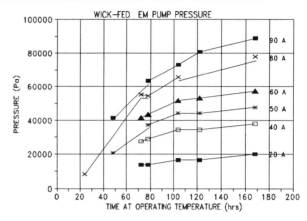

Figure 2 Test Pump pressure as a Function of Time for Different Currents.

For zero-G operation, some form of capillary structure will probably be required at the condenser surface to capture efficiently and control the sodium arriving from the BASE tube/electrode. Transfer of the arriving sodium to the inlet of the EM pump or high pressure wick structure constitutes the other major aspect of the sodium circulation task. For zero-G operation, that transfer requires the effective coupling of two capillary structures without serious gaps which would reduce the transfer efficiency. The work we have conducted to deal with that aspect of the sodium management task is described in the next section.

Wick Return Cell Testing

As discussed above, zero-G operation of an AMTEC system requires integration of the sodium collection system at the condenser with the means to deliver the sodium working fluid to the high pressure side of the AMTEC electrolyte membrane. Methods for achieving this through capillary pumping using fine wicks have been discussed for many years and one example of a capillary pumped system has been built and operated (Sherrit 1988). In a system 'pumped' only by capillary forces in wick structures, the need for a gravity induced 'head' for an EM pump can be eliminated. Operation in zero-G should then be straightforward, providing that adequate control of the sodium inventory can be achieved so that excess liquid is not permitted to accumulate outside of the wick structures.

The major limitation of a capillary pumped system arises from the extremely small pore size required to achieve capillary pressures equivalent to AMTEC operating temperatures in the usual range from 800°C to 1000°C. The pore size/operating temperature relationship was illustrated in Figure 1. Because the hot side operating temperature of AMTEC systems pumped only by capillary action is generally restricted by the pressure capabilities of the wick structure, the peak power density and efficiency of such systems are currently lower than can be achieved when an EM pump is employed. A program has just begun to develop wicks that may be capable of the high temperature operating regime presently limited to EM pumps.

For space applications, one clear advantage of a fully wick returned AMTEC arises from the zero-G capability which appears to be inherent in such designs. In order to investigate this operating concept we have built and tested a simple, single-tube AMTEC cell in which the sodium is returned solely by a wick structure. As will be discussed later, this approach can lead to very useful efficiencies if carefully implemented. Work to reach higher efficiency will be proposed for a Phase II continuation of this program.

Sherrit and Sayer (1988) have conducted experiments on an AMTEC cell in which the sodium is returned against the sodium vapor pressure from the condenser reservoir to the high temperature, high pressure side of the BASE tube by a fine wick. Their cell suffered from a large internal resistance which severely limited the power output and prevented achievement of the efficiency of which the system concept should be capable. The high resistance was apparently due in part to the nature of the electrodes used and in part to the current collection bus system. Further, the sodium handling aspect of their design was not intended to control the sodium inventory throughout its cycle in AMTEC operation as will be required for zero-G operation.

An advanced wick return cell design modified for zero-G operation has been designed, built and tested in the laboratory. A schematic diagram of the overall cell configuration is shown in Figure 3. Thermal energy was provided by a cartridge heater in the heater well. Sodium contained within a wick on the outer surface of the heater well is evaporated and deposited on the inside of the BASE tube. The sodium on the inside of the beta tube ionizes and passes through the tube from the high pressure to low pressure region. The electrons released on the inside of the beta tube are conducted from the base of the cell to the load and returned through the positive feedthrough to the permeable electrode on the outside surface of the BASE tube. Sodium ions emerging from the BASE tube pick up a returning electron, then evaporate and re-condense on the interior cell wall. The liquid is then wicked back to the inside of the BASE tube along a continuous capillary path.

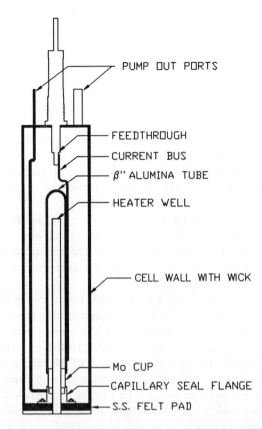

Figure 3 Simple, wick-return AMTEC Cell Design Operable in Any Orientation.

The BASE tube used in the cell is 8 cm long, 1.5 cm in outside diameter with a wall thickness of 1.0 mm. The active electrode is 4.5 cm long with an area of approximately 20 cm^2. The current collector on the outside of the tube is a copper screen cinched down to the tube surface with a helical wrap of 0.025 cm diameter molybdenum wire in the conventional manner. Copper braid bus wires of approximately 4 mm diameter were strapped longitudinally to the tube in contact with the first layer molybdenum wrap and with the copper screen. A reactively sputtered titanium nitride positive electrode was used.

There are several features of this initial test cell which reflect convenience of assembly rather than optimization of performance or efficiency. The positive feedthrough shown at the top of the drawing has a .32 cm copper core brazed though a Kovar sleeve to an alpha alumina insulator. This available feedthrough, is sized for much larger currents than can be delivered by the small area BASE tube in this cell. A smaller feedthrough, optimally sized for the expected power will be used in later cells to reduce parasitic heat loss. 347 SS felt was chosen for the wick material both for the wall of the condenser (where it is shown as integral with the wall here) and for the high pressure wick which feeds sodium to the BASE tube. Advanced wick materials are being developed to allow significantly higher temperature operation.

Current - Voltage Curves Current voltage curves were measured at intervals after the cell temperature was raised. A series of three of these curves taken at different condenser temperatures with a hot zone temperature of 590°C are shown in Figure 4. As expected from the analysis (Hunt 1978), it is apparent from these results that even at the rather low hot zone temperatures achievable in this wick-fed design with the readily available wick materials, the main response of the cell output to changes in condenser temperature occurs at low currents, near open circuit conditions. The ability of AMTEC systems to operate well with high exhaust temperatures is particularly useful for minimizing radiator weight for space applications and for terrestrial, cogeneration applications in which the exhaust heat can contribute to other processes.

The time history of the output current and power during the present testing period is shown in Figure 5. This cell, designated WR-1, is still operating normally after 1300 hours. It can be seen that shortly following start up, the cell was left on open circuit for several days. With stable operating conditions established, the cell was then placed under passive load for durability testing with the load drawing somewhat less than the maximum power (which is approximately 1.04 watts at this temperature). The sodium inventory

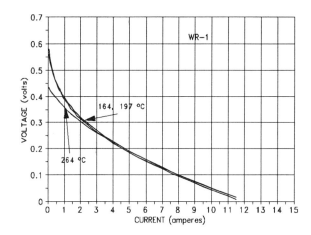

Figure 4 Current-Voltage Curves for Wick Return Cell Showing the Effect of Raising the Condenser Temperature from 164°C to 264°C.

in this cell is 20 grams. This amount makes a complete pass through this cell in about 4 hours at the 4.9 ampere level at which it has been operated.

Inverted Operation   After 430 hours of operation in the standard, upright (90°) orientation, the cell was turned to lie horizontally (0°) while remaining under load. One day later it was rotated further to a position 45° below horizontal, placing it effectively in a "-0.7 G" posture with respect to gravity and conventional AMTEC operating attitude. The timing of these changes is indicated by the vertical lines on the current/-power graph in Fig. 5. It can be seen that there was no decrease in output as the result of operating this cell head down. This cell has now been operating for over 1300 hours.

The current-voltage curves shown in Figure 6 also demonstrate clearly that changes in the orientation of this cell do not significantly degrade its performance. The apparent slight increase in output observed in the later test taken in the inverted position may be attributable to the slight differences in the thermometry between the two conditions. Because there are substantial temperature gradients in this cell, slight shifts in the position of the thermocouples during the rotation operation may have affected their readings and led to a higher effective temperature in the inverted situation. Changes in the heat losses from the exterior of the condenser chamber may also have occurred following the change of orientation. During the course of this initial program phase it was not feasible to turn the cell completely upside down   (-90°) due to the specific arrangement of the diagnostic and heater leads and the pump out ports. The cell operating temperature is 595°C and the condenser temperature is 250°C.

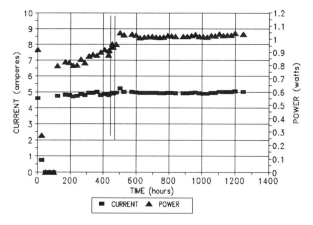

Figure 5 Current and Power Output of WR-1 Cell as a Function of Time Under Constant Passive Load. The Vertical Lines Indicate the Times at which the Cell Orientation with Respect to Gravity was changed.

Analysis of WR Cell   Detailed electrochemical and thermal models were developed to predict the cell performance. The electrochemical models (Williams et al. 1990a,b) include mass transport, interfacial kinetics and ohmic losses in the current collectors. The thermal models developed specifically for this cell geometry (details to be published) included all significant heat loss mechanisms such as conduction between

1321

the BASE and condenser via the heater well, tube support, and feedthrough paths, radiation between the BASE and condenser, and heat transport due to sodium flow.

The model predictions were compared to the experimental results. The predicted current-voltage curves are shown in Figure 6. The solid curve, representing the predicted values, fits the experimental data extremely well using typical performance parameters of G = 100, R_s = 0.08 ohm-cm^2, and j_o taken from Sievers et al. (1989), for a TiN electrode with a copper current collector screen at 600°C. The heat losses are itemized in Table 1. The model predicts a required input thermal power of 55 watts for 1 Watt electrical output and 5 Amperes current. This compares quite well with the 60 watts measured. The 'near term' column gives the values predicted for an improved cell design that may be built with present technology. The advanced cell described in the 3rd column requires a significant advance in electrolyte fabrication.

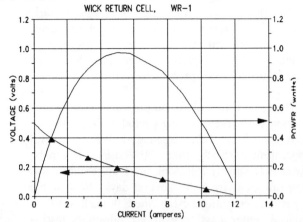

Figure 6 Current-Voltage Relation for Wick Return Cell Showing Analysis and Data Points.

It should be noted that for the experiments reported here, no attempt was made to design this wick return cell for optimum efficiency. The model, which works for this case, can also be used to predict the performance of improved cell designs. This model predicts that efficiencies exceeding 10% at 600°C, and exceeding 25% at 800°C may be achievable.

CELL ENERGY INVENTORY	CELL # 1	NEAR TERM	ADV. CELL
OPERATING TEMPERATURES (°C)			
Electrolyte	595	700	800
Condenser	250	400	400
ELECTROLYTE THICKNESS (mm)	0.8	0.8	0.1
ELECTRODE LENGTH (cm)	5	10	10
THERMAL LOSSES TO CONDENSER (watts)			
Radiation between electrolyte and condenser	32.5	2.0	0.8
Conduction through sodium vapor	0.1	0.0	0.0
Sensible heat transport by vapor flow	0.5	1.1	2.4
Latent heat transport by vapor flow	4.6	10.9	17.5
Conduction through structure	6.6	1.1	1.3
THERMAL LOSSES THROUGH LEADS (watts)	9.6	0.0	0.0
TOTAL THERMAL LOSSES, Q_p (watts)	53.9	15.1	22.0
OUTPUT POWER (watts)	1.1	3.9	9.5
EFFICIENCY	2.0 %	20.5 %	30.1 %

CONCLUSIONS

Significant progress has been made on both component and converter cell development, with particular interest focused here on high efficiency and zero-G operation. Experiments were performed that demonstrate that electromagnetic pumping will operate without a gravity induced head, allowing pumped cell design to operate in zero-G. A cell has also been developed that will operate in zero-G without an EM pump. Wick return cells, without pumps, were also tested and found to be feasible for zero-G operation. Design calculations, using test validated models, indicate that these cells can have efficiencies greater than 25%. These results will strongly influence state of the art and advanced design development.

Acknowledgments

The work described in this paper was supported in part by the United States Department of Energy through the Office of Industrial Technology under contract DE-AC02-90CE40941 and by NASA under an SBIR contract NAS7-1137.

References

Bankston, C.P., T. Cole, S.K. Khanna, and A.P. Thakoor (1984) "Alkali Metal Thermoelectric Conversion (AMTEC) for Space Nuclear Power Systems", Space Nuclear Power Systems, M.S. El-Genk and M.D. Hoover eds., Orbit Book Co., Malabar FL, p. 393-399.

Bankston, C.P., T. Cole, R. Jones and R. Ewell (1983) "Experimental and Systems Studies of the Alkali Metal Thermoelectric Converter for Aerospace Power", J. Energy, 7, 442-448.

Bankston, C.P. (1986) "Alkali Metal Thermoelectric Conversion (AMTEC) Technology Status Review", in Space Nuclear Power Systems 1986, M.S. El-Genk and M.D. Hoover eds., Orbit Book Co., Malabar FL.

Chi, J.W.H., R.K. Sievers and T.K. Hunt (1988) "Space Power AMTEC Reactor Concept (SPARC) Space Power Systems", Proceedings of the 23rd Intersociety Energy Conversion Engineering Conference, Vol. 3, p 691-693.

Cole, T. (1983) "Thermoelectric Energy Conversion with Solid Electrolytes", Science, 221, 915.

Crowley, C.J. and R.K. Sievers (1991) "Preliminary Evaluation of a Space-Solar-AMTEC Power Conversion System" Transactions of the 8th Symposium on Space Nuclear Power p. 1064.

Ewell, R. (1983) "Energy Conversion for Megawatt Space Power Systems", Proceedings of the 18th Intersociety Energy Conversion Engineering Conference, p 87.

Hunt, T.K., N. Weber and T. Cole (1978) "Research on the Sodium Heat Engine", Proc. 13th IECEC, p. 2011-2017.

Hunt, T.K., N. Weber and T. Cole (1981) "High Efficiency Thermoelectric Conversion with Beta"-Alumina Electrolytes, The Sodium Heat Engine", Proc. Conf. on Fast Ionic Transport in Solids, North Holland.

Kummer, J.T. and N. Weber (1968) "Energy Conversion Device Comprising a Solid Crystalline Electrolyte and a Solid Reaction Zone Separator", U.S. Patent 3,404,036 Assigned to the Ford Motor Company, Dearborn, MI.

Markley, R.A., R.K. Sievers, M.D. Carelli, L.R. Hedgecock, J.A. Peoples, J.R. Rasmussen, T.K. Hunt, S. Olsen, N. Weber, C. Germany and S. Limaye (1987) "Nuclear Powered Alkali Metal Thermoelectric Converter System for Multimegawatt Space Power Applications", Phase I Final Report for DOE contract DE-AC03-86SF16503.

McBride, J.R., R.F. Novak, D.J. Schmatz, W.B. Copple, J.T. Brockway, N. Arnon, G.A. Grab and T.K. Hunt (1989) "Advances in Design and Performance of SHE System Components", Proceedings of the 24th Inter society Energy Conversion Engineering Conference, Vol 2. p. 683.

Novak, R.F., J.R. McBride, T.K. Hunt, D.J. Schmatz, W.B. Copple, N. Arnon and J.T. Brockway (1988) "Development of a 1 KWE Sodium Heat Engine", Proceedings of the 23rd Intersociety Energy Conversion Engineering Conference, Vol. 1, p 219-225 1988.

Ryan, M.A., B. Jeffries-Nakamura, R.M. Williams, M.L. Underwood, D. O'Connor and S. Kikkert (1991) "Directly Deposited Current Collecting Grids for AMTEC Electrodes," in Proc. 26th Intersociety Energy Conversion Engineering Conference, 5, 463.

Sievers, R.K. and C.P. Bankston (1988) "Radioisotope Powered Alkali Metal Thermoelectric Converter Design for Space Systems", Proceedings of the 23rd Intersociety Energy Conversion Engineering Conference, Vol. 3, p 159-167.

Sievers, R.K., R.M. Williams, M.L. Underwood, B. Jefferies-Nakamura and C.P. Bankston (1989) "AMTEC System Performance Studies Using the Detailed Electrode Kinetic and Transport Model", Transactions of the 6th Symposium on Space Nuclear Power Systems, Albuquerque, NM.

Sievers, R.K., and M.H. Cooper (1990) "Advanced AMTEC Design Options with Composite Electrolyte Membranes", Transactions of the 7th Space Nuclear Power Symposium.

Sievers, R.K. and R.F. Wright (1990) "High Power Density Alkali Metal Thermal to Electric Converter", Proceedings of the 25th Intersociety Energy Conversion Engineering Conference, Vol. 2, p. 426.

Sherrit, S., M. Sayer and B. Kindl (1988) "Electrode Systems and Heat Transfer in Thermoelectric Generator Design", Proceedings of the 23rd Intersociety Energy Conversion Engineering Conference, Vol. 1, p 241-247.

Underwood, M.L., R.M. Williams, M.A. Ryan (1991) "A Figure of Merit for AMTEC Electrodes," in Proc. 26th Intersociety Energy Conversion Engineering Conference, American Nuclear Society, held in Boston, MA, August 4-9 1991.

Underwood, M.L. (1991) private communication.

Weber, N. (1974), "A Thermoelectric Device Based on Beta-Alumina Solid Electrolyte", Energy Conversion 14, 1-8.

Williams, R.M., M.E. Loveland, B. Jeffries-Nakamura, M.L. Underwood, C.P. Bankston, M. Leduc, and J.T. Kummer (1990) "Kinetics and Transport in AMTEC Electrodes, I. The Interfacial Impedance Model, J. Electrochemical Society (137): 1709-1716.

Williams, R.M., B. Jeffries-Nakamura, M.L. Underwood, C.P. Bankston and J.T. Kummer (1990) "Kinetics and Transport in AMTEC Electrodes, II. Temperature Dependence of the Interfacial Impedance of $Na_{(g)}$/Porous Mo/ Na-B" Alumina, J. Electrochemical Society (137): 1716-1723.

THERMAL MODELING OF AMTEC RECIRCULATING CELL

Jerry W. Suitor, Roger M. Williams, Mark L. Underwood,
Margaret A. Ryan, Barbara Jefferies-Nakamura, and Dennis O'Connor
Jet Propulsion Laboratory
California Institute of Technology
4800 Oak Grove Drive
Pasadena, CA 91109
(818) 354-4917

Abstract

Thermal modeling of the AMTEC energy conversion system was conducted and compared to experimental results. The system studied was an AMTEC Recirculating Test Cell with instrumentation that measured local temperatures. The geometry was tubular and the finite difference routine was cast in cylindrical coordinates. This is the first time a finite difference model has been used to analyze the temperature field within the AMTEC cell. Internal heat generation in the ß" alumina solid electrolyte due to ohmic heating while under load was also included. The results indicated that there was a significant contribution of the condenser film properties on the performance of the test cell. Further work is planned to expand the analysis to include the heat losses through conduction along electrical current carrying leads.

INTRODUCTION

The Alkali Metal Thermal-to-Electric Conversion (AMTEC) system is a static thermal power conversion technology with high efficiency promise. Cells operated in the laboratory have demonstrated high efficiencies and development work on power improvements of the electrodes has suggested high specific energy systems are possible. The device converts heat energy to electrical energy which requires careful attention to heat losses. Heat lost through heat "leaks" reduces the efficiency and total output of the system. Proper design through thermal modeling is needed to identify design characteristics that result in efficient utilization of the thermal energy available. This paper presents the initial work on thermally modeling an AMTEC recirculating test cell (RTC) currently under development at the Jet Propulsion Laboratory (JPL) using finite difference techniques.

DESCRIPTION OF AMTEC SYSTEM

The Alkali Metal Thermoelectric Converter (AMTEC) is a thermally regenerative electrochemical device for the direct conversion of heat to electrical energy with efficiencies potentially near Carnot. It uses ß" alumina solid electrolyte (BASE) as a separator between liquid sodium at 900-1300 K and a low pressure region in which the sodium activity is controlled by a condenser at 400-700 K (see Figure 1). In operation, metallic sodium is oxidized at the sodium/BASE interface allowing sodium ions to enter the BASE. Electrons pass through an external load performing work and then recombine with sodium ions at a porous metal electrode on the low pressure side of the BASE. As a result, the AMTEC converts the work of isothermal expansion of sodium vapor directly to electric power (Weber 1974, Cole 1983). AMTEC has many advantages for terrestrial and space power applications including no moving parts with the resulting potential for low maintenance and high durability, efficiency that is substantially higher than other static power systems, modular construction, and the ability to use high temperature combustion, nuclear, or solar heat sources (Bankston et al. 1983, Sievers and Bankston 1988). Experimentally, AMTEC has demonstrated an energy conversion efficiency of 19 percent (Hunt et al. 1988) and stable electrode performance at 0.4 to 0.8 W/cm^2 for up to 1500 hours (Hunt et al. 1981, Williams et al. 1989). This power density is in the range required by systems projections to deliver >20 percent conversion efficiency.

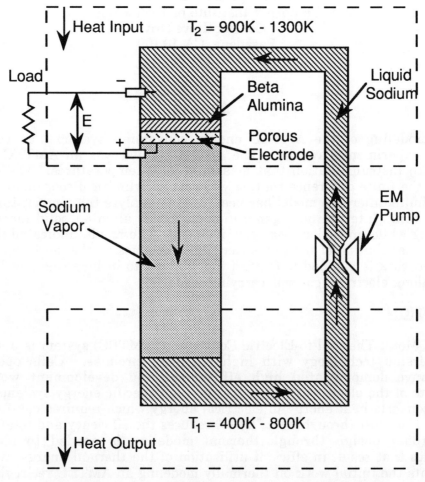

FIGURE 1.   Schematic of the AMTEC Cycle.

The operating temperatures of an AMTEC cell provide a high driving potential for heat transfer by means other than the transport of sodium. Specifically there are thermal losses that occur due to conduction along the current leads from the cell and other structural components of the cell. There is also radiative heat loss from the BASE material to the lower temperature condenser. These heat losses result in lower efficiency and can not be avoided but can be minimized. Proper design of the cells, using thermal modeling, results in substantial heat loss reduction.

The design of the RTC has been described in detail elsewhere (Bankston et al. 1983, Bankston et al. 1987, Underwood et al. 1989). The layout is shown in Figure 2. In operation, a stainless steel heater well is sealed into the interior of the BASE tube. The annular space inside the BASE and outside the heater well is filled with liquid sodium. An electric resistive heater inside the heater well provides the heat input to the cell. The BASE tube is braze sealed to a metal transition piece and a 3.4 cm flange. This flange mates with a metal O-ring seal to the underside of the top flange of the condenser. A 1.6 mm tube, brazed to the heater well, serves as the Na feed line.

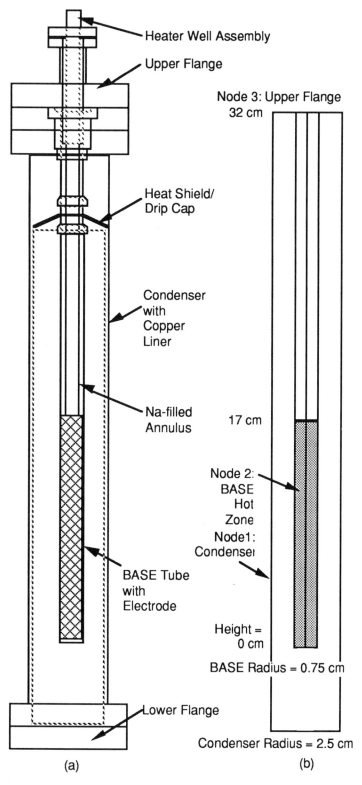

FIGURE 2.   Drawing of Recirculating Test Cell.

On the exterior of the BASE tube, porous Mo or Rh/W electrodes are deposited by magnetron sputtering then wrapped with current collecting grids and attached to electrically isolated feedthroughs (Williams et al. 1989, Bankston et al. 1987). The region between the BASE and the condenser is evacuated to a pressure less than 1 x 10⁻⁵ torr. The condenser is a 5 cm (I.D.) stainless steel tube with a molybdenum mesh liner which promotes temperature uniformity and sodium wetting. An earlier study suggested that a high reflectivity sodium coating was desired on the condenser surface to minimize the radiative heat loss. The mesh was designed to provide wicking of the sodium over the entire condenser so as to provide that highly reflective coating. The exterior of the condenser is wrapped with a heating tape to allow independent control of condenser temperature.

MODELING APPROACH

Previous work on thermal modeling (Underwood et al. 1990) considered the bulk heat losses without detailing temperature profiles within the cell. The modeling approach used in this work considered the basic steady state heat conduction equation, formulated into a series of finite difference equations. These equations were solved using an explicit solution technique to determine the temperature field both in the liquid sodium region as well as in the BASE wall cross section.

Thermal Equation

The steady state heat conduction equation in cylindrical coordinates is given by

$$\frac{\partial}{\partial r}\left(r \frac{\partial T}{\partial r} \right) + \frac{1}{r}\frac{\partial^2 T}{\partial \theta^2} + r \frac{\partial^2 T}{\partial z^2} + \frac{Qr}{k} = 0 \tag{1}$$

where r, θ, and z are the radial, circumferential, and axial coordinates, T is the temperature, Q is the heat generation term, and k is the thermal conductivity of the medium. The cylindrical symmetry of the RTC permits the exclusion of circumferential conduction (the θ term). The steady state heat conduction equation is restated as a finite difference expression based on the mesh nomenclature shown in Figure 3.

$$(r_{i,k} + \frac{\Delta r}{2})\frac{(T_{i+1,k} - T_{i,k})}{(\Delta r)^2} - (r_{i,k} - \frac{\Delta r}{2})\frac{(T_{i,k} - T_{i-1,k})}{(\Delta r)^2}$$

$$+ r_{i,k}\frac{(T_{i,k+1} - T_{i,k})}{(\Delta z)^2} - r_{i,k}\frac{(T_{i,k} - T_{i,k-1})}{(\Delta z)^2} + \frac{(Q_{i,k} r_{i,k})}{k} = 0 \tag{2}$$

Boundary Conditions

The proper identification and formulation boundary conditions of any finite element analysis are critical for success of the analysis. There were four boundary conditions that defined the perimeter of the mesh field. These were heat addition to the interior heater surface and the upper surface, and heat loss from the lower surface and the outer tube surface.

The interior heater heat flux was specified based on experimental operation. The upper surface heat flux was a heat addition to the system due to the presence of heating tape at the top surface to minimize heat loss through that surface. The lower surface heat loss was due to radiation to the condenser shell.

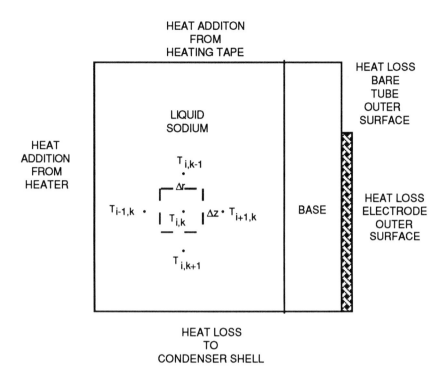

FIGURE 3.   Schematic of Heat Loss Zones and Finite Difference Mesh.

The outer surface heat loss was divided into two zones since not all of the tube had an electrode on it. The top zone had only heat loss due to radiation to the condenser surface. The lower zone, where the electrode was located, had heat loss due to both radiation to the condenser as well as heat loss due to the evaporation of sodium from the surface of the electrode. The evaporative loss is due to the treatment of the sodium as going from the liquid to the vapor state.

RESULTS AND DISCUSSION

The model developed and reported in this paper is the first attempt at modeling the temperature field of an AMTEC cell. The initial results obtained from the model support earlier findings of Underwood et al. (1990) that the condenser emissivity is very important to the efficient operation of the AMTEC cell in this configuration.

Optimization of the cell design is now possible with expansion of this model. Issues that will be addressed include the impact of thermal loss due to structural connections, losses due to the electrical leads, and the interrelationship between operation of the electromagnetic pump and the heat loss/efficiency of the cell. Additional information on the effect of heat conduction on feed throughs will be determined. Feed throughs are typically α alumina, which suffers from chemical attack by sodium at temperatures above 700 C.

Acknowledgement

The research described in this paper was carried out by the Jet Propulsion Laboratory, California Institute of Technology, and was sponsored by the Department of Energy,

Advanced industrial Concepts Division, The Air Force Phillips Laboratory, and the National Aeronautics and Space Administration.

References

Bankston, C.P., T. Cole, R. Jones, and R. Ewell (1983), "Experimental and Systems Studies of the Alkali Metal Thermoelectric Converter for Aerospace Power," *J. Energy*, 7: pp. 442-450.

Bankston, C.P., R. M. Williams, B. Jeffries-Nakamura, and T. Cole (1987), "Alkali Metal Thermoelectric Converter (AMTEC) Electrode Lifetime Studies," in *22nd Intersociety Energy Conversion Engineering Conference*, pp. 1423-1430.

Bankston, C.P., R. M. Williams, B. Jeffries-Nakamura, M. L. Underwood, and T. Cole (1988), "High Power Density Electrode for AMTEC," in *Fifth symposium on Space Nuclear Power Systems*, M.S. El-Genk and M.D. Hoover eds., Orbit Book Co., Malabar, FL, in press.

Cole, T. (1983), "Thermoelectric Energy Conversion with Solid Electrolytes," *Science*, 221: pp. 915-921.

Hunt, T.K., N. Weber, and T. Cole, (1981), "High Efficiency Thermoelectric Conversion with Beta"-Alumina Solid Electrolytes, The Sodium Heat Engine," in *Fast Ion Transport in Solids*, J.B. Bates and G. Farrington eds., North Holland Publishing Co., Amsterdam .

Hunt, T.K., J. V. Lasecki, R. F. Novack, J. R. McBride, and J. T. Brockway (1988), "Sodium Heat Engine/AMTEC System Experiments," in *Space Nuclear Power Systems 1987*, M.S. El-Genk and M.D. Hoover eds., Orbit Book Co., Malabar, FL, pp. 400-412.

Sievers, R.K., and C.P. Bankston (1988), "Radioisotope Powered Alkali Metal Thermoelectric Converter Design for Space Systems," in *Proc. 23rd Intersociety Energy Conversion Engineering Conference, Vol. 3*, American Society of Mechanical Engineers, Denver, Colorado, pp. 159-167.

Underwood, M.L., R. K. Sievers, D. O'Connor, R. M. Williams, B. Jeffries-Nakamura, and C. P. Bankston (1989), "AMTEC Recirculating Test Cell Component Testing and Operation," in *24th Intersociety Energy Conversion Engineering Conference*, (1989), 2833.

Underwood, M. L. D. O'Connor, R. M. Williams, B. Jefferies-Nakamura, M. A. Ryan, and C. P. Bankston (1990), "Thermal Characterization of an AMTEC Recirculating Test Cell," in 25th Intersociety Energy Conversion Engineering Conference, pp. 523-530.

Weber, N. (1974), "A Thermoelectric Device Based on ß" Aluminas," *Energy Conversion*, 14: pp. 1-7.

Williams, R.M., B. Jeffries-Nakamura, M. L. Underwood, B. L. Wheeler, M. E. Loveland, S. J. Kikkert, J. L. Lamb, T. Cole, J. T. Kummer, and C. P. Bankston (1989) "High Power Density Performance of WPt and WRh Electrodes in the Alkali Metal Thermoelectric Converter," *J. Electrochem Soc.* 136(3): pp. 893-901.

AN AMTEC VAPOR-VAPOR, SERIES CONNECTED CELL

Mark L. Underwood, Roger M. Williams, Margaret A. Ryan,
Barbara Jeffries-Nakamura, and Dennis O'Connor

Jet Propulsion Laboratory
California Institute of Technology
4800 Oak Grove Drive
Pasadena, CA 91109
(818) 345-9049

Abstract

The alkali metal thermal-to-electric converter (AMTEC) converts heat energy to electrical energy with efficiencies up to 3 to 4 times the state of the art static converters. However, the high current, low voltage output of a single module requires difficult series connecting of cells, and several life limiting concerns may reduce the achievable efficiency. A concept is proposed that incorporates internal series connecting of cells and sodium supply as a vapor without condensation. This concept overcomes many of the long term materials concerns in liquid anode systems and results in a module with about the same power density, but at a higher voltage and lower current. The vapor-vapor AMTEC concept is described and its performance predicted for typical AMTEC operating conditions.

INTRODUCTION

The alkali metal thermal-to-electric converter (AMTEC) is a device for the conversion of heat energy to electrical energy. As indicated in Figure 1, typical heat input temperatures range from 900 to 1200 K with heat rejection temperatures from 400 to 800 K. In the high temperature region AMTEC uses beta" alumina solid electrolyte (BASE) to separate high activity sodium (Na) from a region in which the Na activity is kept low by a condenser operating near the heat rejection temperature. BASE is a conductor of Na ions (Na^+) and an insulator for electrons. Typical AMTEC designs use liquid Na as the anode on the high activity side (Bankston et al. 1985, Sievers and Bankston 1988, and Weber et al. 1988). At 1160 K, the vapor pressure of Na is 1 atm while the Na vapor pressure at the condenser is 10^{-5} atm or less. This vapor pressure ratio results in an electrical potential across the BASE of up to 1.6 V. Electrodes on either side of the BASE provide sites for an electrochemical reaction at the interfaces. On the high activity side, the anode (typically liquid Na) provides sites for Na oxidation. On the low activity side of the BASE, the cathode (typically a thin, porous metal) provides sites for reduction of Na^+. Electrical power can be drawn from the cell when the anode and cathode are contacted through a load. A complete description and analysis of the AMTEC cycle has been published previously (Weber 1974 and Cole 1983).

AMTEC cells can produce electrical power at high efficiency and power density, but the output per cell is typically about 0.5 V and up to 100 A at maximum power. Power conditioning electronics cannot efficiently convert this low voltage, high current power to a more useful voltage level. Previous AMTEC systems designs typically required series connecting independent modules to produce a more useful voltage for input to the power conditioning circuit. While highly redundant, this approach results in systems with significant mass and volume attributed to non-power producing components. In addition,

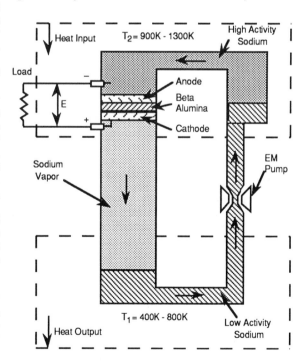

Figure 1: AMTEC Cycle Diagram showing the flow of the sodium working fluid.

previous vapor-fed designs used sodium phase change to supply sodium working fluid and additional heat to the cell. Some of the sodium refluxes over the BASE. As yet unknown is the effect of using very low oxygen content sodium on the BASE during such a reflux process. Over time, the sodium may extract oxygen from the electrolyte and deposit it on the metal walls of the sodium boiler. Finally, in order to achieve very high efficiency operation, previous designs require electrical feedthroughs that operate at high temperature in a Na environment. Under such conditions, the required insulator may degrade and limit the life of the cell.

An AMTEC vapor-vapor cell permits internal series connecting of adjacent cells and addresses the other issues described above. In the vapor-vapor concept, the anode is a porous metal electrode similar to the cathode. The high activity Na is supplied as a vapor from a remote boiler. Several cells using a single Na source can be connected in electrical series to increase the output voltage without the addition of extensive power conditioning electronics. Since the output is at a higher voltage, external series connecting of the cells is less critical, and the electrical feedthroughs can operate at a lower temperature with less of an efficiency penalty. In addition, no Na would reflux over the surface of the BASE, thus reducing the possibility of electrolyte degradation. This paper describes the AMTEC vapor-vapor cell concept and predicts the cell performance based on experimental results with similar systems.

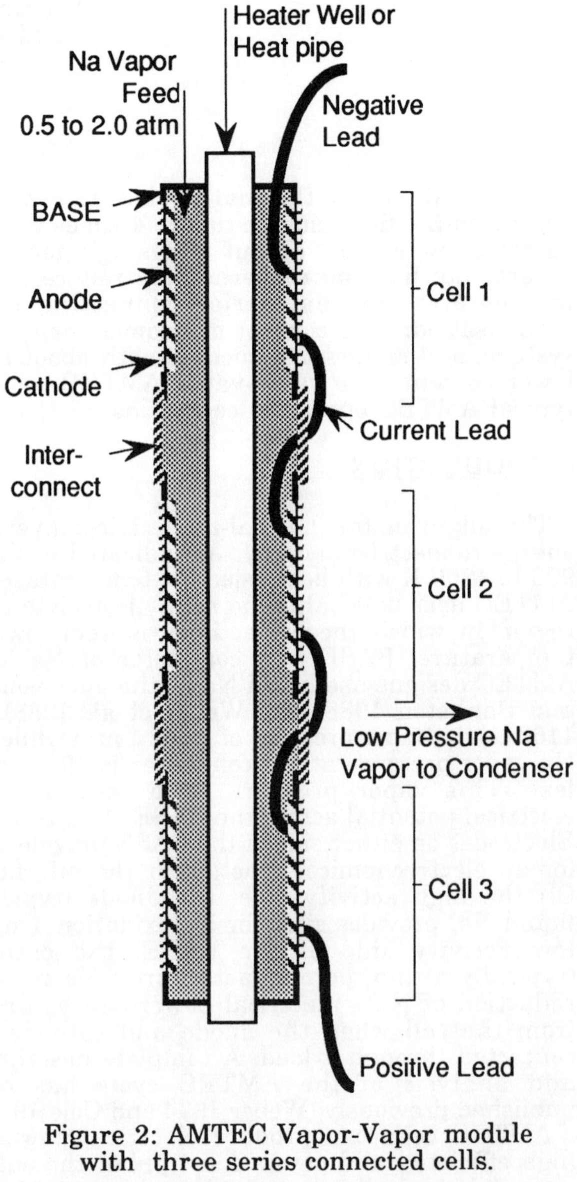

AMTEC VAPOR-VAPOR CONCEPT

A diagram of a possible AMTEC vapor-vapor module is shown in Figure 2. The diagram shows three cylindrical cells internally series connected to form a module. In principle the number of cells is limited only by the ability to supply Na and the breakdown voltage of the materials. To a first approximation, the cell output is the same as the output of a liquid Na anode cell with the same cathode area. Both the anode and cathode are porous metal electrodes similar to cathodes now used in AMTEC cells. These electrodes are sputtered molybdenum (Mo), titanium nitride (TiN), or tungsten alloyed with platinum (Pt/W) or rhodium (Rh/W). The proposed vapor-vapor module is composed of several individual cells that use a single Na source and are connected in electrical series. The connection is made from the anode of one cell to the cathode of the next through a metal interconnect. The leads shown connecting the electrode to the interconnect in the figure could be simply an extension of the electrode and current collector to the interconnect.

Na is delivered as a vapor at 0.5 to 2.0 atm from a remote boiler to the series of anodes. The vapor may be flowing or stagnant except for the flow due to the working fluid. The pressure of the Na vapor is controlled by the temperatures of the boiler and the anode and any flow related pressure drops. The vapor must be super heated to prevent condensation in the anode which might lead to shorting of adjacent anodes. The concept uses a heat pipe in the Na vapor to supply the super heat as well as to overcome thermal losses that would lead to Na condensation. Other methods, such as an external heater could be used to supply the required extra heating.

In order to complete the Na cycle, the Na leaving the cathode is collected by a condenser where waste heat is rejected. The liquid Na is returned to the boiler by an electromagnetic pump, a wick, or other appropriate means. The required flow rate is determined by the current of the module and

Figure 2: AMTEC Vapor-Vapor module with three series connected cells.

the number of cells. A current of 1 A results in an Na flow rate of $0.86N_c$ g/hour where N_c is the number of cells.

Several complications arise in this cell that are not present in the standard liquid Na anode cell. Some, mentioned above, include the need for a large number of BASE to metal seals and the assurance that Na will not condense on the anode. In addition, the series connecting of the cells can result in a large potential difference across a small piece of BASE, and the large number of seals results in several locations for electrical shunt currents to develop. These complications can be controlled.

A large potential across the BASE could occur from the highest potential cell in a stack to the support structure which is probably at ground. The design must prevent this potential difference from becoming large enough to cause electrical breakdown of the BASE. One way to avoid this problem is to set the potential of the cells at the end of the stack as the lowest potential and then series connect the cells toward the center. The two parallel stacks starting from the ends would terminate with the same potential at the center cell. No piece of BASE would then sustain a potential difference larger than the potential of one cell.

A shunt current can develop in an AMTEC cell at the BASE to metal seal. In liquid-vapor AMTEC experiments, the Na shunt current has been estimated as large as 100 mA. For a single cell the shunt current will not greatly affect the cell performance. With the vapor-vapor concept, however, the large number of seals with a shunt current at each one could result in a total current that could significantly reduce the performance. Figure 3 schematically describes the possible shunt currents at an interconnect sealed to the BASE. The shunt current transfers Na across the BASE without producing useful electrical power. One shunt current can develop with Na oxidation at an anode (site a) and a corresponding Na+ reduction at the BASE / interconnect / low activity Na vapor three phase region (site d). Similarly, Na oxidation could occur at the BASE / interconnect / high activity Na vapor three phase region (site b) with corresponding Na+ reduction at the cathode (site c).

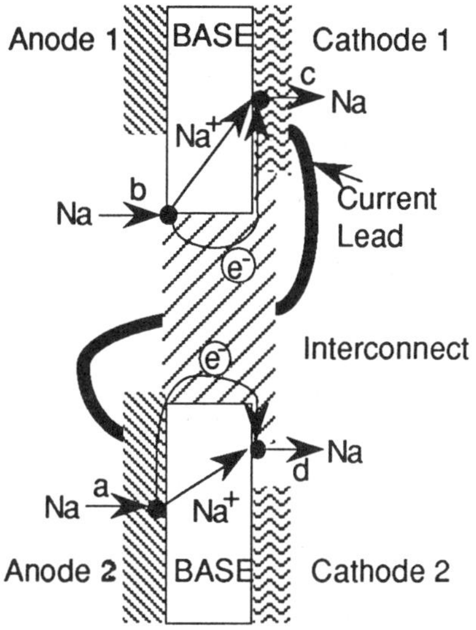

Figure 3: Possible shunt currents at an interconnect between two adjacent cells.

The anode and cathode reactions occur at high rates in the presence of Na vapor. Thus, the magnitudes of the shunt currents are limited by the rates of the reactions at b and d. In order to minimize these reactions, the three phase reaction area at b and d should be covered with a non-porous insulator such as sputtered α-alumina, AlN, or similar material. This covering would increase the vapor-phase mass transport impedance to or from the reaction site to a large value and reduce the resulting shunt current to a small value.

These complications in the vapor-vapor concept can be controlled by proper design and assembly. The resulting module includes several advantages over previous liquid-vapor designs. Primarily, the module output voltage is several times the voltage of a single cell, and the corresponding current is several times smaller than a cell with the same electrode area. Thus, conventional power conditioning electronics can be used without large efficiency losses associated with stepping from a low voltage. In principle, the output can be tailored to the requirements of the user and power conditioning electronics may be eliminated. Secondary advantages of this design include the elimination of the need for high temperature electrical feedthroughs and of Na refluxing over the BASE. Feedthroughs will still be required and high temperature operation would improve efficiency. However, the cell to cell feedthroughs of the vapor-vapor design are inherently operated at high temperature. Thus, the last or final feedthrough from the module to the external electrical circuitry need not operate at high temperature. Additionally, since the total current for the module is small, the size and mass of the conductor can be reduced. This design does not preclude the series connecting of several modules. Instead, since the Na boiler is electrically isolated from the electrodes, series connecting the cells can be accomplished without concern for the potential of the boiler.

Finally, the BASE in the proposed cell is never exposed to liquid Na and no refluxing of the Na takes place. This eliminates the concern that refluxing Na might degrade the BASE by extracting oxygen and depositing it on the walls of the boiler. We should note that BASE has been operated at temperatures as high as 1200 K in contact with liquid Na for many thousands of hours with no apparent degradation. This concern expressed here reflects the realities of thermodynamics and the relative stability of aluminum oxides compared with some other metal oxides.

VAPOR-VAPOR CELL PERFORMANCE PREDICTION

The performance of a vapor-vapor AMTEC cell can be confidently predicted based on the demonstrated performance of present liquid Na anode cells and a non-power producing vapor-vapor cell. The performance of experimental AMTEC cells using a Na liquid anode and a porous metal cathode has been described by an cathode performance model (Williams et al. 1990a and 1990b). The model describes all the major losses of the liquid-vapor AMTEC cell including ohmic losses in the BASE, kinetic losses at the BASE / cathode interface, and mass transport losses through the porous cathode. The model assumes that no mass transport or kinetic polarization occurs at the Na liquid anode / BASE interface. Experimental evidence supports this assumption and has determined typical values for the mass transport and kinetic parameters.

For a vapor-vapor cell, the anode polarization cannot be assumed to be zero. The mass flow impedance of Na from the vapor through the porous anode to the anode / BASE interface can be calculated as a function of anode porosity, temperature, ambient Na pressure, and current density. The mean free path of Na atoms in the vapor at about 1 atm of pressure and 1200 K temperature will be a few tenths of a micrometer. Thus, the Na gas will pass through the porous electrode in viscous flow with the pressure drop proportional to the current density. A typical, thin porous electrode is about 1 μm thick with cylindrical pores about 0.2 μm in diameter. Even for very high currents of 2 A/cm^2, the Reynolds number at 1200 K is only 1.4x10^{-4}, indicating the flow is laminar. The pressure drop can be calculated from the Bernoulli equation with the assumption of an ideal gas. More simply, the delta pressure can be determined from the Poiseuille number which is equal to 32 for laminar flow (Bird et al. 1960 and Weast 1975). The pressure drop through the typical pore is about 400 Pa at 1 A/cm^2 and about 800 Pa at 2 A/cm^2. Since the Na inlet pressure is near the saturated vapor pressure of 1.2x10^5 Pa (at 1200 K), the pressure drop in the anode will have only a small affect on the cell performance. At 1 A/cm^2, the pressure drop corresponds to a depression of the operating voltage of less than 0.03% of the open circuit voltage. Since this mass flow resistance is so small, the cell could also be constructed with a thick, porous anode as structural support for thin BASE which would result in significantly smaller overall impedance for the cell.

An additional component of the anode polarization will result from kinetic losses at the anode / BASE interface. Extensive studies by Williams et al. (1990a and 1990b) of the cathode / BASE interface have demonstrated that at high Na pressures, the overpotential becomes vanishingly small. As the anode will always have high Na pressure, the kinetic loss at this interface will also be insignificant.

Experiments with a non-power producing, vapor-vapor cell have demonstrated the validity of applying the results from the cathode analysis to the performance of the anode. This cell, called the vapor exposure test cell (VETC), uses two electrodes on a section of BASE in a Na vapor environment. By polarizing one electrode relative to the other, the kinetics and mass transport characteristics of each electrode can be determined. The primary difference between the processes at the cathode and anode is the the direction of the Na flow. The current-voltage behavior of the cell as predicted by the cathode model is extended to the anode as well. Thus the applicability of the model to the anode of a vapor-vapor cell is demonstrated (Ryan et al. 1991a and 1990b).

As an extension of these experimental results, the performance of the vapor-vapor cell can be predicted. The cell potential is the equilibrium potential minus the current dependant voltage losses. These losses include cell ohmic losses and the overpotentials at each electrode. Thus

$$E_{cell} = E_{oc} - jr_\Omega - \eta_a - (-\eta_c) \tag{1}$$

where
E_{cell} = Cell Potential, V
E_{oc} = Open Circuit Potential, V
j = Cell Current, A/m^2
r_Ω = Ohmic Resistance, Ω-m^2
η_a = Overpotential at the Anode, V
η_c = Overpotential at the Cathode, V (as defined $\eta_c < 0$)

The overpotentials are determined from the current-overpotential equation with the assumption of no concentration polarization in the BASE:

$$j_b = j_b^o\left[e^{-\alpha f\eta_b} - \left(1 + \frac{K_b j_b}{P_b}\right)e^{(1-\alpha)f\eta_b}\right] \tag{2}$$

where
j_b = Electrode Current, b = a for anode or c for cathode, A/m^2
j_b^o = Exchange Current Density (b = a or c), A/m^2
α = Transfer Coefficient
f = F/RT$_2$, V^{-1}
F = 96485 C/mol
R = 8.314 J/(mol K)

1334

T_2 = Cell Temperature, K
K_b = Mass Transfer Loss Coefficient (b = a or c), Pa m²/A
P_b = Na Pressure with no Current Flow (b = a or c), Pa

At the cathode (b = c), $j_c = j \geq 0$, and $\eta_c \leq 0$. This overpotential has been described in detail (Williams et al. 1990a and 1990b). Under most AMTEC operating conditions, the kinetic parameters (j_c^o and α) and transport parameter (K_c) are important in determining the overpotential. At the anode (b = a), $j_a = -j \leq 0$, and $\eta_a \geq 0$. The overpotential at the anode is small in magnitude relative to the overpotential at the cathode for the reasons described above. In terms of the parameters in Equation 2, the mass transport loss, $K_b j_b / P_b$, is much smaller for the anode than the cathode, and the exchange current density for the anode, j_a^o, is much larger than for the cathode, j_c^o. These differences are described below.

The mass transport, or flow loss, is the ratio of the pressure drop due to Na flow to the equilibrium pressure. The product of the mass transport parameter and the current ($K_b j_b$) is the pressure drop due to Na flow through the electrode and away from (or toward) its surface. This pressure drop can be calculated from the properties of Na vapor and the electrode geometry. As discussed above for the anode, the pressure drop can be determined from the Poiseuille number:

$$P_n = 32 = (P_a - P_i)\left(\frac{4\pi r^4 N F \rho}{\mu(-j_a) M L}\right) \text{ with } -K_a j_a = (P_a - P_i) \tag{3}$$

where
P_n = Poiseuille number
P_i = Na Pressure at the reaction interface, Pa
r = Electrode Pore Radius, m
N = Number density of pores, m⁻²
ρ = Na density, ~ 27 kg/m³ at 1200 K
μ = Na Viscosity, ~ 2x10⁻⁵ kg/(m-s) at 1200 K
M = Na Molecular Weight, 0.02299 kg/mol
L = Pore Length, m

The number density of pores is estimated by assuming the pores are arranged in a square grid with pore centers 3r apart.

For the cathode, gas phase Na transport is by Knudsen flow. The pressure drop has been shown to follow the form:

$$P_j + P_\Delta = K_c j_c = \left(1 + \frac{3G}{8\pi}\right)\sqrt{2\pi M R T_2}\,\frac{j_c}{F} \tag{4}$$

where
P_j = Pressure Drop due to flow away from the Cathode, Pa
P_Δ = Pressure Drop due to flow through the Cathode, Pa
G = Morphological Parameter

The morphological parameter, G, has been described before and typically has a value near 50.

The kinetic parameters have been determined experimentally to be independent of cathode composition, within experimental error. The transfer coefficient, α, is 0.5 also within experimental error. For the vapor-vapor cell, α is set to 0.5 for both electrodes. The exchange current density for the cathode, j_c^o, has been determined by Williams et al. (1990a and 1990b) to be proportional to the collision rate of Na at the interface at unit activity. Thus we define the temperature independent exchange current:

$$B = j_2^o\frac{\sqrt{T_2}}{P_2} \approx 120\,\frac{A\text{-}K^{1/2}}{Pa\text{-}m^2} \tag{5}$$

where
B = Temperature Independent Exchange Current, A-K¹ᐟ²/Pa-m²
P_2 = Saturated Na Vapor Pressure at T_2, Pa

The exchange current densities at the cathode and, by extension, at the anode are calculated from

$$j_2^o = j_c^o\sqrt{P_2 / P_c} \text{ and } j_2^o = j_a^o\sqrt{P_2 / P_a} \tag{6}$$

Because P_a is much larger than P_c, j_a^o is much greater than j_c^o.

The equilibrium pressures at the electrodes, P_a and P_c, are established by T_2 and the Na source temperatures, or they can be evaluated from the open circuit voltage.

For $\alpha = 0.5$, Equation 2 can be solved for the overpotential as a function of current. The solution can be used to determine the overpotentials once the magnitudes of the parameters have been determined. The overpotential as a function of current is:

Table 1: Parameters for Predicting the Performance of a Vapor-Vapor Cell

Cell Parameters		Electrode Parameters		
			Anode	Cathode
Electrode Temperature	1200 K			
Na Boiler Temperature	1170 K		$P_n = 32$	G = 50
Condenser Temperature	500 K	Transfer Coefficient, α	0.5	0.5
BASE Thickness	0.12 cm	Temperature Independent Exchange Current, B	120 A-K$^{1/2}$/Pa-m^2	120 A-K$^{1/2}$/Pa-m^2
Non-BASE Ohmic Resistances	0.1 Ω-cm^2			

$$\eta_b = \frac{2}{f} \ln \left[\frac{-\dfrac{j_b}{j_b^o} + \sqrt{\left(\dfrac{j_b}{j_b^o}\right)^2 + \left(1 + \dfrac{K_b\, j_b}{P_b}\right)}}{2\left(1 + \dfrac{K_b\, j_b}{P_b}\right)} \right] \qquad (7)$$

These equations, then allow the calculation of the vapor-vapor cell performance. Figure 4 is the calculated current-voltage curve under the cell for the conditions summarized in Table 1. The overpotentials and ohmic loss are also plotted as a function of cell current. The peak power of the cell is 0.46 W/cm^2 at 1 A/cm^2. At this point, the overpotential at the anode is only 3.7 mV while the overpotential at the cathode is 700 mV. This demonstrates that the anode does not create a significant impedance for the cell.

The performance of a module is determined as the series connected performance of several cells. Thus a series of 10 identical cells that each have 10 cm^2 of cathode area would produce a current of 10 A at a voltage of 4.6 V. A single liquid fed cell with a 100 cm^2 cathode would produce about 100 A at 0.46 V under similar conditions.

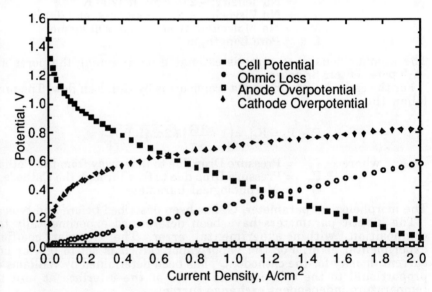

Figure 4: Current-Voltage relationship for the vapor-vapor cell showing the losses due to anode and cathode overpotential as well as the ohmic loss.

CONCLUSION

A new concept has been identified for an AMTEC vapor-vapor series connected cell. The concept uses a porous metal anode with Na delivered as a vapor. This delivery mode allows internal series connecting of several cells with a single Na boiler. The resulting module would produce about the same power density as a liquid Na anode cell, but the power would be at a higher voltage and a lower current. This results in several advantages including: fewer non-power producing parts, easier power conditioning for a single module, and smaller current busses. In addition no Na will reflux over the BASE, and the need for a high temperature feedthrough to achieve high efficiency is reduced. Some complications also arise for this design. The module is complicated by the need to supply additional heat to the electrodes to prevent Na condensation and the use of a large number of BASE to metal seals. Also the shunt currents at these seals need to be controlled, and the design requires that the potential across the BASE at the connection to the structure be held within limits.

The performance of the vapor-vapor cell is calculated from parameters determined from similar experiments. The performance is expected to be about 0.46 W/cm² of cathode area per cell. Thus a module with 10 cells, each with 10 cm², could produce 46 W of power at 4.6 V.

Acknowledgements

The research described in this paper was carried out by the Jet Propulsion Laboratory, California Institute of Technology, and was sponsored by the Energy Conversion and Utilization Technologies Program/Department of Energy, the Air Force Phillips Laboratory, and the National Aeronautics and Space Administration.

References

Bankston, C.P., Cole, T., Khanna, S.K., and Thakoor, A.P. (1985) "Alkali Metal Thermoelectric Conversion (AMTEC) for Space Nuclear Power Systems," *Space Nuclear Power Systems 1984*, Orbit Book Co., Malabar, FL, pp. 393-402.

Bird, R. B., Stewart, W. E., and Lightfoot, E. N. (1960) *Transport Phenomena*, John Wiley and Sons, pp. 460-468.

Cole, T. (1983) "Thermoelectric Energy Conversion with Solid Electrolytes," *Science*, **221**, pp. 915-920.

Ryan, M. A., Jeffries-Nakamura, B., Williams, R. M., Underwood, M. L., O'Connor, D. and Kikkert, S. (1991a) "Directly Deposited Current Collecting Grids for AMTEC Electrodes," in *Proc of the 26th Intersociety Energy Conversion Engineering Conference, Vol. 5*, American Nuclear Society, Boston, MA, August 3-10, 1991, pp. 463–468.

Ryan, M. A., Williams, R. M., Jeffries-Nakamura, B., Underwood, M. L., and O'Connor, D. (1991b) "Vapor Exposure Test Cell for Thin Film Electrodes on Solid Electrolytes," JPL New Technology Report 18620-8166.

Sievers, R. K. and Bankston, C.P. (1988) "Radioisotope Powered Alkali Metal Thermoelectric Converter Design for Space Systems," *Proceedings of the 23rd Intersociety Energy Conversion Engineering Conference*, Vol. 3, pp. 159-167.

Weast, R. C. Editor (1975) *Handbook of Chemistry and Physics*, CRC Press, Cleveland OH, p. F325.

Weber, N. Rasmussen, J. R., Harkins, G., and Olsen, S. L. (1988) "Design and Performance of a Small Circulating Sodium Heat Engine," *Proceedings of the 23rd Intersociety Energy Conversion Engineering Conference*, The American Society of Mechanical Engineers, pp. 215-217.

Weber, N. (1974) "A Thermoelectric Device Based on Beta Alumina Solid Electrolyte," *Energy Conversion*, **14**, pp. 1-8.

Williams, R. M., Loveland, M. E., Jeffries-Nakamura, B., Underwood, M. L., Bankston, C. P., Leduc, H. and Kummer, J. T. (1990a) "Kinetics and Transport at AMTEC Electrodes, I. The Interfacial Impedance Model," *J. Electrochem Soc.* **137**, pp. 1709-1716.

Williams, R. M., Jeffries-Nakamura, B., Underwood, M. L., Bankston, C. P., and Kummer, J. T. (1990b) "Kinetics and Transport at AMTEC Electrodes, II. Temperature Dependance of the Interfacial Impedance of Na$_{(g)}$ / Porous Mo / Na-ß" Alumina," *J. Electrochem Soc.* **137**, pp. 1716-1723.

THERMOACOUSTIC POWER CONVERSION
FOR SPACE POWER APPLICATIONS

William C. Ward and Michael A. Merrigan
Los Alamos National Laboratory
MEE-13, MS J576
Los Alamos, NM 87545
(505)667-6466

Abstract

Thermoacoustic engines are a recent class of devices that convert between heat and sound energy without moving parts. When coupled to a suitable transducer, thermoacoustic prime movers can produce electric power with high reliability and efficiency in lightweight packages that feature low vibration levels. This paper begins with an introduction to thermoacoustics and an overview of design and optimization and then considers analytical results for two transducers that also have no moving parts: piezoelectric transduction for a helium-based engine, and magnetohydrodynamic (MHD) transduction in an engine with liquid sodium working fluid.

BACKGROUND

The origins thermoacoustic engines go back over a hundred years and include the Sondhauss tube, Taconis oscillations, Rijke tubes, and acoustic attenuation in ducts. The foundation for the modern thermoacoustic theory was laid only 20 years ago by Rott and colleagues (Rott, 1969). Wheatley and colleagues (Wheatley et al., 1983) created the first efficient thermoacoustic engines, as we now know them, at the Los Alamos National Laboratory (LANL) less than a decade ago. Swift (Swift, 1988) later derived and compiled an authoritative reference on thermoacoustic engines that is a comprehensive guide to the fundamentals of the field. A dozen or so working thermoacoustic engines have been built in several laboratories, and commercial applications are now beginning to come to light.

Thermoacoustic Principles

We can achieve an informal, qualitative understanding of why thermoacoustic engines work with only a small effort; for a thorough explanation of this (and most points mentioned in this paper), the reader is referred to Swift's review and tutorial article (Swift, 1988).

In a resonator containing an inviscid fluid supporting a standing wave of frequency ω, the sound waves are adiabatic and undergo temperature oscillations of amplitude T_1 that are in phase with the pressure oscillations. For an ideal gas, $T_1 = T_m \beta p_1 / \rho_m c_p$, where T_m is the (absolute) mean temperature, $\beta = 1/T_m$ is the thermal expansion coefficient, p_1 is the acoustic pressure amplitude, ρ_m is the mean density of the fluid, and c_p is its isobaric specific heat. The acoustic displacement amplitude at a point, which is also in phase with the temperature oscillations for a standing wave, is $x_1 = u_1/\omega$, where $u_1 = p_1/\rho_m a$ is the acoustic velocity amplitude and a is the speed of sound in the fluid. If a short plate parallel to the acoustic motion is now introduced into the resonator, the acoustic temperature swings of the fluid very near the plate will be affected by the plate temperature. The thermal penetration depth, $\delta_\kappa = \sqrt{2K/\rho_m c_p \omega}$, is approximately how far heat can diffuse through the fluid during an acoustic period (K is the thermal conductivity of the fluid). If the points at $x = x_1$ and $x = -x_1$ on the plate are at temperatures $T_h = T_m + \nabla T x_1$ and $T_c = T_m - \nabla T x_1$, the temperature gradient ∇T along the plate can be compared with the temperature change $2T_1$ that a gas parcel undergoes over its peak-to-peak displacement, $2x_1$. If the gas parcel is at the hot end of the plate

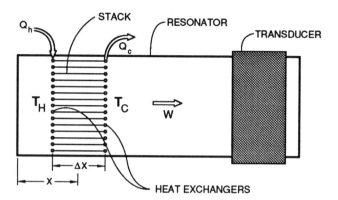

Figure 1. Schematic representation of an acoustic engine (shown with the essential parts required for a complete working system).

when the acoustic pressure is a maximum and $T_m + T_1 < T_h$, the parcel will absorb heat from the plate and expand. If the same particle reaches the cold end of the plate during the acoustic pressure minimum, and $T_m - T_1 > T_c$, the gas will deposit heat in the plate, and contract. Over the cycle, the gas absorbs heat from the plate in the form of acoustic work, and the engine acts as a prime mover. The presence of the plate causes a phase change between the pressure and the temperature oscillations in the wave, so that the sound wave is no longer adiabatic, and work is done on the gas parcels. This work takes the form of a traveling wave superimposed on the standing wave field; in the steady state, the standing wave amplitude is constant, while the travelling wave component is absorbed in resonator losses and in the transducer.

If we changed the temperatures (or moved the plate) so that $T_m + T_1 > T_h$ and $T_m - T_1 < T_c$, the plate would be absorbing acoustic work and cooling the cold end of the plate, acting as a heat pump. The critical mean-temperature gradient is given by

$$\nabla T_{\text{crit}} = T_m \beta \omega P_1 / \rho_m c_p u_1, \tag{1}$$

which is the ratio T_1/x_1. This defines the transition point between prime mover and heat pump behavior. The ratio of the operating mean-temperature gradient to the critical temperature gradient is a useful quantity:

$$\Gamma = \nabla T / \nabla T_{\text{crit}}. \tag{2}$$

For $\Gamma - 1 > 0$, the engine is a prime mover producing work; if $\Gamma - 1 < 0$, it is a heat pump.

Practical thermoacoustic engines use a stack of parallel plates (or a group of channels) that are spaced such that all of the cross-sectional area of the stack is effective for heat flux; the inter-plate spacing may range from 1.5–4 δ_κ. The stack may have a length of many acoustic displacements, and its temperature span is fixed by heat exchangers at either end. A prime mover stack will always be within that $\lambda/8$ of a pressure antinode; the usual spacing is less than $\lambda/16$. The simplest form of a thermoacoustic prime mover is illustrated in Figure 1.

The efficiency of the acoustic engine is given by $\eta = \dot{W}/\dot{Q}_H$, where $\dot{W}$ is the acoustic work output and $\dot{Q}_H$ is the heat input through the hot heat exchanger. Since a thermoacoustic engine's efficiency is substantially proportional to Carnot's efficiency, $\eta_C = 1 - T_C/T_H$, the normalized efficiency η/η_C is a more general indicator of efficiency.

DESIGN AND OPTIMIZATION

Several numerical design tools have been developed at LANL to evaluate thermoacoustic engines. Two codes based on the short-stack approximation give quick preliminary information. A more accurate, but more cumbersome, program is based on numerical integration of the thermoacoustic wave equation as given in the appendix of Swift's tutorial (Swift, 1988). Miscellaneous programs for calculating resonator losses or fluid and material properties may also be used.

The short-stack model given by Swift (Swift, 1988) predicts performance with factor-of-two accuracy, but it is very useful for initial investigations of the broad design parameter space since it offers intuitive insight and quick results. This model assumes that the stack is much shorter than a wavelength, that the temperature span is much smaller than T_m, and that the plate spacing is larger than the penetration depths. The short-stack method is simple enough to allow contour plots of the complete parameter space to be generated on any desktop computer. In the next subsections, we discuss the trends identifiable with the short-stack method and the effect of parameters that may be manipulated in the overall design.

Temperature Span

Because the efficiency goes as Carnot's efficiency, it is always desirable to choose the highest practical hot temperature for thermoacoustic engines. In demanding applications, this may justify the use of advanced refractory materials to maintain strength in the hot sections. Here the thermoacoustic engine has advantages over many heat engines, since there are no moving parts or seals to complicate the high-temperature design, and only the section from the stack center to the closed end (about $\lambda/16$) is at the hot temperature.

Working Fluid and Frequency

To date, thermoacoustic engines have been built with air, helium, argon helium/argon or helium/xenon mixtures, and liquid sodium as working fluids. Binary gas mixtures, such as helium with a small fraction of argon, have a lower Prandtl number (the ratio of viscous to thermal effects) than either pure gas and, hence, have the potential for better efficiency (Susalla, 1988). Liquids can work well as thermoacoustic fluids when $T_m\beta$ is near one, usually approaching their critical point. Mixtures of liquid metals can also be used, but the resulting changes in transport properties are generally undesirable. Swift and colleagues (Swift et al., 1985) built a working liquid sodium engine to test the conversion of acoustic power to electricity by (MHD) transduction.

The choice of working fluid and operating frequency are interrelated, since the frequency and sound speed determine the resonant length for the engine. Also, the conductive and viscous properties influence the maximum frequency for an engine because the stack and heat exchangers become too small to build and the efficiency is reduced at higher frequencies. The choice will also be influenced by the AC operating frequency in the target application and by vibration considerations.

Mean Pressure

The maximum achievable acoustic pressure amplitude is proportional to the mean pressure amplitude, so high mean pressures are important for high volumetric specific power. In gases, interestingly, the specific power per-unit-mass is not dependent on mean pressure because the output power is linear with the mean pressure, and the increased weight for the resonator casing cancels increased power in the engine. In liquids, output power goes as the square of mean pressure, and high mean pressure is important for both measures of specific power.

Stack Design and Placement

The stack is the heart of the thermoacoustic engine, and its design has the most impact on its performance. The plates must have adequate strength and heat capacity, yet they must be thin enough not to disturb or block the flow excessively. If the thermal conductivity of the stack is too high, longitudinal conduction along it will reduce the efficiency significantly. There is also an optimum spacing between plates for each design. Ideally, since this spacing is proportional to the penetration depth, and the penetration depth grows with increased temperature, the spacing should taper from the hot to the cold ends of the stack. Stacks have been made from plates, spiraled sheet material, honeycomb, and blocks with parallel square channels.

More important than the construction and geometry of the stack are its length and placement in the standing wave. The length of the stack changes the temperature gradient, and the position of its center changes the *critical* temperature gradient. Choosing an operating point involves calculating a stack length from Γ and ΔT and a center position from the x/λ value. When the accurate code is used for further tuning, most of the iterations involve changes in stack length and position.

Resonator Design

There are several enhancements that the designer can make to the resonator in order to improve the performance of an engine, both in efficiency and specific power. In many cases, heat exchangers and a thermoacoustic stack can be placed at both ends of the resonator and the specific power can be nearly doubled. The total surface area of the resonator is also reduced, resulting in a smaller internal surface area and a higher effective Q. Whereas the diameter of the stack section is fixed by the required power output, the size and internal surface area can be reduced further by using a shorter, smaller diameter center section. Nonlinear effects prevent the center 'neck-down' from being too small because of the high acoustic velocities involved. If a central spring-mass transducer, such as a linear alternator, is to be used, it can be designed to completely replace the center section, since this tubing serves only to provide the proper phasing of reflected waves. Resonator surface area is not the whole story, however; the large plate surface area in the heat exchangers (that is not active thermoacoustically) is a dominant part of resonator losses, and this must be minimized without degrading heat exchanger performance.

Radiator Temperature

In any space energy conversion application, some Carnot efficiency must be sacrificed by raising the heat rejection temperature so that radiator mass may be reduced. Thermoacoustic engines offer a wide latitude in minimizing total system mass since they can work at high temperatures and across small temperature differences. Unfortunately, only MHD transducers and liquid metal engines can be used at higher temperatures, since transducers will depolarize. Linear alternators (not considered here) can be used for high-temperature gas-based engines with low mass and high efficiency, but reduced reliability.

Vibration

Because the thermoacoustic engine has no moving parts, its vibration levels can be quite low. The only vibratory force present is the acoustic pressure amplitude acting over the surface area of the end of the resonator. Several vibration-canceling designs are possible to eliminate this force; for example, a full-wave resonator has equal opposing acoustic pressures on each end of the resonator at all times. A thin half-wave resonator may be coiled so that its ends meet, or several synchronous, out-of-phase resonators can be rigidly coupled to achieve the same effect. Levels of 0.05 g (rms) at twice the operating frequency of the engine are typical of what can be expected for gas working fluids. Vibrations are an order of magnitude higher (because of the increased acoustic amplitude) in liquid metal engines.

RESULTS

Trends

The design of a thermoacoustic engine usually begins with short-stack calculations to find a starting point for calculations with the full wave equation code. Iterations are performed until the best stack diameter, plate spacing, stack position, and stack length are found for a given set of temperatures, output power, and performance criteria. Like all heat engines, efficiency *vs.* specific power is the primary tradeoff. A qualitative summary of the effects of each design option is given in Table 1.

Table 1. Trade-offs for Gas- and Liquid Metal-based Thermoacoustic Engines.

Increasing Parameter	Efficiency		Power/mass		Power/volume	
	Gas	Liquid	Gas	Liquid	Gas	Liquid
Frequency	−	−	+	+	+	+
Diameter	+	0	−	−	−	−
Doping	+	-	−	−	+	+
Part Load	0	-	na	na	na	na
Resonator Neck-down[1]	+/−	+/−	+	+	+	+
Mean Pressure[2]	0	+	0	+	+	++
Temperature Span[2]	+	+	+	+	+	+

+ =increase   − =decrease   0 =No change   na=Not applicable

[1] Efficiency improves for resonators with moderate neck-down, but it worsens as the secondary resonator diameter is further reduced.

[2] Within materials limits.

Performance of Specific Designs

Two finished designs of a 1-kW helium thermoacoustic engine (stack and resonator, no transducers) are described in Table 2, and a 1-kW liquid sodium engine is given in Table 3. The stack cross-sectional area, Carnot efficiency (η_C), η/η_C, the efficiency reduction (from resonator losses), total efficiency (heat-to-acoustic), mass, and volume are listed. The operating temperatures are $T_H = 1100K$ and $T_C = 330K$. A comparison of the data in Table 2 shows that it is possible to design an engine of the same power with only half the cross-sectional area by choosing a more aggressive stack position and accepting some loss in efficiency.

Table 2. Two 1-kW, 500-Hz, 70-bar Helium Engine Designs.

(cm²) Area	Carnot Efficiency	Thermoacoustic Efficiency	Efficiency Reduction	Overall Efficiency	kg Mass	cm³ Vol
13.6	70%	41%	0.74	21%	1.4	590
6.8	70%	29%	0.74	15%	0.59	330

Transducers

Two options for high reliability transducers with no moving parts have been considered. For the helium engine, piezoelectric elements can be placed in areas of the engine with high pressure amplitudes such that

Table 3. A 1-kW, 1000-Hz, 140-bar Liquid Sodium Engine Design .

(cm^2) Area	Carnot Efficiency	Thermoacoustic Efficiency	Efficiency Reduction	Overall Efficiency	kg Mass	cm^3 Vol
11.4	70%	29%	0.80	16.2%	6.1	277

they will be stressed by the acoustic wave. A lead zirconate titanate (PZT) design could be expected to have a transduction efficiency of 94% considering dielectric and hysteretic losses. Such a transducer would weigh about 8 kg for a 1-kW engine, 2 kg of which is an inductor to offset the capacitance of the PZT at the resonant frequency.

Preliminary designs for an MHD transducer for the liquid sodium engine show a conversion efficiency of 85% and an additional transducer mass of 5 kg at a power of 1 kW. This mass includes internal permanent magnets, copper conductors, and transformers to convert the low-voltage, high-current output into more useful power.

Acknowledgment

Conclusions reached in this paper are drawn from work supported by DOE under the Energy Utilization and Conversion Technologies program.

References

Rott, N. (1969). Damped and thermally driven oscillations in wide and narrow tubes. *Z. Angew. Math. Phys.*, 20:230.

Susalla, M. P. (1988). Thermodynamic improvements for the space thermoacoustic refrigerator (STAR). Master's thesis, Naval Postgraduate School, Monterey, California.

Swift, G. W. (1988). Thermoacoustic engines. *J. Acous. Soc. Am.*, 84:1145–1178.

Swift, G. W., Migliori, A., Hofler, T., and Wheatley, J. C. (1985). Theory and calculations for an intrinsically irreversible acoustic prime mover using liquid sodium as primary working fluid. *J. Acous. Soc. Am.*, 78:767.

Wheatley, J. C., Hofler, T., Swift, G. W., and Migliori, A. (1983). An intrinsically irreversible thermoacoustic heat engine. *J. Acous. Soc. Am.*, 74:153.

PREDICTION OF THE START-UP CHARACTERISTICS
OF THERMIONIC CONVERTER IN A STAR-C REACTOR

David P. Lieb, Carl A. Witt, Gabor Miskolczy, and Celia C.M. Lee
Thermo Electron Technologies Corporation
85 First Avenue
Waltham, MA 02254-9046
617-622-1394

John McVey
Rasor Associates, Inc.
253 Humboldt Court
Sunnyvale, CA 94080
408-734-1622

Abstract

The design for a Space Thermionic Advanced Reactor-Compact (STAR-C) power system with a baseline power of 40 kW(e) consisted of 1230 parallel planar thermionic converters surrounding a space reactor system. The converters were similar to the Solar Energy Thermionic (SET) converters. The collectors were coupled to sodium-filled heat pipes which rejected heat to heat pipe radiators. A cesium intercalated graphite reservoir in each converter supplied cesium vapor. A computer thermal model was used to predict the start-up characteristics of a converter in the STAR-C system. During start-up, the reactor heat was radiated to the emitter. Heat was radiated across the cesium gap to the collector and conducted to the cesium-graphite reservoir located in the niobium of the collector heat pipe. Waste heat was removed by the heat pipe to the radiators. A transient, finite-element computer-model of the thermionic converter was developed to simulate the behavior of the STAR-C converter. The subject of this paper is the use of a computer thermal model Thermal Analysis Code-2 Dimensional, TAC-2D to predict the start-up characteristics of a SET type converter with a cesium-graphite reservoir in the collector heat pipe. When the reactor is started, electron cooling of the emitter will not occur until suffficient cesium vapor is introduced into the interelectrode gap. A transient finite element model of the thermionic converter, fuel, and the cesium-graphite reservoir simulated the operating conditions. The model utilized special boundary conditions at the collector to simulate the behavior of a heat pipe. The heat loss from the radiator is stimulated by heat transfer proportional to the fourth power of the temperature. The start-up time of the TFE is limited by the availability of cesium pressure during heating. The model showed that the converter can be started up in less than 60 minutes without overheating the emitter. The calculation shows there is almost sufficient direct heating of the collector heat pipe system to warm the collector without requiring electron heating. The heat transfer characteristics of the heat pipe are partially responsible for this behavior since the cooling does not become active until the reservoir temperature is high enough to generate cesium pressure. Thus the thermal time constraints of the converter-reservoir system are well within the 60 minute requirements and will not be a limiting factor for rapid start-up of the system.

INTRODUCTION

The STAR-C Thermionic Space Reactor Power System with a baseline power of 40 kWe consists of a series of planar thermionic converters situated between a reactor system and the outer radiation panels (Figure 1). The core is based on the proven Romashka reactor, and the thermionic converters are based on the Solar Energy Thermionic (SET) converters (Brosens 1964). The system is designed for scalability from 10 kW(e) to 100 kW(e).

The design of the STAR-C thermionic converter cell incorporates the experience from past thermionic development and testing programs such as the Advanced Thermionic Technology Program (U.S. DOE 1983), the Thermionics Combustion Module (U.S. DOE 1985), SP-100 Thermionic Technology Program (U.S. DOE 1985), and the Thermionic Fuel Element Verification Program (Bohl et al. 1991). The thermionic configuration most relevant to the STAR-C converters is the Solar Energy Thermionic (SET) Program initiated at the Jet Propulsion Laboratory in 1961. The SET Program involved nine major design iterations in which five generators and over 135 thermionic

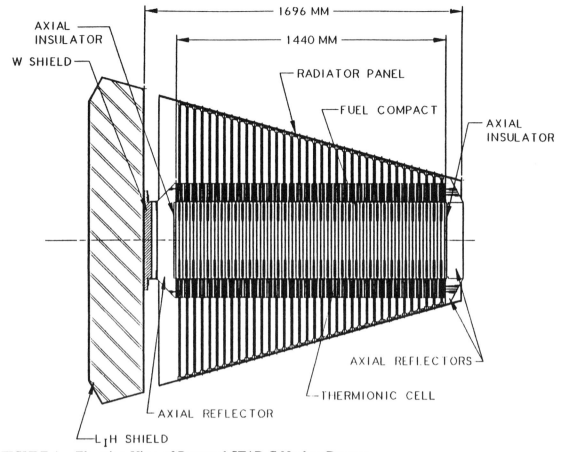

FIGURE 1. Elevation View of Proposed STAR-C Nuclear Reactor.

converters were built and evaluated. This is the only thermionic program in which planar out-of-core thermionic converters provided data on factors governing long-term performance changes and converter lifetimes.

Issues considered in the design of thermionic converters were efficiency, durability, lifetimes, material suitability, minimization of stresses, possible failure modes, and manufacturability. The current design consisted of improvements to the planar converters developed during the SET Program. Different materials for various converter components, such as the emitter sleeve, were investigated to optimize efficiency. Converter efficiency was also optimized by reducing external heat losses from the emitter. The thermionic converter computer program, TECMDL (Rasor Associates 1990), was used to predict converter performance of the current STAR-C cell design. Previously developed thermionic and thermal computer codes, such as EFFIC (Thermo Electron Technologies 1989), were used to evaluate optimal converter efficiencies for varying design parameters such as material selection and emitter sleeve design.

The thermionic cells of the proposed STAR-C Thermionic Space Power System are planar diodes surrounding the reactor. The cell consists of a tungsten heat collector/thermionic emitter, an emitter sleeve, a thermionic collector, a cesium reservoir, a collector heat pipe, multi-foil insulation, and the emitter and collector electrical leads. The components are shown in Figure 2. Heat from the reactor core surface radiates across a 3-mm gap to the tungsten heat collector/thermionic emitter. A 4.9-cm^2 circular section on the back of the heat collector forms the thermionic emitter surface. The gap between the emitter and collector surfaces is 0.1 mm. Cesium vapor is provided in the interelectrode gap by a cesium graphite reservoir located within the walls of the collector heat pipe. The cesium source consists of interlaminar cesium-carbon compounds of the form C_nCs, where the value of n is determined by the fabrication parameters and is predicted by the desired cesium pressure-graphite temperature behavior. This is the reference cesium source used in the Thermionic Fuel Element Verification Program (TFEVP) (Bohl et al. 1991), which successfully conducted irradiation tests in FFTF out to 3 x 10_{22} nvt (E > 0.1 MEV).

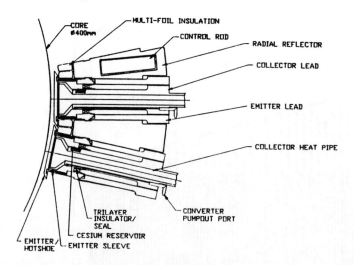

FIGURE 2. STAR-C Converter Assembly.

The thermionic collector forms the evaporator of a niobium-1% zirconium collector heat pipe. The heat pipe transports the waste heat of the thermionic process to the radiator panel for rejection to space. The heat pipe uses sodium as the working fluid and a niobium wick.

The emitter and collector electrical leads are thick annular beryllium sleeves that surround the niobium heat pipe. Beryllium was selected as the lead material because of its low electrical resistivity and high nuclear reflective properties. A trilayer insulator consisting of an insulating ceramic sandwiched between two layers of niobium electrically isolates the emitter and collector leads. A triple emitter sleeve connects the emitter to the beryllium emitter lead. The triple sleeve design was selected to allow a thicker sleeve and maintain converter efficiency by reducing external emitter heat losses. A triple sleeve design is a sleeve that folds back twice forming one continuous sleeve allowing for a longer conduction path and improving the converter efficiency. Heat loss is further reduced by using multi-foil insulation around the emitter sleeve.

The design point operating temperatures were 1860 K for the emitter and 1020 K for the collector. With an electrode spacing of 0.1 mm and a cesium pressure of 4.0 torr, the corresponding current density of the converter was 12.7 A/cm^2. The conceptual design was based on detailed two-dimensional heat transport calculations. Heat loss due to electron cooling was determined using a thermionics correlation program called EFFIC, (Thermo Electron Technologies 1989). Radiative heat loss between the emitter and collector was considered as one-dimensional radiative heat transfer between parallel plates as a function of local surface temperatures. Conductive heat transfer through the cesium vapor was modeled as a function of emitter and collector temperatures, cesium pressure, and characteristic geometry.

During start-up of the STAR-C Space Power System, heat from the reactor core radiated to the emitter hot shoe. The emitter begins heating the collector by radiation. Electron heating of the collector from thermionic conversion does not occur until cesium vapor is introduced into the interelectrode gap. The cesium graphite reservoir, located in the wall of the collector heat pipe must reach a temperature of 900 K as specified for the particular loading of the cesium-graphite reservoir in order to provide enough cesium vapor to ignite the converter. When ignition occurs, the emitter was cooled and the collector heated to the desired operating temperatures.

A finite difference model of one STAR-C converter was developed to investigate the converter behavior during start-up. One concern was that ignition of the converter would not occur soon enough, which would overheat the emitter and the fuel. Excessive temperatures for the emitter were consider be 500 K greater than the desired emitter operating temperature (2360 K) and fuel temperatures over 2700 K. This transient thermal model determined the temperatures of the converter components and fuel for different reactor start-up scenarios.

THERMAL MODEL SUMMARY

A detailed finite difference thermal model of a STAR-C thermionic converter was developed using the TAC-2D thermal analysis code (Boonstra 1976). The axisymmetric model represented the heat flow and temperature distribution from the core through one converter and part of the length of the collector heat pipe. The model included radiant heating of the emitter hot shoe from a block of fuel representing the core, radiation and current density dependent thermionic emission heating of the collector, thermionic emission cooling of the emitter, temperature-dependent graphite cesium reservoir pressure, sodium vapor-pressure dependent thermal transport through the sodium-filled heat pipe, and radiant heat rejection from the end of the collector heat pipe toward the radiators.

Several steady-state cases of the thermal model with increasing fuel power and no simulated electron cooling were run until the cesium graphite reservoir reached a temperature (900 K) previously determined to produce sufficient cesium for ignition of the converter. Converter ignition was then simulated by the addition of electron cooling/heating proportioned to the current density. Fuel power was further increased until full power and desired operating temperatures were obtained. The TAC-2D model showed that it is possible to start-up from cold to operating conditions in about one hour. A start-up of 15 minutes caused excessive peak fuel temperatures (2750 K) and was not recommended.

Description of Thermal Model

Figure 3 shows the axisymmetric thermal model for one STAR-C thermionic converter. The heat was input from the fuel block representing the reactor core. The complex graphite/fuel core was simulated by inserting a thermally resistive layer between the heat source (fuel) and the radiative core surface. Neutronics analysis provided an estimated temperature drop between the hottest part of the fuel and the core surface at operating condition (Begg 1991). The thermal conductivity of the thermally resistive layer was adjusted to provide this same temperature gradient under the same conditions. Gamma heating of the converter components was not considered in this model. All heat was rejected at the radiator end of the simulated heat pipe by means of a coolant with the heat transfer coefficient following T^4 heat loss. The center line of the model, the core end of the model, and the outer radial surface were considered adiabatic. Each of these boundaries corresponded to a region of symmetry or near symmetry and was well approximated as adiabatic. The inputs to the model were geometry, material properties (emissivity and thermal conductivity, for example), heat sources and sinks, and heat transfer coefficients. The output of one quasi-steady state step of TAC-2D provided the time and temperature at each node. A heat balance was available at the completion of each run. This output allowed a temperature distribution at each time step to be obtained.

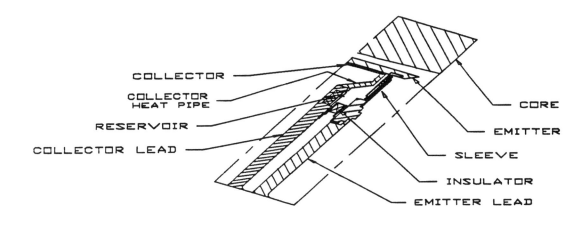

FIGURE 3. STAR-C Converter Thermal Model.

Heat generated in the reactor core heated the emitter, which in turn radiated to the collector surface across the interelectrode gap. Thermionic emission coupling was included in the interelectrode gap to represent the addition of cesium vapor just prior to ignition. This electron heating/cooling was added as positive and negative heating rates within the collector and emitter respectively. The heating cooling rates were proportional to the current density which could be adjusted interactively as the program advanced through time steps. The heat lost through thermionic conversion was approximated by the difference between the emitter cooling and collector heating. Heat was conducted from the emitter through the emitter sleeve out to the emitter lead. The collector heat pipe behavior was simulated with a near isothermal temperature and a high thermal conductivity modeling the thermal transport of the sodium phase changes in an actual heat pipe. This simulation assumed a local thermal conductivity in the heat pipe wall equal to that of niobium plus a term proportional to the sodium vapor pressure at the local temperature. Thus the conductivity of the heat pipe was equal to that of niobium up to about 800 K. The conductivity then increased exponentially with temperature to its final high value at operating temperature. The thermal conductivity at the operating collector temperature was assumed to be 1000 times that of pure niobium. The cesium graphite reservoir, located within the confining niobium of the collector heat pipe, was heated as the collector heat pipe was heated. Heat rejection of the heat pipe was modeled with a fourth power temperature dependence simulating the heat rejection of the radiators to ambient temperature.

A quasi-interactive calculating mode allowed for variable inputs such as fuel power and simulated electron cooling to be changed after each time step. In a typical calculation, the heat input was set with the entire model at 250 K. Fuel heating, simulated by internal heating in the fuel block of the model, was set at a fraction of its operating value and the system allowed to heat until the cesium graphite reservoir reached a temperature which could produce sufficient cesium pressure to support converter current. The converter was then ignited, and simulated electron cooling and heating of the emitter and collector was introduced. The emitter temperature was expected to be somewhat higher than normal emitter operating temperature until electron cooling occurred. The converter current (and thus the electron cooling) could be increased following TECMDL and cesium pressure curves. Heat input and radiator heat transfer coefficients were adjusted to produce the desired final emitter, collector, and reservoir temperatures. These values were then used in a single transient calculation which gave the start-up characteristics.

RESULTS

Figure 4 shows the temperature vs time plots for the converter components during a 15 minute start up. A fifteen minute start-up is one in which the cesium reservoir heats quickly enough to allow ignition within fifteen minutes. Ignition was not simulated in this example since the object was to examine the fuel and emitter temperatures at the

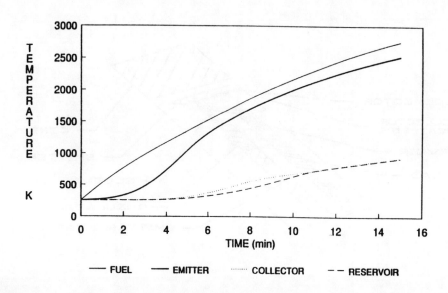

FIGURE 4. Temperature Results for a 15 Minute Start-up.

ignition point. The prescribed power profile for this start-up caused peak fuel temperatures of ~2750 K and emitter temperature of ~2500 K. The temperature vs. time plots show that cesium is not available for electron cooling to cool to the desired temperature levels. A 15 minute start-up could be accomplished with 2750 K maximum fuel temperature. The results of this case suggest that electron cooling must occur sooner in order to prevent emitter and fuel overheating. This requires that cesium vapor be introduced into the interelectrode gap earlier. The cesium graphite reservoir must be heated at a faster rate. Methods of increasing the heating rate of the reservoir are to move the reservoir into the emitter support flange, or to use an internal reservoir heater. An internal reservoir heater would be a block of material such as tungsten located in the cesium/graphite reservoir which would use gamma heating to increase the reservoir heating rate.

Figure 5 is a surface plot of various elements of the STAR-C converter model (shown in Figure 3) at 49 minutes for a 1 hour start up. The fuel heating rate was set at about 65% of its operating value for this initial heating period. Note that the emitter temperature is higher than the desired operating temperature of 1860 K. The initial fuel power allows the system to start up in about an hour while heating the emitter temperature to 2140 K. Higher initial power would heat the cesium reservoir to an ignitable temperature in a shorter time but would increase the peak emitter temperature. This surface plot represents the time before ignition, hence the emitter overheats to 2140 K with respect to the desired operating temperature. The collector and reservoir temperature is just under 900 K in Figure 5. At this point the current density was set at 3 A/cm². The current density was increased to its final value of 12.7 A/cm² over the next five minutes as the graphite reservoir heated and cesium pressure increased. The fuel power was increased to its final operating level at the time of full ignition.

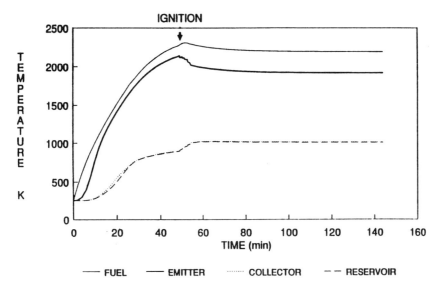

FIGURE 5. STAR-C Temperature Profile at 49 Minutes for a 1 Hour Start-up.

Figure 6 shows the temperature versus. time plots for the fuel, the emitter, collector, and the cesium graphite reservoir for the scenario described above. As expected, the collector and reservoir temperatures follow essentially the same profile. The emitter reaches a peak over temperature of 2140 K just before electron cooling is introduced to the model. At time of ignition, the collector and reservoir temperature is at 900 K, and the fuel temperature is at 2280 K. The fuel temperature decreases only slightly after ignition due to the increase to 100% power.

CONCLUSIONS

The TAC-2D thermal model of the STAR-C thermionic converter demonstrated that a 1 hour start-up is feasible, while a 15 minute start-up leads to overheating of the fuel and emitter. The start-up time is governed by the rate at which the graphite cesium reservoir comes up to temperature. At a fast enough rate, electron cooling was introduced soon enough to reduce the emitter and fuel temperatures before overheating occurred. In order to start-up faster than 1 hour, the cesium graphite reservoir must have a faster means of heating, such as designing the reservoir into the emitter support flange or using an internal reservoir heater.

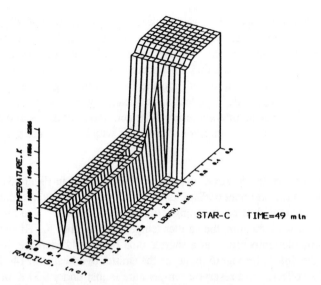

FIGURE 6. Temperature Results for a 1 Hour Start-up.

Acknowledgments

This work was supported by subcontract SC0142291 to General Atomics, the prime contractor. This work has been sponsored and monitored by the United States Air Force, Air Force Systems Command, Phillips Laboratory, Kirtland AFB, NM 87117-6008.

References

Begg, L. (1991)Personal Communication, General Atomics, San Diego, CA, March 1991.

Bohl, R. J., R. C. Dahlberg, D. S. Dutt, and J. T. Wood (1991) "Thermionic Fuel Element Verification Program - Overview," in *Proc. 8th Symposium on Space Nuclear Power Systems,* CONF-910116, M.S. El-Genk and M.D. Hoover, eds., American Institute of Physics, New York, 1991.

Boonstra, R. H., (1976) TAC-2D, A General Purpose Two-Dimensional Heat Transfer Computer Code User's Manual, GA-A14032, General Atomics, San Diego, CA.

Brosens, P. J. (1964) "Solar Energy Thermionic (SET) Converter Design History," in *Trans. High Temperature Conversion Heat to Electricity Symposium,* University of Arizona, Tucson, AZ.

Rasor Associates, Inc. (1990) TECMDL - Ignited Mode Planar Thermionic Converter Model Report, #E563 004-C-082988/NSR-31/90-0775, Rasor Associates, Inc., Sunnyvale, CA.

Thermo Electron Technologies, Corp. (1989) EFFIC Computer Code, Thermo Electron Technologies, Corp., Waltham, MA.

U.S. DOE (1983) Advanced Thermionic Energy Conversion R & D, DEAC02-82-CE40582, U. S. Department of Energy, Washington, D.C.

U.S. DOE (1984) SP-100 Thermionic Technology Program, DEAC03-84-3F12193, U. S. Department of Energy, Washington, D.C.

U.S. DOE (1985) Thermionics Combustion Module, DEAC02-82-CE40582, U. S. Department of Energy, Washington, D.C.

PROJECT THERMION: DEMONSTRATION OF A
THERMIONIC HEAT PIPE IN MICROGRAVITY

Frank J. Redd
Center for Space Engineering
Utah State University
Logan, UT 84322-4130
(801)750-2868

George E. Powell
Department of Electrical Engineering
Utah State University
Logan, UT 84322-4120
(801)750-3554

Abstract

The Idaho National Engineering Laboratory (INEL) is conducting intensive research in the design and development of a small, excore heat-pipe-thermionic space nuclear reactor power system (SEHPTR). Progress in this research effort has identified the need for an in-space flight demonstration of a thermionic heat pipe element. The proposed demonstration will examine the performance of such a device and verify its operation in microgravity. This paper focuses on the design of a microsatellite-based technology demonstration experiment to measure the effects of microgravity on the performance of an integrated thermionic heat pipe device in low earth orbit. Two scenarios, THERMION-I and THERMION-II, emerged from the design process. Selection between the two will depend upon yet undetermined experiment lifetime requirements. THERMION-I is designed for a long-lifetime (greater than one year) investigation of the operations of the thermionic heat pipe element in low earth orbit. Heat input to the element is furnished by a large mirror which collects solar energy and focuses it into a cavity containing the heat pipe device. THERMION-II is a much more simple design which is utilized for short-term (approximately one day) operation. This experiment remains attached to the Delta II second stage and utilizes energy from 253 kg of alkaline batteries to supply thermal energy to the heat pipe device.

INTRODUCTION

The Idaho National Engineering Laboratory (INEL) is conducting intensive research in the design and development of a small excore heat-pipe-thermionic space nuclear reactor power system (SEHPTR). The SEHPTR spacecraft will be able to supply 40 kW of power in any given orbit. The key components in this reactor are the thermionic heat pipes. The heat pipes have two major functions: first, to convert heat energy into electrical energy, and second, to radiate the excess heat to space (Idaho National Engineering Laboratory, 1990).

Very simply stated, thermionic power conversion is a process of converting heat energy into electrical energy with no moving parts. Heat is applied to the cathode surface, as shown in Figure 1. This heat boils off electrons that jump across the gap to the cooler surface of the anode. This causes a potential difference between the two plates and induces a current through the load.

Thermionic power conversion is incorporated into a liquid metal heat pipe in a manner depicted in the Thermacore, Inc. design which is shown in Figure 2 (Thermacore, Inc., 1991). This thermionic heat pipe element is actually a double heat pipe that combines a radial or emitter heat pipe with a linear or collector heat pipe. Thermal energy is transferred radially through the radial heat pipe to the emitter which operates at 1925 K. Lithium is the operating medium. Heat collected on the collector heat pipe is transferred axially from the thermionic device to the spacecraft radiator for radiation into deep space. This heat pipe operates at 1000 K using liquid sodium.

The objective of the THERMION project was to produce a mission design for an in-space demonstration of a thermionic heat pipe device similar to the one illustrated in Figure 2. The project actually produced two designs, THERMION-I and THERMION-II, which are substantially different. These differences derive from differing life-time requirements which, in turn, dictate different heat sources. These heat sources are alternatives to the heat supplied by a nuclear reactor in the actual system which is not available for the in-space demonstration. The details of the alternate designs are described in the sections that follow.

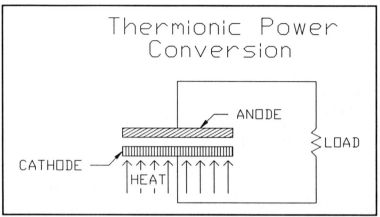

FIGURE 1. Thermionic Power Conversion Converts Heat Energy into Electrical Energy.

REQUIREMENTS

Size constraints dictated by the primary launch configuration permitted only the test of a scaled down version of Thermacore's 40 cm heat pipe. Thus, the collector for the THERMION design is 6 cm long. The 6 cm heat pipe requires 1350 W of thermal energy to operate and hold the emitter at 1950 K and the collector/radiator at 1000 K (Thermacore, 1991).

The demonstration is to be conducted using existing, flight proven components and subsystems as much as possible. Various launch vehicle options were considered, but cost constraints suggested flight as a secondary payload on a launch vehicle with a funded primary payload. After consideration of several "piggyback" opportunities, THERMION was designed to fly as a secondary payload on a McDonnell Douglas Delta II. This option imposed rather severe payload mass and volume constraints on the mission. These are detailed in sections to follow.

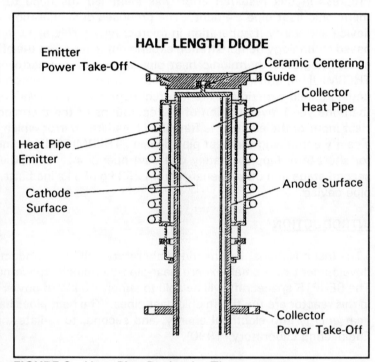

FIGURE 2. Heat Pipe Design by Thermacore.

It was difficult to pin down orbital lifetime requirements. The development of the heat pipe had not progressed far enough to include ground testing; hence, flight test duration requirements were not defined. A long-lifetime mission with continuous testing requires a constant input of thermal energy over the life of the mission. A shorter requirement for only start-up and shut-down transients plus brief continuous operation could be handled with battery power. Since no decision on the lifetime was forthcoming, we produced designs for both scenarios.

THERMION-I supplies thermal energy to the heat pipe device by collecting solar energy and focusing it into an insulating cavity. The requirement to precisely point the collecting mirror at the Sun imposed a pointing requirement of ± 0.75 degrees on the attitude control system. THERMION-II has no

1352

pointing requirement because it uses battery power to generate thermal energy (Utah State University, 1991).

Data from THERMION will be stored and downlinked using a store and forward telemetry system. Operation in this mode requires sufficient computer memory to store 102,857 data points plus housekeeping data for downlink during passage over a single ground station. The trajectory is designed to pass within communications range of the ground station once every eight orbits.

LAUNCH VEHICLE CONSTRAINTS

The decision to design the THERMION mission for launch as a secondary payload on a McDonnell Douglas Delta II launch vehicle set constraints on the spacecraft mass and volume. The volume constraint is illustrated in Figure 3 (McDonnell Douglas, 1989). The Delta II second stage is capable of placing a secondary payload of 270 to 341 kg (600 to 750 lb) into a suitable parking orbit, depending upon required lifetime (McDonnell Douglas, 1989). THERMION-I easily fits within the mass constraint; THERMION-II carries its maximum available mass in batteries to gain maximum experiment lifetime. Despite these constraints, the Delta II allows considerable margin for the achievement of mission objectives. The THERMION-I mission, for example, achieves its optimum orbit by using only 29% of the 727 kg (1600 lb) of excess fuel on the second stage (Utah State University, 1991).

THERMION-I

The THERMION-I design was produced to meet the requirements for long-lifetime, continuous testing requirements. Excess fuel in the Delta II second stage will be used to deploy the spacecraft into a 695 km (375 nm) circular orbit which is inclined at 37 degrees. Deployment at this altitude should insure at least one year of orbital lifetime.

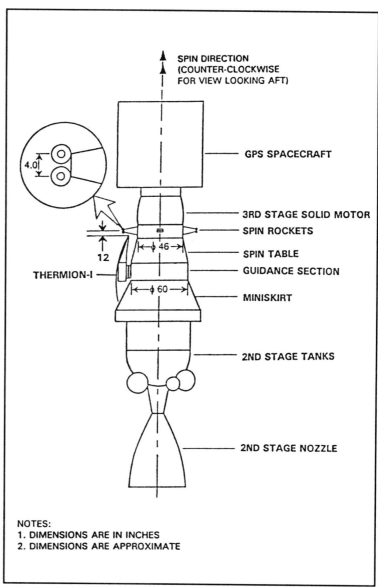

NOTES:
1. DIMENSIONS ARE IN INCHES
2. DIMENSIONS ARE APPROXIMATE

FIGURE 3. 2nd and 3rd Stage Launch Configuration of THERMION-I on the Delta II.

The THERMION-I spacecraft is shown in Figure 4. Primary subsystems include the collecting mirror, the payload insulating cavity and support structure, the attitude control system, the electrical power system, the data management system, and the communications system.

Solar Collection, Concentration, and Measurement

Elements influencing the design of the solar collecting mirror included the following:

- Maximum solar collecting area;
- Ability to fit within Delta II volume constraint;
- Highest possible efficiency;
- Ability to concentrate solar energy into a small area;
- Resistance to contamination; and
- Ability to survive liftoff loads.

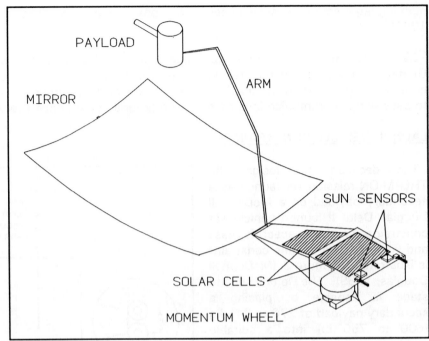

FIGURE 4. THERMION-I in Its Deployed Configuration.

The collecting device is a parabolic mirror that projects a rectangular cross-section of 148.8 x 105.7 cm (58.6 x 41.6 in). An enhanced aluminum alloy will be deposited upon the surface to gain high reflectivity with maximum resistance to atomic oxygen contamination (Smith, 1985; Melles and Griot, 1985). The dimensions of the mirror are set by the Delta II envelope constraints. These constraints also influenced the selection of a focal length of 101.6 cm (40 in) in order to keep the curvature as low as possible.

Solar energy collected by the mirror is concentrated into the opening of the insulating cavity which contains the heat pipe. Because the spacecraft pointing accuracy is $\pm$ 0.75 degrees, the mirror will not focus the energy precisely on the cavity opening. To insure the collection of all radiation into the cavity, a secondary concentrator (Winston, 1991) will be placed at the cavity entrance. The mirror/cavity combination is shown in Figure 5.

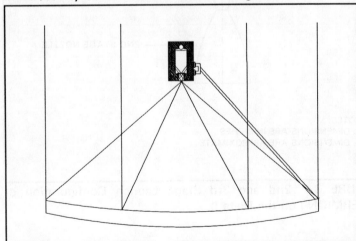

FIGURE 5. Raytrace of Secondary Concentrating Cone and Mirror.

Because the output potential of the thermionic power device is only 0.3 V with a power output of approximately 566 W and a wire temperature of 2330 K, measurement of the output presented an interesting design challenge. The final design includes a well insulated tungsten wire connected to the output leads of the thermionic device. Operation is measured by sensing the heat generated through the wire with a tungsten-rhenium thermocouple. Although precise measurement of potential and current is not possible with this method, measurement of temperature increase does verify its operation.

Attitude Determination and Control

The attitude control system is required to point the spacecraft at the Sun to within $\pm$ 0.75 degrees. The goal was to achieve this pointing accuracy with flight proven components that, when mounted, will not violate the dynamic envelope of the launch vehicle.

After a rigorous examination of various attitude control actuation systems, we decided to use a combination of three magnetic torque rods and a momentum wheel. The spin axis of the momentum wheel is alligned with the vector from the spacecraft to the Sun which gives gyroscopic stiffness about this axis. The torque rods are alligned orthogonally with one axis oriented along the Sun pointing vector.

Attitude determination was accomplished with a combination of two Sun sensors and one horizon sensor. The horizon sensor is mounted to the momentum wheel; the Sun sensors are mounted to the structure with their sensitive axes alligned with the Sun pointing vector. Details on vendor selection may be found in the Utah State University final report (Utah State University, 1991).

Construction of an attitude control simulation allowed the formulation of a control law that would correct a pointing error of 0.4 degrees to within 0.2 degrees in approximately 12 minutes. Although this is not a terribly fast response, it is sufficient for the demonstration mission. The slow response is due to the relatively low torque generated by the torque rods.

Supporting Subsystems

Additional supporting subsystems include the data management and communications subsystem, the electrical power subsystem, and the thermal management subsystem. These subsystems are briefly described below. Additional details may be obtained from the Utah State University final report (Utah State University, 1991).

Data Management and Communications

The data management and communications subsystem is designed to collect information on the operation of the heatpipe and the behavior of the spacecraft bus and then communicate that information to the ground. The data management system is built around a Q88 8088 microprocessor system manufactured by the QSI corporation (Elwell, 1991). This microprocessor contains 128 Kbytes of EPROM, 256 Kbytes of RAM, and 128 Kbytes of EEPROM. It consumes only 270 mW and operates at baud rates from 150 to 38,000 baud. A significant advantage of the Q88 is its allowance for the use of Turbo C code for use in a ROM-based system.

Communications with the ground are accomplished with a store and forward scheme that allows the use of only one ground station. Data is stored until a particular orbital pass is suffiently close to the ground station to permit a clear downlink. On-board memory is suffient to store the data until that occurs. Downlink communications are achieved by use of 2.2 GHz on-board transmitter (Aydin model T-105SE) and a high gain (38.2 dB), sixteen foot parabolic ground receiving antenna. Uplink is conducted using a UHF helix antenna for transmission and an on-board command receiver. The downlink signal margin is 29 dB at 2.2 GHz.

Electrical Power

It is ironic that the power generated by the thermionic conversion system cannot be used for spacecraft power. The output of this device is only 0.7 V (110 Amperes) which is below the threshold of any DC/DC converter. Therefore, we determined that it was necessary to supply spacecraft power through more conventional means. The orbit average power requirement of 6.77 W is met with a solar panel/battery system utilizing off-the-shelf silicon solar cells and 36 D-size Nickel-Cadmium batteries.

Because the same side of the spacecraft always points at the Sun, only this side is covered with solar cells (0.232 m²).

Thermal Management

A computer based thermal analysis revealed that all components were within the desired operating limits when the spacecraft is in eclipse with the exception of the Sun sensors and the solar cells. Because these components are not operating in that condition, the low temperatures present no problems. The results allow us to use standard thermal control methods (coatings and small heaters) to maintain components within temperature limits.

<u>**Launch Vehicle Interface**</u>

The primary goal in the selection of the launch vehicle was low cost. The small size of the THERMION vehicles allowed flight as a secondary payload where the majority of the cost would be picked up by the owner of the primary payload. A number of options were considered, but none was as attractive as the McDonnell Douglas Delta II (Utah State University, 1991).

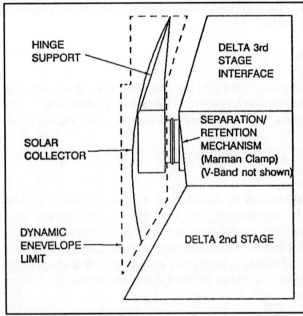

Flight as a secondary payload imposes some rather stern constraints on the design. The THERMION mission cannot interfere with or impose any risk on the primary payload, and it must fit within a rather restricted volume. Its interfaces with the launch vehicle must conform to existing specifications, and it must not exceed strict mass constraints. It should appear transparent to the primary mission.

Both THERMION missions have been designed to meet the constraints described. We found the volume constraint to be the most restrictive. Figure 6 shows how THERMION-I was configured to meet this constraint and how that constraint limited the size and curvature of the collecting mirror. The spacecraft is mated to the Delta II second stage with a Marman clamp that also serves as an ejection mechanism when the explosive bolts are activated.

FIGURE 6. Side View of Launch Configuration and Envelope Constraints.

<u>**THERMION-II**</u>

The complexity of the THERMION-I design is driven by the requirement to provide continuous thermionic heat pipe operation over a one year period. If this requirement is relaxed to require only a short-term verification of operation in a microgravity environment, an entirely different approach to the design is possible. If "short-term" can be on the order of one day, and/or discontinuous, for example, a large bank of batteries can be used to furnish energy to activate the heat pipe element.

We developed the THERMION-II alternative to provide a much more simple design to meet minimum lifetime requirements. It also will fly as a secondary payload on the Delta II; however, THERMION-II will remain attached to the Delta II second stage. Instead of using a solar collector, 253 kg of alkaline cells will be used to provide power to a tungsten grid heater attached to the thermionic heat pipe. The experiment can be run for 21 hours, either continuously or intermittently. The design permits operation of a heat pipe element that is 65% larger than that on THERMION-I and, therefore, much closer to the full scale element.

The THERMION-II configuration is shown in Figure 7. Four battery packs, each containing 480 alkaline cells, are attached to the Delta II second stage structure. (Only two battery packs are visible in Figure 7; the other two packs are on the opposite side.) The payload module is supported between two of the battery packs.

Alkaline cells were selected because they are inexpensive, available, and safe. They also have a high energy density. Lithium cells provide the highest energy density, but they present an unacceptable risk to the primary payload. The alkaline D-cells are packaged in groups of 120 with 4 groups to a pallet or pack. Four pallets are carred on the Delta II.

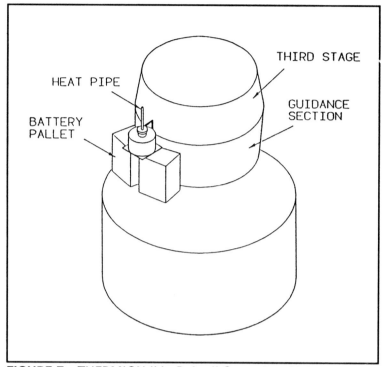

FIGURE 7. THERMION-II in Delta II Secondary Volume.

The power system must nominally supply 1800 W at 60 V to the tungsten heating grid. Additional power requirements raise the average orbital power requirements to 1851 W. The thermionic heat pipe element is surrounded by concentric layers of ceramic insulation material. The design forecasts a heat loss of approximately 80 W, or 4.5% of the total power input. Figure 8 shows an enlarged side view of the battery-payload configuration.

The primary launch vehicle constraint on THERMION-II is the mass limitation which limits the battery capacity. Increasing the number of batteries would allow the test of a larger heat pipe element and/or a longer lifetime mission. Addition of more batteries would soon approach the volume constraint, however.

The simplicity of the THERMION-II approach is evident in all the subsystems. Elimination of the solar collector also eliminates the need to point at the Sun; thus, the pointing requirement is eliminated. The short duration allows battery power to meet spacecraft needs, therefore, solar cells are not needed. Also, the THERMION-II uses the Delta II communications system, eliminating the need for a unique telemetry system.

CONCLUSIONS

The salient conclusion to be drawn from the THERMION effort is the tremendous leverage that the requirements place on the design. In this case, differing experiment lifetime requirements dictated two totally different approaches. The complexity of the THERMION-I approach, when compared to THERMION-II, is driven by the requirement for long-term, continous operation. Inability to define lifetime requirements at this stage of the thermionic heat pipe element compels an examination of both approaches, but it is essential that lifetime and operations requirements be defined before proceeding further with the development effort.

The THERMION design exercise has provided an interesting look at the use of small satellites for the demonstration/validation of emerging space systems technologies. Reluctance to utilize new technology developments in space systems results from the perception that such use will impose an unacceptable risk to mission success. The demonstration of new technologies on relatively inexpensive small satellites provides a nice opportunity to reduce risk through in-flight testing.

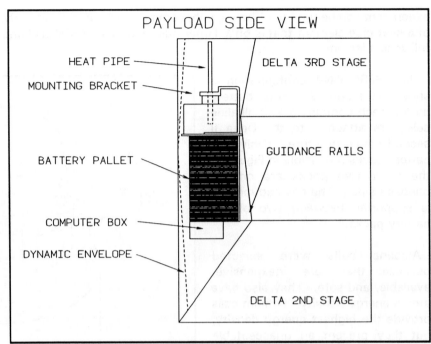

FIGURE 8. Side View of THERMION-II Battery/Payload Configuration in the Delta II Payload Volume.

Acknowledgements

The THERMION design project was conducted by a Utah State University student team funded by NASA under the NASA/USRA Advanced Design Program. A total of 18 students participated in this effort. Advisory and documentation support was provided by faculty and industry personnel from Utah State University and the Idaho National Engineering Laboratory, and by our NASA mentor, Mr. James Burke from the Jet Propulsion Laboratory. Documentation on the thermionic heat pipe element was provided by Thermacore, Inc., and documentation on the Delta II launch vehicle was provided by the McDonnell Douglas Space Systems Company.

References

Elwell, J. (1991) Telephone conversation, QSI Corporation, 1991.

Idaho National Engineering Laboratory (1990) "Small Excore Heat-Pipe-Thermionic Space Nuclear Reactor, SEHPTR," Idaho National Engineering Laboratory, October 1990.

McDonnell Douglas (1989) "Commercial Delta II Payload Planners Guide," (MDC H3224B), McDonnell Douglas Commercial Delta, Inc., December 1989.

Melles and Groit (1985) "Optics Guide 3," Copyright Melles and Groit, 1985.

Smith, W. (1985) *Modern Optical Engineering*, McGraw Hill, New York, 1985.

Thermacore (1991) "Thermionic Heat Pipe Module, Final Report," Thermacore, Inc., 5 April 1991.

Utah State University (1991) "Thermion: Verification of a Thermionic Heat Pipe in Microgravity," Final Report to NASA/USRA Advanced Design Program, Utah State University, June 1991.

Winston, R. (1991) "Nonimaging Optics," in *Scientific America*, March 1991, p.76-81.

CASCADED THERMIONIC CONVERTERS

Gary O. Fitzpatrick and
Daniel T. Allen
Advanced Energy Technology, Inc.
P.O. Box 327
La Jolla, CA 92038
(619) 455-4310

John B. McVey
Rasor Associates, Inc.
253 Humboldt Ct.
Sunnyvale, CA 94089
(408) 734-1622

Abstract

A combination of experimental results and analytical work indicates the feasibility of a cascaded thermionic converter comprised of a barium-cesium stage with an emitter temperature of 2000 K and a cesium close-spaced stage with a collector temperature of 740 K having an overall lead efficiency of 27%. This cascaded converter has potential beneficial application to space nuclear power, space terrestrial solar power and topping cycles in fossil or nuclear central stations.

INTRODUCTION

An efficient cascaded thermionic converter is described. The top stage is a barium-cesium converter, and the bottom stage consists of a close-spaced cesium converter.

Experimental results have been reported for barium-cesium converters operated in Germany (Henne 1975) and the USSR (Kalandarishvili 1986). These results are notable in that useful power densities at good calculated efficiency are demonstrated for high collector temperatures. For emitter temperatures in the range of 2000 to 2200 K, output of 5 We/sq cm at near 17% lead efficiency can be calculated for a 1200 K collector temperature. The average work function used in this calculation is 2.2 eV, which is supported by the experimental data (Henne 1975).

Close-spaced cesium converters in the USA (Dick *et al.* 1983, Nyren *et al.* 1984 and Fitzpatrick & Dahlberg 1989) and USSR (Nikolaev *et al.* 1990) have been shown experimentally to give good output at relatively low emitter temperatures. These data support a calculated lead efficiency of approximately 12% at an emitter temperature of 1200 K and a collector temperature of 740 K.

Since the collector temperature of the barium-cesium converter is the same as the emitter temperature of the close-spaced converter, it is possible to propose a cascade of these two types of thermionic converters. The potential combination of these two experimentally verified converter designs would yield an overall lead efficiency of 27%.

ENABLING TECHNOLOGIES

Barium-Cesium Converters

The cesium in a conventional high-pressure cesium, ignited mode thermionic converter serves two functions. One function is the neutralization of the space charge in the gap between the emitter and the collector. The cesium ions in the inter-electrode plasma accomplish this function. The other function of the cesium in the conventional converter is to cover the surfaces of the refractory metal electrodes in order to reduce their work functions for emission of electrons. In its first function (space charge neutralization) the cesium contributes to the voltage drop between the electrodes, and for high performance converters (which would be those having high temperatures and relatively high inter-electrode current densities) lowering this voltage drop can provide significant improvement in conversion efficiency. Lowering cesium pressure results in lower voltage drop, but it undermines the other function of the cesium, which is the electrode work function modification that brings about higher voltage output.

A solution put forward early in the development of thermionic technology (Psarouthakis 1964) has been the utilization of a mixed vapor where an "additive" provides the electrode work function coverage and the cesium is needed only for the space charge neutralization and its pressure can be reduced. Experiments (Bondarenko 1968) have shown that the heat of absorption of the alkaline earths Ba, Sr and Ca are greater than that of cesium, and therefore they selectively adsorb on the electrodes and dominate the effective work function. On the other hand, they have a larger ionization potential than cesium and therefore contribute little to the processes of the inter-electrode plasma. Thus, the addition of barium allows the operating cesium pressure to be lower, reducing the voltage drop.

Experimental studies in Germany by Henne (1975) demonstrated two important characteristics of the barium-cesium converter. One is that the lower inter-electrode voltage drop (due to the lower cesium pressure) enables good performance at greater inter-electrode spacings, compared to conventional cesium only converters. The second characteristic is that the high collector work function (due to the barium) results in good converter output at collector temperatures much higher than those for which output falls off in conventional cesium only converters.

Henne's experiments quantified the sensitivity to spacing with cesium plus barium. Shown in Figure 1 are the results of the variation of inter-electrode spacing from 0.1 mm to 1.0 mm in the experimental converter. As can be seen, the decline in maximum power is on the order of 12% for the change from 0.1 to 0.5 mm.

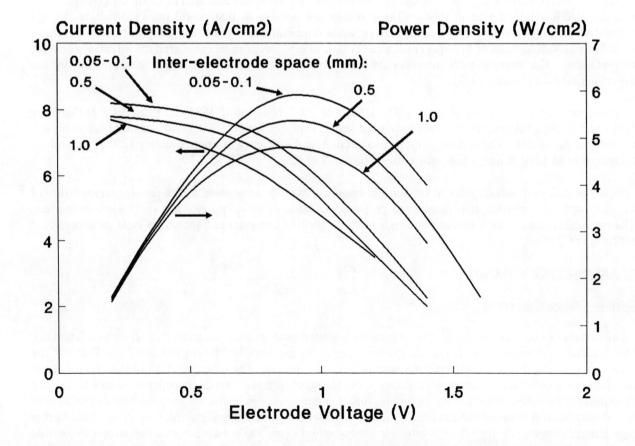

FIGURE 1.   Effect of Variation of Inter-electrode Gap in Experiments by Henne (1975).

Experimental work in the USSR, notably that by Kalandarishvili and co-workers (1986) at Sukhumi, has verified the performance of barium-cesium converters at high collector temperature. As shown in Figure 2, the identical converter with cesium only has optimum performance at a lower collector temperature as compared to operation with barium-cesium at emitter temperatures of 1773, 1923 and 2073 K. These data are from experiments with converters employing oriented tungsten emitters and niobium collectors.

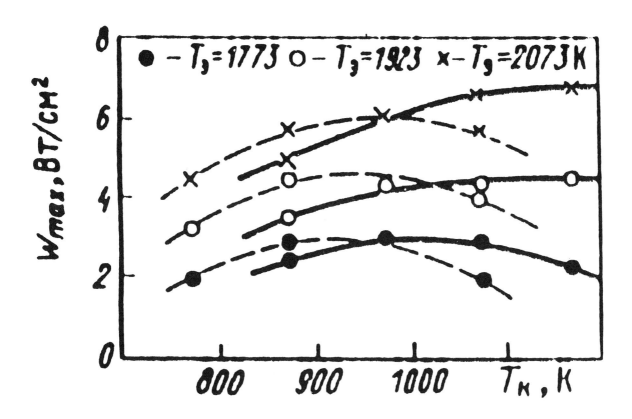

FIGURE 2.   Experimental Results of Maximum Output Variation with Collector Temperature in Experiments Reported by Kalandarishvili (1986).

Close-Spaced Cesium Converters

It was shown by Hatsopuolos and Kaye (1958) that heat could be converted directly to electricity with good efficiency in a converter operating in the quasi-vacuum mode provided, that the inter-electrode space was small. They obtained 13% lead efficiency with a gap of less than 25 μm and emitter and collector temperature of 1533 K and 810 K. Such relatively high efficiency is possible because there is no plasma arc drop. Subsequent efforts to exploit this discovery in engineering devices resulted in short circuit problems at small electrode spacing.

Experiments were sponsored by the U.S. Department of Energy in 1982 through 1984 (Dick *et al.* 1983 and Nyren *et al.* 1984) which proved the feasibility of SAVTEC close-spaced thermionic converters. The key to the SAVTEC technology is that in manufacture the emitter and collector are in physical contact, and the differential thermal expansion of the emitter lead upon heating to power opens and maintains the small gap.

Figure 3 is a J-V plot for experimental cesium close-spaced converter SAVTEC 12A. Tests were done at emitter temperatures from 1100 K to 1750 K. A SAVTEC test at 1750 K had an indicated efficiency of 18%.

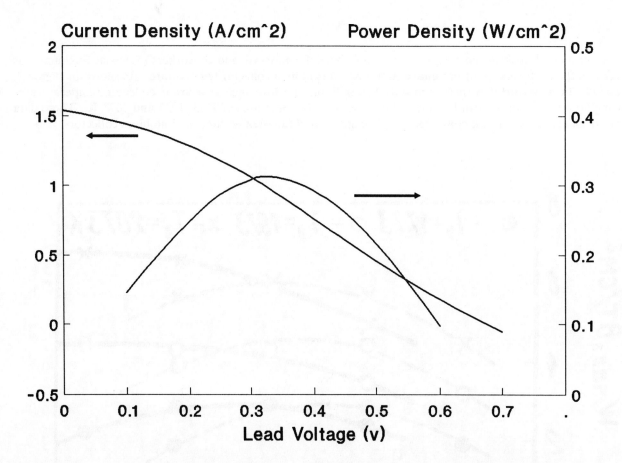

Current Density (A/cm^2)   **Power Density (W/cm^2)**

Lead Voltage (v)

FIGURE 3.   Performance of Converter SAVTEC 12A (Dick *et al.* 1983 and Nyren *et al.* 1984).

Recently published experimental work by the Scientific-Industrial Association *Luch*, located in Podolsk, describes results with a close-spaced converter of a different design and which utilizes single crystal alloys for both emitter and collector (Nicolaev *et al.* 1990).   The emitter was a W-Ta-Re crystal and the collector Mo-Nb-Re.

CASCADED THERMIONIC CONVERTERS

A two-part analysis was conducted to compare the performance of two configurations of a cascaded-cycle thermionic system.   Table 1 is emitter temperature and so on.

TABLE 1.   Converter Operating Conditions.

SAVTEC 12A Cs Converter		Ba-Cs Converter	
Emitter Temperature:	1200 K	Emitter Temperature:	2003 K
Collector Temperature:	743 K	Collector Temperature:	1223 K
Cesium Res. Temperature:	461 K	Cesium Res. Temperature:	453 K
Gap:	10 µm	Barium Res. Temperature:	953 K
		Gap:	0.5 mm

In the first configuration, the Ba-Cs converter and the SAVTEC converter are in series electrically. The emitting areas of the two converters can be different, so that although the same total current flows through each, the current densities in each can vary.

The input heat to the Ba-Cs emitter is equal to the heat leaving:

$$Q_{in} = A_1(q_{ell} + q_{rad1} + q_{Vap1}) + Q_{lead},$$

where q_{el} is the electron cooling, q_{rad} is the thermal radiation heat transfer, q_{Vap} is the vapor conduction, and Q_{lead} is the total thermal loss down the emitter lead. The heat balance for the SAVTEC emitter is:

$$Q_{in} - IV_1 - Q_{lead} = A_2(q_{e12} + q_{rad2} + q_{Vap2})$$

Here IV_1 represents the power converted to electricity in the Ba-Cs cycle. The total lead efficiency for the combined cycle is:

$$\eta = \frac{I(V_1 = V_2) - I^2 R_{lead}}{Q_{in}}$$

where R_{lead} is the lead resistance, and is related to the thermal conductivity used to evaluate Q_{lead}. The relationship between current density and voltage (the J-V curve) for each cycle completes the model. These were taken from experimental data.

The ratio A_2/A_1, the total current I, and the lead resistance R_{lead} were varied in order to optimize η. Results are shown in Figures 4 and 5 and Table 2.

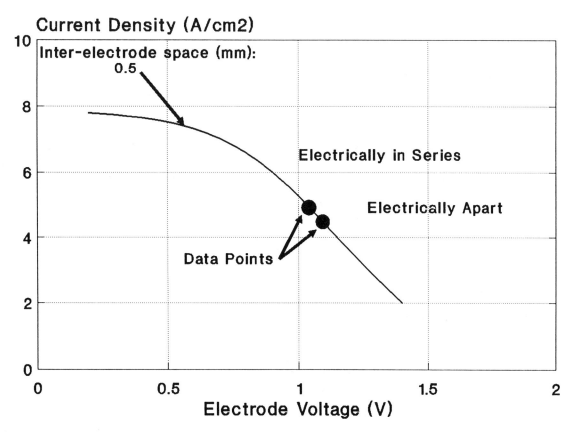

FIGURE 4.   Current Voltage Points of Barium-Cesium Converter.

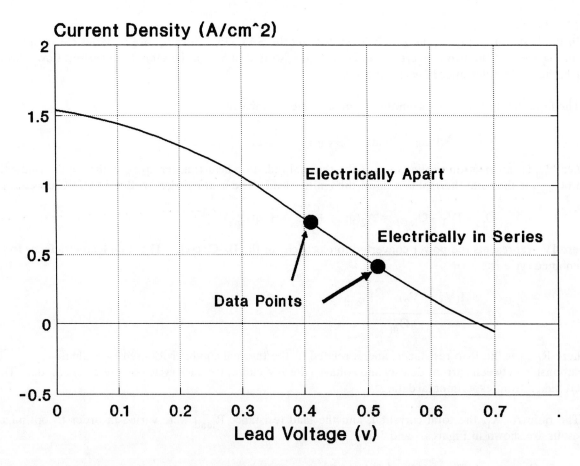

FIGURE 5. Current-Voltage Points of SAVTEC 12.

TABLE 2. Cascaded Electrically Series.

SAVTEC 12A Cs Converter		Ba-Cs Converter	
Current:	48 Amp.	Current:	48 Amp.
Current Density:	0.39 A/sq cm	Current Density:	4.8 A/sq cm
Lead Voltage:	0.52 v	Lead Voltage:	1.07 v
Input Power Density:	1.62 Wth/sq cm	Input Power Density:	25.3

Ratio A2/A1:	12.4
Cascaded Efficiency:	26%

In an alternative electrically apart type of cascaded cycle, the total currents in each converter are separate. The mathematical model for this cycle is:

$$Q_{in} = A_1(q_{ell} + q_{rad1} + q_{Vap1}) + Q_{lead12}$$

$$Q_{in} - I_1 V_1 = A_2(q_{e12} + q_{rad2} + q_{Vap2}) + Q_{lead23}$$

and,

$$\eta = \frac{I_1V_1 + I_2V_2) - I_1^2R_{lead12} - I_2^2R_{lead23}}{Q_{in}}$$

The area ratio, individual total currents, and lead efficiencies are optimized. Results are shown in table 3. Independent optimization of the current densities allows the efficiency to rise to 27%. For this type of cycle it is possible to independently define a lead resistance for each part. The efficiency of the Ba-Cs converter was 12%, and 17% for the SAVTEC.

TABLE 3.   Cascaded Electrically in Parallel.

SAVTEC 12A Cs Converter		Ba-Cs Converter	
Current:	66 Amp.	Current:	44 Amp.
Current Density:	0.7 A/sq cm	Current Density:	4.4 A/sq cm
Lead Voltage:	0.42 v	Lead Voltage:	1.1 v
Input Power Density:	2.0 Wth/sq cm	Input Power Density:	24.0
Lead Efficiency:	12%	Lead Efficiency:	17%
Ratio A2/A1:		9.47	
Cascaded Efficiency:		27%	

It should be especially noted that the "rejected" heat from the Stage 2 converter is at a temperature of 750 K. This heat could be further utilized to produce more output power or to do other useful work, and so there is the possibility that the cascaded converter could be applied in systems of even greater overall efficiency.

ENGINEERING REALIZATION

A difficulty in actually making such a cascaded converter would be the disparity input power density and current density between the stages, which is approximately a factor of ten. Use of a heat pipe is one way to get around the problem. A niobium heat pipe with lithium working fluid could couple the stages. Electric current flows down the heat pipe. A schematic for a two-stage converter is in Figures 6 and 7.

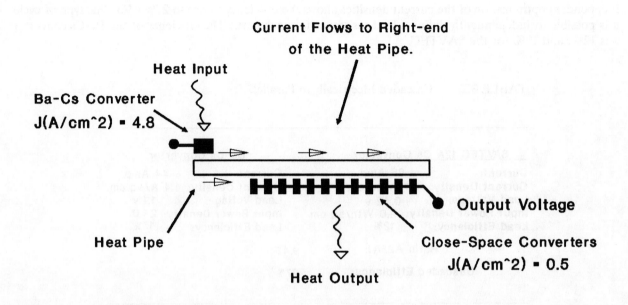

FIGURE 6.   Use of Heat Pipe to Combine Stages.

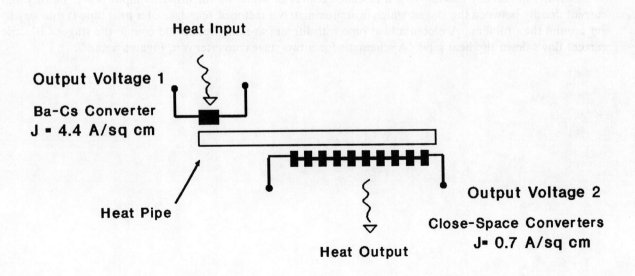

FIGURE 7.   Use of Heat Pipe to Combine Stages.

CONCLUSIONS

A cascaded converter with very attractive efficiency is a real possibility. The converter described is based on experimental converters built and tested. For the barium-cesium stage the converters were built and tested in the USSR and Germany. For the close-spaced stage the converters have been built in the USSR and USA.

This cascaded converter has potential beneficial application in addition to space nuclear power, such as for space and terrestrial solar power conversion and for topping cycles in fossil or nuclear central stations.

References

Bondarenko, V. D., and Y. K. Guskov (1968) "The Characteristics of Thermionic Converter Filled with Vapor Mixture,", Second International Conference on Thermionic Power Generation, Stresa, Italy, 1968.

Dick, R. S., J. B. McVey, G. O. Fitzpatrick and E. J. Britt (1983) "High Performance, Close-Spaced Thermionic Converter," *Proceedings of the Eighteenth Intersociety Energy Conversation Engineering Conference*, held in Orlando, Florida (839030), August 1983.

Fitzpatrick, G. O. and R. C. Dahlberg (1989) "SAVTEC Thermionic Converter with an SP-100 Heat Source," *Trans. Sixth Symposium on Space Nuclear Power Systems*, CONF-890103, Albuquerque, New Mexico, January 1989.

Hatsopoulos, G. N. and J. Kaye (1958) "Measured Thermal Efficiencies of a Diode Configuration of a Thermo Electron Engine," *Journal of Applied Physics*, 1958.

Henne, R. (1975) "Features of a Barium/Cesium Diode with Plane, Polycrystalline Molybdenum Electrodes for Thermionic Energy Conversion," Deutsche Forschungs- und Versuchsanstalt fur Luft- und Raumfahrt, Institut fur Energiewandlung und Elektrische Antriebe, Eureopean Space Agency report TT-171, Stuttgart, June 1975.

Kalandarishvili, A. G. (1986) *Working Medium Sources for Thermionic Power Converters*, Energoatomizdat, Moscow, 1986.

Nikolaev, Yu., V., Ya. R. Kucherov, *et al.* (1990) "Close-spaced Thermionic Energy Converter," *Proc. Seventh Symposium on Space Nuclear Power Systems*, CONF-900109, Albuquerque, New Mexico, January 7-10, 1990.

Nyren, T., M. Korringa, *et al.*, (1984) "Design and Testing of a Combustion-Heated Nineteen Converter SAVTEC Array," *Proceedings of the Nineteenth Intersociety Energy Conversation Engineering Conference*, San Francisco (849308), August 1984.

Psarouthakis, J. (1964) "Thermionic Energy Converter with Barium and Cesium Vapors," Thermionic Conversion Specialists Conference, Cleveland, 1964.

A POWER PROPULSION SYSTEM
BASED ON A SECOND-GENERATION THERMIONIC NPS
OF THE "TOPAZ" TYPE

Georgi M.Gryaznov, Eugene E.Zhabotinski,
Pavel V.Andreev, Gennadie A.Zaritski
Scientific and Production
Association "Krasnaya Zvezda"
Moscow, USSR

Nikolai N.Ponomarev-Stepnoi
Veniamin A.Usov
I.V.Kurchatov Institute
of Atomic Energy
Moscow, USSR

Anatoly S.Koroteev, Viktor M.Martishin,
Vladimir N.Akimov
Research Institute of Thermal Processes
Moscow, USSR

Edward J.Britt
Space Power Inc.
San Jose, California
USA

Abstract

The paper considers the concept of power propulsion systems-universal space platforms (USPs), on the basis of second-generation thermionic nuclear power systems (NPSs) and stationary plasma electric thrusters (SPETs). The composition and the principles of layout of such a system, based on a thermionic NPS with a continuous power of up to 30 kWe allowing power augmentation by a factor of 2-2.5 as long as during a year, as well as SPETs with a specific impulse of at least 20 km/s and a propulsion efficiency of 0.6-0.7 are discussed. The laylouts and the basic parameters are presented for a power propulsion system ensuring cargo transportation from an initial radiation-safe 800 km high orbit into a geostationary one using the "Zenit" and "Proton" launch systems for injection into an initial orbit. It is shown that the mass of mission-oriented equipment in the geostationary orbit in the cases under consideration ranges from 2500 to 5500 kg on condition that the flight time is not longer than a year.

The power propulsion system can be applied to autonomous power supply of various spacecraft including remote power delivery. It can be also used for deep space exploration.

INTRODUCTION

The concept of autonomous propulsion systems, such as universal space platforms (USPs), combining a power unit and a dual-thrust propulsion unit seems today promising. These systems can solve many problems in informatics, production, and technology. At electrical capacities of tens of kilowatts the USP can efficiently function as an interorbital tug, including the task of removing spacecraft having fulfilled their missions from the operational orbits. In future the USPs could be used for remote delivery of power to spacecraft with the help of centimeter-band radiation. Of all available nuclear power systems (NPSs) only thermionic ones seem most useful for USPs in virtue of their compactness and ability of essential power augmentation during 10-15% of the in-service time.

The paper considers the principles of construction of a USP on the basis of a thermionic NPS having a continuous power of 25-30 kWe and allowing the power augmentation by a factor of 2-2.5 within a year, and stationary plasma thrusters (SPETs) with a specific impulse of at least 20 km/s.

ELECTRIC THRUSTERS

Electric thrusters for interorbit flights must possess sufficiently high specific impulse and efficiency. At present stationary plasma electric thrusters (SPETs) with closed drift of electrons appear to be in-flight proved best of all. These thrusters allow to attain a specific impulse of 20 to 40 km/s at a propulsion efficiency of 0.6-0.7. The SPET lifetime depends mainly on sputtering of the discharge chamber walls by accelerated ions, the service life of the chamber being proportional to its characteristic dimension. For the engines 50-70 mm and more in diameter one can surely achieve a half-yearly lifetime, which is confirmed by operation of SPETs under full-scale conditions for several thousands hours.

A 25-kWe SPET module with a 300-mm OD accelerating channel was chosen for the sustainer propulsion system of the USP. The DC supply voltage for the SPE is about 500 V. SPET modules of less dimension can be used for manoeuvres in the operational orbits. Xenon is used as a coolant in the SPET.

THERMIONIC NPS

The NPS comprises a nuclear power unit (NPU), a system of NPS movement away from an instrument compartment (IC) of the USP, equipment of an automatic control system (ACS) and an electricity quality assurance system. The NPU includes a reactor-converter (RC) with a cesium supply system, control devices drives, a radiation shield, a heat removal system with an electromagnetic pump (EMP) and a cooler-radiator (CR), a cable system, and supporting structures.

The core contains a zirconium hydride moderator with holes wherein TFE with cooling channels are placed. The core together with a top and a bottom reflectors is enclosed in a casing, separating the core from a side reflector. The TFEs have a five-layer collector stack allowing their connection in commutation boxes in the atmosphere of inert gas to obtain an RC output voltage of about 120 V. The number of the TFEs in the reactor is 60, their total emission area being about 1.6 m^2. The average specific continuous power is 2-2.5 W/cm^2 at an emitter temperature of below 1870 K. The conversion efficiency at the end of lifetime is no lower than 8%. In the enforced mode at the beginning of lifetime during an interorbit flight the average specific power is about 5W/cm^2 at an efficiency of at least 10% and an emitter temperature of up to 2020K.

Twelve rotating control rods divided into four separate groups are placed in the side reflector. A safety rod is at the center of the core.

The neutron-physical efficiency of the control rods and the safety rod is enough to ensure a proper nuclear safety in any accident situation in the period of spacecraft preparation and launching into orbit.

The NPS has a single heat removal system. The heat-pipe CR has a fixed and a folding parts. The maximum temperature of the coolant (Na-K eutectic) in the enforced mode is 900K.

The angle of radiation shielding is chosen to create a shadow cone encompassing not only the IC, but also all its protrusions.

The ACS and the electricity quality assurance system disposed in the IC comprise an NPS regime control, a temperature regulator for a cesium vapour generator, a fast-response voltage regulator, an EMP supply unit and a converter for power supply of a nuclear thermal propulsion system (NTPS).

The distance of NPS movement away from the IC was determined to meet the requirement of minimum total mass of the NPU and the NPS movement system together with electric power lines at radiation levels of 10^6 rad and 10^{12} neutron/cm^2 on the IC front wall for five years. The optimal distance of the IC from the core center is 15-17m.

USP COMPOSITION, LAYOUT AND PARAMETERS

To ensure reliably radiation safety the USP operating altitude range is limited by 800-km and higher orbits. The electric power consumed by the NTPS in the enforced mode is 65 kW with allowance for auxiliary power and power conversion and transmission losses.

Let us consider the principles of construction and the basic parameters of the USP as applied to the problem of transfer of a spacecraft from a reference (800-km high) orbit into a geostationary orbit (GSO) by the "Zenit" and "Proton" launch systems. Masses Mo in a reference orbit which can be transferred by these launch systems using a preoperating engines with a specific impulse of 3.1 km/s from a low (200-km high) orbit into the reference one are 12000 kg and 17500 kg for the "Zenit" and "Proton" respectively.

In order to fly from an 800-km high circular reference orbit with an inclination of 51° into a GSO the characteristic velocity must be changed by about 7.5 km/s.

Figure 1 shows the flight time t_{tr} dependence on the NTPS specific impulse J and effeciency for M_o=12000 kg and 17500 kg. It follows from Figure 1 that, if t_{tr} cannot be

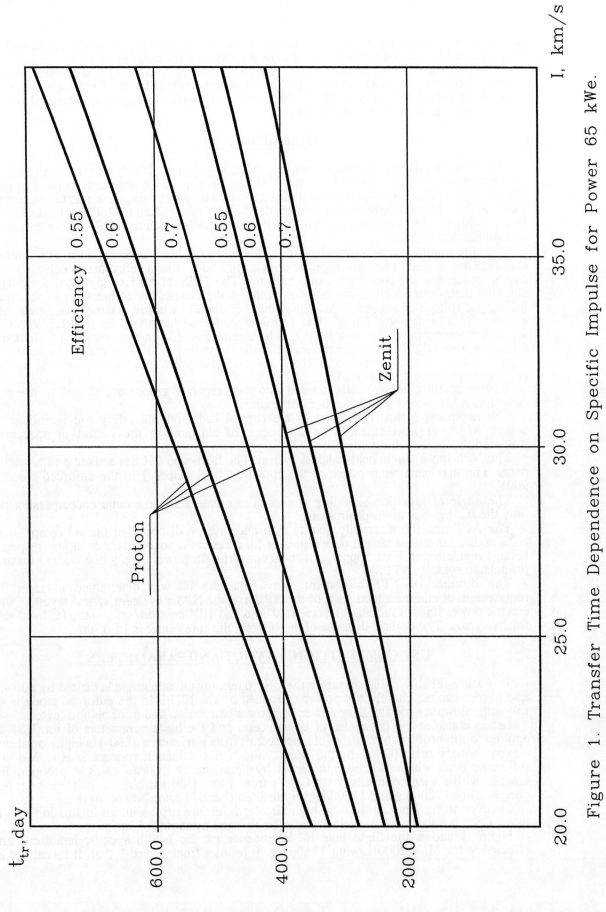

Figure 1. Transfer Time Dependence on Specific Impulse for Power 65 kWe.

longer than a year, J can reach 20-25 km/s at an efficiency ranged from 0.55 to 0.7 for M_0 = 17500 kg and from 28 to 35 km/s for M_0 = 12000 kg.

A two-launch variant of placing a spacecraft into a reference orbit, considered alongside a single-launch variant for the "Zenit" launch system the USP docking with the spacecraft in the reference orbit. In this case at greater than 20 km/s M_0 will be about 18500 kg, which is sufficiently close to the case with the "Proton" launch system.

The USP is composed of
- an NPS;
- a USP movement control system;
- an NTPS ensuring transfer of a spacecraft into a GSO as well as correction and stabilization of the USP position in operation;
- an ET coolant supply system;
- an IC enclosing an ACS, an electricity quality assurance system, a start-up battery, a USP control system, tanks of the feed system; and
- a thermal-control system ensuring a required temperature regime inside the IC.

Depending on the variant of injection into a reference orbit the USP can be equipped by the following units:
- a preoperating engines to transfer the USP from the low orbit into the reference one;
- a unit for docking the USP with a purpose spacecraft in the reference orbit, if they are launched separately; and
- a compartment to dispose either a purpose equipment (the single-launch variant) or additional tanks with ET coolant.

The ET contains nine sustainer engine modules with three of them operating simultaneously.

To reduce the overall dimensions of the USP in the launching position the CR in the NPS has a folding construction. It is developed in the orbital position in moving the NPU away from the IC. The IC is not pressurised, the thermal-control system based on heat tubes including diode ones. The equipment is disposed in the periphery of the IC, the middle part of which is occupied by tanks with xenon. The IC diameter is chosen so, that the IC be placed under the fairing of the "Zenit" launch system and that the sustainer electric thrusters disposed in the IC periphery be retracted in the launching position inside special wells made in the IC.

The length of the USP is 8.1 m in the launching and 17.7 m in the operating position, the maximum diameter of the USP being 3.2 m. The total length of a spacecraft for the single-launch variant with the "Zenit" launch system is 11.8 m in the launching position. The layouts of the USP in the launching and orbital positions are shown in Figure 2. The mass of the USP without the xenon tanks is about 5300 kg including the 3800 kg mass of the NPS with the start-up battery.

The mass of payload (the purpose equipment) in the GSO is determined for the single-launch variant by the following expression

$$M_{п.г} = M_0 \exp\left(-\frac{\Delta V}{J}\right) - M_{укп} - M_{крд} - \Delta M_T^{эрду} - \alpha_{схрт} \cdot M_T^{эрду}$$

where:

$M_{укп}$ is the dry mass of the USP;

$M_{крд}$ is the dry mass of the preoperating engine;

M_T is the mass of xenon for injection into the GSO;

$\Delta M_T^{эрду}$ is the mass of xenon for the manoeuvre in the GSO at a total impulse of at least 5.10^6 н·с

$\alpha_{схрт}$ is the relative mass of the coolant storage and supply system (it was assumed that =0.15).

For the single-launch variant $M_{п.г}$ must be reduced by the mass of the docking units. Moreover, it was accepted that the expenditure of xenon for the orbiting control is about 5% of its mass for injection into the GSO. The calculations at $M_{крд}$=250 kg showed that for the "Proton" launch system can amount to about 5300 kg; for the "Zenit" launch system is about 2500 kg and 5400 kg in the single-launch and two-launch variants respectively. It

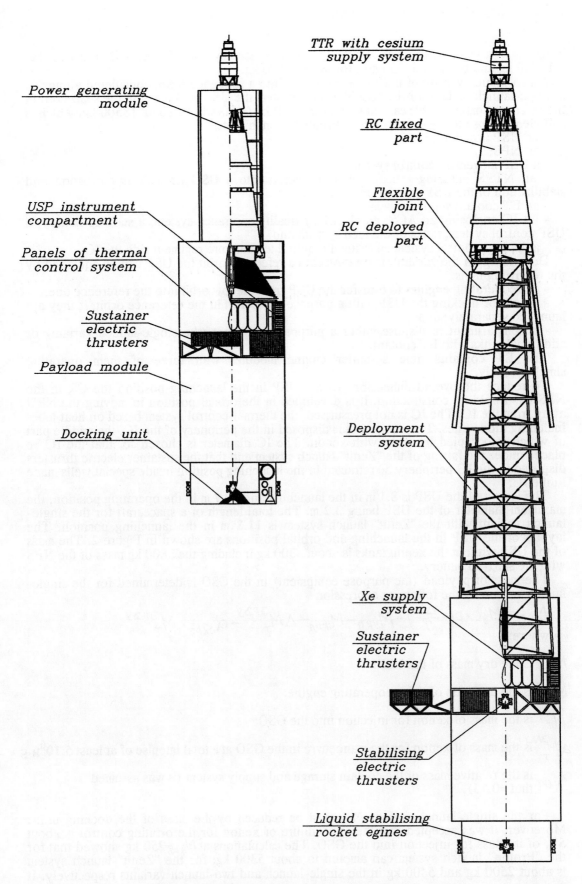

Power generating
module

TTR with cesium
supply system

RC fixed
part

USP instrument
compartment

Flexible
joint

RC deployed
part

Panels of thermal
control system

Sustainer
electric
thrusters

Payload module

Deployment
system

Docking unit

Xe supply
system

Sustainer
electric
thrusters

Stabilising
electric
thrusters

Liquid stabilising
rocket egines

Fig.2 Universal space platform (USP).

should be noted that in addition to the purpose equipment the spacecraft in the GSO has a power system ensuring 25-30 kWe during 5-7 years and a propulsion system for the spacecraft stabilization and maneuvring.

CONCLUSION

The combination of a second-generation thermionic NPS with electric thrusters increases essentially the efficiency of their use. An NPS with a continuous capacity of 25-30 kWe can be used as a basis for creation of a power propulsion system placed into a reference radiation-safe orbit by the existing "Zenit" or "Proton" launch systems and ensuring the delivery of the purpose equipment with a mass of up to 5500 kg. Besides the solution of the transportation problems, such a system can function as an autonomous module of power supply for various spacecraft to which it is docked or ensures the transmission of SHF power. The system can be useful for deep space exploration and, in particular, for tracking of the Vesta asteroid and the Enke comet.

ACKNOWLEDGMENTS

The present work was performed under the agreement with a Small-scale State Enterprise "Power Technologies Scientific Production" by the individuals from Kurchatov's Institute of Atomic Energy, Scientific and Production Assotiation "Krasnaya Zvezda" and Research Institute of Thermal Processes, consulting with SPI (USA) representative Dr. Edward J.Britt.

REFERENCE

V.N.Akimov, P.V.Andreev, M.N.Valt et al.(1991) "Investigation and Optimization of Method of a Payload Launching into a Geostationary Orbit Using Power Propulsion Systems Based on Thermionic NPS and Electric Thrusters", Report of a Small-scale State Enterprise "Power Technologies Scientific Production", NPE-91/005-01 dated June 26, 1991.

AUTHOR INDEX
(Bold Page Numbers Indicate Senior Authorship)

AUTHOR INDEX
(Bold Page Numbers Indicate Senior Authorship)

AUTHOR INDEX
(Bold Page Numbers Indicate Senior Authorship)

AUTHOR INDEX
(Bold Page Numbers Indicate Senior Authorship)

AUTHOR INDEX
(Bold Page Numbers Indicate Senior Authorship)

(Note: Names of the presenters in the Applied Technology sessions are included in the author list, but written versions of their presentations are not available.)